SPECIES INVASIONS
Insights into Ecology, Evolution, and Biogeography

SPECIES INVASIONS

INSIGHTS INTO ECOLOGY, EVOLUTION, AND BIOGEOGRAPHY

Edited by

DOV F. SAX, JOHN J. STACHOWICZ, AND STEVEN D. GAINES

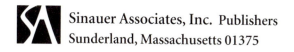

Sinauer Associates, Inc. Publishers
Sunderland, Massachusetts 01375

About the Cover

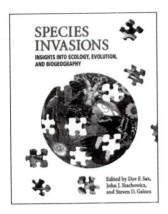

Species invasions come in all shapes and sizes, from tiny, innocuous species that are hardly noticed to large or rapidly expanding species that dramatically alter habitat structure and ecosystem dynamics. All these species offer a series of opportunities to solve complex puzzles in ecology, evolution, and biogeography; the clues these invaders provide come as much from their successes as their failures. Some species invade and have large ecological and economic impacts, while others either fail to become established or do so only marginally. It is this variation in performance taken together with the tremendous number of invaders present everywhere on the planet that makes species invasions at once an issue for significant concern and an opportunity for tremendous insight. In light of this, we have chosen a set of cover images that spans the range of environments invaded, as well as the range of success these invaders enjoy, bearing in mind that all these species help to solve the complex puzzles nature presents.

Cover Photo Credits

Goat (*Capra hircus*): photo by Scott Bauer, ARS/USDA. Sunflowers (*Helianthus annuus*): photo by Bruce Fritz, ARS/USDA. Yellow-banded dart-poison frog (*Dendrobates leucomelas*): photo by Thomas Villegas. Arabian angelfish (*Pomocanthus asfur*): photo by Andrzej Zabawski. Iceplant (*Carpobrotus edulis*): photo by Mark Vellend. Flea beetle (*Aphthona flava*): photo by ARS/USDA.

Species Invasions: Insights into Ecology, Evolution, and Biogeography

Copyright © 2005 by Sinauer Associates Inc. All rights reserved.
This book may not be reproduced without the permission of the publisher.

For information address Sinauer Associates, Inc., 23 Plumtree Road, Sunderland, MA 01375 U.S.A.

 FAX 413-549-1118
 EMAIL order@sinauer.com; publish@sinauer.com
 WEBSITE www.sinauer.com

Library of Congress Cataloging-in-Publication Data

Species invasions : insights into ecology, evolution, and biogeography / edited by Dov F. Sax, John J. Stachowicz, and Steven D. Gaines.

 p. cm.
 Includes bibliographical references and index.
 ISBN: 0-87893-811-7 (paperbound), 0-87893-821-4 (casebound)
1. Biological invasions—Congresses. 2. Ecology—Congresses. 3. Evolution (Biology)—Congresses. 4. Biogeography—Congresses. I. Sax, Dov F. II. Stachowicz, John J. III. Gaines, Steven D.
QH353.S29 2005
577'.18—dc22
         2005013019

Printed in U.S.A.

4 3 2 1

Contents

Part III / Insights into Biogeography 309

Introduction by Julie L. Lockwood

Contributors

Sebastián R. Abades Center for Advanced Studies in Ecology and Biodiversity (CASEB) and Departamento de Ecología, Pontificia Universidad Católica de Chile, Casilla 114-D Santiago, Chile
sabades@bio.puc.cl

Michael Barfield Department of Zoology, 223 Bartram Hall, PO Box 118525, University of Florida, Gainesville, FL 32611-8525 USA mjb01@ufl.edu

Mark D. Bertness Department of Ecology and Evolutionary Biology, Walter Hall Box G-W, Brown University, Providence, RI 02912 USA Mark_Bertness@brown.edu

Tim M. Blackburn School of Biosciences, University of Birmingham, Edgbaston, Birmingham B15 2TT United Kingdom
t.blackburn@bham.ac.uk

Keryn D. Bromberg Department of Ecology and Evolutionary Biology, Walter Hall, Box G-W, Brown University, Providence, RI 02912 USA
Keryn_Bromberg@brown.edu

James H. Brown Department of Biology, University of New Mexico, Albuquerque, NM 87131 USA jhbrown@unm.edu

John F. Bruno Department of Marine Sciences, University of North Carolina, Chapel Hill, NC 27599-3300 USA
jbruno@unc.edu

Ragan M. Callaway Division of Biological Sciences, University of Montana, Missoula, MT 59812 USA
ray.callaway@mso.umt.edu

Phillip Cassey School of Biological Sciences, University of Birmingham, Edgbaston B15 2TT United Kingdom
p.cassey@bham.ac.uk

Carla M. D'Antonio Environmental Studies Program, University of California, Santa Barbara, CA 93106 USA
dantonio@lifesci.ucsb.edu

Andrew P. Dobson Department of Ecology and Evolutionary Biology, Princeton University, Princeton, NJ 08544-1003 USA
dobber@princeton.edu

Jason D. Fridley Department of Biology, University of North Carolina, Chapel Hill, NC 27599-3280 USA fridley@unc.edu

Steven D. Gaines Marine Science Institute and the Department of Ecology, Evolution and Marine Biology, University of California, Santa Barbara, CA 93106-6150 USA gaines@lifesci.ucsb.edu

Kevin J. Gaston Biodiversity and Macroecology Group, Department of Animal and Plant Sciences, University of Sheffield, Sheffield S10 2TN United Kingdom k.j.gaston@sheffield.ac.uk

George W. Gilchrist   Department of Biology, Box 8795, College of William & Mary, Williamsburg, VA 23187-8795 USA gwgilc@wm.edu

Richard Gomulkiewicz   Department of Mathematics and School of Biological Sciences, Washington State University, Pullman, WA 99164 USA gomulki@wsu.edu

Richard K. Grosberg   Center for Population Biology, University of California, Davis, CA 95616 USA   rkgrosberg@ucdavis.edu

Alan Hastings   Department of Environmental Science and Policy, One Shields Avenue, University of California, Davis, CA 95616 USA amhastings@ucdavis.edu

Andrew P. Hendry   Redpath Museum & Department of Biology, McGill University, 859 Sherbrooke St. W., Montréal, QC H3A 2K6 Canada   andrew.hendry@mcgill.ca

José L. Hierro   Division of Biological Sciences, University of Montana, Missoula, MT 59812 USA jose.hierro@umontana.edu

Sarah E. Hobbie   Department of Ecology, Evolution, and Behavior, University of Minnesota, St. Paul, MN 55108 USA shobbie@tc.umn.edu

Robert D. Holt   Department of Zoology, 223 Bartram Hall, PO Box 118525, University of Florida, Gainesville, FL 32611-8525 USA rdholt@zoo.ufl.edu

Raymond B. Huey   Department of Biology, Box 351800, University of Washington, Seattle, WA 98195-1800 USA hueyrb@u.washington.edu

A. Randall Hughes   Section of Evolution and Ecology, University of California, Davis, CA 95616 USA   arhughes@ucdavis.edu

Brian P. Kinlan   Department of Ecology, Evolution, and Marine Biology, University of California, Santa Barbara, CA 93106-9610 USA   kinlan@lifesci.ucsb.edu

Armand M. Kuris   Department of Ecology, Evolution, and Marine Biology, University of California, Santa Barbara, CA 93106-9610 USA   kuris@lifesci.ucsb.edu

Fabio A. Labra   Center for Advanced Studies in Ecology and Biodiversity (CASEB) and Departamento de Ecología, Pontificia Universidad Católica de Chile, Casilla 114-D, Santiago, Chile   flabra@bio.puc.cl

Kevin D. Lafferty   U.S. Geological Survey Western Ecological Research Center, and Marine Science Institute, University of California, Santa Barbara, CA 93106-6150 USA   lafferty@lifesci.ucsb.edu

Julie L. Lockwood   Ecology, Evolution and Natural Resources, 14 College Farm Road, Rutgers University, New Brunswick, NJ 08901 USA   lockwood@aesop.rutgers.edu

Richard N. Mack   School of Biological Sciences, Washington State University, Pullman, WA 99164-4236 USA rmack@mail.wsu.edu

Pablo A. Marquet   Center for Advanced Studies in Ecology and Biodiversity (CASEB) and Departamento de Ecología, Pontificia Universidad Católica de Chile, Casilla 114-D, Santiago, Chile pmarquet@bio.puc.cl

Michael L. McKinney   Department of Earth and Planetary Sciences, University of Tennessee, Knoxville, TN 37996 USA mmckinne@utk.edu

Stephen J. Novak   Department of Biology, Boise State University, 1910 University Drive, Boise, ID 82725-1515 USA snovak@boisestate.edu

William R. Rice   Department of Ecology, Evolution, and Marine Biology, University of California, Santa Barbara, CA 93106-9610 USA   rice@lifesci.ucsb.edu

Robert E. Ricklefs   Department of Biology, 8001 Natural Bridge Road, University of Missouri-St. Louis, St. Louis, MO 63121-4499 USA   ricklefs@umsl.edu

Dov F. Sax   Department of Ecology, Evolution, and Marine Biology, University of California, Santa Barbara, CA 93106 - 9610 USA   sax@lifesci.ucsb.edu

Katherine F. Smith   Department of Ecology, Evolution, and Marine Biology, University of California, Santa Barbara, CA 93106 -9610 USA   k_smith@lifesci.ucsb.edu

John J. Stachowicz   Section of Evolution and Ecology, University of California, Davis, CA 95616 USA   jjstachowicz@ucdavis.edu

Andrea S. Thorpe   Division of Biological Sciences, University of Montana, Missoula, MT 59812 USA andrea.thorpe@mso.umt.edu

David Tilman   Department of Ecology, Evolution, and Behavior, University of Minnesota, St. Paul, MN 55108 USA tilman@umn.edu

Mark E. Torchin   Smithsonian Tropical Research Institute, Apartado 0843-03092, Balboa, Ancon, Republic of Panama torchinm@si.edu

Mark Vellend   Departments of Botany and Zoology, and Biodiversity Research Centre, University of British Columbia, Vancouver, BC V6T 1Z4 Canada mvellend@interchange.ubc.ca

Geerat J. Vermeij   Department of Geology, University of California, One Shields Avenue, Davis, CA 95616 USA gjvermeij@ucdavis.edu

John P. Wares   Department of Genetics, University of Georgia, Athens, GA 30602 USA   jpwares@uga.edu

Ethan P. White   Department of Biology, Utah State University, Logan, UT 84322 USA epwhite@biology.usu.edu

Preface

Charles Darwin, Joseph Grinnell, G. Ledyard Stebbins, and many other luminaries in the fields of ecology, evolution, and biogeography have suggested that species invasions can provide invaluable opportunities to advance our understanding of nature. In spite of their foresight, most studies of species invasions have concentrated on the applied problems posed by non-native species. This book is an attempt to refocus some portion of the attention paid to species invasions back onto basic research questions, with the hope that such efforts can not only advance our understanding of basic issues in the life sciences, but also produce new insights that will improve our ability to deal with the very real applied problems species invasions present.

This book is predominately a product of serendipity and collaboration. None of its three parts, nor indeed the book as a whole, would have been possible without sustained collaboration and a willingness to proceed into areas unknown by any single author or editor. The kernel of the idea for this book formed after presenting material for a job seminar, in the hopes of finding a useful unifying theme. From there the book took on a life of its own. Early on, one contributor suggested holding a workshop that would bring the chapter authors together to discuss, flesh out, and improve their ideas. Such a workshop was held at the National Center for Ecological Analysis and Synthesis (NCEAS) in March of 2004. Many of the book's participants were able to attend, and the forum provided an opportunity to revisit what has been accomplished since the Asilomar Conference hosted by H. G. Baker and G. L. Stebbins some 40 years ago. Indeed, the meeting proved to be of tremendous benefit both to those participating and to the book as a whole. It is our hope that these resulting chapters will be useful to those who read them and to the academic community at large.

Certainly this book would not have been possible if not for the genuine interest of the participants, the support of the agencies that provided funding to the many authors involved, and the wonderful assistance provided by Sinauer Associates throughout the publishing process.

We also wish to acknowledge the participation of Mark Davis, Robin Pelc, and Eric Seabloom, who contributed substantially to the discussions at the NCEAS meeting. This book benefited tremendously from chapter reviews, and we thank all reviewers who assisted in this process. Finally, we thank Andrea Gaines for her suggestions for the book cover.

DOV SAX

JAY STACHOWICZ

STEVE GAINES

Introduction

Dov F. Sax, Steven D. Gaines, and John J. Stachowicz

Few species reside solely in the location where they originated. In fact, the movement of species from one place to another is a predominant feature of life on Earth. Species expand, contract and shift their geographical distributions. These natural dynamics are due to the interplay of ecology and evolution, to changing physical geography and climate. Over time these dynamics lead to the development of regional and local biotas. It is our perception of these biotas and the boundaries in time and space that we draw around them that allows us to assign species as being native or non-native to any given locality or region—a process that is by its nature artificial and arbitrary. In spite of this, it is nearly impossible not to be concerned about recent species invasions and not to agree with Elton (1958) that we live today in a world of "ecological explosions" caused by non-native species.

Elton's book, *The Ecology of Invasions by Animals and Plants*, drew attention to the tremendous damage that non-native species can cause to ecological systems, human health, and economic well-being. Such damage can come in many forms. Introductions of exotic pathogens have led to outbreaks of bubonic plague (*Yersinia pestis*), malaria (*Falciparum malaria*), and AIDS (the HIV virus), as well as to drastic declines of American chestnut (*Castanea americana*), American elm (*Ulmus americana*), and oaks (*Quercus* spp.). Exotic agricultural weeds and garden escapes have contributed cheatgrass (*Bromus tectorum*), black mustard (*Brassica nigra*), kudzu (*Pueraria lobata*), and many other problematic plants. Exotic animals released intentionally for economic or aesthetic reasons have become persistent (and often undesired) components of the landscape, e.g., rainbow trout (*Oncorhynchus mykiss*), feral pigs (*Sus scrofa*), European starlings (*Sturnus vulgaris*); and exotic biocontrol agents, such as the giant toad (*Bufo marinus*), small Indian mongoose (*Herpestes auropunctatus*), and rosy wolf snail (*Euglandina rosea*) have occasionally gone awry, decimating native populations. In the nearly 50 years since the publication of Elton's book, the continuing threats posed by non-native species have received much attention and been well documented (e.g., Ebenhard 1988; D'Antonio and Vitousek 1992; Fritts and Rodda 1998).

While the consequences of invasion by some non-native species are certainly dire, others provide tremendous benefits to human society. In many cases, non-native species have come to form such a large part of our diet, livelihood, and traditions that it is hard to imagine a time without them. For example, consider the many foods (tomatoes, potatoes, etc.) that are not native to Europe but which are now integral to European cultures. A strikingly large portion of the food we eat, the clothes we wear, and the wood we use to build our homes comes from species grown outside of their native range. Most of these benefits come from non-native species that humans have domesticated, but we also derive benefits from many non-domesticated species (e.g., the bees that pollinate our crops and introduced mammals that we hunt for food). Non-native species clearly have a checkered legacy.

The persistent threats posed by non-native species, particularly to ecological systems, have led to a proliferation of research addressing applied questions, including: What are the impacts of invasive species and how can we manage systems to minimize these impacts? (e.g., Dodd 1959; Ebenhard 1988; Vitousek and Walker 1989: Keenan et al. 1997; Crooks 1998; Fritts and Rodda 1998; Kriwoken and Hedge 2000); and, What are the characteristics of species that make them likely to become invasive, and what are the characteristics of communities that make them likely to be invaded? (e.g., Forcella and Wood 1984; Fox and Fox 1986; Newsome and Noble 1986; Crawley 1987; Mack 1995; Rejmánek 1996; Veltman 1996; Blackburn and Duncan 2001; Cassey 2002). Research on these questions (particularly the latter two) was spearheaded and popularized by several influential books published in the late 1980s (Groves and Burdon 1986; Mooney and Drake 1986; Drake et al. 1989). Since that time, a great amount of research on invasions has been performed and the field of "invasion biology" has emerged as a rich discipline focusing on the diverse problems posed by non-native species.

Species Invasions Help Solve Puzzles in Ecology, Evolution, and Biogeography

The chapters in this book take a different course. Instead of focusing on applied research, we consider the insights we can gain about the natural world from the study of species invasions. We posit that studying species invasions can provide a rich source of knowledge about fundamental issues in ecology, evolution, and biogeography. The power of species invasions as a research tool comes from several characteristics. First, invasions allow us to observe processes in real time, rather than having to infer the operation of processes that occurred in the past solely from the patterns they generated. Second, the exact time, place, and characteristics of species introductions are sometimes known. As a result, we can measure many rate processes (such as genetic change and spatial spread) directly; these processes are difficult to study with long-established native species. Third, many different types of species have been introduced in great numbers and in many locations. The scope of these species-addition "experiments" is so large that we can examine ecological and evolutionary processes over spatial and temporal scales that would otherwise be impossible to study. Collectively, we believe these unique windows into the natural world may provide the missing pieces necessary to solve many long-standing puzzles in ecology, evolution, and biogeography.

Discerning observers of the natural world probably have been drawing insights from non-native species for a very long time. The first written accounts that we are aware of come from the writings of Darwin. In explaining how native species could have coped with climate changes during glacial periods, Darwin wrote in 1859 that

> The Glacial period, as measured by years, must have been very long; and when we remember over what vast spaces some naturalised plants and animals have spread within a few centuries, this period will have been ample for any amount of migration.

Such a basic inference—that species could migrate or otherwise modify their geographical ranges by great distances in a relatively short period of time—was later supported by fossil data (e.g., Williams et al. 2001), but would have been difficult if not impossible to discern at the time of Darwin's writing without drawing insight from non-native species.

It was some time after Darwin, however, when the first explicit mention of non-native species as an "experiment" was suggested. Joseph Grinnell, one of the most influential naturalist, ecologist and biogeographers of the early twentieth century wrote a noteworthy paper in 1919, titled "The English House Sparrow has arrived in Death Valley: An experiment in nature." Grinnell's paper suggested that the arrival of English house sparrows in Death Valley (one of the warmest and least humid sites in North America) presented an experiment in which the potential for phenotypic change could be measured over time. Interestingly, this same species, *Passer domesticus*, elicited further attention in this light later in the century. Johnston and Selander in a series of publications (1964, 1967, 1971, and 1973) explored the evolution of "races" of house sparrows in North America. Among their findings was the observation that significant population variation in body size and coloration conforming to Bergmann's and Gloger's Rules, respectively, had formed in as few as 50 years (Johnston and Selander 1964); this finding provided valuable insights into potential rates of ecotypic diversification and evolutionary change.

The seminal volume exploring insights from species invasions was the published proceedings of a conference organized by H. G. Baker and G. L. Stebbins and held in Asilomar, California in 1964. This conference on "The Genetics of Colonizing Species" drew together some of the best minds of the time to ask what could be learned from species invasions. Most of the fertile ground plowed at that historic conference, however, has lain largely fallow since that time. Only relatively recently has this situation started to change.

In the past dozen years there has been a burgeoning of interest in non-native species. With this interest has come a growing appreciation of the role species invasions can play in improving our understanding of ecological and evolutionary questions. For example, David Lodge (1993) wrote an influential paper suggesting that the study of non-native species could "provide clues to long-standing issues" in ecology. Since that time, many investigations have used species invasions to draw inferences about the natural world. A few of these are work by Huey and Gilchrist (2000) studying evolutionary processes in introduced fruit flies; work by Sax (2001) exploring mechanisms for latitudinal gradients in diversity and geographical range size; work by Levin (2003) on the process of speciation; and work by Zacherl et al. (2003) exploring the mechanisms that set species' geographical range limits. We believe that these studies are but the tip of a large iceberg of work that will emerge in the coming decades. This book itself is a good start on this process, but it is not alone. A forthcoming book edited by Cadotte et al. (2005) shares many of the goals we outline here, particularly to draw insights from species invasions to better understand fundamental issues in ecology and evolutionary biology.

In this book, we did not attempt to bring together the leaders in the field of invasion biology (although we have included a few), but tried instead to draw together leaders and emerging leaders in the fields of ecology, evolution, and biogeography. We asked these individuals to review a topic or consider a problem that they typically study, but to do so in light of the advantages or additional information that species invasions might provide. In many cases, this request attracted biologists who typically have nothing to do with species invasions, or at least not with non-native species *per se*. This choice was deliberate. One need not be an "invasion biologist" to draw insights and inferences from species invasions. Indeed, we hope this book will help broaden the range of scientists who incorporate species invasions into their research.

Terminology, Structure, and Intellectual Contributions

There are currently a wide variety of terms used to describe non-native species and their various perceived stages of establishment and affect (see Richardson et al. 2000). The appropriate use of this terminology is a subject of debate and presently is fairly contentious (e.g. Richardson et al. 2000; Chew and Laubichler 2003; Larson et al. 2005). We have done our best, on the one hand, to standardize the total set of terms used to describe non-native species without, on the other hand, constraining authors to such a point that they were forced to use terms they didn't find appropriate. In this light, the terms *non-native, exotic, alien,* and *introduced* are used interchangeably among chapters to refer to those species that are not indigenous to a region in question. These terms do not necessarily imply (even if it is the case) that these species are *established* in that region, in that these populations have not necessarily formed self-sustaining populations. The term *naturalized* refers specifically to those non-native species that are established. We have tried to restrict the use of the term *invasive* to mean those species that are known to cause ecological or economic damage, but in a few cases (in each case indicated by the author) *invasive* has been used instead to denote a species that is naturalized and has spread broadly within its newly occupied region. We use the term *species invasion* to refer to any species that has occupied a region in which it was not present historically or prior (in the case of paleontological studies) to some reference point of interest. It is important to remember that all of these terms are somewhat arbitrary, because nearly all species have spread at one point or another to and from any one geographic space.

The book is organized into three principal sections that focus on insights into (1) ecology, (2) evolution, and (3) biogeography, respectively. While any such divisions in a book of this nature are artificial, we believe this structure helps to enhance an appreciation of the broad set of fields to which such studies can contribute. The chapters themselves are introduced in each section by a short essay. These introductory essays provide a specific context for the work discussed in each section.

The chapters cover a diverse set of topics and take a wide range of approaches. Some produce results that we suspect will be somewhat controversial. Bruno et al. (Chapter 1) suggest that the role of competition has been greatly overestimated and that facilitation and predation are equally important structuring forces in community ecology. Blackburn and Gaston (Chapter 4) suggest that extinction is a largely idiosyncratic process and that we can expect to learn little that will allow us to predict its influence on species demise. Huey et al. (Chapter 6) suggest that evolution can be exceedingly fast, so that strong evolutionary changes can take place in a species within a single decade following introduction. Novak and Mack (Chapter 8) and Wares et al. (Chapter 9) both suggest that genetic bottlenecks are more rare and weaker forces than have previously been appreciated. Vermeij (Chapter 12) suggests that species invasions historically have led to the long-term promotion of species diversity, and that the results of today's invasions should be no different. McKinney and Lockwood (Chapter 14) suggest that species abundance data offers relatively little information in our attempts to understand processes of biotic homogenization. Sax et al. (Chapter 17) suggest that although species diversity is generally increasing as a consequence of species invasions that ultimately there may be a "species capacity," or maximum number of species that any given place can support.

Other chapters offer constructs and theoretical frameworks that should be useful in better understanding both species invasions and fundamental processes. Stachowicz and Tilman (Chapter 2) employ theory of inherent trade-offs in species interactions to explain species invasions and better understand species coexistence. Ricklefs (Chapter 7) suggests that taxon cycle theory can help us understand species invasions and, reciprocally, that species invasions can aid in understanding taxon cycles. Holt et al. (Chapter 10) draw on theory of rapid evolutionary change as a mechanism to rescue populations with suboptimal ecological fits to the introduced environment. Callaway et al. (Chapter 13) explore how transcontinental reciprocal transplants and experiments can shed light on mosaic patterns of coevolution. Labra et al. (Chapter 16) use minimum spanning trees and scaling approaches to study biogeographic patterns of species abundance.

Still other chapters review key topics and draw out directions for future research. D'Antonio and Hobbie (Chapter 3) review ecosystem consequences and insights from invasions. Lafferty et al. (Chapter 5) explore the role of parasites and pathogens in community structure and distribution. Rice and Sax (Chapter 11) consider the functioning of introgression, sexual reproduction, and other evolutionary processes; and Kinlan and Hastings (Chapter 15) consider range expansion and dispersal on both the land and in the sea.

Individually, we believe that each of these chapters has something significant to offer. Collectively, we hope that this book has much to offer to both invasion biology and to our fundamental understanding of ecology, evolution, and biogeography.

Literature Cited

Baker, H. G., and G. L. Stebbins, eds. 1965. *The genetics of colonizing species*. Academic Press, New York.

Blackburn, T. M., and R. P. Duncan. 2001. Determinants of establishment success in introduced birds. Nature 414:195–197.

Cassey, P. 2002. Life history and ecology influences establishment success of introduced land birds. Biological Journal of the Linnean Society 76:465–480.

Chew, M. K., and M. D. Laubichler. 2003. Natural enemies: Metaphor or misconception? Science 301:52–53.

Drake, J. A., H. A. Mooney, F. di Castri, R. H. Groves, F. J. Kruger, M. Rejmánek and M. Williamson, eds. 1989. Biological invasions: A global perspective. Wiley, New York.

Cadotte, M. W., S. M. McMahon, and T. Fukami. 2005. *Conceptual ecology and invasions biology: Reciprocal approaches to nature*. Kluwer. *In press*.

Crawley, M. J. 1987. What makes a community invasible? Symposium of the British Ecological Society. 26:429–453.

Crooks, J. A. 1998. Habitat alteration and community-level effects of an exotic mussel, *Musculista senhousia*. Marine Ecology Progress Series 162:137–152.

D'Antonio, C. M., and P. M. Vitousek. 1992. Biological Invasions by exotic grasses, the grass/fire cycle, and global change. Annual Review of Ecology and Systematics 23:63–87.

Darwin, C. 1859. *On the origin of species*. Murray, London.

Dodd, A. P. 1959. The biological control of prickly pear in Australia. In A. E. M. Nairn, ed. *Biogeography and Ecology in Australia*, pp. 565–577. Dr. W. Junk BV, The Hague.

Ebenhard, T. 1988. Introduced birds and mammals and their ecological effects. Swedish Wildlife Research 13:1–107.

Elton, C. S. 1958. *The ecology of invasions by animals and plants*. Methuen and Co., London.

Forcella, F., and J. T. Wood. 1984. Colonization potential of alien weeds are related to their "native" distributions, implications for plant quarantine.

Journal of the Australian Institute of Agricultural Science 50:35–41.

Fox, M. D., and B. J. Fox. 1986. The susceptibility of natural communities to invasion. In R. H. Groves and J. J. Burdon, eds. *Ecology of biological invasions*, pp. 57–76. Cambridge University Press, Cambridge.

Fritts, T. H., and G. H. Rodda. 1998. The role of introduced species in the degradation of island ecosystems: a case history of Guam. Annual Review of Ecology and Systematics 29:113–140.

Grinnell, J. 1919. The English House Sparrow has arrived in Death Valley: An experiment in Nature. American Naturalist 53:468–472.

Groves, R. H., and J. J. Burdon, eds. 1986. *Ecology of biological invasions*. Cambridge University Press, Cambridge.

Johnston, R. F., and R. K. Selander. 1964. House Sparrows: Rapid evolution of races in North America. Science 144:548–550.

Johnston, R. F., and R. K. Selander. 1967. Evolution of the House Sparrow. I. Intrapopulation variation in North America. The Condor 69:217–258.

Johnston, R. F., and R. K. Selander. 1971. Evolution of the House Sparrow. II. Adaptive differentiation in North American populations. Evolution 25:1–28.

Johnston, R. F., and R. K. Selander. 1973. Evolution of the House Sparrow. III. Variation in size and sexual dimorphism in Europe and North and South America. American Naturalist 107:373–390.

Keenan, R., D. Lamb, O. Woldring, T. Irvine, and R. Jensen. 1997. Restoration of plant biodiversity beneath tropical tree plantations in northern Australia. Forestry Ecology and Management 99:117–131.

Kriwoken, L. K., and P. Hedge. 2000. Exotic species and estuaries: Managing *Spartina anglica* in Tasmania, Australia. Ocean & Coastal Management 43:573–584.

Huey, R. B., G. W. Gilchrist, M. L. Carlson, D. Berrigan, and L. Serra. 2000. Rapid evolution of a geographic cline in size in an introduced fly. Science 287: 308–309.

Larson, B. M. H., B. Nerlich, and P. Wallis. 2005. Metaphors and biorisks: The war on infectious disease and invasive species. Science Communication 26:243–268.

Levin, D. A. 2003. Ecological speciation: Lessons from invasive species. Systematic Botany 28:643–650.

Lodge, D. M. 1993. Biological invasions: Lessons for ecology. Trends in Ecology and Evolution 8:133–137.

Mack, R. N. 1995. Understanding the process of weed invasions, the influence of environmental stochasticity. Weeds in a changing world, BCPC Symposium Proceedings number 64, British Crop Protection Council.

Mooney, H. A., and J. A. Drake, eds. 1986. *Ecology of biological invasions of North America and Hawaii*. Springer-Verlag, New York.

Newsome, A. E. and I. R. Noble. 1986. Ecological and physiological characters of invading species. In R. H. Groves and J. J. Burdon, eds. *Ecology of biological invasions*, pp. 1–20. Cambridge University Press, Cambridge.

Rejmánek, M. 1996. A theory of seed plant invasiveness, the first sketch. Biological Conservation 78:171–181.

Richardson, D. M., P. Pyšek, M. Rejmánek, M. G. Barbour, F. D. Panetta, and C. J. West. 2000. Naturalization and invasion of alien plants: concepts and definitions. Diversity and Distributions 6:93–107.

Sax, D. F. 2001. Latitudinal gradients and geographic ranges of exotic species: implications for biogeography. Journal of Biogeography 28:139–150.

Veltman, C. J., S. Nee, and M. J. Crawley. 1996. Correlates of introduction success in exotic New Zealand birds. American Naturalist 147:542–557.

Vitousek, P. M., and L. R. Walker. 1989. Biological invasion by *Myrica faya* in Hawaii: Plant demography, nitrogen fixation, ecosystem effects. Ecological Monographs 59:247–265.

Williams, J. W., B. N. Shuman, and T. Webb. 2001. Dissimilarity analyses of late-Quaternary vegetation and climate in eastern North America. Ecology 82:3346–3362.

Zacherl, D., S. D. Gaines, and S. I. Lonhart. 2003. The limits to biogeographical distributions: insights from the northward range extension of the marine snail, *Kelletia kelletii* (Forbes, 1852). Journal of Biogeography 30:913–924

PART I

Insights into

ECOLOGY

Phillip Cassey

The transportation and successful introduction of non-native species to new locations and habitats is a pervasive component of human-induced global change, one that presents distinct challenges for the conservation and management of biological diversity. For ecologists, the ongoing transportation of species also provides an unprecedented opportunity to study directly the processes that structure the distribution and abundance of species, their community associations, and their evolutionary relationships. Because relationships between species, and between species and habitats, change through time, the patterns we can observe and manipulate today may bear little resemblance to the historical processes and interactions that created them. Non-native species thus provide a unique opportunity to directly study these processes, and to quantify how they change through time. Importantly, many researchers believe that the geographic scales at which humans transport species (and thus biological invasions occur) help to facilitate the development of significant insights into the complex nature of ecological systems (Holt et al. 1997; Maurer 1999).

The first major synthesis to frame an ecological theory with examples from non-native species was Charles Elton's monograph *The Ecology of Invasions by Animals and Plants* (1958). Elton presented a series of examples supporting his assertion that community structure influences temporal changes in a community and its constituent species—and most notably supporting the premise that simple communities are less stable than complex ones. Moreover, Elton was fascinated by many anecdotal examples provided by non-native species: their population variability; their potential abundance; the rate at which they might increase in a recipient community; the extent of damage they might produce; and how readily communities might be invaded. Elton's interest in what non-native species can tell us about natural systems was typically prescient: 50 years later, studies of non-native species are a core element of ecological research.

The single largest research effort to synthesize both empirical and theoretical studies on biological invasions was the SCOPE (Scientific Committee on Problems of the Environment) program on "The Ecology of

Biological Invasions." The SCOPE initiative ran for more than 10 years and produced at least 15 edited volumes (see references in Williamson 1996). The two core questions asked by the program's participants were, What factors determine whether a species will become an invader or not?, and What site properties determine whether an ecological system will be prone to, or resistant to, invasions? Both of these questions have clear analogs in population and community ecology. However, despite the enormous wealth of primary studies and the large number of researchers who contributed to the SCOPE effort, it was generally acknowledged that the program's results suffered from a heavy reliance on anecdotal examples from single-species studies, and that these studies largely failed to resolve many of the idiosyncrasies that are inherent in the introduction of non-native taxa (e.g., Lodge 1993; Williamson 1999).

In the past 10 years, the greatest advances in studies of non-native species have resulted from researchers who responded to these criticisms and developed quantitative methods for synthesising comparative data that spans biogeographic regions, phylogenetic hypotheses, and long periods of historical transportation (e.g., Blackburn and Gaston, Chapter 4; see also Prinzig et al. 2002; Cassey et al. 2004; Jeschke and Strayer 2005). The chapters in this section describe much of the progress that has taken place to date. They also open multiple doors for future research in both the study of the characteristics shared by non-native species and the variety of ecological effects that result when these species become established.

When non-native species establish new populations, they forge new interactions with species in their recipient communities. This is not a contentious point—every species interacts with at least some other species. Non-native species must therefore affect the species already present in a community in some way, whether the consequences are minor or major. It is intuitive, then, that non-native species are able to facilitate ecosystem change, and that this change is most likely to be detectable when invasions occur rapidly (e.g., Vitousek et al. 1987). Indeed, in locations where non-native species have established repeatedly (such as the Hawaiian Islands), they often spread and become abundant quickly (Williamson 1996). Such situations provide researchers with unique opportunities to evaluate how an ecosystem's characteristics can influence the magnitude of species effects. Notably, we have previously lacked systematic studies that compare the range of effects across either related taxa or different regions. Whereas most studies have focused on single species, the proportional diversity and density of non-native species can, for some taxa, approach 100% within recipient communities (see McKinney and Lockwood, Chapter 14). Studies of these newly assembled communities will likely provide the best opportunities for critically examining detectable changes in ecological relationships (e.g., species abundances, range sizes) as well as trophic and other interactions. Comparisons of non-native species traits versus resident species traits across a range of habitats (controlling for similarities among related species) can help to identify what types of ecosystems are most affected and the magnitude of these ecosystem changes (see Chapter 3 by D'Antonio and Hobbie and Chapter 5 by Lafferty et al.).

In the past it has been difficult for ecologists to quantify the importance of even the most universal interactions (such as competition, predation, and facilitation) using only contemporary measurements, because the strengths of interactions measured today have been shaped by a multitude of unobserved interactions in the past (Connell 1980).

The study of non-native species provides a partial solution to this problem of shared evolutionary history, since interactions among exotic and native species (and among exotics themselves) are influenced largely by contemporary ecological processes. Thus non-native species provide natural "experiments" in the absence of long-term coevolution (Lodge 1993). These repeated "species-addition experiments" in a wide range of habitat types allow their effects in driving community structure and composition to be assessed over much broader spatial scales (landscapes to regions) than typical scientific addition and removal experiments allow, as described by Bruno et al. in Chapter 1.

In Chapter 2, Stachowicz and Tilman show that studying invaded communities offers insights into the fundamental processes that influence community assembly, composition, structure, and diversity. At least in the short term, many introductions of non-native species have resulted in a net gain in the number of species present at the local or regional level (Sax and Gaines 2003). This is similar to what has been observed from geological periods of biotic interchange, during which most regions gain more donor species than they lose through the extinction of resident species. To date it has been difficult to conclude whether overall invasion rates increase or decrease with the continued addition of non-native species. Most notably, it remains to be proven whether or not the differences in the quality and magnitude of human-induced invasions (and the resulting effects on biodiversity and ecological processes at various scales) are different from those observed from geological periods of biotic interchange (see Vermeij, Chapter 12). Future studies of introduction-level factors (e.g., propagule pressure), species-level factors (e.g., reproductive traits), and environment-level factors (e.g., disturbance regimes) are all needed in order to determine the var-

ious influences these factors can have on the net change of invasion rates.

More and more, ecological researchers are using non-native and invasive species as a core for their studies and publications (see Simberloff 2004). This is unsurprising given both the increasing number of non-native species and the ecological and social consequences of the few dramatic invasions that have resulted in radical community changes and strong interactions with resident species. Consequently, current ecological theory and practice is being based (and perhaps even becoming reliant) on non-native species and their recipient communities. Indeed, it is the major premise of the following chapters that we can substantially advance our understanding of longstanding and classical ecological issues through directly observing the consequences of introduced non-native species.

Interestingly, it has been claimed that studies of "natural" systems, in which coevolutionary processes have taken place over a long period of time, offer a more realistic basis for understanding complex issues in evolutionary ecology (e.g., Clegg et al. 2002). Regardless of one's perspective on this point, it is obvious that non-native species are prevalent and increasingly ubiquitous members of global ecological communities. Whether or not non-native species are easier to study than resident native species, or are useful tools for observing changes in ecological processes, they are a major component of human-wrought global change and undoubtedly will remain a persistent feature of global ecosystems in the future (but see Simberloff and Gibbons 2004).

The rationale for studying non-native species is largely driven by the multitude of interactions that non-native species *can* have with resident species. There is every indication that the rate at which locations are accumulating non-native species is accelerating as free trade and globalization proceed. In the great proportion of locations and environments, the interactions between non-native and resident species are poorly studied and their impacts incompletely understood. The roles that ecologists choose to take when studying non-native species can not only address existing controversies regarding the processes that form and structure ecological assemblages, but can also assist society in responding to the changes (and future threats) that non-native species present when they alter the structure and functioning of ecosystems. The following chapters build on the foundations laid by Elton and further developed by the SCOPE program. They mark the way for future research agendas in the ecological study of non-native species.

Literature Cited

Cassey, P., T. M. Blackburn, D. Sol, R. P. Duncan, and J. L. Lockwood. 2004. Global patterns of introduction effort and establishment success in birds. Proceedings of the Royal Society of London, Series B 271: S405–S408.

Clegg, S. M., S. M. Degnan, C. Moritz, A. Estoup, J. Kikkawa, and I. P. F. Owens. 2002. Microevolution in island forms: The roles of drift and directional selection in morphological divergence of a passerine bird. Evolution 56:2090–2099.

Connell, J. H. 1980. Diversity and coevolution of competitors, or ghost of competition past. Oikos 35:131–138.

Elton, C. 1958. *The ecology of invasions by animals and plants*. Methuen, London.

Holt, R. D., J. H. Lawton, K. J. Gaston, and T. M. Blackburn. 1997. On the relationship between range size and abundance: Back to the basics. Oikos 78:183–190.

Jeschke, J. M., and D. L. Strayer. 2005. Invasion success of vertebrates in Europe and North America. Proceedings of the National Academy of Sciences USA. In press.

Lodge, D. M. 1993. Biological invasions: Lessons for ecology. Trends in Ecology and Evolution 8:133–137.

Maurer, B. A. 1999. *Untangling ecological complexity*. University of Chicago Press, Chicago.

Prinzig, A., W. Durka, S. Klotz, and R. Brandl. 2002. Which species become aliens? Evolutionary Ecology Research 4:385–405.

Sax, D. F., and S. D. Gaines. 2003. Species diversity: From global decreases to local increases. Trends in Ecology and Evolution 18:561–566.

Simberloff, D., and L. Gibbons. 2004. Now you see them, now you don't! Population crashes of established introduced species. Biological Invasions 6:161–172.

Simberloff, D. 2004. A rising tide of species and literature: A review of some recent books on biological invasions. Bioscience 54:247–254.

Vitousek, P. M., L. Walker, L. Whiteaker, D. Mueller-Dombois, and P. Matson. 1987. Biological invasion by *Myrica faya* alters ecosystem development in Hawaii. Science 238:802–804.

Williamson, M. H. 1996. *Biological invasions*. Chapman & Hall, London.

Williamson, M. H. 1999. Invasions. Ecography 22:5–12.

1

Insights into Biotic Interactions from Studies of Species Invasions

John F. Bruno, Jason D. Fridley, Keryn D. Bromberg, and Mark D. Bertness

We took advantage of the extensive literature on invasion ecology to gain insights into the nature, frequency and role of biotic interactions in structuring ecological communities. Specifically, we reviewed 120 studies of exotic species that considered how biotic interactions influence populations and community structure and synthesized a number of related reviews and publications on biotic interactions. We make five general conclusions—some unsurprising, others likely contentious—and discuss their implications for ecological theory and conservation. (1) Biotic interactions are common and can have substantial effects on populations and communities. (2) Competition is not the dominant force structuring communities, but is one of several important factors. Competition occurs relatively frequently, but rarely limits immigration or leads to local extinction. (3) Evidence for facilitation exceeds its perceived influence. Facilitation of native species by exotics is common, and some exotic habitat-forming species have increased native species richness, especially in aquatic habitats. (4) A relatively small proportion of facilitative and competitive species interactions are coevolved or species-specific. In contrast, many consumer-prey interactions are somewhat specific and do appear coevolved. (5) Propagule supply strongly influences community structure, and many communities are unsaturated.

Introduction

Understanding the frequency, net effects, and relative importance of biotic interactions has posed a major challenge to generations of ecologists. Early theoretical (Lotka 1925; Volterra 1926) and laboratory (Gause 1934; Park 1948; Huffaker 1958) studies focused on competition and predation as the major processes regulating populations and structuring communities. Subsequent theory has sought to explain community structure via evolved species differences that limit competitive exclusion, particularly with reference to the niche (Hutchinson 1957, 1959; Elton 1958; MacArthur and Levins 1967; MacArthur 1970). The role of trophic interactions in controlling both community- and ecosystem-level processes has also received extensive attention from theoretical and empirical ecologists (Hairston et al. 1960; Paine 1966). However, 30 years ago, ecologists began to question the ultimate importance of competition (Connell 1975; Simberloff 1981; Connor and Simberloff 1983; Lawton 1984). More recently, positive species interactions, or facilitation, received renewed attention as an important driver of community structure (Callaway and Walker 1997; Bruno et al. 2003a), particularly in stressful environments (Bertness and Callaway 1994; Bruno and Bertness 2001).

Currently few generalities about the relative frequency and importance of the three key classes of species interactions—competition, predation, and facilitation—have emerged. For example, two extensive reviews of competition published in 1983 came to somewhat different conclusions. Schoener (1983) found that competition was nearly universally important, while Connell (1983) argued that less than half of field experiments found evidence of interspecific competition and that its importance varied substantially among taxa, trophic levels, and systems. The ability of predators to control prey populations in different contexts is also a perennial topic of debate (Hairston et al. 1960; Menge and Sutherland 1976; Leonard et al. 1998; Silliman and Bertness 2002), and the importance of facilitation across a wide variety of ecosystems is still controversial (Bruno et al. 2003a).

As Connell (1980) recognized, it is notoriously difficult to quantify the importance of a process like competition using only contemporary measurements of interaction strengths because interactions measured today have been shaped by interactions in the past. One way to overcome the problem of shared evolutionary history when determining how species interactions influence community structure is to evaluate interactions between species with no historical geographic overlap and thus limited ecological and evolutionary exposure. Invasions of exotic species allow studies of species interactions that are influenced largely by contemporary ecological processes and thus provide natural "experiments" in the absence of long-term coevolution (Lodge 1993). These conveniently replicated additions of competitors, predators, and facilitators in a wide range of habitat types allow us to assess their effects on communities over much broader spatial scales (landscapes to regions) than typical scientific addition and removal experiments allow. Because many of these invasions began decades or centuries ago, problems related to slow or delayed population responses are minimized. Organisms or propagules are added to resident assemblages through a variety of human activities, overcoming barriers to dispersal and immigration that previously limited the distributions of many species. Studies of exotic invasions thus provide a means to evaluate the relative roles of different types of species interactions in driving community structure and composition while reducing the need to consider the limits on interaction intensities imposed by evolutionary history and immigration processes (Simberloff 1981; Lodge 1993).

We surveyed the literature on species invasions to answer several questions about biotic interactions. Specifically, we asked: (1) What are the relative frequencies and importance of competition, predation, and facilitation as determinants of community structure? (2) How do they vary among taxa and habitats? (3) How specialized are biotic interactions—do interactions between species with no shared evolutionary history differ from those between species pairs with historically overlapping distributions? We also synthesized several related reviews and other recent empirical work on species interactions and invasions to address some broader questions and implications for community ecology and conservation biology. We consider the role of propagule limitation, ask whether communities are saturated, and outline the theoretical predictions and realized effects of increasing the size of regional species pools via species introductions. We outline recent evidence that many exotic habitat-forming species have striking and largely unexpected positive effects on native species. Finally, we discuss the repercussions of the uncritical adoption of the competition paradigm—that communities are saturated and are effectively structured by competitive processes, and that competition often leads to local extinction of subordinate species (Simberloff 1981). We argue that this perspective has severely limited our ability to understand the processes driving species introductions and their subsequent effects on native communities.

A Review of the Literature on Species Invasions

We focused our review on studies of exotic species—species that have colonized or are in the process of colonizing whole regions in which they have not been present historically—that considered how species interactions influence community structure (species abundances, distributions, composition, and diversity). We included unidirectional and bidirectional interactions between exotics and natives and between two or more exotic species (in their non-native habitat). Because all of the papers we considered have been peer-reviewed, we did not reevaluate evidence or question the general findings of the studies. We excluded papers focusing on the biology of a single exotic species (e.g., on stress tolerance, competitive ability, or intraspecific interactions). We considered only studies of "natural" (nonagricultural) communities and thus did not include studies with an agronomic or biocontrol focus. We considered both experimental and descriptive studies on both plants and animals in a wide variety of terrestrial and aquatic ecosystems.

We searched for relevant papers in peer-reviewed ecological journals between 1981 and 2003. We used studies published in 1981 and after to preclude overlap with Simberloff's (1981) review of the community effects of exotic species. Our search results indicated that studies of community ecology considering exotic species became much more common in the late 1990s; the majority of the papers scored in this study (66%) were published between 1996 and 2003. We constructed a pool of relevant papers by a series of refinements to an initial database query. First, we performed a series of database searches using the ISI Science Citation Index Expanded (ISI Web of Science 2003) using the search terms ("invas*" or "exotic") and ["communit*" not (patient* or infect*)]. Our combined initial searches produced approximately 700 papers on exotic species. We then selected papers from this pool for further analysis if they considered each of the following four factors: (1) one or more exotic species, (2) at least two species in total, (3) one or more species interactions, and (4) some type of community-level response variable (e.g., the abundance, distribution, or diversity of native or exotic species). Our final selection pool included 120 papers, 86 of which were from the one of the following journals: *Conservation Biology, Ecology, Ecological Applications, Ecological Monographs, Ecology Letters, American Naturalist, Oikos, Oecologia, Journal of Ecology, Journal of Animal Ecology, Biological Invasions, Nature,* and *Science*. For each of these 120 papers, we scored which biotic interaction each paper explicitly tested and whether or not the paper found evidence for a significant effect of that interaction. We also calculated the frequency of six different categories of effects of exotics on native species and communities and on other exotic species. Although the "vote counting" technique we used has a number of well-known flaws (Gurevitch et al. 1992), we could not perform a more quantitative statistical meta-analysis of the results in these papers because of the large range of response variables we considered (e.g., individual and population growth rates, mortality, richness, distribution, fertilization success, etc.).

What does the literature address?

The 120 papers that did consider the effects of biotic interactions among exotic species and between exotic and native species were somewhat biased toward certain groups, but generally represented a broad range of habitats, taxa, and trophic levels. Most studies were in terrestrial systems, approximately half considered plants and their interactions with other plants or animals, and only 18% focused on vertebrates (Table 1.1). Many of the invertebrate studies were on ants, spiders, or beetles, usually focusing on their role as herbivores or predators. Competition was the most commonly tested interaction. Most studies considered the role of only a single type of interaction, and only one (MacIsaac 1996) examined competition, facilitation, and predation simultaneously. The use of field experimentation was less common than we expected. Only 62 of

TABLE 1.1   *Coverage of topics in the 120 studies included in the literature review*

Study topic	Number (percentage) of studies
Organisms	
Plant	39 (33%)
Plant-Animal	26 (22%)
Vertebrate	21 (18%)
Invertebrate	32 (27%)
Habitat	
Freshwater	22 (18%)
Marine	15 (13%)
Terrestrial	83 (69%)
Interaction type	
Competition	73 (61%)
Predation	40 (33%)
Facilitation	40 (33%)

the 120 studies we analyzed performed any type of controlled and replicated manipulation (the rest were descriptive field studies). Like Simberloff's (1981), our review of more recent literature found that a large proportion of published exotic studies are anecdotal, often describing characteristics of exotic species and the habitats they invade, or speculating about the traits and environmental factors that facilitate or prevent invasion.

The review had a number of obvious limitations. As mentioned previously, the studies were biased toward some taxa, habitats, and categories of interactions. Like all literature reviews, our sample is biased toward those studies finding a significant positive result, particularly with respect to detecting a species interaction (i.e., studies that fail to find significant effects are less likely to be submitted and published). Few of the original pool of about 700 studies were experimental, measured community-level parameters, or considered more than one factor. Finally, we had to take for granted some dubious interpretations, such as the assumption that negative abundance correlations between two species were the result of competitive exclusion.

To begin to address some of the inherent problems with this type of literature analysis, we calculated two measures of the frequency of each main interaction type: (1) the proportion of studies in a given category (e.g., aquatic studies) reporting evidence of a significant interaction effect, and (2) the proportion of studies reporting evidence of a significant interaction effect that explicitly purported to test for that effect. The first measure was probably biased by differences in how frequently each interaction type was actually considered (see Table 1.1), while the second measure was possibly inflated because some authors, after failing to find a significant effect, may not have mentioned that test or hypothesis. Examining both measures probably improved our estimates of the true frequency and importance of each factor.

What do species invasions tell us about the role of biotic interactions?

Overall, we found strong evidence that species interactions play an important role in structuring natural communities; 82% of the 120 studies reported that an interaction had a significant effect (for a breakdown by interaction type, see Figure 1.1). However, in nearly half of the cases we examined, the exotic species

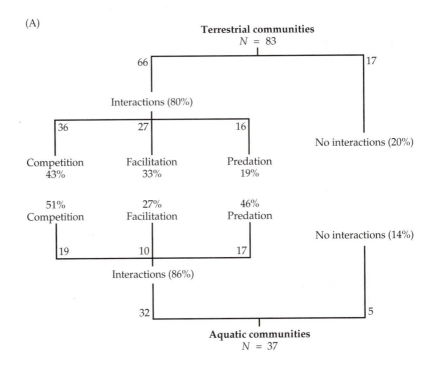

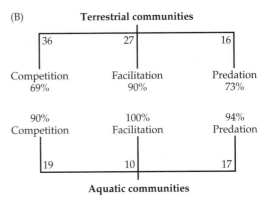

Figure 1.1   Classification trees partitioning the 120 studies into the processes that were found to influence community structure in terrestrial (top of panels A and B) and aquatic (bottom of panels A and B) communities. (A) The proportion of studies that found a significant interaction effect. (B) The proportion of studies that explicitly tested for an interaction and found a significant effect. Studies are included in multiple categories where appropriate (i.e., percentage sums may exceed 100%). Numbers at branches indicate the number of studies represented by each category.

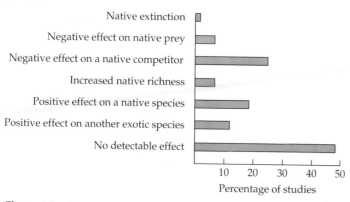

Figure 1.2   Percentages of the 120 papers from a literature review of studies that found evidence for each category of exotic impact.

had no detectable effect on any measured parameter (Figure 1.2). Simberloff (1981) estimated that 80% of species introductions have no effect, but we specifically targeted studies that tested for biotic interactions and so were more likely to find significant effects. Additionally, many effects of exotic species are indirect, which makes it difficult to quantify or even estimate their net effects. For instance, exotic grasses can compete with native plants yet facilitate native birds (McAdoo et al. 1989). Similarly, Grosholz et al. (2000) found that invasive green crabs (*Carcinus maenas*) have directly caused a decline of native clam abundance in soft-sediment marine habitats, but have simultaneously facilitated an increase in the abundance of native crustaceans and polychaetes disturbed by the burrowing of the native clam. Of the studies in our review that explicitly considered each interaction type (see Table 1.1), 75%, 83%, and 93% found a significant effect of competition, predation, and facilitation, respectively (46%, 28%, and 31% of all 120 studies).

The role of interspecific competition

We defined "competition" as a negative interaction between two species of the same trophic level that occurred as a result of competition for a shared resource (Connell 1983). We did not consider negative relationships mediated by trophic interactions with a third species (e.g., apparent competition) to be competition. Of the 73 studies that tested for competition, 55 found significant effects (see Figure 1.1). For example, exotic species including marine snails (Race 1982), freshwater fishes (Townsend 1996; Taniguchi et al. 2002), and terrestrial plants (Meyer and Florence 1996) were found to reduce the abundance and, in some cases, the diversity of native species, especially on islands (Lavergne et al. 1999). There were also examples of native species suppressing exotic colonizations, at least at small spatial scales. For instance, in the absence of disturbance, native

TABLE 1.2   *Percentage[a] of studies explicitly testing for the effects of predation or competition that found a significant effect of competition among or predator control of autotrophs and heterotrophs*

	Competition	Predator control
Autotrophs	78% (36)	67% (12)
Heterotrophs	73% (36)	89% (28)

[a] The sample size (number of studies reviewed) is given in parentheses.

grasses prevented invasion by the exotic plant *Conyza sumatrensis* (Case and Crawley 2000).

Our review found a higher frequency (75%) of relevant studies demonstrating competition than Connell (1983) did. He reviewed 72 papers published between 1974 and 1982 and reported that 43% of all experiments (on a *per species* basis) demonstrated interspecific competition. This conclusion was perhaps due in part to Connell's strict criteria for credible tests of competition and his exclusion of nonexperimental studies. Schoener (1983) employed similarly rigid criteria in his review of competition, but reported that 90% of 164 studies found competitive effects. Like Schoener, we did not correct for the number of response species being considered (Connell also found a frequency of 93% when he included studies of a single response species). Very few papers in the invasion literature have tested reciprocal effects between two competing species. Thus, like Schoener's (1983), our review overestimates the frequency of competitive effects on a *per species* basis (i.e., our calculated frequencies are *per study*). Like Connell (1983), we did not detect major differences in the frequency of competition between autotrophs and heterotrophs (Table 1.2) or between sessile and mobile species [80% ($n = 34$) and 69% ($n = 30$), respectively; for both comparisons, data were pooled across habitat types]. A more recent and restricted meta-analysis of competition in field experiments (Gurevitch et al. 1992) found that interspecific competition had a large effect on biomass and, surprisingly, that the effects were significantly greater among herbivores and filter feeders than among carnivores or plants.

The role of predation

We defined "predation" as a negative effect on one species occurring as the direct result of being consumed (including parasitism) by another. The effects of exotic predators and parasites on native prey in a wide variety of terrestrial and aquatic communities highlight the dominant role consumers play in regulating prey populations and community structure. There are many striking case studies, including that of the invasive ctenophore, *Mnemiopsis leidyi*, that substantially altered Black Sea zooplankton communities (Shiganova 1998), the current rapid decline of Canada hemlock (*Tsuga canadensis*) populations in the eastern United States due to hemlock wooly adelgid (*Adelges tsugae*) parasitism (McClure and Cheah 1999), and the effects of the Argentine ant (*Linepithema humile*) on native insects in Hawaii (Cole et al. 1992). Exotic fishes and other vertebrates seem to have especially large effects on native prey (Lodge 1993; Townsend 1996; Blackburn et al. 2004). Native prey species in insular ecosystems that have evolved in the absence of whole taxonomic classes of predators seem particularly vulnerable following the establishment of those predators (Blackburn et al. 2004). One of the best-known examples is the decimation of the avifauna on Guam by the exotic brown tree snake (*Boiga irregularis*) (Savidge 1987). Several studies have also found compelling evidence that native consumers have limited exotic populations or prevented invasion (Lodge 1993). Native herbivores can limit the spread and abundance of invasive exotic plants (D'Antonio 1993; Case and Crawley 2000), and native fishes can limit exotic bivalve populations through predation (Robinson and Wellborn 1988; Magoulick and Lewis 2002). The frequency of studies in our survey indicating predator control of exotic or native prey species was about 20% higher for heterotrophic prey than for autotrophic prey (see Table 1.2). We were not able to use our survey to test the prediction that herbivores are more frequently controlled by consumers than are plants or carnivores (Hairston et al. 1960) because omnivory was common and in many cases we could not reliably categorize species within specific consumer trophic levels. The effects of exotic predators are generally thought to be greater than those of exotic competitors (Connell 1975; Simberloff 1981; Ricklefs 1987), although recent analyses question this belief (Gurevitch and Padilla 2004). In fact, of the studies in our review that tested for it, 41% found evidence for deleterious exotic competitor effects on natives, while only 20% demonstrated negative exotic consumer effects (25% and 7% of all 120 studies, respectively; see Figure 1.2).

The role of interspecific facilitation

We defined "facilitation" broadly as direct positive interactions between two organisms that benefit at least one (Hacker and Gaines 1997; Stachowicz 2001), not including trophic interactions in which one species benefits from consuming another. This definition includes commensalisms (in which one species benefits and the other is unaffected by the interaction) and mutualisms (in which both species benefit from the interaction) as well as mutually obligate and facultative relationships (Bronstein 1994). Facilitations generally occur when one organism makes the local environment more favorable for another by reducing environmental stress or the negative effects of enemies. We also included reproductive mutualisms that increase pollination success or propagule dispersal.

In general, facilitation has received much less attention, and is perceived to be less important in structuring communities, than competition and predation. Facilitation is not included in most ecological theories, is only briefly mentioned in most textbooks, and is the focus of few empirical studies (Callaway 1995; Bruno and Bertness 2001). For example, at the August 2004 meeting of the Ecological Society of America (ESA), 175 talks or posters were on competition, 141 covered predation, and only 43 focused on some aspect of facilitation ($n = 2709$ abstracts). Likewise, in the peer-reviewed journals, *Ecology* and *Ecological Monographs,* published by the ESA, only 108 papers during the last 25 years were on facilitation, while 814 and 804 were on competition and predation, respectively (Figure 1.3).

Overall, our review suggests that direct facilitative interactions are at least as common and important as competition and predation in structuring communities (see Figure 1.1). In several cases, native species facilitated exotic colonization and spread via habitat modification or reproductive mutualisms (e.g., native vertebrates and insects consumed and dispersed exotic fruits and seeds: Glyphis et al. 1981; Bossard 1991; Vilà and D'Antonio 1998). Exotics also frequently facilitated native species and other exotics (see Figure 1.2). In 8 of the 120 studies we examined, and in 53% of the studies that explicitly tested

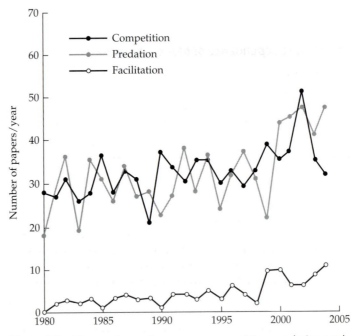

Figure 1.3 The numbers of papers on competition, predation, and facilitation in the peer-reviewed journals *Ecology* and *Ecological Monographs* during the last 25 years. Data are from an electronic search for the terms *competition, predation,* and *facilitation* or *mutualism* in the abstracts of all published papers in these journals between 1980 and 2004.

for it, exotic species increased native richness. In contrast, only 2 studies (Savidge 1987; Holway et al. 1998) found evidence of local (i.e., at the scale of the experiment, usually within small plots < 1 hectare) native extinction caused by exotic predators or competitors (see Figure 1.2). For example, the invasive freshwater snail *Potamopyrgus antipodarum* facilitates several native invertebrates and increases benthic species richness (Schreiber et al. 2002). Similarly, Wonham et al. (2005) found that the invasive mud snail *Batillaria attramentaria* increases the abundance of several other species, including native hermit crabs and exotic anemones, snails, and eelgrass, through habitat provision, bioturbation, and grazing. Two recent reviews of exotic-exotic facilitation (Simberloff and Von Holle 1999; Richardson et al. 2000) found that the facilitation of newly arriving exotic species by established exotics was ubiquitous. Richardson et al. (2000) outlined a variety of facilitative mechanisms through which both native and exotic resident species facilitated exotic plant invasions, including pollination mutualisms, mycorrhizal associations, and countless examples of dispersal by native birds, reptiles, mammals, and invertebrates. They argued that a key lesson of invasion ecology was that facilitation by numerous resident species was often required for successful colonization and that positive interactions are at least as important as competition in structuring communities.

The context dependence of biotic interactions

Ecologists have long recognized that the relative importance of ecological factors varies in space and time (Connell 1983). The explicit recognition of context dependence became incorporated into ecological theory in the 1970s when several models were developed to predict when and where competition, for instance, was expected to be strong. For example, Grime (2001) predicted that competition should be more intense in highly productive systems. In 1976 Menge and Sutherland proposed their original prey stress models, which made predictions about the relative importance of predation, competition, and abiotic stress in limiting the abundance of basal prey populations across a gradient of environmental stress.

The structure and dynamics of aquatic and terrestrial ecosystems differ in a number of aspects (Carr et al. 2003), and it seems reasonable to expect that the relative importance of different types of species interactions should vary between these two broad habitat types. A greater proportion of the studies we reviewed found significant biotic interaction effects in aquatic communities (see Figure 1.1). Of the studies that considered predation, 94% found significant effects in aquatic systems, compared with 73% in terrestrial ecosystems (Figure 1.1), most of which were cases of carnivory. Like Connell (1983), we found that the frequency of experiments showing interspecific competition was about 20% higher in marine and freshwater habitats than in terrestrial communities. In contrast, Schoener (1983) found nearly identical frequencies in all three habitat types (94%, 91%, and 89%, respectively). All tests of herbivory in aquatic ecosystems in our review ($n = 3$) found significant effects, compared with only 5 of 9 in terrestrial systems. These differences are small given the error and biases involved, but they are concordant with the prevailing view that herbivory is generally more important in aquatic ecosystems (Cyr and Pace 1993; Duffy and Hay 2001).

While we did not necessarily expect large differences in the overall occurrence of facilitation within aquatic and terrestrial systems, we did expect that dispersal-based facilitations would be more frequent in terrestrial than in aquatic habitats. The greater density of water compared with air enables propagules to disperse much greater distances in water, even in the absence of morphological or behavioral traits that increase dispersal. As a result, maximum propagule dispersal distances are two orders of magnitude greater for sessile marine species than for terrestrial plants (Kinlan and Gaines 2003). Most marine species disperse at least tens or hundreds of meters, whereas the seeds of terrestrial plants frequently disperse only between 1 and 10 meters (Chambers and MacMahon 1994; Kinlan and Gaines 2003), which can drive dispersal limitation even at very small spatial scales (Chambers and MacMahon 1994). Native-exotic dispersal enhancement was the mechanism in more than half of the 30 documented cases of facilitation in terrestrial ecosystems (Figure 1.4). In contrast, we found only one case of dispersal enhancement in an aquatic system.

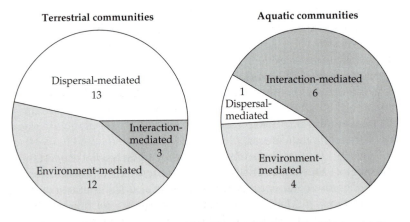

Figure 1.4   Differences in modes of facilitation between the 28 terrestrial and 11 aquatic studies in which facilitation was an important determinant of community structure. Dispersal-mediated facilitation occurred when one species facilitated the entry of another species into a community through dispersal of propagules. Environment-mediated facilitation occurred when one species benefited from the presence of another species because of the neighbor's effect on the environment. Interaction-mediated facilitation occurred as a result of a third species reducing the effects of a negative species interaction (e.g., provision of a predation refuge).

The Predicted, Perceived, and Realized Significance of Native-Exotic Competition

There is currently a fairly widespread perception that species invasions represent a major crisis for the conservation of native communities on par with climate change and habitat fragmentation (Vitousek et al. 1987; Czech and Krausman 1997; Wilcove et al. 1998; but see Gurevitch and Padilla 2004). A significant portion of the threat has been attributed to exotic competitors, despite the limited evidence for competitive exclusion by exotics (Simberloff 1981; Slobodkin 2001; Davis 2003; Houlahan and Findlay 2004). Because the competition paradigm permeates conservation and exotic invasion theory, the logical expectations are that communities are saturated and that species additions will result in local extinctions and declines in native richness. There are indeed several well-documented examples of suppression and local displacement of native species by exotics. For example, the invasion of Norway maple (*Acer platanoides*) in the central and northeastern United States has been associated with decreased neighborhood species diversity, probably due to its ability to tolerate deeply shaded conditions (Martin 1999). Argentine ants (*Linepithema humile*) are highly aggressive and competitively displace native ant species by monopolizing shared food resources (Porter and Savignano 1990; Suarez et al. 1999). Green crabs have reduced the abundance of native crabs in benthic marine habitats and appear to compete with native fish and bird species as well (Grosholz et al. 2000). Exotic snails have well-documented negative effects on native competitors in soft-sediment habitats (Byers 2000).

However, the volume of evidence does not seem commensurate with the predicted and perceived threat of native species declines caused by competitive displacement by exotics (Slobodkin 2001; Gurevitch and Padilla 2004; Houlahan and Findlay 2004). Furthermore, most available evidence shows negative effects on a single or a few species, mainly based on abundance reductions and not local extinctions (Gurevitch and Padilla 2004). This is not because nonexperimental evidence has been ignored; we included correlative studies in our review and found that descriptive data concordant with native exclusion via competition (e.g., post-invasion declines of native richness) are scarce. In fact, most field surveys have found a positive relationship between exotic success (i.e., abundance or richness) and native richness (Chaneton et al. 2002; Sax 2002; Brown and Peet 2003; Stohlgren et al. 2003; Bruno et al. 2004; Houlahan and Findlay 2004), although such patterns do not rule out the possibility of native declines (Fridley et al. 2004). Even at fairly small scales (e.g., 1 m^2) there are few examples of negative correlations between native and exotic species richness (but see Stachowicz et al. 2002; Stohlgren et al. 2002; Brown and Peet 2003) and several examples of positive correlations (Sax 2002; Keeley et al. 2003). Furthermore, null models without any species interactions predict that at very small spatial scales, native and exotic richness should be negatively correlated, largely because of spatial constraints and sampling artifacts (Fridley et al. 2004; Herben et al. 2004). It is true that some exotic plant species appear capable of producing near-monocultures at small scales (e.g., Chinese tallow tree, *Sapium sebiferum*, in the southern United States; garlic mustard, *Alliaria petiolata*, in eastern and central U.S. forests; ice plant, *Carpobrotus chilensis*, along the U.S. Pacific coast; giant hogweed, *Heracleum mantegazzianum*, in riparian areas of Europe and North America). Although such monocultures clearly displace some native species within small areas, there are few documented cases of competitive displacement of natives across sites or landscapes.

Contradicting the expectation of mass native extinctions via exotic competitors, recent evidence indicates that invasions are often additive. For instance, Sax and colleagues (Sax et al. 2002; Sax and Gaines 2003) found that native extinctions are rare and much less common than exotic immigration; consequently, plant diversity has nearly doubled on oceanic islands. In the Hawaiian Islands, for example, there have been 71 recorded native plant extinctions (largely caused by deforestation, other land use changes, and disturbances associated with exotic vertebrates) and 1090 known exotic plant invasions (Sax et al. 2002). In New Zealand the pattern is even more extreme: 3 recorded native plant extinctions and 2069 successful exotic plant colonizations. Similarly, Gido and Brown (1999) found that exotic fish invasions have increased the total number of fish species in most southwestern U.S. streams and concluded that stream ecosystems in general may not be saturated. For some groups (e.g., birds), richness has remained relatively constant, with native extinctions, which are largely driven by exotic vertebrate predators (Blackburn et al. 2004), roughly matched by exotic colonizations (Sax et al. 2002). On balance, most evidence indicates that species richness has increased at landscape to regional scales as a result of recent introductions (Lodge 1993; Gido and Brown 1999; Sax and Gaines 2003). There are certainly exceptions (e.g., the effects of some ant introductions), and these results do not suggest that invasions by new potential competitors will not affect the abundance of native species or the evenness or composition of native communities—they probably will (Gido and Brown 1999). Our point is simply that available evidence indicates that in many systems, the introduction of new competitors rarely causes native extinctions. This is a key lesson of exotic invasions. Species frequently compete, and in some cases competition with exotics reduces the fitness, distribution, or abundance of native species. Nonetheless, the resulting negative effects on community diversity are faint and much less profound than predicted by the competition paradigm and the group of related ideas that spawned it (e.g., the competitive exclusion principle, mathematical competition models, and concepts related to niche and species packing).

Several other reviews of biotic interactions and the effects of species introductions also have suggested that a large proportion of exotic species have no detectable competitive effects on their neighbors and that cases of outright competitive exclusion are scarce (Simberloff 1981; Lodge 1993; Hager and McCoy 1998; Davis 2003; Sax and Gaines 2003; Gurevitch and Padilla 2004; Houlahan and Findlay 2004). Simberloff (1981) found that solid evidence of exclusion by exotic competitors was rare (only 3 of 854 introductions he reviewed resulted in extinction via competition) and, in contrast to our results, that exotic predators generally had greater community effects (50 instances in the same pool of 854 studies). Invasion ecologists have long recognized that only a small proportion (perhaps only 1%–10%) of successful invaders will have highly deleterious effects on native communities (Lodge 1993; Mack et al. 2000; Byers et al. 2002). Yet recent analyses indicate that there is little scientific evidence that even some of the most abundant exotic invasive species, widely cited as leading threats to native communities, actually caused the problems attributed to them. For example, several studies have concluded there is little if any evidence that purple loosestrife (*Lythrum salicaria*), an exotic wetland plant, competitively excludes native plants (Hager and McCoy 1998; Farnsworth and Ellis 2001; Houlahan and Findlay 2004). Yet it is widely perceived as a major cause of native extinction and has even been the target of biological control programs through the release of exotic predators in North America (Hager and McCoy 1998; Houlahan and Findlay 2004). Given the lack of evidence of a substantial effect on native communities, this perception and reaction seem unwarranted. Hager and McCoy (1998) outlined the costs of acting on the acceptance of such unproven beliefs, including the loss of limited conservation resources that could have been spent on problems supported by scientific evidence, the unintended consequences of biological control (Louda et al. 1997), and the loss of scientific credibility. A similar case has been made against the evidence (or lack thereof) of deleterious effects of other well-known problem exotics. Gurevitch and Padilla (2004) argued that most native plants threatened by exotic competitors are also negatively affected by exotic vertebrates (via consumption or disturbance), habitat degradation, and other factors. Only 2%–4% of these plants are considered threatened exclusively by exotic competition.

What explains the lack of evidence for such a widely predicted and perceived phenomenon? We suggest that it may be an outcome of one or more of three general explanations. First, it is possible that few studies have attempted to describe the negative effects of exotic species on native richness (i.e., the effects are real but not quantified). This may seem surprising, given the large number of published papers on species introductions and the fact that exotic invasions and native extinctions are among the most important motivating forces in conservation biology. Still, invasion ecologists have recently called for more research documenting the negative effects of problem exotic species (Byers et al. 2002), and this possibility should not be discounted lightly. Second, perhaps exotic species are usually not abundant enough to cause severe impacts or extinctions of native species. We suspect this is true in many systems. In Rhode Island estuarine strandline plant communities, more than half of resident species are exotic (Bruno et al. 2004). However, the most abundant exotic, *Rosa rugosa*, covers only 4.3% of the landscape, and a large proportion of exotics cover less than 1% (Bruno et al. 2004). Third, there may be a lag of tens or hundreds of years after invasions before even local displacement occurs. In other words, competition may be just as important as predation, facilitation, and recruitment limitation, but its effects could take much longer to be realized. If this is true, current increases in diversity could be ephemeral (Rosenzweig 2001a). This is a very difficult hypothesis to test or refute, especially at very small scales (i.e., 1 m², for which there are few long-term data sets of native richness and composition), and cannot be dismissed. Yet, evidence from two types of investigations suggests this is not the case: (1) small-scale seed addition experiments (e.g., Tilman 1997), in which exclusion is expected to occur in months to years, and (2) global-scale tests of the effect of the size of regional species pools on local richness (e.g., Witman et al. 2004), which effectively incorporate local species interactions that have taken place across relatively long periods of ecological time. As we discuss in a later section, the results of both types of studies are generally in agreement with the view that communities are unsaturated and recruitment-limited and that species introductions will generally increase local diversity, causing proportionately few native extinctions (Lodge 1993; Gido and Brown 1999; Rosenzweig 2001a).

The Unanticipated Role of Exotic Foundation Species in Aquatic Communities

The invasion of habitat-forming or foundation species (*sensu* Dayton 1972; a.k.a. keystone resource species, keystone facilitators, and autogenic ecosystem engineers) into aquatic habitats has highlighted their large effect on community structure and ecosystem processes (Crooks 2002; Castilla et al. 2004). Because competition is generally assumed to be more influential than facilitation and invasive exotics are expected to have negative effects, increases in native richness via facilitation by exotic species was almost entirely unanticipated and has largely gone unnoticed. Measuring the influence of foundation species is difficult because it is usually not practical to add or remove them. However, the introduction and spread of exotic foundation species have created a unique opportunity for ecologists to compare invaded and uninvaded landscapes as well as particular locations before and after invasion (Crooks 2002).

A remarkable example is the invasion of rocky mid-intertidal habitats in northern Chile by the ascidian *Pyura praeputialis* (Castilla et al. 2004). This large ascidian forms dense aggregations and a complex matrix on an otherwise homogeneous substrate and probably facilitates native species by providing a refuge from predation and reducing thermal stress (Bertness and Leonard 1997). The presence of *Pyura* aggregations increases local invertebrate richness by a factor of four (Castilla et al. 2004). Another South American example is the invasive marine tube worm *Ficopomatus enigmaticus*, which creates small reefs within estuaries in Argentina, transforming the landscape and increasing the abundance and diversity of benthic invertebrates (Schwindt and Iribane 2000). Crooks (1998) found that the exotic Japanese mussel *Musculista senhousia* increased the density and richness of invertebrates by increasing the heterogeneity of subtidal mudflats in San Diego Bay, California. But like many other foundation species, the Japanese mussel can negatively affect other primary space holders, in this case the native eelgrass *Zostera marina* (Thorsten and Williams 1999). Even the notorious zebra mussel, *Dreissena polymorpha*, a virtual poster child for ecosystem-altering exotic species, has a net positive effect on native benthic invertebrate density and richness and also facilitates native fishes and benthic macrophytes (Lodge 1993; Stewart and Haynes 1994; MacIsaac 1996; Simberloff and Von Holle 1999; Ricciardi 2003). Furthermore, as predicted by facilitation theory (Bruno and Kennedy 2000; Bruno and Bertness 2001), its facilitative effects on other benthic species are strongly positively related to its density (Ricciardi 2003). However, the zebra mussel literally settles on the backs of freshwater unionid mussels (a highly threatened group) and causes a variety of economic and environmental problems (Ricciardi 2003), so its reputation may be well deserved.

It is important to point out that like native habitat-forming species, exotic foundation species generally do not dominate to the degree that they exclude other species at large spatial scales. Foundation species are often more susceptible to disturbance and consumers than their associated species, generating a mosaic of modified and unmodified habitat patches in most landscapes, and thereby maximizing β-diversity (Castilla et al. 2004). However, some species that modify entire habitats can have negative effects on associated species and total community diversity. Some autogenic engineers increase environmental stress and displace their neighbors (Crooks 2002), such as the exotic ice plant *Mesembryanthemum crystallinum*, which increases surface soil salinity (Vitousek 1986). Sea stars and other native keystone predators can reduce α-diversity (although they probably increase β-diversity) in rocky intertidal habitats by removing habitat-generating mussels that facilitate dozens of associated species (Witman 1985; Lohse 1993; Seed 1996). Crooks (2002) pointed out that exotic species that reduce native diversity (e.g., rabbits, nutria, isopods, and snails) usually do so by consuming foundation species, thereby reducing the complexity of the physical environment. Most exotic species that significantly increase diversity—that is, foundation species or autogenic engineers—facilitate other species by increasing heterogeneity (Crooks 2002). Biotic habitat provision and destruction are two of the most powerful types of species interactions: both can transform landscapes and seascapes and facilitate or remove entire communities. The magnitude and scales of these effects are generally much greater than most forms of competition, predation, and facilitation.

How Specific Are Biotic Interactions?

We used the papers in our review and several other recent articles to determine to what degree different categories of biotic interactions are specialized or species-specific. By comparing interactions between native species with native-exotic encounters, we can begin to assess whether having a shared history or evolutionary exposure affects the frequency or intensity of interactions. Assuming that natural selection can maximize differences among competing species, it is expected that pairs of species with historically overlapping distributions should compete less frequently or intensely than pairs with no or little shared evolutionary history (Connell 1980; Abrams 1983). However, a crude comparison of the results of our review of the exotic species literature with previous reviews largely restricted to competition between natives suggests that the frequency of detectable competitive effects is not greater between exotic and native species than between pairs of natives. Our review and several other recent reviews also indicate that, at least qualitatively, the effects of native-exotic competition are not noticeably greater than those of competition among natives. Removing the so-called "ghost of competition past" (Connell 1980) did not appear to change the role of competition in structuring natural communities. There are several serious problems with this approach, however, most notably that in some cases, exotics evolve surprisingly quickly (Sakai et al. 2001; Callaway and Hierro 2005; Callaway et al., this volume; and Huey et al., this volume) and could already be "coevolved" by the time a given study was performed. Therefore, these results suggest that either that (1) coevolution is not particularly important in mediating competition between species or that (2) coevolution can be important, but generally begins to occur very quickly (i.e., in years to decades) following species co-occupation of an area.

It is widely believed that exotic species prosper in their new ranges in part because many of their natural enemies are absent, an implicit assumption being that predators and parasites are less likely to consume or infect novel prey species (Keane and Crawley 2002). There is some support for this idea (e.g., Mitchell and Power 2003; Colautti et al. 2004), which is known as the "enemy release hypothesis" (Keane and Crawley 2002). There is also contradictory evidence indicating that native consumers *prefer* exotic prey (Agrawal and Kotanen 2003; Colautti et al. 2004), probably because those prey lack effective consumer deterrents, an idea called the "increased susceptibility hypothesis" (Colautti et al. 2004). Evidence that some native predators do not consume exotic prey and that some prey defenses are less effective against novel generalist consumers indicates that many predator-prey interactions are somewhat specialized and probably coevolved. The devastation of native prey populations by introduced mammalian predators on oceanic islands that historically had few vertebrate consumers (Blackburn et al. 2004) also supports this point.

We found that exotic and native species frequently facilitate each other. An extensive review by Richardson et al. (2000) also found that native pollinators, habitat modifiers, nitrogen fixers, and so forth can usually fill the role of former facilitators, suggesting that tightly coevolved facilitative interactions are rare. However, we are not aware of any experimental analyses of the strengths of exotic-native and native-native facilitation analogous to those comparing consumer prey preferences, and it is possible that native-native facilitations are stronger even if they are not more frequent. Furthermore, our sample is biased, as species with highly specific mutualists seem less likely to be able to invade a new region. Simberloff and Von Holle (1999) have suggested that facilitation and positive feedbacks among exotic species could lead to an "invasional meltdown" (i.e., to negative effects on native communities and ecosystems). For example, in Bodega Bay, California, recently arriving exotic European green crabs mainly consume the competitively dominant native clams, thereby facilitating an increase in the density of the exotic eastern gem clam (Grosholz 2005). The gem clam invaded the system decades earlier, but until the green crab invasion it was uncommon and generally benign (Grosholz 2005).

Despite this and many other documented cases of exotic-exotic facilitation, there is minimal theoretical basis or empirical evidence for the expectation that exotics are less likely to facilitate native species or that exotic-native facilitation is weaker. The results of our literature review indicate that exotic-exotic and exotic-native facilitation are nearly equally frequent (Figure 1.2). The frequency and strength of facilitation among exotic species could increase over time with increasing exotic richness due to the "sampling" of exotic facilitators (Levine and D'Antonio 1999; Stachowicz et al. 2002; Bruno et al. 2003a) as predicted by the invasion meltdown hypothesis (Simberloff and Von Holle 1999). This could potentially cause an acceleration of exotic impacts. Yet exotic species will also compete with and consume each other and facilitate native species, resulting in positive and negative effects on exotics as a group that should be roughly equivalent to effects on native species. However, an "invasional meltdown" could occur even if exotic species have a net positive effect on natives if positive feedbacks among exotics drive dramatic ecosystem-level changes and degradation.

The Role of Propagule Limitation: Are Communities Saturated?

Saturated communities are not open to species immigration without resident extinction. Despite a variety of negative interactions among native and exotic competitors, hosts and pathogens, and predators and their prey, it is notable that invasions have increased species richness in many communities and at nearly all spatial scales (Gido and Brown 1999; Rosenzweig 2001a; Sax and Gaines 2003; Bruno et al. 2004). Assuming the absence of a significant time lag in native species declines, such net increases suggest that many ecological communities are not entirely saturated at the species level (Lodge 1993; Gido and Brown 1999). The lesson for community ecology is that community composition and richness are driven at least in part by regional processes and dispersal limitation (MacArthur and Wilson 1967; Ricklefs 1987; Hubbell 2001) and are not dominated by the local processes of competition and niche relations. Similarly, the Eltonian paradigm that diverse communities are more stable and "niche-packed" (Elton 1958) is accurate only in certain situations or at very small scales where local processes dominate (Levine 2000; Kennedy et al. 2002; Brown and Peet 2003). As a general rule, the global spread of exotic species suggests that communities are not saturated and will continue to exhibit local enrichment as the size of regional species pools increases (Rosenzweig 2001a).

The well-developed theory and empirical investigations of the relative roles of local and regional processes in controlling local richness are an ideal framework for predicting and analyzing the effects of invasions on native communities (Ricklefs 1987; Srivastava 1999). The regulation of local richness via local processes (e.g., competition) is expected to produce a nonlinear and asymptotic relationship between regional and local species richness; that is, a saturated relationship (Ricklefs 1987; but see Loreau 2000). Conversely, if regional processes (e.g., speciation) also are important in determining local richness, then the relationship should be positive and linear with no local saturation, even at sites within very diverse regional pools. The local-regional hypothesis has been tested most rigorously in marine ecosystems, where regional (Karlson et al. 2004) and global scale analyses (Witman et al. 2004) have documented a positive linear relationship between regional and local invertebrate species richness. Witman et al. (2004) found that approximately 75% of the variation in the local (among-site) richness of subtidal, sessile invertebrate communities is explained by regional species richness. Likewise, Karlson et al. (2004) found that fivefold increases in regional coral species richness resulted in a nearly identical increase in local coral richness. Both studies demonstrated strong regional enrichment of local diversity and indicate that even some of the world's most diverse marine communities are not saturated. This does not indicate that local biotic interactions are weak or unimportant, but instead that regional processes and propagule supply also strongly regulate local community diversity and structure: the two forces are not mutually exclusive and could act simultaneously (Levine and D'Antonio 1999; Chaneton et al. 2002). For example, on highly diverse Pacific reefs, competition among coral species can be intense, and in some situations can lead to local exclusion and reductions in diversity (Connell et al. 2004), despite clear evidence of regional enrichment (Karlson et al. 2004). Conversely, in habitats where competition can be intense and biotic resistance from native competitors is known to be substantial (e.g., shallow subtidal habitats in New England), exotic species can still invade and become established (Stachowicz et al. 2002).

Small-scale experimental tests of the recruitment limitation hypothesis have also demonstrated the effect of propagule richness on community composition and diversity (Tilman 1997; Brown and Fridley 2003). In a system where competition is widely believed to largely control community structure (Tilman 1999), Tilman (1997) nearly doubled local plant richness by experimentally reducing recruitment limitation—an effect documented 4 years after the single seed addition and one that is still evident 7 years later (D. Tilman, personal communication). Recruitment limitation may also be the actual cause of native species declines. Most California grasslands are dominated by exotic annuals, and the formerly widespread native perennial grasses are now rare (Seabloom et al. 2003). Superficially, this appears to be another case of natives being excluded by exotic invaders. However, Seabloom et al. (2003) demonstrated that the native grasses are actually competitively dominant and that their abundances were initially depressed by cattle gazing and other land use changes; therefore, their recovery is essentially limited by seed supply. In plots where seeds were added, the native species flourished and displaced the exotic annuals (Seabloom et al. 2003).

By effectively increasing the sizes of regional species pools (Rosenzweig 2001a; Sax and Gaines 2003), species invasions have provided accidental experiments that are extremely useful to ecologists: replicated broad-scale tests of the recruitment limitation and local-regional richness hypotheses. As described above, the results to date suggest that increases in regional pool sizes have caused increases in local richness (Sax and Gaines 2003). But is the relationship linear, or are there signs of saturation? Because invasions have not been viewed or analyzed from the local-regional richness perspective, this question cannot yet be answered. However, the experimental (small-scale and relatively short-term) and descriptive (large-scale and effectively very long-term) studies we have just described (i.e., Tilman 1997; Karlson et al. 2004; Witman et al. 2004) indicate that both saturation and a delay in substantive extinctions via competition are unlikely.

The human activities that have increased rates of exotic invasions (e.g., global shipping) have effectively reduced the degree of isolation of islands and continents by increasing the dispersal ability of thousands of plants, animals, and microbes. Rosenzweig (2001a) argued that removing the distance barriers that have historically isolated biogeographic provinces will increase the equilibrium level of species diversity at subglobal scales (Sax et al. 2002). This prediction is based in part on the theory of island biogeography: An elevated immigration rate will increase equilibrium diversity even when there is significant competition among exotic and native species (i.e., because the extinction curve is not altered—see Sax et al., Chapter 17, this volume). The theory of island biogeography (MacArthur and Wilson 1967) also predicts a degree of species turnover and extinctions of local populations, which have been and will continue to be documented as a result of exotic invasions. But the important point is that these extinctions will not compensate for increased immigration, and thus the predicted and documented diversity increases are theoretically permanent. In reality, diversity increases may be ephemeral, but probably not because exotic enemies will eventually cause mass native extinctions, but instead because of other anthropogenic factors, especially habitat fragmentation and possibly climate change. Reducing the size of real and virtual habitat islands will lower equilibrium diversity levels in many communities and locations (Rosenzweig 2001b). Ironically, species introductions will tend to mitigate these losses in diversity, but will also greatly alter local and regional composition at many taxonomic levels—creating changes in native biotic communities of grave concern to conservation biologists.

Conclusions

The primary purpose of this chapter was to review the literature on species introductions to gain insights into the role and nature of biotic interactions. Based on this review, we made five broad generalizations about biotic interactions that we believe to be supported by a preponderance of available scientific evidence (Box 1.1). We purposely attempted to do this from a neutral perspective and focused on the general lessons that could be gleaned from species introductions rather than on their negative effects. Nevertheless, we documented many of these effects (e.g., declines in native abundances and the alteration of ecosystem processes), and we recognize the magnitude of the problems they cause (reviewed in Mack et al. 2000). Species introductions have reduced global biodiversity and will homogenize the composition of the earth's communities. Some exotic species alter ecosystem processes, landscapes, and biogeochemical cycles, and mitigating these changes can be very expensive. Exotic predators have caused the loss of some fauna, particularly fishes and birds, and species introductions have altered a number of characteristics of native communities. In some systems, these and additional changes caused by exotic species are clearly significant, yet in others the threat from exotics is trivial, especially in comparison to a variety of more pressing problems. For example, in coral reef ecosystems, stressors such as climate change, disease outbreaks, overfishing, and pollution are having much greater effects than species introductions (Hoegh-Guldberg 1999; Bruno et al. 2003b; Hughes et al. 2003).

BOX 1.1   *Biotic Interactions: Evidence, Insights, and Conclusions*

1. Biotic interactions are common and can have substantial effects on populations and communities. The relative frequencies of competition, predation, and facilitation are similar.

2. Ecologists and conservation biologists in particular have overestimated the importance of competition and the prevalence of competitive exclusion. Competition is not *the* dominant force structuring communities, but is one of many important factors. Competition occurs relatively frequently, but rarely limits immigration or leads to local extinction.

3. Evidence for facilitation exceeds its perceived influence. Facilitation of native species by exotics is common, and some exotic habitat-forming species have increased native species richness, especially in aquatic habitats.

4. A relatively small proportion of facilitative and competitive species interactions are coevolved or species-specific. In contrast, many consumer-prey interactions are somewhat specific and do appear coevolved.

5. Propagule supply strongly influences community structure, and many communities are unsaturated.

In many other ecosystems, habitat degradation, fragmentation, and loss are much greater concerns (Rosenzweig 2001a; Gurevitch and Padilla 2004).

One of the great achievements of modern conservation biology has been quantifying, and in some cases mitigating, the threats from species introductions. However, a notable failure is the inaccuracy of many predictions and assumptions concerning invasions and their effects on native species and communities. This inaccuracy has resulted in part from the uncritical acceptance of the competition paradigm, which has had cascading effects on the theory of invasion ecology, just as it did in basic ecology 40 years ago (Bruno et al. 2003a). The assumption that communities are largely structured by competitive interactions deemphasized the roles of recruitment limitation and facilitation. This assumption led to the now largely falsified predictions that species introductions would drive widespread native extinctions via competition leading to compensatory reductions in native richness, and that diverse native communities could "repel" or prevent invasions (Levine and D'Antonio 1999; Kennedy et al. 2002). Additionally, since communities were considered saturated, successful invaders were assumed to be especially competitive, "released" from natural enemies, and thus more successful and abundant than comparable native species (Agrawal and Kotanen 2003; Vilà et al. 2003). Recent work challenges all three assumptions. For example, exotic plants do not appear to be more competitive than co-occurring natives (Daehler 2003; Vilà et al. 2003), and herbivory can be more intense on exotic plants than on native congeners in the introduced range (Agrawal and Kotanen 2003; Colautti et al. 2004). Additionally, exotic plants are generally not larger in introduced than in native ranges (Thébaud and Simberloff 2001; Vilà et al. 2003; but see Grosholz and Ruiz 2003 for a contrary marine invertebrate example). Furthermore, the abundance distributions of exotic and native plants can be essentially identical, with exotic species representing a random subset of the regional species pool (Bruno et al. 2004). Competition clearly plays an important role in structuring communities (Gurevitch et al. 1992) and in invasion dynamics (Levine and D'Antonio 1999; Stachowicz et al. 2002; Levine et al. 2004), but so do a number of other factors. Disturbance, competitive hierarchies, the context dependence of competitive outcomes, and a variety of other factors appear to limit the incidence and relative importance of competitive exclusion (Buss and Jackson 1979; Petraitis et al. 1989). A major challenge for conservation biology and invasion ecology is to consider recent empirical advancements and all of modern ecological theory in future efforts to predict and understand the processes and effects of species introductions.

Acknowledgments

First we thank the editors for developing the overall theme of this book and the focus of this chapter and for their extensive editing and advice. We also thank L. Hazen, M. O'Connor, J. Parker, T. Rand, E. Selig, C. Shields, J. Wall, two anonymous reviewers, and the members of the NCEAS "Exotic Species: A Source of Insight into Ecology, Evolution, and Biogeography" Working Group for their helpful comments and insightful discussions. This work was funded in part by a National Science Foundation grant to JFB (OCE-0327191), the Andrew W. Mellow Foundation, and the National Center for Ecological Analysis and Synthesis, funded by NSF (grant DEB-0072909), the University of California, and the Santa Barbara campus.

Literature Cited

Abrams, P. 1983. The theory of limiting similarity. Annual Review of Ecology and Systematics 14: 359–376.

Agrawal, A. A., and P. M. Kotanen. 2003. Herbivores and the success of exotic plants: a phylogenetically controlled experiment. Ecology Letters 6:712–715.

Bertness, M. D., and R. Callaway. 1994. Positive interactions in communities. Trends in Ecology and Evolution 9:191–193.

Bertness, M. D., and G. H. Leonard. 1997. The role of positive interactions in communities: lessons from intertidal habitats. Ecology 78:1976–1989.

Blackburn, T. M., P. Cassey, R. P. Duncan, K. L. Evans, and K. J. Gaston. 2004. Avian extinction and mammalian introductions on oceanic islands. Science 305:1955–1958.

Bossard, C. C. 1991. The role of habitat disturbance, seed predation and ant dispersal on establishment of the exotic shrub *Cytisus scoparius* in California. American Midland Naturalist 126:1–13.

Bronstein, J. L. 1994. Our current understanding of mutualism. Quarterly Review of Biology 69:31–51.

Brown, R. L., and J. D. Fridley. 2003. Control of plant species diversity and community invasibility by species immigration: seed richness versus seed density. Oikos 102:15–24.

Brown, R. L., and R. K. Peet. 2003. Diversity and invasibility of southern Appalachian plant communities. Ecology 84:32–39.

Bruno, J. F., and M. D. Bertness. 2001. Positive interactions, facilitations and foundation species. In M. D. Bertness, S. D. Gaines, and M. Hay, eds. *Marine community ecology*, pp. 201–218. Sinauer Associates, Sunderland, MA.

Bruno, J. F., and C. W. Kennedy. 2000. Patch-size dependent habitat modification and facilitation on New England cobble beaches by *Spartina alterniflora*. Oecologia 122:98–108.

Bruno, J. F., J. J. Stachowicz, and M. D. Bertness. 2003a. Inclusion of facilitation into ecological theory. Trends in Ecology and Evolution 18:119–125.

Bruno, J. F., L. Petes, C. D. Harvell, and A. Hettinger. 2003b. Nutrient enrichment can increase the severity of two Caribbean coral diseases. Ecology Letters 6:1056–1061.

Bruno, J. F., C. W. Kennedy, T. A. Rand, and M. Grant. 2004. Landscape-scale patterns of biological invasions in shoreline plant communities. Oikos 107: 531–540.

Buss, L. W., and J. B. C. Jackson. 1979. Competitive networks: nontransitive competitive relationships in cryptic coral reef environments. American Naturalist 113:223–234.

Byers, J. E. 2000. Competition between two estuarine snails: implications for invasions of exotic species. Ecology 81:1225–1239.

Byers, J. E., S. Reichard, J. M. Randall, I. M. Parker, C. S. Smith, W. M. Lonsdale, I. A. E. Atkinson, T. R. Seastedt, M. Williamson, E. Chornesky, and D. Hayes. 2002. Directing research to reduce the impacts of nonindigenous species. Conservation Biology 16:630–640.

Callaway, R. 1995. Positive interactions among plants. The Botanical Review 61:306–349.

Callaway, R. M., and L. R. Walker. 1997. Competition and facilitation: a synthetic approach to interactions in plant communities. Ecology 78:1958–1965.

Carr, M. H., J. E. Neigel, J. A. Estes, S. Andelman, R. R. Warner, and J. L. Largier. 2003. Comparing marine and terrestrial ecosystems: implications for the design of coastal marine reserves. Ecological Applications 13:S90–S107.

Case, C. M., and M. J. Crawley. 2000. Effect of interspecific competition and herbivory on the recruitment of an invasive alien plant: *Conyza sumatrensis*. Biological Invasions 2:103–110.

Castilla, J. C., N. A. Lagos, and M. Cerda. 2004. Marine ecosystem engineering by the alien ascidian *Pyura praeputialis* on a mid-intertidal rocky shore. Marine Ecology Progress Series 268:119–130.

Chambers, J. C., and J. A. MacMahon. 1994. A day in the life of a seed—movements and fates of seeds and their implications for natural and managed systems. Annual Review of Ecology and Systematics 25:263–292.

Chaneton, E. J., S. B. Perelman, M. Omacini, and R. J. C. León. 2002. Grazing, environmental heterogeneity, and alien plant invasions in temperate Pampa grasslands. Biological Invasions 4:7–24.

Colautti, R. I., A. Ricciardi, I. A. Grigorovich, and H. J. MacIsaac. 2004. Is invasion success explained by the enemy release hypothesis? Ecology Letters 7:721–733.

Cole, F. R., A. C. Medeiros, L. L. Loope, and W. W. Zuehlke. 1992. Effects of the Argentine ant on arthropod fauna of Hawaiian high-elevation shrubland. Ecology 73:1313–1322.

Connell, J. H. 1975. Producing structure in natural communities. In M. L. Cody and J. M. Diamond, eds. *Ecology and evolution of communities*, pp. 460–490. Belknap, Cambridge, MA.

Connell, J. H. 1980. Diversity and coevolution of competitors, or ghost of competition past. Oikos 35:131–138.

Connell, J. H. 1983. On the prevalence and relative importance of interspecific competition: evidence from field experiments. American Naturalist 122:661–696.

Connell, J. H., T. P. Hughes, C. C. Wallace, J. E. Tanner, K. E. Harms, and A. M. Kerr. 2004. A long-term study of competition and diversity of corals. Ecological Monographs 74:179–210.

Connor, E. F., and D. Simberloff. 1983. Interspecific competition and species co-occurrence patterns on islands: null models and the evaluation of evidence. Oikos 41:455–465.

Crooks, J. A. 1998. Habitat alteration and community-level effects of an exotic mussel, *Musculista senhousi*. Marine Ecology Progress Series 162:137–152.

Crooks, J. A. 2002. Characterizing ecosystem-level consequences of biological invasions: the role of ecosystem engineers. Oikos 97:153–166.

Cyr, H., and M. L Pace. 1993. Magnitude and patterns of herbivory in aquatic and terrestrial ecosystems. Nature 361:148–150.

Czech, B., and P. R. Krausman. 1997. Distribution and causation of species endangerment in the United States. Science 277:1116–1117.

Daehler, C. C. 2003. Performance comparisons of co-occurring native and alien invasive plants: implications for conservation and restoration. Annual Review of Ecology, Evolution and Systematics 34:183–211.

D'Antonio, C. M. 1993. Mechanisms controlling invasion of coastal plant-communities by the alien succulent *Carpobrotus edulis*. Ecology 74:83–95.

Davis, M. A. 2003. Biotic globalization: does competition from introduced species threaten biodiversity? BioScience 53:481–489.

Dayton, P. K. 1972. Toward an understanding of community resilience and the potential effects of enrichments to the benthos at McMurdo Sound, Antarctica. In B. C. Parker, ed. *Proceedings of the colloquium on conservation problems in Antarctica*, pp. 81–96. Allen Press, Lawrence, KS.

Duffy, J. E. and M. E. Hay. 2001. The ecology and evolution of marine consumer-prey interactions. In M. D. Bertness, M. E. Hay, and S. D. Gaines, eds. *Marine community ecology*, pp. 131–157. Sinauer Associates, Sunderland, MA.

Elton, C. 1958. *The ecology of invasions by animals and plants*. Methuen, London.

Farnsworth, E. J., and D. R. Ellis. 2001. Is purple loosestrife (*Lythrum salicaria*) an invasive threat to freshwater wetlands? conflicting evidence from several ecological metrics. Wetlands 21:199–209.

Fridley, J. D., Brown, R. L., and J. F. Bruno. 2004. Null models of exotic invasion and scale-dependent patterns of native and exotic species richness. Ecology 85:3215–3222.

Gause, G. F. 1934. *The struggle for existence*. Williams and Williams, Baltimore, MD.

Gido, K. B., and J. H. Brown. 1999. Invasion of North American drainages by alien fish species. Freshwater Biology 42:387–399.

Glyphis, J. P., S. J. Milton, and W. R. Siegfried. 1981. Dispersal of *Acacia cyclops* by birds. Oecologia 48:138–141.

Grime, J. P. 2001. Plant Strategies, Vegetation Processes, and Ecosystem Properties. Chichester: Wiley, West Sussex, New York.

Grosholz, E. D. 2005. Recent biological invasion may hasten invasional meltdown by accelerating historical introductions. Proceedings of the National Academy of Sciences USA 102:1088–1091.

Grosholz, E. D., and G. M. Ruiz. 2003. Biological invasions drive size increases in marine and estuarine invertebrates. Ecology Letters 6:700–705.

Grosholz, E. D., G. M. Ruiz, C. A. Dean, K. A. Shirley, J. L. Maron, and P. G. Connors. 2000. The impacts of a nonindigenous marine predator in a California bay. Ecology 81:1206–1224.

Gurevitch, J., and D. K. Padilla. 2004. Are invasive species a major cause of extinctions? Trends in Ecology and Evolution 19:470–474.

Gurevitch, J., L. L. Morrow, A. Wallace, and J. S. Walsh. 1992. A meta-analysis of competition in field experiments. American Naturalist 140:539–572.

Hacker, S. D. and S. D. Gaines. 1997. Some implications of direct positive interactions for community species diversity. Ecology 78:1990–2003.

Hager, H. A., and K. D. McCoy. 1998. The implications of accepting untested hypotheses: a review of the effects of purple loosestrife (*Lythrum salicaria*) in North America. Biodiversity and Conservation 7:1069–1079.

Hairston, N. G., F. E. Smith, and L. B. Slobodkin. 1960. Community structure, population control, and competition. American Naturalist 94:421–425.

Herben, T., B. Mandák, K. Bímová, and Z. Münzbergová. 2004. Invasibility and species richness of a community: a neutral model and a survey of published data. Ecology 12:3223–3233.

Hoegh-Guldberg, O. 1999. Climate change, coral bleaching and the future of the world's coral reefs. Marine and Freshwater Research 50:839–866.

Holway, D. A., A. V. Suarez, and T. J. Case. 1998. Loss of intraspecific aggression in the success of a widespread invasive social insect. Science 282:949–952.

Houlahan, J. E., and C. S. Findlay. 2004. Effect of invasive plant species on temperate wetland plant diversity. Conservation Biology 18:1132–1138.

Hubbell, S. P. 2001. *The unified neutral theory of biodiversity and biogeography.* Princeton University Press, Princeton, NJ.

Huffaker, C. B. 1958. Experimental studies on predation: dispersion factors and predator-prey oscillations. Hilgardia 27:343–383.

Hughes, T. P., A. H. Baird, D. R. Bellwood, M. Card, S. R. Connolly, C. Folke, R. Grosberg, O. Hoegh-Guldberg, J. B. C. Jackson, J. Kleypas, J. M. Lough, P. Marshall, M. Nyström, S. R. Palumbi, J. M. Pandolfi, B. Rosen, and J. Roughgarden. 2003. Climate change, human impacts and the resilience of coral reefs. Science 301:929–933.

Hutchinson, G. E. 1957. Concluding remarks. Cold Spring Harbor Symposium on Quantitative Biology 22:415–457.

Hutchinson, G. E. 1959. Homage to Santa Rosalia, or why are there so many kinds of animals? American Naturalist 93:145–159.

Karlson, R. H., H. V. Cornell, and T. P. Hughes. 2004. Coral communities are regionally enriched along an oceanic biodiversity gradient. Nature 429:867–870.

Keane, R. M., and M. J. Crawley. 2002. Exotic plant invasions and the enemy release hypothesis. Trends in Ecology and Evolution 17:164–170.

Keeley, J. E., D. Lubin, and C. J. Fotheringham. 2003. Fire and grazing impacts on plant diversity and alien plant invasions in the southern Sierra Nevada. Ecological Applications 12:1355–1374.

Kennedy, T. A., S. Naeem, K. M. Howe, J. M. H. Knops, D. Tilman, and P. Reich. 2002. Biodiversity as a barrier to ecological invasion. Nature 417:636–638.

Kinlan, B. P., and S. D. Gaines. 2003. Propagule dispersal in marine and terrestrial environments: a community perspective. Ecology 84:2007–2020.

Lavergne, C., J. Rameau, and J. Figier. 1999. The invasive woody weed *Ligustrum robustum* subsp. *walkeri* threatens native forests on La Reunion. Biological Invasions 1:377–392.

Lawton, J. H. 1984. Non-competitive populations, non-convergent communities, and vacant niches: the herbivores of braken. In D. R. Strong, D. Simberloff, L. G. Abele, A. B. Thistle, eds. *Ecological communities: conceptual issues and the evidence.* Princeton University Press, Princeton, NJ.

Leonard, G. H., J. M. Levine, P. R. Schmidt, and M. D. Bertness. 1998. Flow-driven variation in intertidal community structure in a Maine estuary. Ecology 79:1395–1411.

Levine, J. M. 2000. Species diversity and biological invasions: relating local processes to community pattern. Science 288:761–763.

Levine, J. M., and C. M. D'Antonio. 1999. Elton revisited: a review of evidence linking diversity and invasibility. Oikos 87:15–26.

Levine, J. M., P. B. Adler, and S. G. Yelenik. 2004. A meta-analysis of biotic resistance to exotic plant invasions. Ecology Letters 7:975–989.

Lodge, D. M. 1993. Biological invasions: lessons for ecology. Trends in Ecology and Evolution 8:133–137.

Lohse, D. P. 1993. The importance of secondary substratum in a rocky intertidal community. Journal of Experimental Marine Biology and Ecology 166:1–17.

Loreau, M. 2000. Are communities saturated? On the relationship between alpha, beta and gamma diversity. Ecology Letters 3:73–76.

Lotka, A. J. 1925. *Elements of physical biology.* Williams and Wilkins, Baltimore.

Louda, S. M., D. Kendall, J. Connor, and D. Simberloff. 1997. Ecological effects of an insect introduced for the biological control of weeds. Science 277:1088–1090.

MacArthur, R. H. 1970. Species-packing and competitive equilibrium for many species. Theoretical Population Biology 1:1–11.

MacArthur, R. H. and R. Levins. 1967. The limiting similarity, convergence, and divergence of coexisting species. American Naturalist 101:377–385.

MacArthur, R. H. and E. O. Wilson. 1967. *The theory of island biogeography.* Princeton University Press, Princeton, NJ.

MacIsaac, H. J. 1996. Potential abiotic and biotic impacts of zebra mussels on the inland waters of North America. American Zoologist 36:287–299.

Mack, R. N., D. Simberloff, W. M., Lonsdale, H. Evans, M. Clout, and F. A. Bazzaz. 2000. Biotic invasions: causes, epidemiology, global consequences, and control. Ecological Applications 10:689–710.

Magoulick, D. D., and L. C. Lewis. 2002. Predation on exotic zebra mussels by native fishes: effects on predator and prey. Freshwater Biology 47:1908–1918.

Martin, P. H. 1999. Norway maple (*Acer platanoides*) invasion of a natural forest stand: understory consequences and regeneration pattern. Biological Invasions 1:215–222.

McAdoo, J. K., W. S. Longland, and R. A. Evans. 1989. Nongame bird community responses to sagebrush invasion of crested wheatgrass seedlings. Journal of Wildlife Management 53:494–502.

McClure, M. S., and C. A. S.-J. Cheah. 1999. Reshaping the ecology of invading populations of hemlock woolly adelgid, *Adelges tsugae* (Homoptera: Adelgidae), in eastern North America. Biological Invasions 1:247–254.

Menge, B. A. and J. P. Sutherland. 1976. Species diversity gradients: synthesis of the roles of predation, competition, and temporal heterogeneity. American Naturalist 110:351–369.

Meyer, J., and J. Florence. 1996. Tahiti's native flora endangered by the invasion of *Miconia calvescens* DC. (Melastomataceae). Journal of Biogeography 23:775–781.

Mitchell, C. E., and A. G. Power. 2003. Release of invasive plants from fungal and viral pathogens. Nature 421:625–627.

Paine, R. T. 1966. Food web complexity and species diversity. American Naturalist 100:65–75.

Park, T. 1948. Experimental studies of interspecies competition. I. Competition between populations of flour beetles, *Tribolium confusum* Duval and *Tribolium castaneum* Herbst. Ecological Monographs 18:265–308.

Petraitis, P. S., R. E. Latham, and R. A. Niesenbaum. 1989. The maintenance of species diversity by disturbance. The Quarterly Review of Biology 64:393–418.

Porter, S. D., and D. A. Savignano. 1990. Invasion of polygyne fire ants decimates native ants and disrupts arthropod community. Ecology 71:2095–2106.

Race, M. S. 1982. Competitive displacement and predation between introduced and native mud snails. Oecologia 54:337–347.

Ricciardi, A. 2003. Predicting the impacts of an introduced species from its invasion history: an empirical approach applied to zebra mussel invasions. Freshwater Biology 48:972–981.

Richardson, D. M., N. Allsopp, C. M. D'Antonio, S. J. Milton, and M. Rejmanek. 2000. Plant invasions—the role of mutualisms. Biological Reviews 75:65–93.

Ricklefs, R. E. 1987. Community diversity: relative roles of local and regional processes. Science 235:167–171.

Robinson, J. V., and G. A. Wellborn. 1988. Ecological resistance to the invasion of a freshwater clam, *Corbicula fluminea*: fish predation effects. Oecologia 77:445–452.

Rosenzweig, M. L. 2001a. The four questions: What does the introduction of exotic species do to diversity? Evolutionary Ecology Research 3:361–367.

Rosenzweig, M. L. 2001b. Loss of speciation rate will impoverish future diversity. Proceedings of the National Academy of Sciences USA. 98:5404–5410.

Sakai, A. K., F. W. Allendorf, J. S. Holt, D. M. Lodge, J. Molofsky, K. A. With, S. Baughman, R. J. Cabin, J. E. Cohen, N. C. Ellstrand, D. E. McCauley, P. O'Neil, I. M. Parker, J. N. Thompson, and S. G. Weller. 2001. The population biology of invasive species. Annual Review of Ecology and Systematics 32:305–32.

Savidge, J. A. 1987. Extinction of an island forest avifauna by an introduced snake. Ecology 68:660–668.

Sax, D. F. 2002. Native and naturalized plant diversity are positively correlated in scrub communities of California and Chile. Diversity and Distributions 8:193–210.

Sax, D. F., and S. D. Gaines. 2003. Species diversity: from global decreases to local increases. Trends in Ecology and Evolution 18:561–566

Sax, D. F., S. D. Gaines, and J. H. Brown. 2002. Species invasions exceed extinctions on islands worldwide: a comparative study of plants and birds. American Naturalist 160:766–783.

Schoener, T. W. 1983. Field experiments on interspecific competition. American Naturalist 122:240–285.

Schreiber, E. S. G., P. S. Lake, and G. P. Quinn. 2002. Facilitation of native stream fauna by an invading species? Experimental investigations of the interaction of the snail, *Potamopyrgus antipodarum* (Hydrobiidae) with native benthic fauna. Biological Invasions 4:317–325.

Schwindt, E., and O. O. Iribarne. 2000. Settlement sites, survival and effects on benthos of an introduced reef-building polychaete in a SW Atlantic coastal lagoon. Bulletin of Marine Science 67:73–82.

Seabloom, E. W., W. S. Harpole, O. J. Reichman, and D. Tilman. 2003. Invasion, competitive dominance, and resource use by exotic and native California grassland species. Proceedings of the National Academy of Sciences USA 100:13384–13389.

Seed, R. 1996. Patterns of biodiversity in the macro-invertebrate fauna associated with mussel patches on rocky shores. Journal of the Marine Biology Association of the United Kingdom 76:203–210.

Shiganova, T. A. 1998. Invasion of the Black Sea by the ctenophore *Mnemiopsis leidyi* and recent changes in pelagic community structure. Fisheries Oceanography 7:305–310.

Silliman, B. R., and M. D. Bertness. 2002. A trophic cascade regulates salt marsh primary production. Proceedings of the National Academy of Sciences USA 99:10500–10505.

Simberloff, D. S. 1981. Community effects of introduced species. In M. H. Nitecki, ed. *Biotic crises in ecological and evolutionary time*, pp. 53–81. Academic Press, New York.

Simberloff, D., and B. Von Holle. 1999. Positive interactions of nonindigenous species: invasional meltdown? Biological Invasions 1:21–32.

Slobodkin, L. B. 2001. The good, the bad, and the reified. Evolutionary Ecology Research 3:1–13.

Srivastava, D. S. 1999. Using local-regional richness plots to test for species saturation: pitfalls and potentials. Journal of Animal Ecology 68:1–16.

Stachowicz, J. J. 2001. Mutualism, facilitation, and the structure of ecological communities. Bioscience 51:235–246.

Stachowicz, J. J., H. Fried, R. W. Osman, and R. B. Whitlatch. 2002. Biodiversity, invasion resistance, and marine ecosystem function: reconciling pattern and process. Ecology 83:2575–2590.

Stewart, T. W., and J. M. Haynes. 1994. Benthic macroinvertebrate communities of southwestern Lake Ontario following invasion of *Dreissena*. Journal of Great Lakes Research 20:479–493.

Stohlgren, T. J., G. W. Chong, L. D. Schell, K. A. Rimar, Y. Otsuki, M. Lee, M. A. Kalkhan, and C. A. Villa. 2002. Assessing vulnerability to invasion by nonnative plant species at multiple spatial scales. Environmental Management 29:566–577.

Stohlgren, T. J., D. T. Barnett, and J. T. Kartesz. 2003. The rich get richer: patterns of plant invasions in the United States. Frontiers in Ecology and the Environment 1:11–14.

Suarez, A. V., N. D. Tsutsui, D. A. Holway and T. J. Case. 1999. Behavioral and genetic differentiation between native and introduced populations of the Argentine ant. Biological Invasions 1:43–53.

Taniguchi, Y., K. D. Fausch, and S. Nakano. 2002. Size-structured interactions between native and introduced species: can intraguild predation facilitate invasion by stream salmonids? Biological Invasions 4:223–233.

Thébaud, C., and D. Simberloff. 2001. Are plants really larger in their introduced ranges? American Naturalist 157:231–237.

Thorsten, T. B. H., and S. L. Williams. 1999. Macrophyte canopy structure and the success of an invasive marine bivalve. Oikos 84:398–416.

Tilman, D. 1997. Community invasibility, recruitment limitation, and grassland biodiversity. Ecology 78:81–92.

Tilman, D. 1999. The ecological consequences of changes in biodiversity: a search for general principles. Ecology 80:1455–1474.

Townsend, C. R. 1996. Invasion biology and ecological impacts of brown trout *Salmo trutta* in New Zealand. Biological Conservation 78:13–22.

Vilà, M., and C. M. D'Antonio. 1998. Fruit choice and seed dispersal of invasive vs. noninvasive *Carpobrotus* (Aizoaceae) in coastal California. Ecology 79:1053–1060.

Vilà, M., A. Gomez, and J. L. Maron. 2003. Are alien plants more competitive than their native conspecifics? A test using *Hypericum perforatum L.* Oecologia 137:211–215.

Vitousek, P. M. 1986. Biological invasions and ecosystem properties: can species make a difference? In H. A. Mooney and J. A. Drake, eds. *Ecology of biological invasions of North America and Hawaii*, pp. 163–176. Springer-Verlag, New York.

Vitousek, P. M., L. R. Walker, L. D. Whiteaker, D. Mueller-Dombois, and P. A. Matson. 1987. Biological invasions by *Myrica faya* alters ecosystem development in Hawaii. Science 238:802–804.

Volterra, V. 1926. Fluctuations in the abundance of a species considered mathematically. Nature 118:558–560.

Wilcove, D. S., D. Rothstein, J. Dubow, A. Phillips, and E. Losos. 1998. Quantifying threats to imperiled species in the United States. Bioscience 48:607–615.

Witman, J. D. 1985. Refuges, biological disturbance and rocky subtidal community structure. Ecological Monographs 55:421–445.

Witman, J. D., R. J. Etter, and F. Smith. 2004. The relationship between regional and local species diversity in marine benthic communities: a global perspective. Proceedings of the National Academy of Sciences USA 101:15664–15669.

Wonham, M. J., M. O'Connor, and C. D. G. Harley, 2005. Positive effects of a dominant invader on introduced and native mudflat species. Marine Ecology Progress Series 289:109–116.

2

Species Invasions and the Relationships between Species Diversity, Community Saturation, and Ecosystem Functioning

John J. Stachowicz and David Tilman

In this chapter, we examine how the ecology of invasions helps clarify the relationships among community saturation, diversity, and ecosystem functioning. We begin with a theoretical approach that suggests a common underlying mechanism for the negative effect of diversity on community invasibility and other ecosystem functions, such as productivity. This stochastic model of community assembly predicts that, within a given habitat, increasing species richness should reduce resource availability and decrease invasion success. A brief review of evidence from the fossil record of biotic interchanges reveals that the less diverse region is typically more invaded by species than the more species-rich region. The reason for this could be either that species-poor regions have greater resource availability (there are empty niches), or that species from more diverse regions are competitively superior. Experimental studies with recent invaders suggest that, all else being equal, increasing diversity decreases invasion success by decreasing resource availability. More complete or efficient utilization of resources at higher diversity extends beyond invasion resistance in its contribution to other ecosystem processes, such as productivity, nutrient recycling, and stability/consistency. In accord with theory, data from both grassland plants and marine invertebrates suggest that successful invaders are those with niche requirements most different from those of species in the existing community, providing a degree of determinism to an otherwise stochastic community assembly process. We also suggest some other ways in which the study of introduced species could increase our understanding of diversity and its impacts on ecosystem processes.

Introduction

The geologic history of life on earth has been one of long periods of isolation of biogeographic realms interrupted by periods of immigration when continents collided or land bridges were exposed by lowered sea levels (e.g., Flannery 2001). More recently, human transport has caused a massive increase in introductions of nonindigenous species. For instance, about half of the plant species of Hawaii are exotics, as are about 20% of those of California (Sax et al. 2002; Sax and Gaines 2003). Similarly, San Francisco Bay contains at least several hundred exotic invertebrates, which constitute up to 99% of the biomass at some sites (Cohen and Carlton 1998; Ruiz et al. 2000). Because the same processes that influence invasion by exotic species should also influence the assembly, composition, structure, and diversity of natural communities, the study of invasions—whether human-driven or natural—should provide insight into fundamental ecological processes. In this regard, invasions can be viewed as inadvertent and often unreplicated, but large-scale and long-term, "experiments" that, in total across a wide variety of situations, may provide ecological insights that small-scale, short-term deliberate experiments might never provide.

From a societal perspective, species invasions can pose serious threats to human economic interests or to native species. Introduced agricultural pests such as the Mediterranean fruit fly and various agricultural weeds and diseases impose significant costs on agriculture. Introductions of rats, snakes, and other predators led to a massive wave of bird extinctions in the Pacific islands (Pimm 1987). Invasions of the North American Great Lakes by zebra mussels and sea lampreys have dramatically altered the biota of those lakes via both alteration of the standing stock of phytoplankton and effects on commercially important fish species. Similarly, on land, Argentine ants have displaced native ant fauna in the western United States and imposed significant pest control costs. Regardless of the potential consequences of invasions, it is undeniable that new species are being added to regional biotas by human-mediated transport and that increased understanding of the causes and consequences of these invasions would be valuable.

In this chapter, we discuss the application of the study of invasions in both geologic and recent times to understanding the relationship between community saturation, diversity, and ecosystem functioning. Theoretical discussions of the relationship between the number of species in a system and the productivity or stability of the system have a long history (see Tilman and Lehman 2001 for review). With growing threats to global species diversity from human activities such as habitat modification, concern has mounted as to the consequences of species losses for the functioning of ecosystems. This concern has renewed interest within ecology in the relationship between the number or diversity of species in a community and the extent to which that community maintains its functioning. Indeed, experimental tests of relationships between diversity and ecosystem processes such as productivity, stability, and invasibility have increased rapidly in the last 10 years (e.g., see chapters in Kinzig et al. 2001 and in Loreau et al. 2002).

One of the older hypotheses about the relationship between diversity and ecosystem processes is that more diverse communities should be more resistant to invasion (Elton 1958). The most commonly cited mechanism behind this phenomenon is that as species accumulate, competition intensifies and fewer resources remain available for new colonists (Elton 1958; Case 1990). Experimental and observational approaches to this problem have sometimes yielded conflicting results (see the excellent review by Levine and D'Antonio 1999; Fargione et al. 2003; Taylor and Irwin 2004). Studies that employ both approaches simultaneously show that, all else being equal, diversity does reduce invasion success, although this mechanism is sometimes overwhelmed by other forces, such as propagule supply or disturbance (Levine 2000; Stachowicz et al. 2002a). But the issue of more complete or efficient utilization of resources extends beyond invasion resistance in its contribution to ecosystem functioning, with links to other ecosystem processes such as productivity, nutrient recycling, and stability/consistency. Our aim in this chapter is to assess how studies of invasions inform us about the relationship between the number of species and the utilization of resources in a community and to apply this knowledge to enhance our understanding of the relationship between diversity and ecosystem functioning.

We begin by considering some simple theory, based on trade-offs in resource use, that outlines how increasing species diversity is expected to decrease the amount of available resources and thus decrease invasion success. We then proceed to assess the mechanistic experimental evidence in support of the theory and how this evidence has already contributed to our broader understanding of diversity-saturation-ecosystem functioning relationships. We then evaluate experimental and theoretical predictions in light of evidence from invasions across spatial and temporal scales. To do this, we begin by briefly reviewing evidence from natural biotic interchanges that occur among biogeographic regions on geologic time scales and evaluate possible hypotheses for the highly asymmetrical exchange among biotas that typically occurs. We then compare these findings with those from more recent, human-mediated, biotic exchanges. Next, we compare observational and experimental approaches to studying the diversity-invasibility relationship within a single biogeographic region, reconciling the often disparate results obtained by the two approaches. These sections paint a picture of how changing species richness affects resource use across spatial and temporal scales. Because total resource use is coupled to so many other ecosystem functions, such as productivity and nutrient cycling, we then assess what diversity-invasibility studies can tell us more generally about the relationship between diversity and ecosystem functioning. We end with suggestions for additional ways in which invasions may be exploited to learn about these relationships.

Theory

Insight into the factors controlling the success or failure of invasive species is provided by considering the mechanisms of interaction among established and invading species. Because there are many such mechanisms, including competition for resources, interference competition, mutualism, and top-down forces such as herbivory, predation, and disease, there is no simple, general theory of invasion. Here we will focus on the role that competitive interactions can play as controllers of invasion dynamics (e.g., Case 1990; Tilman 1999a, 2004), but we stress that top-down forces are likely to be of at least equal importance, as witnessed by the major effects that invasive diseases and predators have had on their host and prey species (e.g., Lafferty et al. and Bruno et al., this volume).

Resource competition theory predicts that numerous species can coexist if species have trade-offs in their traits and if the habitat is spatially or temporally heterogeneous (Tilman 1982, 1988). Consider a case in which plant species compete for two limiting resources and in which the habitat has spatial heterogeneity in these two resources (Tilman 1982). The trade-off needed for species to coexist is that species that are better at competing for one of the resources are necessarily poorer at competing for the other resource. This causes each species to have a particular ratio of resource 1 (R_1) and resource 2 (R_2) at which it is the superior competitor. These ratios all fall on a curve, the interspecific trade-off curve (Figure 2.1A). Every point on this curve represents the traits of a potential species that would be able to stably coexist with any and all other potential species if the habitat had a continuous gradient in the supply rates of these resources.

How, then, might established species influence the success or failure of potential invaders? Insight might be provided by stochastic niche assembly theory, which explores the effects of resource competition and stochastic birth-death processes on the probability of establishment of rare invaders (Tilman 2004). Let us start by considering cases in which the invaders have traits that are drawn from the same trade-off curve as the established species. A new invader has to survive, grow, and reproduce using the resources left unconsumed by established species. Consider the pattern of unconsumed resources left by the three-species community in Figure 2.1B. These species create four distinct "patches" of usable resources (shaded areas). The probability that an invader could survive and grow to maturity and then reproduce would be highly dependent on its traits relative to the traits of the established species. These probabilities (calculated as in Tilman 2004) are shown in Figure 2.2B as a function of the optimum R_1:R_2 ratios of potential invaders. Invaders that are

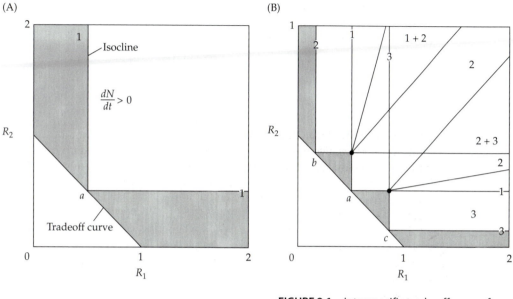

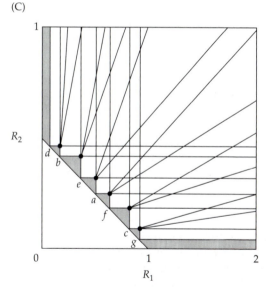

FIGURE 2.1 Interspecific trade-off curves for assembled communities with (A) one, (B) three, or (C) seven species. Axes are the rates of supply of two resources (R_1 and R_2). Isoclines are zero net growth isoclines for each species. Shaded areas indicate the level of unconsumed resources in each community.

similar to the three established species would have very low chances of successful establishment because of their extremely low growth rates. Their low growth rates would cause such invaders to require a long period to become reproductive adults, and thus expose them to mortality for a long time. This greatly decreases the probability of establishment by an invader that is similar to the established species. In contrast, potential invaders would have much higher growth rates, reach maturity more quickly, and have a higher proba-

(A) Invasion into single-species habitat

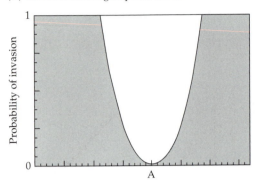

FIGURE 2.2 Probability of invasion as a function of the invader's position on the interspecific trade-off curve for each of the communities in Figure 2.1. Species with a position on the trade-off curve most distant from those of the species in the existing community have the highest probability of successful invasion. Thus, while there is no absolute limit to the number of species in a community, the probability of successful invasion becomes smaller with increasing numbers of species in the community.

(B) Invasion into 3-species habitat

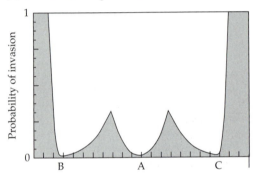

(C) Invasion into 7-species habitat

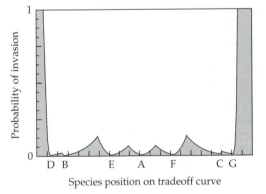

Species position on tradeoff curve

bility that they and their progeny would become established if they had resource requirements that better matched the peaks of resources left unconsumed by the established species (Tilman 2004).

The probabilities of invasion into the three-species community can be compared with those of invasion into the seven-species community in Figure 2.1C. In this more diverse community, there is more complete utilization of the limiting resources, creating much smaller peaks of unconsumed resources. Compared with the less diverse community, this greatly reduces the probabilities of invasion (Figure 2.2C) by species drawn from the same trade-off curve. Note, though, that there would still be a strong patterning to any invasion, with potential invaders that are maximally different from existing species having the greatest chance of establishment.

This brief summary of stochastic niche theory (Tilman 2004), which combines the effects of competition and demographic stochasticity on rare invaders, shows that the chance of successful invasion should decrease markedly as the diversity of the established community increases (Figure 2.3; see also Tilman 2004). Indeed, the log of the probability of further invasion is an approximately linearly decreasing function of the number of established species. In the example of Figure 2.3, each additional species that becomes established in a community decreases by about 30% the chance that another invader

FIGURE 2.3   Results of simulations showing that the probability of invasion decreases with increasing numbers of species in the community. See text and Tilman 2004 for details.

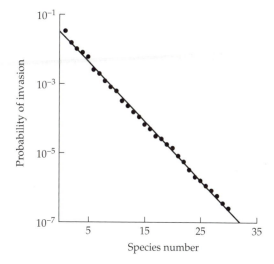

will become established. Once the habitat of Figure 2.3 had accumulated 32 species, for instance, only 1 propagule out of every 10^7 would tend to be successful at becoming established. This extremely low probability of establishment is a result of the low and relatively uniform levels of resources left unconsumed by communities assembled by stochastic niche processes.

This theory (Tilman 2004), as summarized above, assumes that established and invading species have competitive trade-offs that are drawn from the same underlying trade-off surface. With this assumption (which we will later relax), the theory predicts that (1) high-diversity naturally assembled communities should be highly resistant to invasion, (2) successful invaders should decrease the abundances mainly of species that are competitively similar to themselves (that are close to them on the trade-off surface), and (3) successful invading species should rarely, if ever, drive established competitors extinct. From these three predictions, it also should follow that (4) invasions by exotic species should lead to a net increase in regional diversity (e.g., as shown experimentally by Tilman 1997 and via survey by Sax et al. 2002) and (5) exotic species should be much more common and abundant in disturbed habitats because of higher resource levels or lower diversity due to recent extinctions.

Some of these predictions may seem, on their surface, to be contradicted by the many examples of high abundances reached by exotic invasive species in contemporary communities. For example, exotic annuals reach high abundances in California grasslands, and the exotic grass *Bromus tectorum* forms abundant stands in the Intermountain West of the United States, as does (or did) the *Opuntia* cactus in Australia and *Hypericum* (St. John's wort) in Oregon and northern California. However, in each of these cases, exotic species were invading and spreading across habitats that had already been disturbed by cattle grazing. Similarly, the Asian clam *Potamocorbula* in San Francisco Bay and the ascidian *Didemnum lahelliei* in the Gulf of Maine have reached high abundances in environments disturbed by mechanized fishing and other human activities. These examples demonstrate that resident species richness is by no means the only factor affecting resource availability and that the combined and interactive effects of resident species richness, disturbance regimes, and site fertility on resource levels and invasions must be considered (e.g., Davis et al. 2000). For example, disturbance, by freeing up resources, could facilitate invasion simply by increasing the probability that a non-native will become established if introduced or by favoring invaders with life histories that are better adapted to disturbance than those of residents.

Now that we have laid out some basic theory and predictions, let us begin by examining the experimental evidence for the proposed linkages between diversity, resource use, and invasibility and what it can contribute to our understanding of diversity-ecosystem function relationships.

Invasions and the Mechanistic Underpinnings of Diversity-Ecosystem Functioning Relationships

Investigations of the role of diversity in invasion success have already contributed significantly to our understanding of the relationship between diversity and ecosystem processes in the broader sense by pushing for more rigorous examination of the mechanism(s) underlying the potential effect of diversity on invasibility. The studies that have examined this issue have supported the idea that, all else being equal, higher species richness leads to a reduction in the availability of resources (Stachowicz et al. 1999, 2002a; Tilman 1999a; Fargione et al. 2003), in agreement with the theory presented above. We discuss below specifically what these findings contribute to our understanding of the consequences of changing diversity for ecosystem structure and functioning.

By what mechanisms might increasing diversity enhance resource use and thus lead to increased ecosystem functioning? First, diversity may lead to enhanced ecosystem functioning simply because more species-rich communities have a higher probability of containing a "strong interactor"; that is, a species with a dominant effect on resource levels. This effect, termed the "sampling effect" (Aarssen 1997; Huston 1997; Tilman et al. 1997), would seem to be a plausible explanation for observed diversity-invasibility and diversity-ecosystem functioning relationships. However, a variety of tests have rejected the sampling hypothesis as a significant explanation for the observed diversity-productivity relations, suggesting instead that complementary use of resources by species with different niche requirements, such that more diverse communities more completely use available resources, may best explain the effect of diversity on productivity (e.g., Loreau and Hector 2001; Tilman et al. 2001, 2002; Hille Ris Lambers et al. 2004). Although sampling effects may overwhelm complementarity in some studies, this appears likely to occur only in the short term (e.g., Tilman et al. 2002). Similarly, sampling rarely seems to be the major explanation underlying diversity-invasibility relationships (e.g., Knops et al. 1999; Stachowicz et al. 1999, 2002; Naeem et al. 2000; Fargione et al. 2003). Rather, these studies suggest that niche differences and resulting complementary interactions better explain the observed effects of diversity on productivity and invasibility.

In addition to complementary use of resources at any one time, species might differ in their seasonal phenologies such that they occupy different "temporal" niches. As Davis et al. (2000) pointed out, fluctuations in resource availability can have a major effect on the susceptibility of the local community to invasion. Resident species and their temporal patterns of resource use (e.g., seasonal phenologies) are one possible cause of predictable variation in resource availability. For example, in sessile marine invertebrate communities, invasion resistance increases with diversity because individual species are complementary in their temporal patterns of space occupation (Stachowicz et al. 1999, 2002a). Established communities with many species maintain relatively high cover and low resource availability over time, despite high variation in the temporal abundance of individual species (Stachowicz et al. 2002a). In contrast, species-poor communities undergo large fluctuations in resource availability as a result of the seasonal boom-and-bust cycles of their dominant species (Figure 2.4A). Thus, as suggested by Davis et al. (2000), invasion is more likely in areas of fluctuating resource availability (species-poor communities) than in areas of predictably low resource availability (species-rich communities) (Stachowicz et al. 2002a).

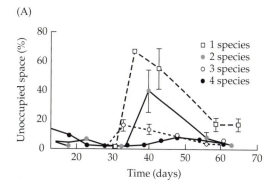

(A)

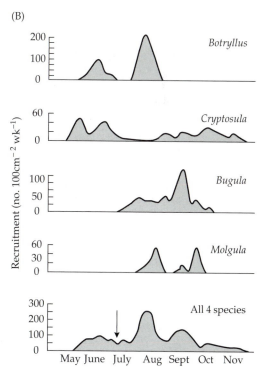

(B)

FIGURE 2.4   (A) Results of experimental manipulations of diversity, showing large peaks in the availability of a limiting resource (available space) in low-diversity communities and consistently low resource availability (i.e., high space occupation) in more diverse communities. This pattern resulted in a negative relationship between native diversity and invader survival and abundance. (B) Temporal variation in the recruitment patterns of sessile marine invertebrates, showing temporal complementarity in seasonal patterns of abundance among native species. Total recruitment is consistently high over the course of the entire season when multiple native species are added to the system, and there are few periods of low recruitment. The arrow indicates the peak in the timing of recruitment for one of the most successful introduced species in this community, *Botrylloides violaceus.* Note that the peak in seasonal abundance of this invader corresponds to a period of low recruitment by natives. (A after Stachowicz et al. 1999, 2002a; B after J. Stachowicz, unpublished data.)

Observational studies across multiple years show similar correlations between local species richness and occupation of space in rocky subtidal habitats (Osman 1977). Despite differences in growth forms, species seem to compete primarily for space, the likely limiting resource in this system. Because species differ in growth rates, in the timing, duration, and magnitude of reproductive output, in the degree to which they are capable of local recruitment, and in their response to biotic and physical stresses, the seasonal fluctuations in abundance of each species tend to be out of phase with one another. In Figure 2.4B we provide one example from records of settlement of marine invertebrates onto 10 × 10 cm bare substrates placed within an established community (detailed methods are given in Stachowicz et al. 2002b). Such substrates reach 100% cover within a few months after deployment, with the dominant species being those that recruited in the first few weeks of deployment. Thus, one can view these settlement records as a proxy for which species would be likely to become established in gaps within established communities throughout the course of the season. As a result of seasonal complementarity in settlement patterns, communities with more species will have residents recruiting to gaps for a greater percentage of the season, resulting in fewer resources (less space) available for new species to exploit. Additionally, those invaders that are successful (see the legend of Figure 2.4B) are often those that have seasonal recruitment periods that coincide with temporal minima in the recruitment of natives.

This temporal niche partitioning is completely analogous to the sorts of resource partitioning presented above, and a more complete theory of invasion and community assembly would account for both. The temporal dimension may be incorporated into models of resource competition like that presented above by adding a third axis, time, to the trade-off plot. In the simplest case, even if all species were at a similar point on the $R_1{:}R_2$ trade-off surface, they might coexist if each were distinct on the temporal axis. As a corollary, invasions can occur into communities in which temporal use of resources is not complete (e.g., Figure 2.4).

A similar pattern of temporal variation in the use of resources may occur among grassland plants (McKane et al. 1990; Tilman and Wedin 1991; Wedin and Tilman 1993). For example, C_4 grasses are the best competitors for nitrogen during the hottest time of year (summer), whereas C_3 grasses use nitrogen during cooler parts of the year (spring and autumn). Different groups also appear to use different spatial pools of N, with forbs using primarily deeper soils for N, C_4 grasses using N from shallower soils, and legumes utilizing atmospheric N via nitrogen-fixing symbionts in their roots. Thus, total N utilization is greater in communities with a greater diversity of these functional groups (Fargione et al. 2003). A prediction that follows from these sorts of functional group arguments (and the theory outlined above) is that invading species most different from those in the established community will have the highest probability of establishment. In other words, communities will be most susceptible to invasion by species with functional traits that are currently rare or absent. In the temporal niche framework, this would suggest that invaders with seasonal phenologies most unlike those of any of the natives would be the most successful. Fargione et al. (2003) did find, in an experiment in which seeds of 27 different plant species were added to plots that differed in diversity and functional guild composition, that the strongest inhibitory effects of resident plants were on introduced plants of the same functional guild.

These patterns suggest a degree of determinism to the community assembly process that may apply broadly to the processes of colonization and establishment: established species should most strongly inhibit invaders that are most similar in resource requirements to them (Tilman 2004). Although which species become part of a new community depends on the composition of the propagule pool and on chance events (e.g., who arrives first, which seed happens to fall in a more favorable spot), this phenomenon could cause communities to assemble toward specific relative abundances of different functional guilds (Tilman 2004). Whether communities are assembled randomly or via a repeatable process has important implications for ecological theory and application, and these few studies from invasion biology suggest that niche-based models may more adequately describe community assembly processes than neutral models (e.g., Fargione et al. 2003). They also suggest that there is some hope for developing predictive models of community assembly.

These insights also have implications for the ways in which we study and manage invasive species. Two major questions of invasion biology are what makes a community more or less invasible than others, and what makes some species better invaders than others. If resident species do more strongly inhibit invading species that are functionally similar to them, then these two questions would be best studied interdependently rather than separately. That is, the invasibility of a community may depend not only on community properties, but also on the characteristics of the invader under consideration. Similarly, the invasive potential of a species may depend not only on its intrinsic properties, but also on the diversity and composition of the community into which it is being introduced.

Having presented theory and experimental evidence suggesting that increasing community diversity can decrease invasion by decreasing resource availability, we now examine the degree to which these mechanisms and theoretical predictions actually correlate with patterns in both natural and human-mediated invasions.

Biotic Interchange

Natural invasions occur when biogeographically isolated biotas are brought into contact via major geologic events. This can occur when, for example, long-separated continents are connected via land bridges, such as happened when the Isthmus of Panama formed during the Pliocene, joining the terrestrial floras and faunas of North and South America. Similarly, the opening of new ocean passages, as occurred with the opening of the Trans-Arctic seaway 3 million years ago, can connect formerly separate marine floras and faunas. What happens when these large groups of independently evolved species come into secondary contact can be quite informative for the study of current ecological and evolutionary scenarios. The study of these interchanges and their contribution to our more general understanding of community assembly is complementary to the study of more recent, human-mediated introductions, because only for the former do we know the ultimate outcome with respect to which species were successful invaders and which, if any, went extinct—as opposed to current invasions, in which consequent species extinctions may not be realized for a long time (Tilman et al. 1996a).

Paleontological data suggest that the flow of species within these interchanges was often unidirectional, or at least highly asymmetrical, from areas of higher species richness to areas of lower species richness. In addition, as predicted from theory, and in agreement with data from recent invasions (Sax et al. 2002), these invasions led to only modest extinction in the recipient biota and generally caused a net increase in diversity over time (e.g., Vermeij 1991a). For example, when the Isthmus of Panama joined North America to South America, mammals were more diverse in North America (where grassland and savanna habitats dominated), and these mammals were very successful in invading South America, where such animals were far less diverse (e.g., Marshall 1981; Marshall et al. 1982; Webb 1991). Conversely, South America, being more tropical and wet and thus having a greater area of rainforest, was more diverse than North America with respect to rainforest plants and associated species, which were very successful invaders into tropical habitats in North America (Gentry 1982). Similarly, movement of marine organisms (particularly mollusks) was predominantly from the Pacific to the Atlantic in the Trans-Arctic interchange, with the species-rich North Pacific contributing far more species to the species-poor North Atlantic than vice versa (Vermeij 1991a). Vermeij (this volume) discusses these and other examples in greater detail and concludes that, although there are exceptions, the pattern of invasion from areas of high diversity to areas of low diversity is robust across a range of taxa, latitudes, and biomes.

The simplest possible explanation for these patterns is that they reflect the statistical expectation: species-rich systems have more species, and thus would be expected to contribute more invaders, than species-poor systems. One way of phrasing this is that if the native diversity of region A outnumbers that of region B by 5:1, then one should expect 5 times more invaders from region A to region B than from region B to region A. Because pre-exchange diversity can be difficult to estimate, relatively few studies have explicitly tested this null hypothesis. However, Vermeij (1991b) found that the exchange of marine organisms across the Trans-Arctic seaway during the Pliocene was more asymmetrical than expected based on the ratios of either the current or the pre-exchange faunal diversities of the North Pacific and North Atlantic. Similarly, prior to the Great American interchange, terrestrial mammal genera in North America outnumbered those in South America by 125:71 (1.76:1), whereas the number of recent genera of northern origin in the south outnumbered southern invaders in the north by 80:24 (3.33:1) (data from Figure 1 in Marshall et al. 1982). A more formal test of the null model for the Great American interchange that differentiated patterns of "invasion" from subsequent patterns of "diversification" would be more complex, particularly as this invasion occurred in stages, but this casual look at the data suggests that an explanation beyond the null may be required. Vermeij (in press) also reports that marine invasions from the Caribbean to Florida are more asymmetrical than predicted by a null hypothesis. These examples do not discount the possibility that some uneven biogeographic exchanges may have a simple null explanation; however, they suggest that in at least some cases, the exchange among regions is even more uneven than expected based on pre-exchange diversity levels.

Multiple ecological and evolutionary mechanisms could underlie this asymmetry in biotic exchanges. First, as first suggested by Darwin in *The Origin of Species* (see McNaughton 1993), biotas with fewer species might utilize available resources less completely or consistently, leaving more opportunities for speciation and subsequent increases in productivity. Such "empty niches" in low-diversity realms could also be readily occupied by invaders from other realms. In this situation, it is possible that successful invaders have traits that fall on the same trade-off curve as those of established species, and are successful merely because their traits allow them to exploit resources left unconsumed by the more species-poor resident community. This is a plausible explanation, for example, for the Trans-Arctic marine interchange, in which the invaded Atlantic biota had recently suffered a series of extinctions, whereas the more diverse source biota of the Pacific had not (Vermeij 1991b). Thus, at least one ancient invasion is in agreement with predictions from theory and mechanistic experiments that decreasing resident diversity should increase susceptibility to invasion.

Alternatively, species from more diverse biotas might possess superior competitive abilities as a result of evolutionary innovation. Areas with more diverse floras or faunas tend to have a greater area of available suitable habitat. Larger regions might also lead to larger total population sizes and thus higher genetic variation. With greater variation, there is greater potential for natural selection to act to maximize resource use efficiency and thus lower the resource levels required for persistence (i.e., lowering the overall trade-off surface). Overall, then, all else being equal, individual species found in regions with a larger physical area and greater diversity might thus be able to subsist on lower levels of resources. With a superior trade-off curve, invaders would have positive growth rates even if they had a $R_1:R_2$ ratio identical to that of an established species. This would increase their chance of establishment, and would mean that, once established, they would competitively displace the original species that were similar to them. On the other hand, if invaders came from a biogeographic realm that had an inferior trade-off curve, they would be highly unlikely to invade new habitats unless those habitats had low diversity and lacked species with R_1:R_2 requirements similar to their own. Additionally, trade-off surfaces can be more complex than the simple two-dimensional approach of the model above: adding a third (or greater) axis could allow one to specify the rates of resource supply required in different habitats or seasons. Regions with a greater range of available habitat types might then have more species specialized for each habitat type and be more resistant to invasion by species from biogeographic realms with less habitat differentiation and specialization. While it would be unreasonable to expect that all species from one continent would be superior to all others from another, it might prove informative and a useful test of the theory outlined above to compare, in a common garden approach, the competitive abilities of a broad range of taxa from biogeographic provinces of differing species richness and/or invasibility.

There are, of course, other ways for invading species to have evolved superior competitive abilities. In particular, key innovations or "novel weapons," such as allelopathic chemicals (see Callaway et al., this volume), may allow invading species to circumvent the normal competition for resources, effectively allowing them, once established, to persist with or exclude competitors with lower resource requirements in the biogeographic realm into which they are being introduced. Root exudates from the European weeds *Centaurea maculosa* and *C. diffusa* strongly inhibit potential competitors in North America, allowing these weeds to become competitive dominants, despite the minimal effect their root exudates have on similar European species with which these species share a long coevolutionary history (Callaway and Aschehoug 2000; Callaway et al., this volume). More generally, superior interference competitive abilities of animals, via aggressive or territorial behavior, might allow them to invade areas even if they require higher resource supply rates than native species to persist.

Alternatively, an invader could be temporarily competitively superior to established species if the invader entered a new habitat that lacked natural enemies (diseases, pathogens, predators, herbivores, etc.) like those that kept it in check in its native habitat (e.g., Torchin et al. 2003; Lafferty et al., this volume, and Ricklefs, this volume). Lower loss rates could allow a species to maintain itself on lower levels of resources and thus increase its competitive ability. Such an increase could allow the invader to become readily established, and possibly dominant, even in a highly diverse community. The successful biological control of species such as *Opuntia* and *Hypericum* suggests that the absence of natural enemies probably contributed at least to their initial success as exotic invasive species. Missing parasites might, however, be a less plausible explanation for invasions following the formation of land bridges or the opening of seaways between once-separated realms because diseases or predators would seem likely to spread with, or soon after, their hosts or prey. Additionally, given enough time, diseases and parasites are likely to either be acquired from the new fauna (e.g., Strong et al. 1977) or arrive in subsequent invasion events, reducing the abundance of the invader to levels more similar to those in its native habitat.

Regardless of the underlying mechanisms, these patterns from large-scale natural invasions are consistent with the hypothesis that habitats with greater species diversity are less readily invaded than habitats with lower diversity. The reported patterns of major biotic interchanges thus seem, at least on first inspection, to be consistent with the hypothesis that communities with a greater diversity of competing species may have lower resource levels and/or fewer "unfilled niches" and thus inhibit invasion. If lower resource levels lead to more intense competition, and thence to evolution of greater competitive ability, it seems plausible that a region with more species would be both harder to invade and more likely to produce successful invaders. Invasions thus present an opportunity to examine biogeographic patterns in competitive ability and resource depletion that have been largely unaddressed.

Does this pattern of asymmetrical exchange hold for more recent, human-mediated interchanges between biogeographic regions? We know of no direct tests in recent invasions that explicitly address the null hypothesis that the number of invasions between two regions is a simple function of the ratios of the pre-contact diversities of the two regions. Of course, unlike ancient natural interchanges, recent invasions involve more than two previously separated regions, complicating any such test. Nevertheless, it is useful here to review a few commonly cited examples of asymmetrical biotic exchange in contemporary invasions.

Elton's original treatise formalizing the idea of asymmetrical biotic exchange drew on comparisons of islands and continental environments, pointing out that islands, with smaller areas and more depauperate biotas, are more frequently invaded than mainlands, which are typically much more diverse. However, the number of attempted introductions of mainland species to islands is often far greater than the number of island species introduced to mainlands; thus, in many analyses, propagule supply appears more likely to be the cause of high rates of island invasion (e.g., Blackburn and Duncan 2001a). Alternatively, it might be that island floras and faunas are inherently different from those of the mainland because only a small proportion of species were capable of being dispersed to islands before humans began transporting species freely. If, for instance, there is a general trade-off between dispersal ability and competitive ability, then the flora and fauna of islands would be biased toward being poor competitors. This could mean that islands are readily invaded not because of their low diversity, but because they are populated with poor competitors relative to the species being introduced to them. Humans often preferentially introduce species that are abundant in their native range; to the extent that abundance correlates with competitive ability, human-introduced species may represent a nonrandom sample of the species from the native range that have high competitive ability but low dispersal ability (e.g., Blackburn and Duncan 2001b).

Because a large fraction of introductions to novel marine ecosystems are the accidental result of transoceanic shipping, these might allow a less biased assessment of the relative invasibility of regions of differing resident diversities. For example, estuaries along the Pacific coast of North America are few in number, small in area, and spaced relatively widely on this geologically active continental margin. These estuaries are geologically quite young and relatively species-poor and are among the most invaded in the world (Cohen and Carlton 1998; Ruiz et al. 2000). Many of the invaders in these regions come from the western North Atlantic—a region with a far greater area of estuarine habitat and a greater diversity of estuarine biota. Of course, there are alternative explanations for why Pacific coast estuaries are so highly invaded, including the volume of ship traffic that passes through those estuaries (Ruiz et al. 2000) and the relatively benign climate of the region.

The available data from human-mediated biotic interchanges appear to be unsuitable, then, for testing theories regarding diversity effects on resource use and invasibility, although they do remind us that diversity and its effect on resource use represent just one of many possible factors that may interact to determine invasibility.

Reconciling Pattern and Processes in Community Invasibility

We now turn to the question of whether diversity, and its effects on resource use, are useful predictors of the relative invasibility of a particular community *within* a given biotic province. A review of the literature highlights the lack of consensus on the strength and direction of the effects of diversity on community resistance to invasion (Levine and D'Antonio 1999). A major characteristic of this debate is the discordant results of studies employing observational versus experimental approaches. Many observational studies find that more diverse native communities support more invaders (Knops et al. 1995; Planty-Tabacchi et al. 1996; Rejmanek 1996; Wiser et al. 1998; Lonsdale 1999; Stohlgren et al. 1999). These studies are often conducted at large scales (e.g., 10–1000 km^2 or more; but see Sax 2002), involve many species, and are favored by some because they involve "natural" communities. But many other factors correlated with diversity can affect the establishment and spread of invaders, confounding such studies and urging caution in their interpretation (e.g., Rejmanek 2003). In particular, a recent reanalysis of the data used by Stohlgren et al. (1999) found that a high level of real estate development was a much better predictor of high rates of invasion by exotic species across the United States than was high regional diversity (Taylor and Irwin 2004).

In contrast to most of the simpler analyses of broad-scale invasion patterns, most experimental manipulations (necessarily conducted at smaller scales) support the idea that species richness decreases invasion success (McGrady-Steed et al.1997; Knops et al. 1999; Lavorel et al. 1999; Stachowicz et al. 1999, 2002; Levine 2000; Naeem et al. 2000; Symstad 2000; Kennedy et al. 2002; Fargione et al. 2003). Such experiments reveal the potential of diversity to reduce invasion success, but rarely assess whether diversity is important relative to other factors, such as propagule supply, disturbance, or predation, for generating patterns of invasion in the field. Two studies have employed a combination of field experiments and observational approaches: both find that when all other things are equal (experimentally controlled), diversity has a negative effect on invasion success, but one shows a positive (Levine 2000) and one a negative (Stachowicz et al. 2002) relationship between native and invader diversity in field surveys. Thus, these results suggest that diversity *can* reduce invasion success in the field, but that it should be visualized as one of several potentially correlated factors (including disturbance and propagule supply) that may affect the invasibility of a community (Tilman 1999a).

Once it is recognized that diversity (through its effects on resource use) is one of many factors, including human-driven habitat disturbance (Taylor and Irwin 2004), that may affect invasion success, a resolution of the apparent contradiction between observational and experimental results emerges. Positive native-invader diversity relationships may often be driven by inherent spatial variation in biotic and abiotic conditions that overwhelms the effects of competition between species for resources. Thus, the positive correlation between native and exotic richness is not causal, but rather results from both natives and exotics responding similarly to some extrinsic factor, such as propagule supply, predation, degree of disturbance, or temperature. Thus, a positive native-exotic relationship will result if the area sampled includes sufficient spatial heterogeneity in these extrinsic factors. But when samples are taken from locations that vary little in these factors (and thus have a smaller range of native diversity), a negative native-exotic relationship is more likely, but not guaranteed. Note that the scale of partitioning of spatial heterogeneity can lead to situations in which negative native-exotic relationships exist at particular spatial scales while a positive relationship exists across spatial scales (Figure 2.5; see also Shea and Chesson 2002). Experimental approaches explicitly control for and attempt to minimize such sources of variability, so they should be viewed as asking what the causal effect of diversity on invasion is "when all else is equal." Thus, even when a positive correlation between native and exotic diversity exists, losses of native species should still lead to an increase in invasion success.

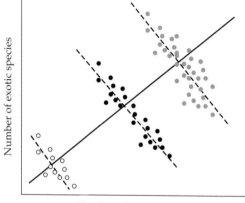

Number of native species

FIGURE 2.5 Hypothetical relationship between native and exotic richness across spatial scales. Clusters of points from a particular location across which extrinsic (abiotic) factors vary little form negative correlations, but when data are lumped across sites that vary considerably in extrinsic factors, an overall positive relationship results, because these factors affect the diversity of native and exotic species in similar ways and overwhelm any effects of diversity on biotic resistance. (After Shea and Chesson 2002.)

So why, then, do observational studies of biotic exchange over geologic time at the largest spatial and temporal scales support theoretical ideas about the negative effects of diversity on invasibility while the evidence from more recent surveys suggests, if anything, that a positive relationship between the number of natives and the number of invaders may be more common? We present three possible explanations. One possibility is that differences between biotic interchanges and current patterns of change may be due to differences in the number of sources for species invasions. In biotic interchanges historically, there were generally just two pools of species involved, whereas in modern invasions, species are often invading from many different regions of the world. This may allow many more species, with a much greater range of traits, to get to species-rich regions than would have been possible historically. Such biotic homogenization should increase the probability of arrival of invaders that occupy novel niche spaces or possess superior trade-off surfaces, as it is unlikely that any single regional biota will be superior in all respects to all other biotas on the planet, even if it is relatively resistant to one particular biota. A second difference between recent and geologic invasions is that recent interchanges are associated with a major source of disturbance—human activities—that can increase the patchiness and spatial heterogeneity of habitat types, potentially allowing more species to coexist (especially species adapted to disturbed habitats, as many invaders are). It may thus be that under the current disturbance regime, the niche space available for opportunist species is unsaturated by native biota. A final possibility is that there has not been enough time since these most recent invasions for exclusion to occur on a global or even a regional scale, and that communities currently carry an extinction debt (Tilman et al. 1996a) that will eventually be paid. Competitive exclusion over large spatial scales may often be a slow process, and it could be that the current patterns are transient and reflect other, faster-acting forces (see discussion in Bruno et al., this volume). Each of these possibilities suggests interesting lines of future research—more in-depth comparisons between natural biotic interchange and our current situation should prove illuminating.

Synthetic approaches that compare experimental results and field surveys in the same system have provided insight into the diversity-invasibility debate, but are generally rare in the diversity-ecosystem functioning literature. This approach was used for native grasslands and for a related biodiversity experiment, which both showed higher productivity and more complete nitrogen use at higher plant diversities (Tilman et al. 1996b). In marine systems, in which experiments are still relatively rare, there have been some promising attempts to assess the relationship between diversity and ecosystem functioning using a correlational approach (Emmerson and Huxham 2002). Pairing these approaches with simultaneous manipulative experiments (as in Levine 2000 or Stachowicz et al. 2002) would allow for significant advances in our understanding of the conditions under which diversity plays a major role in controlling ecosystem functioning and of how their relationship might change across spatial scales.

Broader Implications and Future Directions

The theme of this chapter has been how the study of invasions has contributed to the understanding of the relationship between species diversity, resource utilization, and the functioning of ecosystems. Resource use itself can be considered an important measure of ecosystem functioning, and it can be linked to other processes. For example, if more diverse communities have higher resource utilization, then they are also likely to have greater biomass or productivity and lower losses of resources, such as to leaching, than less diverse communities. The suggestion that diversity enhances the consistency of resource use and space occupancy (e.g., Figure 2.4) lends further evidence to the idea that diversity is related to community stability. For instance, in terrestrial studies and in theoretical work, as diversity increases, the stability of individual species' populations tends to decrease slightly (May 1974), whereas the stability of aggregate community properties, such as biomass or percentage of cover, tends to increase (Tilman 1996; Tilman and Lehman 2002). Here we discuss a few additional ways in which the study of invasions may contribute significantly to the diversity and ecosystem functioning debate.

Multivariate complementarity

The effect of diversity on ecosystem functioning is often measured with a single response variable. For example, the influence of diversity on productivity may be measured in one study, while its influence on nutrient cycling is measured in another. However, while the influence of any species or suite of species may be critical to the performance of any single ecosystem process (e.g., productivity), other species may exert a dominant influence on other ecosystem variables. In many communities, then, any suite of species might seem unlikely to simultaneously control all ecosystem processes. Therefore, an incorporation of multiple ecosystem processes into a multivariate index of ecosystem functioning might prove useful to the study of diversity and ecosystem functioning. For instance, a common critique of the application of the results of diversity-ecosystem functioning experiments to conservation is that the ecosystem process examined often saturates at low levels of diversity, such that, for example, only some small fraction of diversity is required to obtain maximum productivity (e.g., Schwartz et al. 2000; but see Tilman 1999b). However, if a different fraction of that diversity is required to maintain some other function (e.g., nutrient cycling), a larger amount of the diversity may be required to saturate the multivariate index of ecosystem functioning. Alternatively, in spatially or temporally heterogeneous habitats, different species may maximize productivity in different locations in space or time (e.g., Hector et al. 1999; Tilman 1999b; Stachowicz et al. 2002a).

Intriguing results in this vein have been reported for a community of mobile invertebrate grazers in coastal seagrass meadows (Duffy et al. 2003). Different grazer species maximized values of grazer production, grazing impact, and sediment organic carbon, such that the high value of each of these processes at maximum diversity levels is explained by the fact that diverse communities contain individual species with a dominant effect on a particular ecosystem process (a.k.a. the sampling effect, *sensu* Tilman et al. 1997). However, the multispecies mixture was the only combination to achieve high values for all three of these response variables simultaneously. Although this "multivariate complementarity" has been relatively unrecognized explicitly, the work on the relationship between diversity and invasions (in which no single species seems able to exclude all potentially invading species) suggests that it may in fact be quite common. This insight, combined with the joining of experimental and observational approaches, may help broaden our understanding of the role of diversity in ecosystem functioning.

Does increasing diversity enhance ecosystem functioning?

At least in the short term, many invasions have resulted in a net gain in the number of species present at the local or regional level (Sax and Gaines 2003). This is not surprising, and it is similar to what is seen in invasions at the geologic scale: most regions gain in the total number of species, though some resident species eventually go extinct (e.g., Marshall et al. 1982). The diversity-ecosystem functioning debate has mostly been framed in the context of what will happen to ecosystem functioning as species richness declines due to extinctions. But in cases in which invasions cause a net increase in diversity at the scale at which ecosystem processes are measured, it is worth considering what effects this might have on ecosystem functioning. There are virtually no data to address this question, although it is clear that invasive species can affect ecosystem structure and function. For example, aquatic ecosystems lacking dominant filter-feeding organisms (either because natives have been driven ecologically extinct or because there were none to begin with) are dramatically altered when such species are introduced (e.g., Alpine and Cloern 1992). The introduction of a rapidly growing N-fixing legume into a low-nutrient ecosystem lacking such species can also have dramatic effects (Vitousek and Walker 1989; see also D'Antonio and Hobbie, this volume). Assessing the role of increasing diversity in other ecosystem processes might contribute to a fuller understanding of the degree to which ecosystem functioning saturates at high levels of diversity, particularly across multidimensional measures of ecosystem functioning.

Can communities ever be saturated?

If invasions lead to increased local diversity, then, all else being equal, communities might be expected to become less invasible over time as more species accumulate. However, there are many instances in which invasion rates appear to hold constant or even increase after the addition of new species (e.g., Cohen and Carlton 1998). One confounding factor is that many invasive species enter habitats that are highly disturbed. Because increased disturbance can allow more species to coexist, the net effect of disturbance and invasion may be communities that are proportionately less saturated with species than the original undisturbed habitats. Additionally, when species at multiple trophic levels are considered, it is possible that the addition of new species adds some resources while consuming others. Basal species, for example, might consume nutrients but provide food and habitat for many species at higher trophic levels. Adding a species takes away resources for species similar to that one (i.e., in the same functional group), but may add food or shelter resources for other species or modify habitat in a way favorable to some species. For example, while the invasion of a new C_3 grass undoubtedly decreases the resources available to other C_3 grasses (Fargione et al. 2003), it may add a new host plant that can be exploited by different species of herbivorous insects. While it is broadly appreciated that invaders can have cascading effects on species at the same trophic level via competition or at lower trophic levels via predation, the potential importance of cascading effects from the bottom up via facilitation or resource provision is less appreciated, although such effects may be common (Simberloff and Von Holle 1999; Richardson et al. 2000; Bruno et al., this volume). Recognition of these effects would lead to the prediction that as invasion within a functional group occurs, further invasion within that group should decline, while the potential for invasion within other functional groups might stay the same or even increase.

Whether overall invasion rates increase or decrease with the addition of exotics is hard to say, and the influence of propagule supply, disturbance, and other factors may play a larger role in determining the net change in invasion rates. This problem was recognized by Davis et al. (2000), who argued that a more predictive approach to invasions may come from focusing more explicitly on resource availability rather than on the specific factors that affect resource availability. Resource availability does appear to be the proximate determinant of the success of establishment of many invaders, but it is still of interest to understand the ultimate chain of causation that has led to recent changes in resource availability and hence to greater invasion.

Concluding Thoughts

For decades, ecological research has focused on how various ecosystem-level properties (such as productivity and stability) affect the maintenance of diversity on local scales. The emerging research agenda of the past decade has turned this around and has begun to rigorously examine the effects of diversity on these very same ecosystem processes. The data thus far support both directions of causation, revealing that diversity and ecosystem variables mutually influence one another. A serious challenge for ecology lies in assessing the potential for positive and negative feedbacks among diversity and various ecosystem processes to understand more broadly how these two critical ecosystem properties interact. In this chapter, we have outlined some ways in which the study of introduced species has provided significant insights into this area and other ways in which it may contribute further. There is significant potential for such research on introduced species to contribute to the synthesis of traditionally isolated fields ranging from evolution to ecosystem ecology.

Acknowledgments

The authors would like to thank the National Center for Ecological Analysis and Synthesis (funded by NSF grant DEB-0072909), the University of California, and the Santa Barbara campus for funding the "Exotic Species: A Source of Insight into Ecology, Evolution, and Biogeography" Working Group, which greatly benefited the development of this chapter. JS would also like to thank the National Science Foundation (grants OCE 00-02251 and OCE 03-51778) and NOAA-Sea Grant for support of the work on invasions that made some of the insights described within possible. DT also thanks the National Science Foundation (grant DEB/LTER 0090382) and the Andrew Mellon Foundation for support. Comments from the NCEAS Working Group, as well as D. Sax and two anonymous reviewers, improved previous versions of this chapter.

Literature Cited

Aarssen, L. W. 1997. High productivity in grassland ecosystems: affected by species diversity or productive species? Oikos 80:183–184.

Alpine, A. E., and J. E. Cloern. 1992. Trophic interactions and direct physical control of phytoplankton biomass in an estuary. Limnology and Oceanography 37:946–955.

Blackburn, T. M., and R. P. Duncan. 2001a. Determinants of establishment success in introduced birds. Nature 414:195–197.

Blackburn, T. M., and R. P. Duncan. 2001b. Establishment patterns of exotic birds are constrained by non-random patterns in introduction. Journal of Biogeography 28:927–939.

Callaway, R. M., and E. T. Aschehoug. 2000. Invasive plants versus their new and old neighbors: a mechanism for exotic invasion. Science 290:521–523.

Case, T. J. 1990. Invasion resistance arises in strongly interacting species-rich model competitive systems. Proceedings of the National Academy of Sciences USA 87:9610–9614.

Cohen, A. N., and J. T. Carlton. 1998. Accelerating invasion rate in a highly invaded estuary. Science 279:555–558.

Davis, M. A., J. P. Grime, and K. Thompson. 2000. Fluctuating resources in plant communities: a general theory of invasibility. Journal of Ecology 88:528–534.

Duffy, J. E., J. P. Richardson, and E. A. Canuel. 2003. Grazer diversity effects on ecosystem functioning in seagrass beds. Ecology Letters 6:637–645.

Elton, C. S. 1958. The ecology of invasions by animals and plants. Methuen, London.

Emmerson, M., and M. Huxham. 2002. How can marine ecology contribute to the biodiversity–ecosystem functioning debate? In Loreau, M., S. Naeem, and P. Inchausti, eds. *Biodiversity and ecosystem functioning: Synthesis and perspectives*, pp 139–146. Oxford University Press, Oxford.

Fargione, J., C. S. Brown, and D. Tilman. 2003. Community assembly and invasion: An experimental test of neutral versus niche processes. Proceedings of the National Academy of Sciences USA 100:8916–8920.

Flannery, T. 2001. *The eternal frontier: An ecological history of North America and its peoples*. Atlantic Monthly Press, New York.

Gentry, A. H. 1982. Neotropical floristic diversity: phytogeographical connections between Central and South America, Pleistocene climatic fluctuations, or an accident of the Andean orogeny? Annals of the Missouri Botanical Garden 69:557–593.

Hector, A., and 33 others. 1999. Plant diversity and productivity experiments in European grasslands. Science 286:1123–1127.

Hille Ris Lambers, J., W. S. Harpole, D. Tilman, J. Knops, and P. B. Reich. 2004. Mechanisms responsible for the positive diversity-productivity relationship in Minnesota grasslands. Ecology Letters 7:661–668.

Huston, M. A. 1997. Hidden treatments in ecological experiments: re-evaluating the ecosystem function of biodiversity. Oecologia 110:449–460.

Kinzig, A. P., S. W. Pacala, and D. Tilman. 2002. The functional consequences of biodiversity. Princeton University Press, Princeton, NJ.

Kennedy, T. A., S. Naeem, K. M. Howe, J. M. H. Knops, D. Tilman, and P. Reich. 2002. Biodiversity as a barrier to ecological invasion. Nature 417:636–638.

Knops, J. M. H., D. Tilman, N. M. Haddad, S. Naeem, C. E. Mitchell, J. Haarstad, M. E. Ritchie, K. M. Howe, P. B. Reich, E. Siemann, and J. Groth. 1999. Effects of plant species richness on invasion dynamics, disease outbreaks, insect abundances and diversity. Ecology Letters 2:286–293.

Knops, J. M. H., J. R. Griffin, and A. C. Royalty. 1995. Introduced and native plants of the Hastings reservation, central coastal California, a comparison. Biological Conservation 71:115–123.

Lavorel, S., A.-H. Prieur-Richard, and K. Grigulis. 1999. Invasibility and diversity of plant communities: from patterns to processes. Diversity and Distributions 5:41–49.

Levine, J. M. 2000. Species diversity and biological invasions: relating local process to community pattern. Science 288:852–854.

Levine, J. M., and C. M. D'Antonio. 1999. Elton revisited: a review of evidence linking diversity and invasibility. Oikos 87:15–26.

Lonsdale, W. M. 1999. Global patterns of plant invasions and the concept of invasibility. Ecology 80:1522–1536.

Loreau, M., and A. Hector. 2001. Partitioning selection and complementarity in biodiversity experiments. Nature 412:42–76.

Loreau, M., S. Naeem, and P. Inchausti. 2002. *Biodiversity and ecosystem functioning: Synthesis and perspectives.* Oxford University Press, Oxford.

Marshall, L. G. 1981. The Great American interchange—an invasion induced crisis for South American mammals. In M. H. Nitecki, ed. *Biotic crises in ecological and evolutionary time*, pp. 133–229. Academic Press, New York.

Marshall, L. G., S. D. Webb, J. J. Sepkoski, and D. M. Raup. 1982. Mammalian evolution and the Great American Interchange. Science 215:1351–1357.

May, R. M. 1974. *Stability and complexity in model systems*, 2nd Ed. Princeton University Press, Princeton, NJ.

McGrady-Steed J., P. M. Haris, and P. J. Morin. 1997. Biodiversity regulates ecosystem predictability. Nature 390:162–165.

McKane, R. B., D. F. Grigal, and M. P. Russelle. 1990. Spatiotemporal differences in ^{15}N uptake and the organization of an old-field plant community. Ecology 71:1126–1132.

McNaughton, S. J. 1993. Biodiversity and function of grazing systems. In E. D. Schulze, and H. A. Mooney, eds. *Biodiversity and ecosystem function*, pp. 361–383. Springer-Verlag, Berlin.

Naeem, S., J. M. H. Knops, D. Tilman, K. M. Howe, T. Kennedy, and S. Gale. 2000. Plant diversity increases resistance to invasion in the absence of covarying extrinsic factors. Oikos 91:97–108.

Osman, R. W. 1977. The establishment and development of a marine epifaunal community. Ecological Monographs 47:37–63.

Pimm, S. L. 1987. The snake that ate Guam. Trends in Ecology and Evolution 2:293–295.

Planty-Tabacchi, A-M., E. Tabacchi, R. J. Naiman, C. Deferrari, and H. Decamps. 1996. Invasibility of species-rich communities in riparian zones. Conservation Biology 10:598–607.

Rejmanek, M. 1996. Species richness and resistance to invasions. In G. H Orians, R. Dirzo and J. H. Cushman, eds, *Biodiversity and ecosystem processes in tropical forests*, pp. 153–172. Springer-Verlag, Berlin.

Rejmanek, M. 2003. The rich get richer—Response. Frontiers in Ecology and the Environment 1:122–123.

Richardson, D. M., N. Alsopp, C. M. D'Antonio, S. J. Milton, and M. Rejmanek. 2000. Plant invasions–the role of mutualisms. Biological Reviews 75:65–93.

Ruiz, G. M., P. W. Fofonoff, J. T. Carlton, M. J. Wonham, and A. H. Hines. 2000. Invasion of coastal marine communities in North America: Apparent patterns, processes and biases. Annual Review of Ecology and Systematics. 31:481–531.

Sax, D. F. 2002. Native and naturalized plant diversity are positively correlated in scrub communities of California and Chile. Diversity and Distributions 8:193–210.

Sax, D. F., S. D. Gaines, and J. H. Brown. 2002. Species invasions exceed extinctions on islands worldwide: a comparative study of plants and birds. American Naturalist 160:766–783.

Sax, D. F., and S. D. Gaines. 2003. Species diversity: from global decreases to local increases. Trends in Ecology and Evolution 18:561–566.

Schwartz, M. W., C. A. Brigham, J. D. Hoeksema, K. G. Lyons, M. H. Mills, and P. J. van Mantgem. 2000. Linking biodiversity to ecosystem function: implications for conservation ecology. Oecologia 122:297–305.

Shea, K., and P. Chesson. 2002. Community ecology as a framework for biological invasions. Trends in Ecology and Evolution 17:170–176.

Simberloff, D., and B. Von Holle. 1999. Positive interactions of nonindigenous species: invasional meltdown? Biological Invasions 1:21–32

Stachowicz, J. J., H. Fried, R. B. Whitlatch, and R. W. Osman. 2002a. Biodiversity, invasion resistance and marine ecosystem function: reconciling pattern and process. Ecology 83:2575–2590.

Stachowicz, J. J., J. R. Terwin, R. B. Whitlatch, and R. W. Osman. 2002b. Linking climate change and biological invasions: Ocean warming facilitates nonindigenous species invasions. Proceedings of the National Academy of Sciences USA 99:15497–15500.

Stachowicz, J. J., R. B. Whitlatch, and R. W. Osman. 1999. Species diversity and invasion resistance in a marine ecosystem. Science 286:1577–1579.

Stohlgren, T. J., D. Binkley, G. W. Chong, M. A. Kalkhan, L. D. Schnell, K. A. Bull, Y. Otsuki, G. Newman, M. Bashkin, and Y. Son. 1999. Exotic plant species invade hot spots of native plant diversity. Ecological Monographs 69:25–46.

Strong, D. R., E. D. McCoy, and J. R. Rey. 1977. Time and the number of herbivore species: the pests of sugar cane. Ecology 58:167–175.

Symstad, A. 2000. A test of the effects of functional group richness and composition on grassland invasibility. Ecology 81:99–109.

Taylor, B. W., and R. E. Irwin, 2004. Linking economic activities to the distribution of exotic plants. Proceedings of the National Academy of Sciences USA 101:17725–17730.

Tilman, D. 1982. *Resource competition and community structure*. Princeton University Press, Princeton, NJ.

Tilman, D. 1988. *Plant strategies and the dynamics and structure of plant communities*. Princeton University Press, Princeton, NJ.

Tilman, D. 1996. Biodiversity: population vs. ecosystem stability. Ecology 77:350–363.

Tilman, D. 1997. Community invasibility, recruitment limitation, and grassland biodiversity. Ecology 78:81–92.

Tilman, D. 1999a. The ecological consequences of changes in biodiversity: a search for general principles. Ecology 80:1455–1474.

Tilman, D. 1999b. Diversity and production in European grasslands. Science 286:1099–1100.

Tilman, D. 2004. A stochastic theory of resource competition, community assembly and invasions. Proceedings of the National Academy of Sciences USA 101:10854–10861.

Tilman, D., and C. Lehman. 2001. Biodiversity, composition and ecosystem processes: theory and concepts. In A. P. Kinzig, S. W. Pacala, and D. Tilman,

eds. *The functional consequences of biodiversity*, pp. 9–41. Princeton University Press, Princeton, NJ.

Tilman D., P. B. Reich, J. M. H. Knops, D. Wedin, T. Mielke, and C. Lehman. 2001. Diversity and productivity in a long-term grassland experiment. Science 294:843–845.

Tilman, D., R. M. May, C. L. Lehman, and M. A. Nowak. 1996a. Habitat destruction and the extinction debt. Nature 371:65–66.

Tilman, D., and D. Wedin. 1991. Plant traits and resource reduction for five grasses growing on a nitrogen gradient. Ecology 72:685–700.

Tilman, D., D. Wedin, and J. Knops. 1996b. Productivity and sustainability influenced by biodiversity in grassland ecosystems. Nature 379:718–720.

Tilman, D., C. Lehman, and K. Thompson. 1997. Plant diversity and ecosystem productivity: theoretical considerations. Proceedings of the National Academy of Sciences USA 94:1857–1861.

Tilman, D., J. Knops, D. Wedin, and P. Reich. 2002. Plant diversity and composition: effects on productivity and nutrient dynamics of experimental grasslands. In M. Loreau, S. Naeem, and P. Inchausti eds. *Biodiversity and ecosystem functioning: Synthesis and perspectives*, pp. 21–35. Oxford University Press, Oxford.

Torchin, M. E., K. D. Lafferty, A. P. Dobson, V. J. McKenzie, and A. M. Kuris. 2003. Introduced species and their missing parasites. Nature 421:628–630.

Vermeij, G. J.1991a. When biotas meet: understanding biotic interchange. Science 253:1099–1104.

Vermeij, G. J. 1991b. Anatomy of an invasion: the Trans-Arctic interchange. Paleobiology 17:281–307.

Vermeij, G. J. In press. One-way traffic in the Atlantic: causes and consequences of Miocene to early Pleistocene molluscan invasions in Florida and the Caribbean. Paleobiology.

Vitousek, P. M., and L. R. Walker. 1989. Biological invasion by Myrica faya in Hawaii: Plant demography, nitrogen fixation, and ecosystem effects. Ecological Monographs 59:247–265.

Webb, S. D. 1991. Ecogeography and the Great American interchange. Paleobiology 17: 266–280.

Wedin, D. A., and D. Tilman. 1993. Competition among grasses along a nitrogen gradient: initial conditions and mechanisms of competition. Ecological Monographs 63:199–299.

Wiser, S. K., R. B. Allen, P. W. Clinton, and K. H. Platt. 1998. Community structure and forest invasion by an exotic herb over 23 years. Ecology 79:2071–2081.

3

Plant Species Effects on Ecosystem Processes

INSIGHTS FROM INVASIVE SPECIES

Carla M. D'Antonio and Sarah E. Hobbie

Many non-native invasive species invade rapidly, create distinct monospecific patches with spreading borders, occupy a range of habitats, and transform ecosystems quickly. These characteristics make invasive species ideal tools for asking questions about the varying roles and pathways through which plant species affect energy flow, nutrient transformations, and geomorphic and hydrological processes. Studies of invasive species effects on disturbance regimes (a primary control over ecosystem processes) have demonstrated that a species does not have to become dominant to drive dramatic state changes in an ecosystem. For example, subtle changes in fuel distribution can lead to dramatic and long-term changes in fire regime and ecosystem structure. The extent of those changes, however, may depend on whether the prior resident species have experienced an evolutionary history involving fire. Studies of invasive plant species effects on nitrogen cycling have also demonstrated both dramatic and subtle species effects. Dramatic change tends to occur when species with discretely distinct traits invade ecosystems, whereas invaders with traits that overlap with those of residents do not appear to cause rapid change. Invasive plant species offer many further opportunities to test ecological theories regarding linkages between invasiveness and species effects, alternative states, controls over ecosystem change, controls over competitive interactions, and the reversibility of species effects.

Introduction

In the 1960s and 1970s, several now-classic papers detailed the roles that individual animal species could play in altering community structure, particularly in aquatic ecosystems (e.g., Paine 1966; Estes and Palmisano 1974; Menge 1976). These early studies laid the groundwork for further research examining the roles of individual animal species in driving community and ecosystem change. A flurry of work on animal effects then led to published syntheses of conditions controlling when individual species might be considered keystone species or ecosystem engineers (Jones and Lawton 1995; Power et al. 1996).

Although early twentieth-century plant ecologists noted that individual plant species could affect ecosystem development and the course of succession by ameliorating abiotic stress and thereby facilitating the establishment of later-arriving species (Cowles 1899; Clements 1916), little effort was made for most of the twentieth century to evaluate how changes in plant composition and, in particular, individual plant species could affect ecosystem structure and function. The recent rise in interest in plant species effects can be attributed to recognition that the composition of "natural" ecosystems is changing dramatically in response to anthropogenic activities and that these species compositional changes may play a feedback role in global change (Vitousek 1994). Among the widespread ongoing anthropogenic changes is the breakdown of barriers to species dispersal that has occurred with increased global travel and trade, contributing both directly and indirectly to community compositional change. The realizations that plant species could facilitate ecosystem change, rather than simply responding to it, and that plant invasions and subsequent ecosystem change could occur rapidly were brought to light by Vitousek and Walker (Vitousek et al. 1987; Vitousek and Walker 1989), who quantified strong effects of a single invading tree species (*Myrica faya*) on ecosystem development and nitrogen (N) accumulation in Hawaii. Since these publications, numerous other investigators have utilized invasive species (for this chapter, defined as non-native species with established or spreading populations) to understand how plant species control changes in ecosystem processes (e.g., D'Antonio and Vitousek 1992; see reviews in Levine et al. 2003; D'Antonio and Corbin 2003; Ehrenfeld 2003).

At an individual level, all living organisms are involved in the flux and storage of matter, water, and energy and are therefore a part of what we define as ecosystem processes. So, in some sense, it is not at all surprising or new to discover that individual plant species can have important consequences for ecosystem functioning. Yet the details of these effects are less clear. What is the cumulative effect of individual species' populations on rates of energy flow, water fluxes, and nutrient transformations? To what extent are species readily distinguishable from one another in their contributions to such processes? How abundant must a species become before its effects are realized—is the relationship between a species' abundance and its effects on ecosystem processes linear, or is there a threshold density below which its effects are negligible? Because substantial changes in ecosystem processes can affect the direction of succession and the sustainability of valued ecosystem functions (e.g., services), the ability to predict changes caused by species invasions (or extinctions) has practical as well as scientific value. In addition, the potential that individual species might create biospheric feedbacks to regional or even global environmental change has contributed to the growing interest in understanding the mechanisms through which compositional change might alter ecosystem processes.

Because non-native invading plant species often become abundant rapidly and can create spreading monospecific stands in a range of habitats, they offer clear opportunities to investigate questions that have otherwise proved elusive because of the slow rates of change in many "natural" communities and the difficulty of singling out species effects from potentially covarying environmental factors. If one invader can become established across a range of environmental conditions, particularly in the absence of other ongoing environmental changes, it offers the opportunity to evaluate how environmental characteristics such as climate, soils, and topography influence the magnitude of species effects. Likewise, a comparison of invader traits versus resident traits across a range of habitats controlling for invader identity can help us to identify conditions that determine when trait differences contribute to ecosystem change.

Conceptual Background

Predicting the effects of individual species requires defining the traits that are important to ecosystem processes and identifying how similar a species is to members of the community to or from which it is being added or deleted. Vitousek (1990) proposed that individual species cause measurable changes in ecosystem processes if they differ from natives in traits that affect (1) rates of utilization and storage of resources, (2) trophic structure, or (3) the frequency and magnitude of disturbance. He did not explicitly address the traits contributing to these characteristics or the issue of how different a species needs to be from residents to cause changes that would result in substantially altered ecosystem functions or services.

Chapin et al. (1996), building on this theme, suggested that species that were being added or deleted would cause substantial change only if they were "discretely" different from residents in key traits. One obvious example for plants is a species that fixes N entering a community lacking N fixers, as in the case of *Myrica faya* entering successionally young Hawaiian forests lacking other symbiotic N-fixing trees (Vitousek et al. 1987; Vitousek and Walker 1989). By contrast, species whose traits overlapped greatly with those of residents, even if they had differing mean trait values, would probably interact competitively with residents in a largely compensatory manner and not cause substantial ecosystem change. Chapin et al. referred to these invaders as "continuous trait" invaders. They essentially predicted that the plant species that are the most similar to residents may be strongly competitive with residents, but will not necessarily cause ecosystem change, whereas those that are the most different, or have discretely different traits, will cause the largest changes.

Predicting whether a species addition or deletion will affect ecosystem processes requires identifying which species traits have the largest ecosystem consequences, as well as the distribution of those traits among the resident species and the invader. In the case of species additions, we must be able to predict whether an invader can become abundant enough to shift the mean value for an important trait far enough to observe a consequence at the ecosystem scale. In the case of species deletions (e.g., local extirpations), we must be able to evaluate whether the remaining residents will shift their trait values as the declining species disappears, thereby compensating for the missing species. If species are interacting competitively for a resource, then it makes sense that the disappearance of a species should free up resources for its resident competitors. Thus, depending on the relative abundances of the various species, compensation might result in a lack of ecosystem change.

A hypothetical example of invasion and shifting of community mean trait values is presented in Figure 3.1. We based this example on leaf litter quality because it is an important trait controlling decomposition (Swift et al. 1979) and is broadly correlated with rates of N cycling in natural systems (e.g., Scott and Binkley 1997). Yet even for leaf litter quality, it is not well known how different an invader has to be for overall rates of nutrient cycling to shift in an ecosystem, or how abundant the invader has to become before change is apparent. The relationship between abundance and relative trait values is

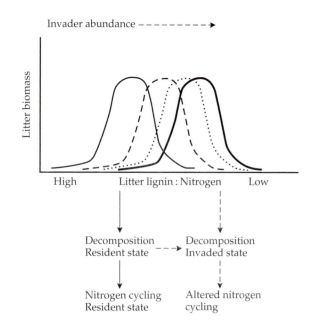

Figure 3.1 Hypothetical shift in tissue quality (lignin : N ratio) of litter inputs into the community litter pool (assuming some sort of relative distribution of a community's litter over time). In this example, the invader has higher N and/or lower lignin, so the lignin:N ratio of the community litter pool begins to shift to the right. Dashed and dotted curves represent transient states of litter distribution as the invader becomes more abundant. A high-biomass-producing invader with a lower lignin:N ratio could presumably shift the community faster by swamping the existing litter pool and shifting the curve to the right rapidly. The example here assumes that the total amount of litter produced by the community is constant after invasion, but an invader could also increase the total amount of litter at the site.

largely unknown. Furthermore, species can simultaneously alter nutrient cycling in multiple ways (e.g., via effects on litter chemistry or on the abiotic environment) that can sometimes have offsetting effects—litter chemistry will be the trait of importance only if substrate characteristics (rather than some aspect of the physical environment) limit microbial activity. Hence graphic presentations such as Figure 3.1 may oversimplify such scenarios, which more realistically involve multiple predictor variables.

As in the case of keystone species (Power et al. 1996), the effect of an invader is likely to be context-dependent. D'Antonio and Corbin (2003) suggest that predicting whether a species will affect an ecosystem process requires knowing (1) the likelihood that the species will be able to invade an ecosystem and become abundant, (2) which key species traits affect the ecosystem processes of interest and how those traits interact with other changes that might be occurring, (3) the relative distribution of important trait values of the invader compared with the residents, (4) the way in which residents will respond to the invader (e.g., can they persist by altering their own trait values?), and (5) the potential for abiotic factors to buffer biological processes driven by the suspected traits of interest. If residents do not readily drop out of the system as the invader becomes abundant, ecosystem fluxes may change more slowly than if residents die or disappear quickly. Likewise, abiotic factors such as soil texture or chemistry (e.g., carbonate content, pH, etc.) may mitigate the extent to which invaders can change ecosystem processes. While little quantitative modeling has been done to address the relative importance of these various processes, we believe that invasive species offer an enormous opportunity to gain insight into these questions, through both modeling and empirical studies.

In the following sections we review some examples of single-species effects on ecosystem processes and what we have learned about controls over ecosystem change through these studies, with an emphasis on invasive plants. We focus first on species effects on disturbance regimes because disturbance represents one of the primary controls over temporal and spatial variation in rates of energy flux, primary production, and nutrient cycling in terrestrial ecosystems. We also focus on species effects on N cycling because N is a critical limiting nutrient across an array of systems and because N storage and cycling can affect succession, carbon storage, N leaching, and other important ecosystem functions. Controls over N cycling have been a focus of ecosystem ecologists for many decades.

Insights Gained from Studying Invasive Species

Species effects on disturbance regimes

Disturbances create spatial and temporal variation in the flux of energy, water, and nutrients through ecosystems. Changes in disturbance regimes, such as changes in the timing, frequency, size, or severity of disturbances, can therefore drive changes in ecosystem processes that influence community composition and structure (Vitousek 1990). Species that alter disturbance regimes are therefore important controllers of successional trajectories and qualify as ecosystem engineers (*sensu* Lawton and Jones 1995; see also Mack and D'Antonio 1998).

The central role of individual animal species (whether native or not) in creating disturbances in communities is well known. From insects to elephants, native animals have been demonstrated to create disturbances that in turn alter a range of processes, including rates of erosion and nutrient cycling. By contrast, our understanding of the effects of individual *plant* species on disturbance regimes has been more limited, but has benefited enormously from the study of invasive species.

Perhaps the most important lesson we have learned by evaluating the effects of invasive plants on disturbance regimes is how quickly dramatic shifts in ecosystem states can come about. Community ecologists have studied and modeled conditions allowing the existence of alternative states for decades (e.g., Scheffer et al. 2001; Scheffer and Carpenter 2003). While shifts between internally reinforced states have been shown to occur rapidly in aquatic systems (e.g., Scheffer and Carpenter 2003), they have been less well studied and rarely quantified in terrestrial ecosystems. Fire-enhancing introduced grasses have aided our understanding of how and when alternative states can arise on the same piece of ground and what ecological constraints contribute to positive internal feedbacks. Further insights into community theory could be gained by careful evaluation of critical thresholds of species abundances, alone or in combination with environmental changes that might contribute to rapid state shifts. While these insights could be gained from systems with "native species invasions," the prevalence of introduced species, often with distinct growth forms or physiological traits, has made tracking of such thresholds easier.

While invading species have been demonstrated to both suppress and enhance aspects of disturbance regimes (Mack and D'Antonio 1998; Brooks et al. 2004), we focus here largely on enhancement of fire disturbance because this area has so far yielded the greatest insights. Plant characteristics and their cumulative effects on vegetation morphology, chemistry, and biomass have long been known to be fundamental to fire regimes and are a part of all commonly used fire behavior models, such as BEHAVE (Rothermel 1972) and its more recent derivatives such as FARSITE (Finney 1995) and BEHAVE-PLUS (http://fire.org). Indeed, parameterization of initial models was based largely on fuel characteristics of single species (e.g., Rothermel and Philpot 1973). Well-studied changes in fire regime driven by humans include those resulting from the elimination of understory fuels through livestock grazing and from increases in fuel biomass as a result of fire suppression. The effects of species as they invade a landscape were, until recently, much less studied. Several pale-oecological studies have documented changes in fire regime that appear to correlate with species invasions, but the extent to which the changes are driven by species invasions and their associated traits versus climate change is less well understood. The rapid spread of invasive plants that alter fuel conditions has provided an excellent opportunity to evaluate the role of individual species traits compared with that of climate and other characteristics as they affect fire regime (D'Antonio 2000; Brooks et al. 2004).

Invasive species research has demonstrated that certain invasive plants, particularly grasses, have had enormous effects on fire regimes in many regions of the world (see reviews in D'Antonio and Vitousek 1992; D'Antonio 2000; Brooks et al. 2004). In some cases, grass invasion and subsequent fires have been the triggers leading to rapid state changes ultimately including feedbacks that perpetuate the new state. A well-studied example that illustrates the importance of species-driven changes in disturbance regime to ecosystem condition is the case of perennial grasses from both the New and Old World that have invaded seasonally dry Hawaiian shrublands and woodlands (Smith 1985). These grasses have tripled the frequency of fire and increased by 60 times the average area burned in a single fire (Tunison et al. 2001). Their presence across an elevation gradient has also allowed investigators to evaluate causes of context dependency in species effects. The grasses have fueled fires across a range of habitats from sea level to 1200 meters. D'Antonio et al. (2000) found that the outcome of these fires depended on both the identity of the invader and pre-fire community composition, including presence of fire-adapted species. In the higher-elevation sites, the grass-fueled fires promoted a state change from forest to grassland that appears to be relatively stable (Hughes et al., 1991; D'Antonio et al. 2001). The internal feedbacks that maintain and reinforce this new state appear to be the homogeneous grass canopy, which promotes more rapid fire spread (Freifelder et al. 1998), and the rapid and dense regrowth form of the invading grasses, which precludes reestablishment of prior dominants (Hughes and Vitousek 1993). Changes in nutrient dynamics with these altered ecosystem conditions (e.g., Mack et al. 2001; Mack and D'Antonio 2003) are discussed in more detail in the section on nutrient cycling below.

We have also learned from the study of invasive species that fire *frequencies* have been altered in habitats where introduced grasses have changed the *distribution* of fine fuels across space and time (e.g., Whisenant 1990; Tunison et al. 2001). This change is usually simultaneous with a change in biomass of fine fuels, but not always. Studies in the Great Basin suggest that subtle changes in fuel distribution without changes in fuel biomass (Whisenant 1990) are all that is needed to trigger dramatic ecosystem change via altered fire frequency. Thus the invader, in this case *Bromus tectorum* (cheatgrass), did not have to become a "dominant" species in order to drive change in the system. Where invasive species change the *biomass* of fine fuels in habitats with historic fire, they tend to change the *intensity*, but not the frequency, of fire (van Wilgen and Richardson 1985; Bilbao 1996; Lippincott 2000; Platt and Gottshalk 2001; Rossiter et al. 2003). Changes in fire frequency under these circumstances may develop over the course of decades (van Wilgen and van Hensbergen 1992). The community and ecosystem consequences of fire intensity changes appear to be less dramatic, at least initially. Thus, when a species changes the quantity of fuel present in sites with historic fire, intensity is the most important disturbance regime characteristic that is altered.

By contrast, when a species invasion results in a qualitatively different type or distribution of fuel, as when the invader has the ability to occupy space that had not previously contained fine fuels (shrub interspaces) or when it shifts the fuel size class distribution dramatically, fire frequency and spatial extent change. A testable hypothesis emerging from this literature is that changes in fire frequency should have more dramatic consequences in terms of being able to trigger rapid changes in ecosystem state than changes in fire intensity.

What we do not yet know is why species that change fire regimes in some places do not cause alterations in the many other places where they are successful invaders. For example, in Idaho's Snake River plains (in the Intermountain West of the United States), the invasive grass *Bromus tectorum* has increased the frequency of fire from once every 60–110 years to once every 3–5 years, with dramatic and long-term consequences for ecosystem conditions and processes (Whisenant 1990; Obrist et al. 2003, 2004). *Bromus tectorum* creates a layer of fine fuels around and between sagebrush shrubs, leading to increased probabilities of ignition and greater probabilities of fire spread after lightning strikes during summer storms. But *B. tectorum* is present in at least 40 states and is abundant in other western states such as Wyoming where sagebrush is also abundant. Its effects on fire frequency have been confined to the northern and western Great Basin, raising the question of what the interplay is between invader demography, fuel characteristics, climate, and ignition sources that has kept *B. tectorum* from having the same effects in other regions. Within the range where fire frequency has changed due to *B. tectorum* invasion, we do not yet know what site or resident community characteristics define *thresholds* above which native or desirable species do not recover on the site. It is likely that there is recovery toward the prior community in some regions or at some sites, but as of yet there are no models or regional studies to evaluate environmental constraints on internally reinforcing conditions.

LIMITATIONS TO USING INVASIVE SPECIES TO STUDY SPECIES-LEVEL CONTROLS OVER DISTURBANCE REGIMES   One of the most obvious limitations on using invasive species to study controls over disturbance regimes is that many invasive species invade as a result of disturbance in the first place (Hobbs and Huenneke 1992; D'Antonio et al. 1999). Thus changes perceived to be driven by species invasion are essentially driven by prior disturbances that may in themselves have contributed to altered fuel conditions. For example, the prevalence of *B. tectorum* in the Great Basin is proposed to be due in part to the removal of native perennial grasses by intensive sheep grazing in the late 1800s (Kennedy and Doten 1901; Kennedy 1903). The elimination of perennial grasses contributed to the creation of large senescent stands of sagebrush with little understory, setting the stage for more catastrophic wildfires. Since most of the fire-mediated cheatgrass-grassland transformations have occurred in "mature" or "senescent" sagebrush, initial conversion in many areas cannot be attributed solely to *B. tectorum*. In other words, the series of triggers or changes that ultimately led to the dramatic ecosystem transformations that we are currently witnessing may be more complex than invading grasses alone.

In many regions, invasion by suspected fire-promoting species is so complete that it is difficult to find control areas for comparisons of changes in fire regime. Often historical fire regimes are poorly known, so comparisons of pre-invasion and post-invasion fire regimes are difficult. Pre- and post-invasion fuel conditions are also sometimes difficult to discern clearly because uninvaded sites in the same climate zone are often on different soils and thus probably supported different vegetation composition and structure. This probably explains the paucity of real numbers to cite when evaluating changes in fire regime or fuel conditions caused by invasive species (see D'Antonio 2000).

Likewise, changes in climate, hydrological conditions, and atmospheric gases are occurring in many regions simultaneously with the spread of invasive species. Changes in fire regimes may result from any of these changes alone or in combination with species invasions (e.g., Dukes 2000). Sorting out the importance of these multiple drivers may be difficult. For example, in desert riparian habitats in western North America, fire frequencies appear to be increasingly correlated with the presence of invasive salt cedar, *Tamarix ramossisima* (e.g., Busch 1995). This fire-responsive species invades sites where the water table and flooding regimes have been altered by dams and levees. At the same time, native species have declined in vigor, most likely due to declining water tables, competition with salt cedar, and lack of floods that remove older plant tissue and provide fresh substrates for regeneration (Busch and Smith 1995; Taylor et al. 1999). Although salt cedar clearly accumulates fine stems that can fuel fires, the contribution of these multiple factors to fire frequency and severity is not clearly known.

Species effects on nitrogen cycling

Nitrogen has been a focus of ecosystem ecologists since the emergence of the discipline because N limits or co-limits net primary production in most temperate ecosystems (e.g., Vitousek and Howarth 1991). More recently, human alterations of the N cycle, sometimes with dramatic effects on terrestrial ecosystems, have been well documented. Studying the effects of invasive species on N cycling (and the cycling of other nutrients) has provided, and can continue to provide, important insights into three areas of ecology. First, these studies potentially provide a way to link plant traits affecting invasion success (invasiveness) with community and ecosystem effects. For plants, the relationship between invasiveness and ecosystem effects has not been well explored. Second, invasive species have served as case studies for exploring mechanisms by which plant species influence biogeochemical cycles. The diversity of growth forms of invaders and the rapidity of change have diversified studies of species effects and broadened the exploration of pathways through which biogeochemical change can come about. Finally, the active control or removal of many invaders from wildland sites offers opportunities to test the reversibility of different types of nutrient cycling changes.

LINKING INVASIVENESS AND SPECIES EFFECTS   Parker et al. (1999) made one of the first attempts to seek generality in describing how species effects arise. Their simple formulation of species impact as the result of a species' abundance, range, and per capita effects emphasizes the importance of demographic traits along with ecophysiological and morphological traits in creating species effects (Figure 3.2). With regard to linking invasiveness and species effects on N cycling, if invasive species are capable of using sources of limiting nutrients that are not available to native species (e.g., via N fixation), they should have a competitive edge over native species, which should contribute to increases in their abundance and eventually in their range. At the same time, if their success is tied to their ability to use a previously unavailable source of N, then their per capita effect on that resource should be high relative to that of resident species. The accumulation of per capita effects as more and more individuals become established should cause large changes in the resource. This in turn should determine successional trajectories following invasion (Adler et al. 1998; Maron and Jefferies 1999; Levine et al. 2003) or during restoration (Yelenik et al. 2004). Despite the simplicity of this logic, the chain of reasoning has rarely been tested. Most of the relevant work has involved invasion by N-fixing woody species into N-poor habitats. By using atmospheric N, N-fixing species can overcome N limitation, potentially promoting their own invasion (*A* and *R* in Figure 3.2), at least initially. Their ability to use atmospheric N and thereby affect N accumulation and cycling is not new to ecosystem ecology (e.g., Boring et al. 1988). Linking this "per capita" trait to invasion and ultimately to species effects, however, is new.

Alleviation of N limitation has probably facilitated the invasion by the N fixers *Myrica faya* and *Acacia* spp. into N-poor lava flows in Hawaii (Vitousek and Walker 1989) and the Cape floristic region, South Africa (Witkowski 1991). Although N fixers generally should have an advantage in N-limited ecosystems, subtle interactions between the N-fixing capabilities of potential invasive species and the N status of potentially invasible ecosystems may ultimately

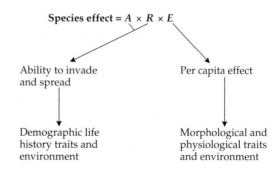

Figure 3.2 General equation describing species effects. *A* = abundance, *R* = range, *E* = per capita effects. (After Parker et al. 1999.)

Species effect = $A \times R \times E$

Ability to invade and spread

Per capita effect

Demographic life history traits and environment

Morphological and physiological traits and environment

determine which N-fixing species are successful invaders. For example, there can be tremendous variation within and among legume crop species and their wild relatives in N fixation rates (Provorov and Tikhonovich 2003). Likewise, Haubensak (2001) has demonstrated variation in N fixation among sites by *Genista monspessulana* (French broom) invading central California. In addition, in the Cape floristic region, Stock et al. (1995) have demonstrated that reliance on N fixation by two invasive *Acacia* species was related to the N status of the invaded site, with *A. saligna*, which invades the relatively N-poor fynbos, relying more heavily on N fixation than *A. cyclops*, which invades the relatively N-rich strandveld. Interestingly, however, the effects of the two invaders on annual rates of N mineralization were opposite to predictions: the species entering the N-poor site had a less pronounced effect on N cycling rates than the one entering the N-rich site (Stock et al. 1995), at least over the short term. Such nonintuitive findings open up new avenues of inquiry into the conditions controlling when species traits translate into ecosystem process effects.

Because of the difficulties associated with quantifying N fixation rates in situ, interactions between the N-fixing activity of invasive species, the N status of invaded sites, and the success of N fixers as invaders have rarely been explored. K. A. Haubensak and C. M. D'Antonio (unpublished data) experimentally evaluated the importance of N limitation of grassland vegetation to successful establishment of the N fixer *G. monspessulana*. Contrary to expectations, N limitation of the resident vegetation played only a minor role in French broom establishment. Future research could prove helpful in explaining the relative success of different invasive N-fixing species and how their ability to fix N relates to their spread, abundance, and effects.

Besides overcoming nutrient limitation via N fixation, invasive species may increase nutrient stocks if they partition nutrients differently from native species, using nutrients in complementary ways. For example, invasive species might use nutrients from different soil depths or at different times of the year than native species (Fargione et al. 2003). Such complementary resource use should increase total nutrient use, perhaps helping to explain the overall increase in net primary production that often occurs with plant invasion (Ehrenfeld 2003). Indeed, in a biodiversity manipulation of grassland species, resident species were more susceptible to invasion by species that were dissimilar to them in their nutrient use (Fargione et al. 2003; J. Fargione and D. Tilman, unpublished data), suggesting that invasive species increased overall nutrient use by the community. Did this increase, in turn, result in increased rates of nutrient accumulation and turnover? Such linkages await further research.

PATHWAYS THROUGH WHICH PLANTS AFFECT ECOSYSTEM N BIOGEOCHEMISTRY   Any plant species, whether native or not, could conceivably influence rates of N cycling. Pathways through which this might happen could be direct, as through litter quality, or indirect, whereby species might initiate changes in the animal, plant, or microbial communities, and these changes in turn might cause N cycling rates to change (Figure 3.3). The most obvious direct effect is that of

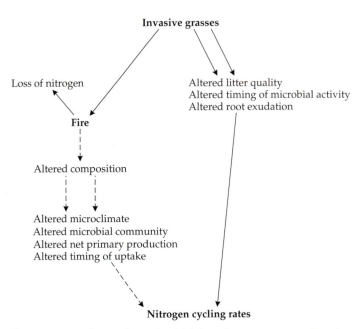

Figure 3.3 Pathways through which invasive grasses may alter nitrogen cycling in an ecosystem. Solid arrows depict direct pathways; dashed arrows depict indirect pathways.

N-fixing species increasing the amount of N in an ecosystem, as discussed above. The Vitousek and Walker (1989) example is so often cited because these N-fixing invasive species probably represent an endpoint of large and unambiguous effects of invasive species on nutrient cycling. However, the rapidity of change caused by many fast-spreading species also offers the opportunity to evaluate more subtle species effects on N loss and internal rates of N cycling.

To our knowledge, few studies have examined the effects of invasive species on nutrient losses. Invasive species could, however, decrease nutrient losses from ecosystems. If invasive species differ from natives in where and when they access soil nutrient pools (Fargione et al. 2003), they may reduce soil inorganic N pools and thus leaching losses, as demonstrated in studies manipulating plant diversity (i.e., more diverse communities exhibit lower soil inorganic N pools: Tilman et al. 1997; but see Hooper and Vitousek 1998). However, if invasive species promote nitrification, as suggested by Ehrenfeld (2003), they could promote N losses via leaching and/or denitrification. Invasive species that increase fire frequencies, as described above (D'Antonio and Vitousek 1992; Whisenant 1990), may promote losses of N directly through volatilization. They may also have indirect effects on N losses via fire-induced changes in species composition that alter N mineralization and/or the amount and timing of plant N uptake (e.g., Mack et al. 2001). For example, C_4 grass invasions in submontane woodlands in Hawaii have increased fire frequency, promoting their own dominance—as discussed previously (Tunison et al. 2001). These grasses directly cause N loss via fire (Mack et al. 2001) and reduce N accretion after fire by suppressing recruitment of species associated with N fixation (Ley and D'Antonio 1998; Mack et al. 2001). They also change N cycling indirectly by promoting net N mineralization rates in the wet season but reducing them in the dry season (Mack and D'Antonio 2003). Their competitive suppression of native woody species (Hughes and Vitousek 1993) results in a greatly simplified post-fire community. As a consequence, plant N demand occurs over a narrower portion of the year than in uninvaded communities, and annual N availability exceeds plant demand (Mack et al. 2001), probably promoting gaseous and leaching N losses. These species have provided an ideal opportunity to trace the importance of these different direct and indirect pathways of species-induced changes. Similar invasions in other semiarid ecosystems offer opportunities to compare the magnitude of direct and indirect pathways.

In addition to altering nutrient inputs and losses, invasive species can alter rates of nutrient cycling within communities, both by altering the abiotic environment (e.g., Mack and D'Antonio 2003) and by changing the quantity, chemistry, and placement of plant inputs into soils (see Figure 3.3; e.g., Evans et al. 2001; Windham and Ehrenfeld 2003). Interest in the influence of invasive species on nutrient cycling grew out of comparative studies of native species that linked variation in species traits such as litter chemistry (e.g., lignin or N concentration) to ecosystem processes such as decomposition and litter nutrient dynamics (Gosz et al. 1973; Melillo et al. 1982), and in turn to variation in soil nutrient processes (e.g., Wedin and Tilman 1990; Scott and Binkley 1997). Observations of such linkages among native species have spurred interest in whether invasive species could similarly affect nutrient cycling. In addition, interest has broadened from studying the effects of variation in litter chemistry to consideration of other mechanisms by which invasive species might affect nutrient cycling. Ehrenfeld (2003) reviewed a number of invasive species traits that might influence nutrient cycling, such as size, canopy and root architecture, phenology, mycorrhizal status, photosynthetic pathway, growth form, and tissue chemistry, but concluded that little evidence exists to link specific traits of invasive species to observed changes in nutrient cycling at this time.

REVERSIBILITY OF SPECIES EFFECTS ON SOIL NUTRIENTS The extent to which species effects are readily reversed is of theoretical as well as practical interest. From a practical point of view, where an invasive species does affect soil nutrient cycling, managers who oversee removal of these species assume that after removal, they will be able to return the site to a more desirable condition. Yet legacies of the invader, such as elevated labile soil N pools or altered rates of cycling, may interfere with the return of the vegetation to a more desirable condition. From a theoretical point of view, interests lie with trying to understand what mechanisms are operating that would tend to keep the system from going back to a presumed alternative state. Since several invasive species appear to cause soil changes (e.g., Ehrenfeld 2003), there are many opportunities to test hypotheses regarding species legacies, the strength or importance of internal feedbacks and the reversibility of alternative states.

The more rapid N cycling observed in many sites invaded by non-native species has spurred both basic and applied research. After removal of the undesirable species, many studies have utilized a technique of adding labile carbon in the form of sucrose or sawdust to soils in order to stimulate soil microbial growth and immobilization of N. This in turn should temporarily decrease N availability to plants (see review in Corbin et al. 2004). Ecological theory predicts that lower N cycling should decrease the competitive edge of faster-growing species (i.e., the undesirable invader), allowing slower-growing native species to increase in abundance (see Gurevitch and Scheiner 2002 for a discussion of N effects on plant competition). Several studies have found that adding labile carbon can indeed reduce the biomass of undesirable species. Sometimes the biomass of desirable species increases (e.g., Blumenthal et al. 2003; Perry et al. 2004), but other times the reduced N levels also reduced the growth of more desired species (see review in Corbin et al. 2004), both confirming the original predictions under some conditions but raising further questions about the assumptions that underlie use of this technique.

Overall, we believe that the prevalence of exotic species removal projects in natural landscapes offers an excellent opportunity to evaluate many questions in both basic and applied soil and plant ecology. From testing the reversibility of species effects to practical tests of how to assist succession toward more desired plant composition and ecosystem structure and function, invasive species that influence soil properties offer many opportunities for future research.

LIMITATIONS TO STUDYING INVASIVE SPECIES AND NITROGEN CYCLING   Invasive species present both limitations and opportunities as model systems for understanding how plant species affect nutrient cycling. The approach is limited because often one or two invasive species are compared with a handful of native species (or with an entire native community); such a small number of species makes attributing effects on nutrient cycling to specific plant traits difficult. For example, Ehrenfeld et al. (2001) found that two invasive species of eastern deciduous forests, a shrub and a C_4 grass, both accelerated nitrification rates, despite opposite effects on the quantity and N concentration of detritus produced, making the mechanisms responsible for the changes in nitrification difficult to discern. Her studies make clear that simple predictions based on tissue quality, the dominant paradigm in ecosystem research prior to the recent spate of invasive species studies, are inadequate to predict species effects.

Although invasive species may differ from natives in traits known to alter nutrient cycling, the magnitude of trait difference or the abundance of the invasive species needed to produce biologically significant effects at the ecosystem level is difficult to predict. In addition, invasive species can have multiple, sometimes opposing, effects on nutrient cycling, such that predictions of their net effects based on single traits are difficult to make and may be context-dependent. For example, *B. tectorum* invasion into grasslands of Canyonlands National Park, Utah, decreased rates of N availability (measured using ion-exchange resin bags), a result that is consistent with its having litter that was higher in lignin and lower in N than the native grasses (Evans et al. 2001). However, this depression of N availability was apparently dependent on precipitation, occurring in a wet year, but not in a dry year. Likewise, Svejcar and Sheley (2001) observed no effect of *B. tectorum* on rates of N cycling at their sites in eastern Oregon when it was compared with native bunchgrasses. Sweet-clovers (*Melilotus* spp.), despite being N fixers, depressed rates of net N mineralization in Rocky Mountain National Park, Colorado, apparently by reducing soil moisture (Wolf et al. 2004). Such studies demonstrate that simple assumptions about trait effects on N cycling need reevaluation.

Nevertheless, comparisons across multiple sites invaded by a single species, or among studies of multiple invaders, provide opportunities for synthesizing results that may offer new insights into the nature of species effects on ecosystem processes. In a review of effects of invasive species on biogeochemical processes, Ehrenfeld (2003) concluded that invasive species more frequently accelerate rates of N cycling (mineralization, nitrification) than would be expected by chance alone, but for unclear reasons. Whether this pattern arises because positive effects are reported more frequently than lack of effects, because studies of invasive species effects often involve N fixers whose high-N tissue is likely to accelerate N cycling (but see Wolf et al. 2004), or because invasive species tend to be rapidly growing species with high photosynthetic rates and high tissue N concentrations (e.g., see review in D'Antonio and Corbin 2003) is unknown.

Attributing causation to invasive species in all studies requires being able to separate the effects of the invader from the effects of events or spatial heterogeneity that might have stimulated invasion in the first place. In the case of species effects on N cycling, for example, disturbance often stimulates N availability, thereby favoring fast-growing species that can capitalize on pulses of high N and high light. These may be non-native or native invaders, but changes in N cycling due to their presence are difficult to separate from changes in N cycling due to the initial disturbance. Planting plots with different species after disturbance can help to separate the causes of accelerated or altered N cycling, but simple comparisons of "invaded" versus "uninvaded" patches of habitat may be misleading. Selection of control areas for comparisons with invaded sites must always be done with such cautions in mind. Nonetheless, since many invading species form slowing spreading fronts, measurements can be taken from the middle of infestations and along gradients of infestation age. Uninfested areas outside of spreading centers can be carefully checked for disturbance, preexisting soil differences, or other complicating factors.

Although the consequences of altered nutrient cycles for both invasive and noninvasive species have often been suggested (Vitousek and Walker 1989; Ehrenfeld and Scott 2001), few studies have demonstrated all of the causal relationships that link invasive species effects on nutrient cycling to changes in community structure. Furthermore, it is unclear whether invasive species' effects on nutrient cycling are always analogous to those of native species, given that invasive species may have different relationships with soil biota in their home ranges than in the ecosystems they invade (Callaway et al. 2004; Callaway et al., this volume). Studies of invasive plant species may provide an opportunity to evaluate the importance of familiarity between plants and soil microbes in influencing N cycling.

Future Challenges

Invasive plant species have provided opportunities to evaluate how and when individual species or groups of closely related species can affect controls over ecosystem processes. Most of the best-studied examples are those in which plant species have altered disturbance regimes or soil N cycling; we have reviewed several of them here in order to elucidate what we have learned from these examples about species traits as drivers of ecosystem change. Notable cases are largely those situations in which residents differ greatly from the invaders: in these cases, invasion and change occur rapidly, sometimes resulting in the creation of "alternative states." Nonetheless, we believe that much more insight can be brought to ecosystem ecology by studying invaders across a broader range of habitats, not just those sites where their effects are already dramatic and easy to measure. One unanswered question is to what extent species have to become dominant in order to exert a measurable effect on the rates of key processes.

Simultaneously with the rise in interest in studying invasive species effects on ecosystem processes, a second line of research developed that addressed the role of species composition, independent of origin (e.g., with no concern over whether species were native or not), in affecting ecosystem processes (Schulze and Mooney 1994). Some investigators focused on the role of species diversity, while others focused on how the loss of individual species might affect the timing and rates of nutrient and energy flux in natural ecosystems (e.g., Wedin and Tilman 1990; Hobbie 1992; Chapin 1993; Naeem et al. 1994; Hooper and Vitousek 1997; Wardle et al. 1999; Loreau and Hector 2001; Tilman et al. 2001; Hector et al. 2002; Loreau et al. 2002; Diaz et al. 2003). This work has been complementary to the line of research involving invasive species. Both approaches have utilized removal experiments, or synthetic communities in planted plots or mesocosms, to evaluate species effects, and both have been largely trait-based (see Stachowicz and Tilman, this volume). These parallel lines of research should be compared to achieve greater synthesis and a clearer understanding of the strengths and weaknesses of both approaches. A greater attempt at comparison between these approaches might also shed light on whether studies of non-native invaders tell us anything different from studies of general compositional shifts in plant communities.

Using naturally occurring invasions can be problematic for interpreting cases of change in which invaders do not have discretely different traits from residents because rates or spatial patterns of invasion may be driven by covarying environmental factors that may be difficult to discern and control. Hence careful measurements and experimentation should be used to tease apart potentially covarying factors or eliminate their importance. Experimental manipulation of resident communities irrespective of species origin (e.g., removal experiments in "natural communities") or the creation of mesocosm plots in fields or plantations largely circumvents this problem if plots are randomly arrayed on the landscape. But this approach too has a disadvantage because it starts with a major soil disturbance, and because composition may rely on maintenance of potentially artificial conditions as compared with a "natural community." Measuring changes caused by invaders across patch boundaries where invaders are spreading locally allows the measurement of "natural change" and transient states.

Because dramatic changes are taking place in many "natural" ecosystems around the globe today, trying to understand and predict when and where shifts will occur is important for successful management and restoration of these systems. Identifying thresholds of species abundances, or combinations of environmental thresholds and invasive species abundances, that trigger change or impede restoration is critical. While invasive species have already given us insights into the mechanisms responsible for ecosystem change, we believe they have still been underutilized in such studies.

Acknowledgments

This work was conducted as a part of the "Exotic Species: A Source of Insight into Ecology, Evolution, and Biogeography" Working Group supported by the National Center for Ecological Analysis and Synthesis, funded by the National Science Foundation (grant DEB-0072909), the University of California, and the Santa Barbara campus. We thank Dov Sax, Jay Stachowicz, the NCEAS Working Group participants, and two anonymous reviewers for feedback on an earlier draft of this chapter.

Literature Cited

Adler, P. B., C. M. D'Antonio, and J. T. Tunison. 1998. Understory succession following dieback of *Myrica faya* in Hawai'i Volcanoes National Park. Pacific Science 52:69–78.

Bilbao, B. 1995. Impacto del regimen de quemas en las caracteristicas edaficas, produccion de material organica, y biodiversidad de sabanas tropicales en Calabozo, Venezuela. PhD dissertation. Instituto Venezolano de Investigaciones Cientificas (IVIC), Caracas, Venezuela. 215 pp.

Blumenthal, D. M., N. R. Jordan, and E. L. Svenson. 2003. Soil carbon addition controls weeds and facilitates prairie restoration. Ecological Applications 13:605–615.

Boring. L. R., W. T. Swank, J. B. Waide, and G. S. Henderson. 1988. Sources, fates and impacts of nitrogen inputs to terrestrial ecosystems: review and synthesis. Biogeochemistry 6:119–158.

Brooks, M. L., C. M. D'Antonio, D. M. Richardson, J. M. DiTomaso, J. B. Grace, R. J. Hobbs, J. E. Keeley,

M. Pellant, and D. Pyke. 2004. Effects of invasive alien plants on fire regimes. Bioscience 54:677–688.

Busch, D. E. 1995. Effects of fire on Southwestern riparian plant community structure. Southwestern Naturalist 40:259–267.

Busch, D. E., and S. D. Smith. 1995. Mechanisms associated with declines of woody species in riparian ecosystems of the Southwestern USA. Ecological Monographs 65:347–370.

Callaway, R. M., G. C. Thelen, A. Rodriguez, and W. E. Holben. 2004. Soil biota and exotic plant invasion. Nature 427:731–733.

Chapin, F. S., III. 1993. Functional role of growth forms in ecosystem and global processes. In J. R. Ehleringer and C. B. Field, eds. *Scaling physiological processes: Leaf to globe*, pp. 287–312. Academic Press, San Diego.

Chapin, F. S., H. Reynolds, C. M. D'Antonio, and V. Eckhart. 1996. The functional role of species in terrestrial ecosystems. In B. Walker and W. Steffen, eds. *Global change in terrestrial ecosystems*, pp. 403–428. Cambridge University Press, New York.

Clements, F. E. 1916. Plant succession: analysis of the development of vegetation. Carnegie Institute of Washington Publication, No. 242. Washington, DC.

Corbin, J. C., C. M. D'Antonio, and S. J. Bainbridge. 2004. Tipping the balance in the restoration of native plants: Experimental approaches to changing the exotic: native ratio in California grassland. In M. Gordon and L. Bartol, eds. *Experimental approaches to conservation biology*, pp. 154–179. University of California Press, Los Angeles.

Cowles, H. C. 1899. The ecological relations of the vegetation on the sand dunes of Lake Michigan. The Botanical Gazette 27:95–391.

D'Antonio, C. M. 2000. Fire, plant invasions and global changes. In H. Mooney and R. Hobbs, eds. *Invasive species in a changing world*, pp. 65–94. Island Press, Covela, CA.

D'Antonio, C. M., and J. D. Corbin. 2003. Effects of plant invaders on nutrient cycling: Using models to explore the link between invasion and development of species effects. In C. D. Canham, J. J. Cole, and W. K. Lauenroth, eds. *Models in ecosystem science*, pp. 363–384. Princeton University Press, Princeton, NJ.

D'Antonio, C. M., and P. Vitousek. 1992. Biological invasions by exotic grasses, the grass-fire cycle and global change. Annual Review of Ecology and Systematics 23:63–88.

D'Antonio, C. M., T. Dudley, and M. Mack. 1999. Disturbance and biological invasions. In L. Walker, ed. *Ecosystems of disturbed ground*, pp. 429–468 Elsevier, New York.

D'Antonio, C. M, J. T. Tunison, and R. Loh. 2000. Variation in impact of exotic grass fueled fires on species composition across an elevation gradient in Hawai'i. Austral Ecology 25:507–522.

Diaz, S., A. J. Symstad, F. S. Chapin, III, D. A. Wardle, and L. F. Huenneke. 2003. Functional diversity revealed by removal experiments. Trends in Ecology and Evolution 18:140–146.

Dukes, J. S. 2000. Will the increasing atmospheric CO_2 concentration affect the success of invasive species? In H. Mooney and R. Hobbs, eds. *Invasive species in a changing world*, pp. 95–114. Island Press, Covela, CA.

Ehrenfeld, J. G. 2003. Effects of exotic plant invasions on soil nutrient cycling processes. Ecosystems 6:503–523.

Ehrenfeld, J. G., and N. A. Scott. 2001. Invasive species and the soil: effects on organisms and ecosystem processes. Ecological Applications 11:1259–1260.

Ehrenfeld, J. G., P. S. Kourtev, and W. Z. Huang. 2001. Changes in soil functions following invasions of exotic understory plants in deciduous forests. Ecological Applications 11:1287–1300.

Estes, J. A., and J. F. Palmisano. 1974. Sea otters: their role in structuring nearshore communities. Science 185:1058–1060.

Evans, R. D., R. Rimer, L. Sperry, and J. Belnap. 2001. Exotic plant invasion alters nitrogen dynamics in an arid grassland. Ecological Applications 11:1301–1310.

Fargione, J., C. S. Brown, and D. Tilman. 2003. Community assembly and invasion: An experimental test of neutral versus niche processes. Proceedings of the National Academy of Sciences USA 100:8916–8920.

Finney, M. A., 1995. FARSITE: A fire area simulator for mangers. In *The Biswell symposium: fire issues and solutions in urban interface and wildland ecosystems*. General technical report PSW-158 USDA Forest Service, Berkeley, CA.

Freifelder, R., P. Vitousek, and C. M. D'Antonio. 1998. Microclimate effects of fire-induced forest/grassland conversion in seasonally dry Hawaiian woodlands. Biotropica 30:286–297.

Gosz, J. R., G. E. Likens, and F. H. Bormann. 1973. Nutrient release from decomposing leaf and branch litter in the Hubbard Brook forest, New Hampshire. Ecological Monographs 43:173–191.

Gurevitch, J., S. Scheiner, and G. Fox. 2002. *The ecology of plants*. Sinauer Associates, Sunderland, MA.

Haubensak, K. A. 2001. The ecology and impacts of French and Scotch broom in coastal prairie environments. PhD dissertation. University of California, Berkeley.

Hector, A., E. Bazeley-White, M. Loreau, and S. Otway. 2002. Overyielding in grassland communities: testing the sampling effect hypothesis with replicated biodiversity experiments. Ecology Letters 5:502–511.

Hobbie, S. E. 1992. Effects of plant species on nutrient cycling. Trends in Ecology and Evolution 7:336–339.

Hobbs, R. J., and L. F. Huenneke. 1992. Disturbance, diversity, and invasion: implications for conservation. Conservation Biology 6:324–337.

Hooper, D. U., and P. M. Vitousek. 1997. The effects of plant composition and diversity on ecosystem processes. Science 277:1302–1305.

Hooper, D. U., and P. M. Vitousek. 1998. Effects of plant composition and diversity on nutrient cycling. Ecological Monographs 68:121–149.

Hughes, R. F., and P. M. Vitousek. 1993. Barriers to shrub reestablishment following fire in the seasonal submontane zone of Hawai'i. Oecologia 93:557–563

Jones, C. G., and J. H. Lawton, eds. 1995. *Linking species and ecosystems*. Chapman and Hall, New York.

Kennedy, P. B. 1903. Summer ranges of eastern Nevada sheep. Nevada Agricultural Experiment Station Bulletin 55, Reno, NV.

Kennedy, P. B., and S. B. Doten. 1901. A preliminary report on the summer ranges of western Nevada sheep. Nevada Agricultural Experiment Station Bulletin 51, Reno, NV.

Lawton, J. H., and C. G. Jones. 1995. Linking species and ecosystems: organisms as ecosystem engineers. In C. G. Jones, and J. H. Lawton, eds. *Linking species and ecosystems*, pp. 141–150. Chapman and Hall, New York.

Levine, J. M., Vilà, M., D'Antonio, C. M., Dukes, J. S., Grigulis, K., and S. Lavorel. 2003. Mechanisms underlying the impacts of exotic plant invasions. Proceedings of the Royal Society of London B 270:775–781.

Ley, R., and C. M. D'Antonio. 1998. Exotic grasses alter rates of nitrogen fixation in seasonally dry Hawaiian woodlands. Oecologia 113:179–187.

Lippincott, C. L. 2000. Effects of *Imperata cylindrica* (cogongrass) invasions on fire regime in Florida sandhill. Natural Areas Journal 20:140–149.

Loreau, M., and A. Hector. 2001. Partitioning selection and complementarity in biodiversity experiments. Nature 412:72–76.

Loreau, M., S. Naeem, P. Inchausti, and B. Schmid, eds. 2002. *Biodiversity and ecosystem functioning: Synthesis and perspectives*. Oxford University Press, Oxford.

Mack, M. and C. M. D'Antonio. 1998. Impacts of biological invasions on disturbance regimes. Trends in Ecology and Evolution 13:195–198.

Mack, M. C., and C. M. D'Antonio. 2003. Exotic grasses alter controls over soil nitrogen dynamics in a Hawaiian woodland. Ecological Applications 13:154–166.

Mack, M., C. M. D'Antonio, and R. Ley. 2001. Pathways through which exotic grasses alter N cycling in a seasonally dry Hawaiian woodland. Ecological Applications 11:1323–1335.

Maron, J. L., and R. L. Jefferies. 1999. Bush lupine mortality, altered resource availability, and alternative vegetation states. Ecology 80:443–454.

Melillo, J. M., J. D. Aber, and J. F. Muratore. 1982. Nitrogen and lignin control of hardwood leaf litter decomposition dynamics. Ecology 63:621–626.

Menge, B. A. 1976. Organization of the New England rocky intertidal community: role of predation, competition and environmental heterogeneity. Ecological Monographs 46:355–393.

Naeem, S., L. J. Thompson, S. P. Lawler, J. H. Lawton, and R. M. Woodfin. 1994. Declining biodiversity can alter the performance of ecosystems. Nature 368:734–737.

Obrist, D., E. H. Delucia, and J. A. Arnone III. 2003. Consequences of wildfire on ecosystem CO_2 and water vapour fluxes in the Great Basin. Global Change Biology 9:563–574.

Obrist, D., D. Yakir, and J. A. Arnone III. 2004. Temporal and spatial patterns of soil water following wildfire-induced changes in plant communities in the Great Basin in Nevada, USA. Plant and Soil 262:1–12.

Paine, R. T. 1966. Food web complexity and species diversity. American Naturalist 100:65–75.

Parker, I. M., D. Simberloff, W. M. Lonsdale, K. Goodell, M. Wonham, P. M. Kareiva, M. H. Williamson, B. Von Holle, P. B. Moyle, J. E. Byers, and L. Goldwasser. 1999. Impact: toward a framework for understanding the ecological effects of invaders. Biological Invasions 1:3–19.

Perry, L. G., S. M. Galatowitsch and C. J. Rosen. 2004. Competitive control of invasive vegetation: a native wetland sedge suppresses *Phalaris arundinacea* in carbon-enriched soil. Journal of Applied Ecology 41:151–162.

Platt, W. J. and R. M. Gottschalk. 2001. Effects of exotic grasses on potential fine fuel loads in the groundcover of south Florida slash pine savannas. International Journal of Wildland Fire 10:155–159.

Power, M. E., D. Tilman, J. A. Estes, B. A. Menge, W. Bond, S. L. Mills, G. Daily, J. C. Castillo, J. Lubchenco and R. T. Paine. 1996. Challenges in the quest of Keystones: identifying keystone species is difficult, but essential to understanding how loss of species will affect ecosystems. Bioscience 46:609–620.

Provorov, N. A., and I. A. Tikhonovich. 2003. Genetic resources for improving nitrogen fixation in legume-rhizobia symbiosis. Genetic Resources and Crop Evolution 50:89–99.

Rossiter, N. A., S. A. Setterfield, M. M. Douglas, and L. B. Hutley. 2003. Testing the grass-fire cycle: alien grass invasion in the tropical savannas of northern Australia. Diversity and Distributions 9:169–176.

Rothermel, R. C. 1972. A mathematical model for predicting fire spread in wildland fires. USDA Forest Service Techn. Paper, INT-115. Intermountain Research Station, Logan Utah. 40 pp.

Rothermel, R. C. and C. W. Philpot. 1973. Predicting changes in chaparral flammability. Journal of Forestry 71:640–643.

Scheffer, M. and S. R. Carpenter. 2003. Catastrophic regime shifts in ecosystems: linking theory to observation. Trends in Ecology and Evolution 18:648–656.

Scheffer, M., S. R. Carpenter, J. A. Foley, C. Folke, and B. Walker. 2001. Catastrophic shifts in ecosystems. Nature 413:591–596.

Schulze, E.-D., and H. A. Mooney, eds. 1994. *Biodiversity and ecosystem function.* Springer-Verlag, Berlin.

Scott, N. A., and D. Binkley. 1997. Foliage litter quality and annual net N mineralization: comparison across North American forest sites. Oecologia 111:151–159.

Smith, C. W. 1985. Impact of alien plants on Hawaii's native biota. In C. P. Stone and J. M. Scott, eds. *Hawaii's terrestrial ecosystems: preservation and management,* pp. 180–250. Cooperative National Park Resources Study Unit, Honolulu.

Swift, M. J., O. W. Heal, and J. M. Anderson. 1979. *Decomposition in terrestrial ecosystems.* Blackwell Scientific, Oxford.

Stock, W. D., K. T. Wienand, and A. C. Baker. 1995. Impacts of invading N_2-fixing Acacia species on patterns of nutrient cycling in two Cape ecosystems: evidence from soil incubation studies and ^{15}N natural abundance values. Oecologia 101:375–382.

Svejcar, A. and R. Sheley. 2001. Nitrogen dynamics in perennial and annual dominated arid rangeland. Journal of Arid Environments 47:33–46.

Taylor, J. B., D. B. Webster, and L. M. Smith. 1999. Soil disturbance, flood management, and riparian woody plant establishment in the Rio Grande floodplain. Wetlands 19:372–382.

Tilman, D., J. Knops, D. Wedin, and P. Reich. 1997. The influence of functional diversity and composition on ecosystem processes. Science 277:1300–1302.

Tilman, D., P. B. Reich, J. M. H. Knops, D. A. Wedin, T. Mielke, and C. L. Lehman. 2001. Diversity and productivity in a long-term grassland experiment. Science 294:843–845.

Tunison, J. T., R. Loh, and C. M. D'Antonio. 2001. Fire, grass invasions and revegetation of burned areas in Hawaii Volcanoes National Park. In K. E. Galley and T. P. Wilson, eds. *Proceedings of the invasive species workshop: The role of fire in the controls and spread of invasive species,* pp. 122–131. Tall Timbers Research Station Publication No. 11, Allen Press, Lawrence, KS.

Van Wilgen, B. W., and D. M. Richardson. 1985. The effects of alien shrub invasions on vegetation structure and fire behaviour in South African fynbos shrublands: A simulation study. Journal of Applied Ecology 22:955–966.

Van Wilgen, B. W., and G. G. van Hensbergen. 1992. Fuel properties of vegetation in Swartboskloof. In B. W. van Wilgen, D. M. Richardson, F. J. Kruger and H. J. van Hensbergen, eds. *Fire in South African mountain fynbos,* pp. 37–53. Springer-Verlag, Berlin.

Vitousek, P. M. 1990. Biological invasions and ecosystem process-towards an integration of population biology and ecosystems studies. Oikos 57:57:7–13.

Vitousek, P. M. 1994. Beyond global warming: ecology and global change. Ecology 75:1861–1876.

Vitousek, P. M., and R. W. Howarth. 1991. Nitrogen limitation on land and in the sea: How can it occur? Biogeochemistry 13:87–115.

Vitousek, P. M., and L. R. Walker. 1989. Biological invasion by *Myrica faya* in Hawai'i: Plant demography, nitrogen fixation and ecosystem effects. Ecological Monographs 59:247–265.

Vitousek, P. M., L. Walker, L., Whiteaker, D. Mueller-Dombois, and P. Matson. 1987. Biological invasion by *Myrica faya* alters ecosystem development in Hawai'i. Science 238:802–804.

Wardle, D. A., K. I. Bonner, G. M. Barker, G. W. Yeates, K. S. Nicholson, R. D. Bardgett, R. N. Watson, and A. Ghani. 1999. Plant removals in perennial grassland: vegetation dynamics, decomposers, soil biodiversity, and ecosystem properties. Ecological Monographs 69:535–568.

Wedin, D. A., and D. Tilman. 1990. Species effects on nitrogen cycling: a test with perennial grasses. Oecologia 84:433–441.

Whisenant, S. G. 1990. Changing fire frequencies on Idaho's Snake River Plain: ecological and management implications. USDA Forest Service Intermountain Research Station, General Technical Report INT-276. pp. 4–10.

Windham, L., and J. G. Ehrenfeld. 2003. Net impact of a plant invasion on nitrogen-cycling processes within a brackish tidal marsh. Ecological Applications 13:883–897.

Witkowski, E. T. F. 1991. Effects of invasive alien acacias on nutrient cycling in the coastal lowlands of the Cape Fynbos. Journal of Applied Ecology 28:1–15.

Wolf, J. J., S. W. Beatty, and T. R. Seastedt. 2004. Soil characteristics of Rocky Mountain National Park grasslands invaded by *Melilotus officinalis* and *M. alba.* Journal of Biogeography 31:415–424.

Yelenik, S. G., W. D. Stock, and D. M. Richardson. 2004. Ecosystem level impacts of invasive *Acacia saligna* in the South African fynbos. Restoration Ecology 12:44–51.

4

Biological Invasions and the Loss of Birds on Islands

INSIGHTS INTO THE IDIOSYNCRASIES OF EXTINCTION

Tim M. Blackburn and Kevin. J. Gaston

The spread of humans around the globe in the last few millennia has led (and is leading) to the alteration of virtually all ecological systems as exotic species become established and native species are driven to extinction. These two processes are connected: indeed, since an exotic species can be defined as one transported and introduced to a new location by human agency, which definition clearly includes Homo sapiens, *the vast majority of all recent extinctions can probably be attributed to the action of exotics. However, obviously not all native species go extinct following human arrival. In this chapter, we review the characteristics that typify species driven extinct by invaders, focusing on extinctions of birds, with the aim of using these invasion-driven events to further understanding of the extinction process. We show that avian extinction is not a random process with respect to either species or location. Patterns in extinction can be explained at least in part by the characteristics of exotic invaders and by environmental changes wrought by them. However, these conclusions must be tempered by the certainty that of the bird species that have disappeared in the recent extinction crisis, only a fraction are known.*

Introduction

It is now generally accepted that humans (*Homo sapiens*) evolved in sub-Saharan Africa, from where their geographic range has subsequently expanded to encompass essentially every land area on the planet free of permanent ice or snow. The dates of this expansion are controversial, but it may have begun approximately 65,000 years before the present (BP) with colonization of Asia (Quintana-Murci et al. 1999). Archaeological and genetic evidence places humans in East Asia at about this time, on mainland Australia about 60,000–55,000 years BP, and in North America perhaps about 15,000 years BP. Colonization of the many islands in the Pacific Ocean has been much more recent, with humans reaching Melanesia and Micronesia in the last 5000–4000 years, Fiji and Samoa perhaps 3500 years BP, and New Zealand in the last 1000 years (e.g., Anderson 1991; Milberg and Tyrberg 1993; Higham et al. 1999; Roberts et al. 2001; Bortolini et al. 2003; Hurles et al. 2003).

The first arrival of humans on every landmass colonized has broadly coincided with a marked, and sometimes sustained, increase in the extinction rate of the native biota (Martin 1984, 2001; Diamond 1989; Flannery 1994, 2001; Roberts et al. 2001). The net result is that current global extinction rates are estimated to be 1000–10,000 times higher than typical background rates revealed by the fossil record (May et al. 1995). For this reason, it is frequently argued that the earth is currently experiencing a human-induced biodiversity "crisis" or mass extinction event (Wilson 1985; May et al. 1995; Gaston and Spicer 2004). Although the exact causes of these extinctions are the subject of much debate (see Martin and Klein 1984; MacPhee 1999; Holdaway and Jacomb 2000; Alroy 2001; Grayson 2001; Martin 2001; Johnson 2002; Lyons et al. 2004), human colonization is typically associated with three processes that have the potential, at least, to alter the population sizes of native species dramatically and hence cause extinctions (Diamond 1989; Gaston 2002): (1) overexploitation of populations, most obviously for food, but also for building, clothing, or ceremonial purposes; (2) habitat destruction and fragmentation, for example, as a result of clearing for agriculture, collecting for firewood, or burning as a method of hunting; and (3) colonization by species other than humans that are not native to the location, which may affect native populations through processes similar to those already attributed to humans. These three processes may lead to a fourth, "extinction cascades," as the disappearance of certain key species can lead in turn to extinctions of species not directly affected by human exploitation, habitat destruction, or non-native species (e.g., obligate parasites). The length of time these processes have had to act means that hardly any ecological systems have escaped the imprint of human activities.

The third of these processes, colonization by exotic invasive species, is one that has recently received increasing attention from ecologists and conservation biologists. As humans have spread around the globe, they have deliberately or accidentally transported a huge variety of plant and animal species to locations beyond their natural geographic ranges (Elton 1958; Long 1981; Lodge

1993; Williamson 1996, 1999; Grayson 2001). Many of these species have established self-sustaining wild populations, some of which have had profound effects on the ecosystems they have invaded, including substantial economic and health costs to human societies (Elton 1958; Ebenhard 1988; Lever 1994; Simberloff 1995; Williamson 1996; Vitousek et al. 1997; Dalmazzone 2000; Mack et al. 2000; McNeely 2001). The highly deleterious effects of exotic invaders also extend to the native flora and fauna, which they may affect through competition for space, energy, or nutrients; by predation; or by harboring and spreading exotic diseases (reviewed by Williamson 1996; Mack et al. 2000).

The term "exotic invader" is typically used to refer to species that have been deliberately or accidentally transported by humans to a new location beyond their normal geographic range limits (i.e., to exclude species that are expanding their geographic ranges without human assistance). In fact, humans themselves are accommodated by this definition. Although, as noted above, a distinction is usually drawn in studies of extinction between the effects of human and nonhuman colonists, from the point of view of the native biota such a distinction is arbitrary. Severe population reductions by exotic predators can be classified as overexploitation whether or not those predators are human. Habitat may be destroyed by humans or by other exotic species; in the former case, the destruction can often be viewed as competition for space, light, and nutrients between native species and those with which humans wish to replace them. In the remainder of this chapter, we adopt the view that humans are just another exotic invader, exposing native species to new sources of competition and predation. Humans are special only in that they act as the transport mechanism for nonhuman exotics. The explicit treatment of humans as an exotic species is not new; Ehrlich (1986) compared humans with other great apes to further understanding of what distinguishes successful from unsuccessful invaders. Here, we consider the next step on the invasion pathway and incorporate humans into our assessment of the effects invaders have following successful establishment. We review patterns of loss across native species caught up in the current extinction crisis and consider how the entire assemblage of exotic invaders might drive these extinctions. Our aim is to gain insights from extinctions that have already occurred that will help us lessen future losses.

The current extinction crisis has affected many plant and animal taxa (Groombridge 1992; Smith et al. 1993a,b; May et al. 1995). However, we focus here on the relationship between exotic species and extinctions in birds. The reasons for this are entirely pragmatic: birds are probably the best-known major taxon in terms of both extant and extinct species, allowing us to make more rigorous statements about patterns of extinction in relation to invasions in this group than in any other. Nevertheless, even for birds, there are many problems with the data on extinction. These problems are perhaps best illustrated by considering evidence for the scale of the recent extinction crisis in birds. Thus, that is where we begin.

To assist our review, we have assembled a data set on avian extinctions by searching the literature for information on prehistoric (but post-human colonization) and historic species extinctions and combining it with information on current avifaunas. Most of these data refer to oceanic islands, for reasons that will soon become evident, and so for islands of this type we also collated information relating to the island: land area (km²), maximum elevation (m), isolation from nearest mainland (km), absolute latitude (degrees), and time since first human colonization (years). We also collated life history information on extinct species where possible, on bird species threatened with extinction, and on island endemic species. The data largely came from BirdLife International (2000), J. E. M. Baillie (unpublished Ph.D. thesis), Biber (2002), Blackburn and Gaston (2002), and a wide range of other sources listed in those works.

The Scale of Avian Extinctions

BirdLife International (2000) listed 128 bird species that they considered on the basis of the evidence then available to have become extinct since AD 1500 (although some of these species may still persist and others listed as threatened may actually have become extinct: Diamond 1987; Flood 2003; Saville et al. 2003). Of these species, 92% were endemic to islands, even though only about 12% of extant species can be considered island endemics. Holocene fossil deposits provide specimen evidence that human colonization of islands has been followed by the extinction of at least an additional 200 island endemic bird species (Cassels 1984; Milberg and Tyrberg 1993; Steadman 1995; Biber 2002). These fossils must nevertheless constitute only a small fraction of all the bird species that went extinct following human spread across the globe. The scale of the true losses has been estimated by Steadman (1995) using data from Pacific islands. He noted that Pacific islands that have Holocene bird fossils that have been studied in detail are typically found each to have lost several bird populations to extinction in the period following human colonization. Taking a conservative estimate of 10 population extinctions per island and a total of about 800 Pacific islands, Steadman (1995) extrapolates a loss of at least 8000 bird populations in this region in response to human expansion. Since many of the fossils are those of flightless rails endemic to single islands, he suggested that human expansion may have resulted in the extinction of 2000 species of rails in the Pacific alone, although Curnutt and Pimm (2001) suggest that this is a large overestimate. These losses are set against a backdrop of about 10,000 currently extant bird species in the region. They also mean that many other extant bird species with very restricted current distributions, including several island endemics, were in the past distributed much more widely (Olson and James 1982; Milberg and Tyrberg 1993; Steadman 1995, 1997; BirdLife International 2000).

Steadman's (1995) logic is mirrored exactly in our worldwide island data sample. These 220 islands have lost more than 700 bird populations from over 500 species since human colonization. However, extinctions have been recorded from only about a hundred of the islands; thus, those islands where losses are recorded have lost an average of close to 7 populations per island. In fact, the loss per island is likely to be even greater than this. Only 40 of the islands in our sample show fossil evidence of extinctions. Some islands may have been surveyed for fossils and no extinctions found, but this set will be small in comparison with those islands that have not been surveyed. Those islands that do have fossils of species that disappeared there following human colonization have lost an average of 10 populations. These 700 populations losses involve the global extinction of more than 200 species.

Steadman (1995) also provided evidence of the difference in the rate of population loss before and after human colonization, using data on about half a million Holocene fossil bird bones collected from lava tubes on the Galápagos Islands. These bones document the loss of 21 to 24 populations in the few hundred years since human colonization of the islands, but of no more than 3 populations in the preceding 4000–8000 years. More than 90% of the bones predate human arrival, meaning that the difference is unlikely to be a result of preservation bias between the two periods. The arrival of humans on the Galápagos Islands was associated with a hundredfold increase in the avian extinction rate.

The extent of the population losses documented by Steadman's (1995) and our data is enormous, but even so could still be an underestimate. The reason is that most fossil faunas will not include representatives of all species lost: many populations will have vanished without a trace. Pimm et al. (1994, 1995) suggested a method analogous to mark-recapture studies to estimate this number. Some extinct species are known from both specimen skins and fossil bones. Others are known from only one of these two sources. The number of unknown extinctions in certain Pacific island bird faunas can be estimated as follows: number of species with neither skin nor bones = (number with bones but no skins × number with skins but no bones) / number with bones plus skins. Applying this equation to Hawaii, Pimm et al. (1995) obtained an estimate of 30 bird species that have disappeared from these islands without a trace following human colonization. They considered this estimate to be conservative. These extinctions are in addition to the more than 40 species known only from fossils (Olson and James 1991; James and Olson 1991; Pimm et al. 1995) and the 20 that have disappeared since AD 1500 (BirdLife International 2000). Biber (2002) applied Pimm et al.'s approach to 22 islands or archipelagoes with recorded prehistoric bird extinctions. His calculations suggest that unknown extinctions on these islands are almost as numerous as known extinctions (209 vs. 216). Biber (2002) and Steadman (1995) use different samples of islands, and Biber's analysis deals with "endemic bird areas," which often cluster islands that Steadman treats separately. Nevertheless, if Biber's estimates applied more widely, they would suggest that Steadman's (1995) figure of 8000 bird population extinctions in the Pacific may actually be too low.

The species that have disappeared without a trace are likely to be a biased subset of losses. Species with poor potential for preservation and small-bodied species less likely to be sampled by collectors will be underrepresented in fossil data (Milberg and Tyrberg 1993; Pimm et al. 1995). For example, an initial collecting trip by Olson (1985) to a fossil site on Puerto Rico revealed only one hummingbird bone, which was stuck to the bone of a larger species. A second trip to the same site yielded more than 75 hummingbird bones when the sediment was sieved through a finer mesh. Species hunted by humans or by owls, and hence present in middens, will be overrepresented (Milberg and Tyrberg 1993; Pimm et al. 1995). Bones from midden sites may overestimate the extent to which humans were directly responsible for extinctions.

Perhaps the only pre-human avifauna that we can reconstruct in its entirety with reasonable certainty is that of the archipelago of New Zealand (Holdaway et al. 2001). New Zealand was first colonized by humans within the last millennium (Anderson 1991; Higham et al. 1999), making it the last major inhabited landmass to be colonized. One consequence of this recent colonization date is that abundant subfossil and specimen information exist for the avifauna. These data are summarized by Holdaway et al. (2001), who list a core fauna on the two main islands in the archipelago (North and South) of 132 species at the time of first human colonization. Of these, 46 species are now globally extinct, while each of the two main islands has lost more than 50 native species (out of 100 and 113 species, respectively). The global losses include entire bird families (Dinornithidae and Aptornithidae), terrestrial, arboreal, freshwater, and sea birds, and species from 30 g to 200 kg in mass.

Of course, as Pacific islands go, New Zealand is relatively large, and hence likely to be relatively species-rich; therefore, the number of species available to be lost is likely to be relatively high. On the other hand, its size also makes the complete extirpation of a species over the entire archipelago a significant feat, especially since humans and their associated invaders have had less than a millennium in which to achieve it. Smaller islands colonized longer ago have lost proportionately more species. Easter Island, for example, has lost all native land birds and all seabird species bar one (Steadman 1995).

In sum, the true magnitude of the extinction event that has affected the planet following the expansion of the human geographic range beyond Africa is unknown, and perhaps will never be known. Even for birds, probably the best-documented group in terms of lists of extant and extinct species, the number of species for which we have evidence of extinction is certainly a huge underestimate of the total number that have gone extinct. This lack of knowledge is a major limitation when it comes to assessing the true form of extinction patterns, and hence the processes that have caused those patterns, as it is impossible to know for certain the magnitude of any given bias. Although the trends in such biases ought to be broadly predictable on the basis of likely biases in fossil preservation and discovery and in natural history observations, this particular nonrandomness nevertheless must be remembered when considering other forms of nonrandomness in extinctions. With that in mind, we next consider patterns in avian extinctions, first in terms of the characteristics of locations that have lost species and then in terms of the characteristics of the species lost.

Nonrandomness in Extinction

Characteristics of location

As noted earlier, most of the bird extinctions since AD 1500 (BirdLife International 2000) have involved island species (see also King 1985; Johnson and Stattersfield 1990). Pacific islands are heavily represented on this list. Neither of these observations will come as a surprise: the latter observation is a simple consequence of the fact that most birds have been lost from islands, and most islands are in the Pacific. The vulnerability of island birds would be expected given the relatively small geographic range sizes and population sizes of most such species (Diamond 1984; Purvis et al. 2000; but see Manne et al. 1999). This vulnerability would be increased by the lack of predators of birds on islands prior to human arrival. The well-known lack of appropriate escape behaviors of island individuals consequent on their inexperience with predators (e.g., Darwin 1839, cited in Milberg and Tyrberg 1993; Flannery 1994) would be likely to increase mortality rates in island forms. Most known prehistoric Holocene bird extinctions also involve island species.

However, the extent to which the preponderance of island species is a true reflection of the pattern of recent human-caused bird extinctions is unclear. Certainly the colonization by humans of continental landmasses was also accompanied by extinctions among the native megafauna, including birds (Martin and Klein 1984; Martin and Steadman 1999; Miller et al. 1999; Roberts et al. 2001). Moreover, all of the continents were colonized much earlier (10,000 years or more) than any of the islands for which we have extinction data. Continental extinctions are likely to be much less well known because losses of species other than megafauna could conceivably have passed unnoticed. The history of the Australian mammalian fauna since European colonization, with the near or total extinction of several species by associated exotic predators (e.g., cats, foxes) (e.g., Smith and Quin 1996), clearly demonstrates that more than megafauna can disappear when humans invade even geographically extensive regions. North Island and South Island of New Zealand each cover more than 110,000 km^2, yet humans and other invading mammalian predators have between them been able to eradicate more than 50 bird populations from each island in under 1000 years, including many small-bodied species (Holdaway 1999; Holdaway and Jacomb 2000; Holdaway et al. 2001; Duncan et al. 2002; Roff and Roff 2003). It does not seem far-fetched to us that exotic species could be responsible for a large number of unknown bird extinctions on continental landmasses.

The known extinctions of island birds allow us to investigate characteristics of locations that may relate to the number of species lost in this group. We explored the extent to which the number of bird species driven extinct, and the proportion of the fauna lost, since human colonization were related to features of the islands concerned.

The most consistent predictor of the number of species lost was the isolation of an island from the mainland, which was significant in all univariate

TABLE 4.1 *Univariate relationship between five variables and the number of extinctions on islands*

Variable	N^a	Including New Zealand			Excluding New Zealand		
		Estimate	S.E.	F	Estimate	S.E.	F
All islands							
Abs. latitude	220	0.021	0.010	6.09*	0.006	0.011	0.51
Log area	186	0.691	0.101	64.34***	0.527	0.120	28.17***
Log elevation	175	1.379	0.354	33.61***	0.943	0.305	19.06***
Log isolation	211	0.710	0.305	10.34**	0.726	0.274	11.41***
Log colonization	197	−0.290	0.325	1.33	−0.106	0.310	0.18
Only islands with fossils							
Abs. latitude	40	0.043	0.014	9.07**	0.011	0.019	0.34
Log area	40	0.256	0.106	5.92*	0.061	0.119	0.27
Log elevation	40	0.681	0.298	6.35*	0.358	0.264	2.05
Log isolation	40	0.625	0.275	7.05*	0.795	0.222	11.68**
Log colonization	40	−1.062	0.346	8.92**	−0.703	0.357	3.74

Note: The number of extinctions was analyzed as a count variable using Poisson models corrected for overdispersion (specifically, using a general linear model with quasi-Poisson errors and a log link function, implemented by the R statistical package version 1.8 (Ihaka and Gentleman 1996; http://www.r-project.org).

[a]N = number of islands; with New Zealand excluded, sample size = $N − 2$. (The North and South Islands of New Zealand are outliers in most analyses; see, e.g., Figure 4.1.)

*$P < 0.05$, **$P < 0.01$, ***$P < 0.001$

(Table 4.1) and multivariate (Table 4.2) analyses: more isolated islands have lost more bird species (Figure 4.1). Note that the two obvious outliers from this plot are North and South Island, New Zealand: indeed, they are outliers from several of the relationships in our analyses. However, the effect of isolation is still significant whether or not these data are included. Note also that no fossil information is available for most of the islands for which we were able to obtain current species lists. The number of extinctions experienced by such islands is likely to be a severe underestimate (see above). For this reason, we also repeated our analyses only for islands with fossil data. This strategy had the additional benefit of reducing pseudoreplication in our data that arises from the presence of data from several islands in the same archipelago. Isolation was still a significant correlate of the number of species lost in this restricted set (Tables 4.1 and 4.2).

When we analyzed data for all islands, the number of extinctions also increased with island area in both univariate and multivariate analyses (Tables 4.1 and 4.2). For islands with fossil data, multivariate analysis identifies latitude and elevation, in addition to isolation, as significant predictors of the number of extinctions (Table 4.2): more isolated and elevated islands at higher latitudes have lost more species. These results are not due to the influence of New Zealand. Repeating the multivariate analyses in Table 4.2 with number of native

TABLE 4.2 *Multivariate relationship between five variables and the number of extinctions on islands*

Variable	Including New Zealand			Excluding New Zealand		
	Estimate	S.E.	*F*	Estimate	S.E.	*F*
All islands						
Abs. latitude						
Log area	0.83	0.09	106.9***	0.71	0.11	54.0***
Log elevation						
Log isolation	1.36	0.28	41.2***	1.24	0.28	32.1***
Log colonization						
Only islands with fossils						
Abs. latitude	0.06	0.01	28.7***	0.06	0.02	8.3**
Log area						
Log elevation	0.62	0.19	12.4**	0.62	0.21	9.6**
Log isolation	1.09	0.22	34.9***	1.09	0.24	25.9***
Log colonization						

Note: Nonsignificant explanatory variables were sequentially removed from a full model that included all five variables listed. We then tested the significance of each explanatory variable in the resulting minimum adequate model by comparing models with and without it, using an *F*-test (Crawley 2002). Parameters are given only for significant explanatory variables. Other details of the model are as noted in Table 4.1.
*$P < 0.05$, **$P < 0.01$, ***$P < 0.001$

species instead of number of extinctions showed that bird species richness increased with area and decreased with isolation, regardless of the set of islands included (see also Adler 1992, 1994), as island biogeography tells us it should

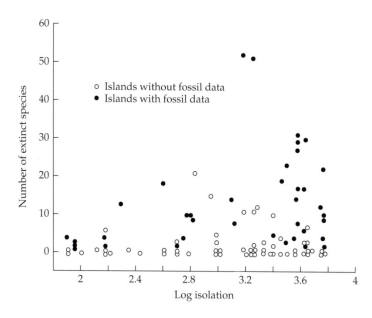

Figure 4.1 The relationship between the number of species known to have gone extinct on an island and the logarithm (base 10) of that island's isolation (in kilometers) from the nearest mainland region. Note that extinction here refers to the loss of a species from an island and so does not necessarily imply global extinction. The two outliers are North and South Island, New Zealand.

TABLE 4.3 *Univariate relationship between five variables and the proportion of an island fauna going extinct following human colonization*

Variable	Including New Zealand			Excluding New Zealand		
	Estimate	S.E.	F	Estimate	S.E.	F
All islands						
Abs. latitude	0.021	0.010	5.33*	0.006	0.012	0.31
Log area	0.147	0.109	2.25	−0.059	0.124	0.30
Log elevation	0.696	0.270	10.29**	0.456	0.261	4.65*
Log isolation	1.640	0.237	61.18***	1.633	0.242	60.13***
Log colonization	−0.728	0.262	8.91**	−0.573	0.278	5.27*
Only islands with fossils						
Abs. latitude	0.039	0.025	2.62	0.010	0.041	0.07
Log area	−0.323	0.149	4.83*	−0.577	0.158	14.65***
Log elevation	−0.193	0.400	0.23	−0.466	0.414	1.30
Log isolation	2.201	0.303	68.45***	2.218	0.302	71.00***
Log colonization	−2.023	0.547	19.19***	−1.925	0.631	10.84**

*Note:*The proportion of extinctions was analyzed using logistic models corrected for overdispersion, with the dependent variable being the number of extinct species out of the total native bird fauna on the island (specifically, we used a general linear model with quasibinomial errors and a logit link function, implemented by the R statistical package, version 1.8).
*$P < 0.05$, **$P < 0.01$, ***$P < 0.001$

TABLE 4.4 *Multivariate relationship between five variables and the proportion of an island fauna going extinct following human colonization*

Variable	Including New Zealand			Excluding New Zealand		
	Estimate	S.E.	F	Estimate	S.E.	F
All islands						
Abs. latitude						
Log area	0.31	0.09	12.6***			
Log elevation	0.45	0.23	4.3*			
Log isolation	1.83	0.28	58.5***	1.64	0.27	49.5***
Log colonization	−0.74	0.26	8.0**			
Only islands with fossils						
Abs. latitude	0.04	0.01	9.9**			
Log area						
Log elevation						
Log isolation	2.27	0.28	85.9***	1.98	0.29	57.9***
Log colonization	−1.05	0.42	5.7*			

*Note:*We sequentially removed nonsignificant explanatory variables from a full model including all five listed in the table, then tested the significance of each explanatory variable in the resulting minimum adequate model by comparing models with and without it, using an *F*-test (Crawley 2002). Parameters are given only for significant explanatory variables. Other details of the model are as listed in Table 4.3.
*$P < 0.05$, **$P < 0.01$, ***$P < 0.001$

(MacArthur and Wilson 1967). Thus, the results in Table 4.2 are not a simple consequence of the number of species present: more isolated islands tend to have fewer species, but to have lost more following colonization.

Patterns in extinction independent of species richness can be further explored by analyzing the proportion of species lost to extinction. Now the most consistent predictors of loss are isolation and time since colonization (Tables 4.3 and 4.4): more isolated islands, and more recently colonized islands, have lost a higher proportion of known native species (Figure 4.2). Isolation and time since colonization are not tightly correlated in these data ($r = -0.126$, $N = 161$, $P = 0.11$). Isolation is significant in all univariate and multivariate analyses (see also Biber 2002; Blackburn et al. 2004). Species inhabiting more isolated islands have been suggested to be more susceptible to exotic invaders because the smaller number of natural immigrants that such locations receive makes the residents particularly ill equipped to face novel competitors and predators, especially if these start to arrive at a high rate (Elton 1958; Pimm et al. 1994; Biber 2002; but see Simberloff 2000). There is also evidence that human population density may have been higher on more isolated islands, which would not have helped matters (Hurles et al. 2003).

We can also test for patterns within the set of bird populations that have been lost from the islands in our sample. Of particular note here is that pre-historic extinctions constitute a higher proportion of all extinctions from an island if it was colonized longer ago (estimate ± S.E. = 2.43 ± 0.56, $F = 23.8$, $P < 0.001$, in a minimum adequate model that also includes positive effects of island area and isolation), as has been shown elsewhere for island birds (Pimm et al. 1995; Biber 2002; Blackburn et al. 2004). This finding is consistent with a "filtration effect" in extinctions (Pimm et al. 1995; Balmford 1996): islands colonized longer ago have lost fewer species in historic times because they had already lost more species in prehistory.

The repeated significance of time since colonization as a predictor of proportional extinction in island avifaunas suggests an artifact. It seems unlikely that more recently colonized islands should actually have lost a higher proportion of their bird fauna, especially as the significance of time since colonization in the multivariate models for all islands and for islands with fossil data (New Zealand excluded) is independent of isolation. Rather, this significance implies that we simply know better which species were lost from islands colonized more recently, and therefore that islands colonized earlier have lost far more species than we realize. These results simply reinforce the view expressed earlier that what we know of recent avian extinctions does not come close to representing the true scale of the losses.

Characteristics of species

Although we have direct evidence for the loss of many species in the wake of human spread across the globe, and indirect evidence for the loss of many more, we can for the most part only guess at the characteristics that these

Figure 4.2 The relationship between the proportion of native species on an island that are known to have gone extinct and (A) the logarithm (base 10) of that island's isolation (in km) from the nearest mainland region and (B) the time since colonization by humans (years BP). Note that extinction here refers to the loss of a species from an island, and so does not necessarily imply global extinction.

(A)

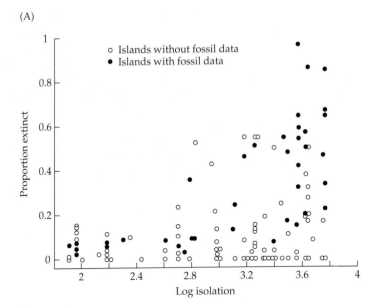

(B)

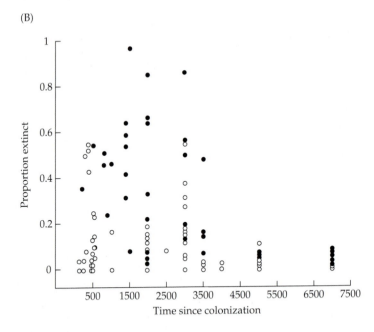

species may have possessed—even the biased set for which specimen data exist. This is a problem in that the characteristics of the species lost can be informative about the processes that caused them to disappear (Walker 1967; Cassels 1984; Gaston and Blackburn 1995; Owens and Bennett 2000; Gaston and Blackburn 2003).

Nonetheless, we may still be able to make reasonable inferences about the characteristics of extinct species, for two reasons. First, we can obtain information on some characteristics, such as body size and flightlessness, from fossil or other specimens. Second, closely related bird species share many features of life history and ecology because of their extensive shared evolutionary history. Most variation in avian life history is located between taxonomic families (Owens and Bennett 1995), meaning that even confamilial species will share many characteristics. The degree of similarity will be even greater for congeneric species. Thus, even simple taxonomic information about extinct species allows inferences to be made about their characteristics.

We employed the logic embodied in this second approach in two ways. First, we explored the taxonomic composition of extinct and threatened island endemic bird species to assess what this composition might tell us about patterns of extinction. We tested for unexpectedly high or low numbers of island endemic bird species extinctions within families using a simple randomization routine (Table 4.5; see Lockwood et al. 2000; Blackburn and Duncan 2001). We

TABLE 4.5 *Total number of island endemic species in a family and the number of those species globally extinct (or threatened with extinction) following human colonization*

Family	No. species	No. species extinct or threatened	P
More extinctions than expected			
Aepyornithidae	7	7	< 0.0001
Dinornithidae	11	11	< 0.0001
Anatidae	38	25	< 0.0001
Rallidae	82	56	< 0.0001
Fewer extinctions than expected			
Zosteropidae	61	2	< 0.0001
Corvidae	161	14	< 0.0001
Fewer threatened species than expected			
Meliphagidae	49	8	< 0.0001

Note: Extinctions were randomly allocated to species on the island endemic list up to the total observed in the actual data, and the number of extinctions per family was then calculated. This process was repeated 50,000 times, and the number of times each family had more, the same, or fewer extinctions than observed was calculated. These numbers were used to calculate the two-tailed probability that the number of extinctions observed was no different from that expected by chance as described by Blackburn and Duncan 2001 (for example, if a family contained more extinct species than all except one of the simulations, the probability that the family suffered no more extinctions than expected by chance = 0.00004). Families are shown only if their probability is significantly lower than expected ($\alpha = 0.05$) once a sequential Bonferroni correction for multiple statistical tests (Rice 1989) has been applied. *P* is the two-tailed probability of observing that number of extinct or threatened species, given the number of species in the family and the proportion of the world's island endemic bird species that have either gone extinct or are threatened. *P* was calculated using simulation.

repeated this process to test for unexpectedly high or low numbers of threatened species across the extant island species within families. Second, we assumed that any unknown body mass for an island endemic bird species was the mean of all other species in the same genus (using data from Dunning 1992). Of course, the precise taxonomic affinities of extinct species are often harder to assess than those of extant species, but if an extinct species is classified in an extant genus, it is presumably similar enough to other members to make the use of the generic mean a reasonable assumption. Any errors that incorrect genus assignments introduce will be small in relation to the (six orders of magnitude) span of body masses in our data. This method allowed us to compile body mass data for 85% of all such species, including 68% of all the extinct species on our list, and examine the relationship between threat status and mass in this set. Body mass is a useful variable for interspecific comparisons because many aspects of a species' life history are related to its size (Peters 1983; Calder 1984; Harvey and Pagel 1991). Note, however, that knowledge of size alone does not allow finer-scale explorations of the influence on extinction of the various life history characteristics correlated with size (see Johnson 2002; Roff and Roff 2003).

TAXONOMIC PATTERNS OF EXTINCTION   Four bird families with island endemic species have experienced significantly higher than expected numbers of species extinctions (Table 4.5). For the Aepyornithidae and Dinornithidae, these extinctions have led to the loss of the entire family. A further five families had significantly more extinctions than expected before a Bonferroni correction for multiple comparisons was applied, albeit not after: the Ardeidae, Threskiornithidae, Scolopacidae, Acanthisittidae, and Fringillidae. Cassels (1984) notes that rates of extinction in Pacific island birds were particularly high for "ratites, rails, waterfowl, birds of prey, and crows." Milberg and Tyrberg (1993) similarly identified Anatidae, Rallidae, and Drepanididae (Hawaiian honeycreepers, now classified as the tribe Drepanidini within Fringillidae) as being overrepresented in Holocene fossils. However, neither of these studies based their conclusions on quantitative methodology.

Two families have had significantly fewer extinctions than expected among their island endemic species (see Table 4.5). These are both passerine families, and hence include mainly small-bodied, arboreal species. One of these families, Corvidae, which includes crows, has had fewer extinctions than expected despite having lost an absolute number of species that is relatively high, as noted by Cassels (1984). A further six families had significantly fewer extinctions than expected before the Bonferroni correction was applied. These are also all families mainly composed of small-bodied species and species that are primarily arboreal or aerial (Apodidae, Tyrannidae, Muscicapidae, Sylviidae, Nectariniidae, and Passeridae).

The proportion of extant island endemic species that is threatened is so high (> 40%) that no family has a higher number than expected by chance once a Bonferroni correction is applied. However, levels of threat within both Rallidae and Procellariidae are the closest (and close) to significance. High levels of current threat are also experienced by island endemic Anatidae, Psittacidae, Spheniscidae, Pittidae, and Phasianidae (uncorrected $P < 0.05$ in every case). The only family with a significantly lower than expected number of island endemic species at risk of extinction is the Meliphagidae. In comparison, a study of the taxonomic distribution of threat across all bird species (Bennett and Owens 1997) identified high levels of threat in the Megapodiidae, Cracidae, Columbidae, and Gruidae, as well as in the families listed above, and low levels in the Picidae.

SPECIFIC CHARACTERISTICS AND DRIVERS OF EXTINCTION   The patterns of extinction and threat across families for island birds suggest a number of features of taxa at high risk. These factors include relatively large body size (e.g., Aepyornithidae, Dinornithidae, Anatidae, Phasianidae), flightlessness (e.g., Aepyornithidae, Dinornithidae, Rallidae, Spheniscidae), and ground-dwelling or ground-nesting habits (e.g., Rallidae, Procellaridae, Acanthisittidae, Pittidae). Those families with fewer than expected extinct or threatened species tend to be small-bodied species living or breeding off the ground. One obvious exception to these generalizations is the Fringillidae, whose high level of extinction is largely due to extinctions of Hawaiian Drepanidini. To the above list of characteristics of extinct species, Cassels (1984) added high endemism (i.e., species that are endemic to an island at a high taxonomic level) and small clutch size (with larger eggs). He also noted that most surviving flightless species are nocturnal and inhabitants of primary forest, and that extinct species capable of flight were often large-bodied. Our data show that a high proportion of the largest-bodied island endemic bird species are extinct or at risk of extinction (Figure 4.3) and that large body mass is significantly associated with extinction and threat in these species (Table 4.6).

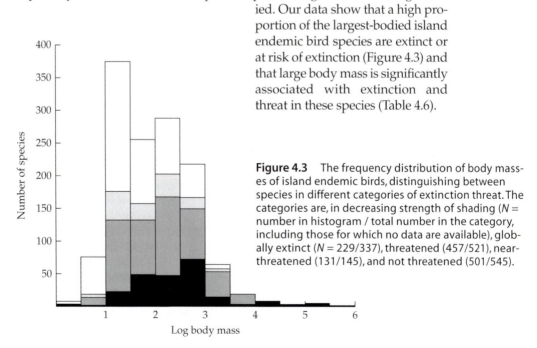

Figure 4.3   The frequency distribution of body masses of island endemic birds, distinguishing between species in different categories of extinction threat. The categories are, in decreasing strength of shading ($N =$ number in histogram / total number in the category, including those for which no data are available), globally extinct ($N = 229/337$), threatened (457/521), near-threatened (131/145), and not threatened (501/545).

TABLE 4.6 *Summary of the body masses of island endemic bird species*

Status	No. species (N)	Mean	S.D.	P
Total	1318	1.958 (91)	0.788	—
Not threatened	501	1.629 (43)	0.646	< 0.0001
Near-threatened	131	1.857 (72)	0.658	0.12
Threatened	457	2.103 (127)	0.750	< 0.0001
Vulnerable	272	2.145 (140)	0.781	—
Endangered	104	2.019 (104)	0.694	—
Critical	81	2.068 (117)	0.708	—
Extinct	229	2.444 (278)	0.883	< 0.0001

Note: Mean = mean $\log_{10}$ body mass in grams (the back-transformed value is in parentheses). S.D. = standard deviation of the $\log_{10}$ mass distribution for each category. P is the two-tailed probability that the geometric mean mass for any category could derive from a random sample of island endemic species (calculated using a randomization test drawing N species from the total of 1318, with 50,000 iterations; see Gaston and Blackburn 1995 for details of the method).

Four processes have been repeatedly suggested to have driven Holocene extinctions in island birds: climate (or some other natural environmental) change, overexploitation by human colonists, the effects of other exotic species, and habitat destruction (Cassels 1984; Holdaway 1999; Duncan et al. 2002; Roff and Roff 2003). Climate change is readily dismissed, as it is difficult to see why its effects should have occurred at different times on different islands (and why humans should always have colonized just before its effects were felt), or why it should differentially have affected flightless or ground-dwelling species (Cassels 1984; Worthy 1999; Roff and Roff 2003; see also Lyons et al. 2004). This leaves the effects of exotic invaders (human or otherwise) as the only likely drivers of these extinctions. The only outstanding question is the relative contribution of the processes associated with them.

The characteristics of extinct island bird species point strongly to human hunting as a primary driver of these extinctions (Cassels 1984; Diamond 1984, 1989; Steadman and Kirch 1990; Milberg and Tyrberg 1993; Steadman 1995; Holdaway 1999; Hurles et al. 2003; Roff and Roff 2003). Large-bodied, flightless, ground-dwelling species would have been particularly vulnerable to humans. Indeed, Cassels (1984) notes that an effect of hunting is "not contradicted by any of the evidence." However, this evidence does suffer from a general lack of quantitative analyses (Roff and Roff 2003), and much of it is circumstantial (Duncan et al. 2002). Processes associated with human arrival other than hunting, including habitat destruction and the introduction of mammalian predators, could also account for the pattern of bird losses on most Pacific islands (Olson and James 1982; Cassels 1984; Steadman 1995). Habitat destruction may differentially affect large-bodied species, as these may require larger tracts of continuous habitat to persist. Polynesians often employed burning to clear habitat (McGlone 1978, 1983, 1989; Cassels 1984; Bussell 1988), from which flightless and ground-dwelling species may be ill equipped to flee. Moreover, although the bones of prehistorically extinct birds are often found in middens, so too are the bones of species that survived prehistoric human occupation. Prehistoric humans hunted a wide variety of birds, only some of which became extinct (Worthy 1997). Factors other than hunting, including habitat destruction, may thus have been important.

These issues were recently addressed by Duncan et al. (2002) for avian extinctions in New Zealand. They compared the abundance of fossil bird species in Maori middens at an archaeological site at Marfell's Beach, South Island, with the abundance of the same species in nearby natural dune deposits. They argued that species that were overrepresented in the middens in comparison to their abundance in dune deposits were species that had been specifically targeted by Maori hunters, and they showed that extinction was more likely in the targeted species than in those not favored by the hunters. The same pattern was seen when controlling for differences in body mass and taxonomic association among the species. Thus, human hunting can be directly linked to extinction in this avifauna (Duncan et al. 2002).

Two further results from Duncan et al.'s analysis are of interest here. First, they showed that larger-bodied taxa were more likely to be selected by Maori hunters, but were also more prone to extinction for a given level of hunting pressure. This finding suggests a "double whammy" for large-bodied species: not only are they more likely to be hunted, but life history traits associated with their size (such as lower population growth rates) increase their risk of extinction for a given level of harvesting.

Second, extinction probability among these birds was also related to their habitat use: terrestrial birds were more likely to go extinct than freshwater or coastal species. Several large-bodied terrestrial taxa became extinct despite relatively low hunting pressure. Duncan et al. suggested that species in this group would have suffered population declines as a result of widespread forest clearance by Maori (McGlone 1983; McGlone and Basher 1995). Thus, their results are consistent with both hunting and habitat destruction being agents of population decline in prehistoric New Zealand (see also Roff and Roff 2003). Habitat destruction may well be more important on smaller islands where total destruction of the native habitat is more easily achievable (see also Anderson 2001), such as Easter Island (Flenley and King 1984).

Nevertheless, it is unlikely that all Holocene island bird extinctions can be attributed directly to the action of humans. Our taxonomic analyses identify high levels of loss in the Acanthisittidae and Fringillidae and high levels of threat in the Pittidae. These families include a number of small-bodied, ground-dwelling species with poor powers of flight. Moreover, many extinctions in the Rallidae involved species with these characteristics (Cassels 1984). Indeed, although large body mass is associated with extinction and threat in island endemic species (see Table 4.6), the frequency distribution of body masses for island endemic birds (see Figure 4.3) shows that many small-bodied species are extinct or at risk of extinction, and that extinctions are spread across all body mass classes.

Although it is possible that these species could have been affected by habitat destruction, an explanation for their loss more consistent with these characteristics is the influence of exotic predators other than humans (Cassels 1984; Milberg and Tyrberg 1993; Holdaway 1999; Roff and Roff 2003). In particular, humans were accompanied in their initial spread across the Pacific by the Pacific rat (*Rattus exulans*), which is known from every island colonized by Polynesians (Wodzicki and Taylor 1984; Atkinson 1985; Roberts 1991; Holdaway 1996; Hurles et al. 2003). The Pacific rat is known to be a potential predator of invertebrates, small-bodied vertebrates, and small eggs, but to restrict its foraging largely to ground level (Atkinson 1985; Holdaway 1999). The coincidence of the spread of a small-bodied mammalian predator across Pacific islands in the last 4000 years with the loss of many small-bodied potential prey species among island resident birds is highly suggestive (Wragg 1995). Moreover, the arrival of rats on Midway Island (Hawaii) in 1943 and on Big South Cape Island (New Zealand) around 1962 (albeit a different species, *Rattus rattus*) was followed within a matter of months by several bird extinctions (Atkinson 1985). Other examples of exotic species that have heavily depleted island avifaunas include the brown tree snake (*Boiga irregularis*) on Guam (Savidge 1987; Fritts and Rodda 1998; Wiles et al. 2003) and domestic cats (*Felis catus*) on Ascension Island (Ashmole et al. 1994).

Good quantitative evidence for the effects of nonhuman exotic predators on an avifauna is again provided by data for New Zealand. Humans colonized New Zealand in two waves: Maori settlement began about AD 1300 (Anderson 1991; Higham et al. 1999), and European settlement followed arrival in AD 1769. The extinction of species in the New Zealand avifauna can accurately be assigned to one of these two periods of human colonization. The major introduced predators in the prehistoric period were humans and the Pacific rat. Following European arrival, this list was expanded to include the brushtail possum (*Trichosurus vulpecula*), brown rat (*Rattus norvegicus*), black rat (*R. rattus*), house mouse (*Mus domesticus*), domestic cat (*Felis catus*), and three mustelid species. Thus, the effects associated with exotic predators ought to differ between the two periods of human colonization, and so too should the characteristics of species driven extinct if these predators are a primary cause.

Holdaway (1999) examined the characteristics associated with extinction in six groups of New Zealand birds (flightless, petrels, coastal/freshwater, arboreal volant, ground-dwelling volant, predators). He was able to show that the characteristics of species lost from each group following each wave of human colonization were those of species that ought to be susceptible to the suite of exotic predators present. Although Holdaway did not provide formal statistics, general linear modeling of his extinction data across the avifauna as a whole supports his conclusions (Duncan and Blackburn 2004). First human colonization was followed by the extinction of large-bodied species, which disappeared regardless of their other characteristics (see also Cassey 2001). Small-bodied species also disappeared in prehistory, but only if they were ground-nesting species. These characteristics are associated with susceptibility to predation by humans and Pacific rats, respectively. Following European arrival, extinction risk was elevated for medium-sized species and also for arboreal taxa. These species would have been susceptible to the larger-bodied mammalian predators, which included good climbers, that Europeans introduced (Duncan and Blackburn 2004). Note that extinctions occurred across a wide range of New Zealand birds in terms of body size and habitat use, but different waves of predators took species with different characteristics. This is one reason why the effect of predation is difficult to identify unequivocally in other locations with poorer data.

A different approach to the reanalysis of Holdaway's extinction data was taken by Roff and Roff (2003). They used regression tree analysis to identify specific characteristics that predict extinction from the native New Zealand avifauna following the first wave of human colonization by Maori. This approach revealed four main groupings among extinct species. The first group consisted of species with a body mass greater than 3.75 kg, all bar one of which were flightless. All of these species went extinct, again implicating human hunting. The second group consisted of flightless species less than 3.75 kg in mass; within this group, small-bodied species had higher extinction probabilities. Roff and Roff (2003) argue that this finding implicates a size-limited predator. The only real candidate is the Pacific rat. The third group included burrow-nesting volant species, all of which were petrels (Procellariidae). Roff and Roff again argue that the data implicate the Pacific rat as the driver of extinction in this group. The final group consisted of a somewhat disparate collection of volant species that nested on or above the ground, for which Roff and Roff (2003) could identify no obvious extinction driver.

Thus, all the evidence points to the importance of exotic predators, be they humans, rats, or other mammals, as drivers of Holocene bird extinctions in New Zealand. Since the Pacific rat is widely distributed across Pacific islands, the influence of this species on small-bodied, ground-dwelling bird species was presumably equally disastrous on other islands; witness the large numbers of endemic extinctions of small-bodied Rallidae.

Finally, we would note here one more potential bias in the extinction data. Most of the data on avian extinctions come from islands, as consequently does most of the evidence for the importance of exotic species other than humans as extinction drivers. However, nonhuman exotics may primarily affect island species. BirdLife International (2000) lists 242 species as being at risk of extinction from invasive species: 225 of these (93%) are island endemics. In contrast, only 45% of species listed as threatened by human exploitation, and only 37% of species listed as threatened by habitat destruction, are island endemics. That does not negate the effect of nonhuman exotic predators on island avifaunas, but it does suggest that nonhuman exotics play a lesser role on continental mainlands, where the primary exotic driver of extinction threat is *H. sapiens*.

What Have We Learned about the Process of Extinction?

The data and studies reviewed here lead us to several conclusions about the current Holocene extinction crisis.

Every landmass that has been colonized by humans as they have spread out of Africa has seen species extinctions soon afterward. Just how many extinctions have occurred will probably never be known, although it is certainly vastly more than have been documented to date. What is also certain is that the coincidence between these losses and the arrival of humans is not a coincidence at all: the extinctions are predominantly anthropogenic in cause. Indeed, some would go further:

> 'As far as I know, no biologist has documented the extinction of a continental species of plant or animal caused solely by non-human agencies such as competition, disease or environmental perturbation in situations unaffected by man' (Soulé 1983, p. 112). I am even less informed. I cannot recollect hearing of a non-anthropogenic extinction of an island species (as against an island population) occurring within the last 8000 years. (Caughley 1994)

The case for anthropogenic island extinctions is, in our opinion, beyond doubt. That for continental extinctions is more controversial, as it is harder to believe that early human societies could have caused the large-scale loss of species over such vast geographic areas. However, recent models have shown that such effects are well within the capabilities of these people (Holdaway and Jacomb 2000; Alroy 2001). In our view, the principal question now is which process(es) associated with human colonizations drove the extinctions.

An obvious corollary of the importance of colonization by humans and their associated exotics is that patterns and mechanisms pertaining to the current extinction crisis will differ from those for other mass extinction events, in which humans certainly were not involved. Vermeij (2004) suggests that previous mass extinctions were driven by processes, such as asteroid impacts or volcanic activity, that disrupted primary productivity. He terms these "supply-side extinctions." In contrast, he argues that current extinctions are driven by the effects of competition, predation, invasion, and habitat destruction and are thus "demand-side" extinctions. Vermeij's assessment of the current crisis concurs with our own, albeit that we would argue that most of the deleterious competition, predation, and habitat destruction is driven by invaders. It follows from this that while the study of the effects of invaders will undoubtedly provide important information on the current extinction crisis, it will not provide insights into the "general" process of extinction. The current crisis is special in being driven from the top down, rather than the bottom up.

Most recent bird extinctions can be attributed to predation by exotic mammalian predators. It has been suggested that the greatest impacts of invasive species occur in situations in which they perform an entirely novel role in the community (Parker et al. 1999); mammalian predators introduced to islands are a classic example (Elton 1958). The principal extinction driver for large-bodied species was *H. sapiens*. There is now good evidence that humans differentially targeted large-bodied species and that for a given level of hunting, large-bodied species were more likely to go extinct (Duncan et al. 2002). Note that reproductive rate, rather than body size per se, may have been the most important ultimate factor determining which species disappeared, as in certain circumstances large-bodied species could survive if their reproductive rate was high enough (Johnson 2002). A wider range of exotic predators has driven extinctions of small-bodied species, but the Pacific rat can be singled out as likely to have been particularly damaging. That is not because this species is notably voracious, but rather because it had the right characteristics (body size, diet), in the right place (Pacific islands) at the right time (first).

Habitat destruction by one exotic species in particular, *H. sapiens*, has also caused bird extinctions. In most systems this process probably acts in concert with predation, rather than alone (e.g., Olson and James 1982; Steadman and Kirch 1990; Steadman 1995; Duncan et al. 2002). Indeed, habitat destruction may ease the passage of other invaders and elevate rates of predation. A combination of these processes is perhaps most likely to explain the prehistoric disappearance of Hawaiian Drepanidini (Olson and James 1982; Burney et al. 2001), although it has undoubtedly been helped along in historic times by apparent competition with introduced birds mediated via the action of avian diseases (e.g., van Riper 1991).

We have focused on avian extinctions because of the scale of losses (both geographic and taxonomic) and the quality of information for this taxon. These extinctions have been driven largely by exotic predators. However, birds are but one taxon suffering losses in the current extinction crisis. This observation raises the question of the relative contributions of different extinction drivers across all plant and animal groups. This question is probably impossible to answer in any quantitative way because the many biases in knowledge of bird extinctions that we have discussed here apply even more strongly to other taxa. Nevertheless, studies that explore extinctions or declines in other groups are beginning to accumulate.

Many fish extinctions can be attributed to predation by nonhuman exotic predators (e.g., Craig 1992; Rahel 2002), while human predation continues to drive down the populations of many others (e.g., Casey and Myers 1998). Habitat destruction, such as drainage or damming, is also important for this group (Pringle et al. 2000) and may facilitate invasions (Moyle and Light 1996). Global declines in amphibian populations have been well documented, although the cause has yet to be identified from a wide range of possible hypotheses (Collins and Storfer 2003). Prehistoric mammalian megafaunal extinctions are widely, if not unanimously, attributed to exotic human hunters, and the lack of such extinctions in the African "native range" of *H. sapiens* is also notable in that light (see recent summary by Lyons et al. 2004). A detailed consideration of size biases in the mammalian fossil record would help elucidate the extent to which other extinction drivers, such as habitat destruction or nonhuman predators, might play a role in this group. They certainly drive historic mammalian extinctions (Smith and Quin 1996). Many extinctions of island snails have followed the introduction of predatory exotic snails (Hadfield 1986; Cowie 2000). Plant populations may be much more susceptible to habitat destruction or to competition with introduced species than to predation (herbivory), although in many cases these effects may be mediated via disturbance or dispersal by exotic herbivores (Simberloff and Von Holle 1999). Taken together, these studies suggest that the precise set of drivers varies across (and within) groups, but that exotic species are implicated in most cases, given that humans are exotic species in much of the world. Deeper exploration of these and other groups will no doubt clarify the generality of the conclusions we draw here for birds.

The evidence that is accumulating on the current extinction crisis leads us to one final and important point. We return to birds to make it. Bird species are more likely to have been driven extinct if their characteristics predisposed them to predation by exotic mammals. Although this is true for all species, the naïveté of island species, and of inhabitants of isolated islands in particular, to predators would not have helped their cause. The wholesale destruction of island faunas, even on islands retaining native habitat, is testament to this. However, it follows that the precise set of species that disappear in any given context is largely idiosyncratic and depends on the set of predators that are introduced. Large-bodied species will always tend to disappear, as humans are a constant of these invasions. What else they bring with them determines what else will go. Imagine a Pacific across which the Polynesians had not spread *R. exulans*. The later spread of mammalian predators by European explorers means that we would probably still have few extant island rails, but we would have many more mounted specimens of the species lost. This idiosyncrasy may explain why it has proved difficult in the past to make general statements about the characteristics of species lost: without knowledge of the species introduced, the characteristics of the species lost are unlikely to make much sense. This is true for understanding past extinctions in the current crisis, but it is also relevant for predicting future ones. We have the potential to lose almost anything, depending on which exotics we introduce.

Acknowledgments

We thank Phillip Cassey and Richard Duncan for many stimulating discussions on the subjects of extinction and introduction and Dov Sax, Steve Gaines, and two anonymous referees for their helpful comments on this chapter. This work was conducted as a part of the "Exotic Species: A Source of Insight into Ecology, Evolution, and Biogeography" Working Group supported by the National Center for Ecological Analysis and Synthesis, a Center funded by the National Science Foundation (Grant # DEB-0072909), the University of California, and the Santa Barbara campus.

Literature Cited

Adler, G. H. 1992. Endemism in birds of tropical Pacific islands. Evolutionary Ecology 6:296–306.

Adler, G. H. 1994. Avifaunal diversity and endemism on tropical Indian Ocean islands. Journal of Biogeography 21:85–95.

Alroy, J. 2001. A multispecies overkill simulation of the end-Pleistocene megafaunal mass extinction. Science 292:1893–1896.

Anderson, A. 1991. The chronology of colonization in New Zealand. Antiquity 65:767–795.

Anderson, A. 2001. No meat on that beautiful shore: the prehistoric abandonment of subtropical Polynesian islands. International Journal of Osteoarchaeology 11:14–23.

Ashmole, N. P., N. J. Ashmole, and K. E. L. Simmons. 1994. Seabird conservation and feral cats on Ascension Island, South Atlantic. In D. N. Nettleship, J. Burger, and M. Gochfeld, eds. *Seabirds on islands: threats, case studies and action plans*, pp 94–121. BirdLife International, Cambridge, U.K.

Atkinson, I. A. E. 1985. The spread of commensal species of Rattus to oceanic islands and their effects on island avifaunas. In P. J. Moors, ed. *Conservation of island birds: Case studies for the management of threatened island species*, pp. 35–81. ICBP Technical Publication No. 3.

Balmford, A. 1996. Extinction filters and current resilience: the significance of past selection pressures for conservation biology. Trends in Ecology and Evolution 11:193–196.

Bennett, P. M., and I. P. F. Owens. 1997. Variation in extinction risk among birds: chance or evolutionary predisposition? Proceedings of the Royal Society of London B 264:401–408.

Biber, E. 2002. Patterns of endemic extinctions among island bird species. Ecography 25:661–676.

BirdLife International. 2000. *Threatened birds of the world*. Lynx Ediciones and BirdLife International, Barcelona and Cambridge.

Blackburn, T. M., and R. P. Duncan. 2001. Establishment patterns of exotic birds are constrained by non-random patterns in introduction. Journal of Biogeography 28:927–939.

Blackburn, T. M., and K. J. Gaston. 2002. Extrinsic factors and the population sizes of threatened birds. Ecology Letters 5:568–576.

Blackburn, T. M., P. Cassey, and R. P. Duncan. 2004. Extinction in island endemic birds reconsidered. Ecography 27:124–128.

Bortolini, M. C., F. M. Salzano, M. G. Thomas, S. Stuart, S. P. K. Nasanen, C. H. D. Bau, M. H. Hutz et al. 2003. Y-chromosome evidence for differing ancient demographic histories in the Americas. American Journal of Human Genetics 73:524–539.

Burney, D. A., H. F. James, L. P. Burney, S. L. Olson, W. Kikuchi, W. L. Wagner, and M. Burney. 2001. Fossil evidence for a diverse biota from Kaua'i and its transformation since human arrival. Ecological Monographs 74:615–641.

Bussell, M. R. 1988. Mid and late Holocene pollen diagrams and Polynesian deforestation, Wanganui district, New Zealand. New Zealand Journal of Botany 26:431–451.

Calder, W. A. 1984. *Size, function, and life history*. Harvard University Press, Cambridge, MA.

Casey, J. M., and R. A. Myers. 1998. Near extinction of a large, widely distributed fish. Science 281:690–692.

Cassels, R. 1984. The role of prehistoric man in the faunal extinctions of New Zealand and other Pacific islands. In P. S. Martin, and R. G. Klein, eds. *Quaternary extinctions: a prehistoric revolution*, pp. 741–767. University of Arizona Press, Tucson.

Cassey, P. 2001. Determining variation in the success of New Zealand land birds. Global Ecology and Biogeography Letters 10:161–172.

Caughley, G. 1994. Directions in conservation biology. Journal of Animal Ecology. 63:215–244.

Collins, J. P., and A. Storfer. 2003. Global amphibian declines: sorting the hypotheses. Diversity and Distributions. 9:89–98.

Cowie, R. H. 2000. Non-indigenous land and freshwater molluscs in the islands of the Pacific: conservation impacts and threats. In G. Sherley, ed. *Invasive species in the Pacific: A technical review and draft regional strategy*, pp. 143–172. South Pacific Regional Environment Programme, Apia, Samoa.

Craig, J. F. 1992. Human induced changes in the composition of fish communities in the African Great Lakes. Review in Fish Biology and Fisheries 2:93–124.

Crawley, M. J. 2002. *Statistical computing: An introduction to data analysis using S-Plus*. John Wiley and Sons Ltd., Chichester, U.K.

Curnutt, J., and S. L. Pimm. 2001. How many bird species in Hawai'i and the central Pacific before human contact? Studies in Avian Biology 22:15–30.

Dalmazzone, S. 2000. Economic factors affecting vulnerability to biological invasions. In C. Perrings, M. Williamson, and S. Dalmazzone, eds. *The economics of biological invasions*, pp. 17–30. Edward Elgar, Cheltenham, U.K.

Darwin, C. 1839. *Journal of researches into the geology and natural history of the various countries visited by the H.M.S. Beagle, under the command of Captain Fitzroy, R.N. from 1832 to 1836*. Henry Colburn, London.

Diamond, J. M. 1984. "Normal" extinctions of isolated populations. In M. H. Nitecki, ed. *Extinctions*, pp. 191–246. Chicago University Press, Chicago.

Diamond, J. M. 1987. Extant unless proven extinct? or, extinct unless proven extant? Conservation Biology 1:77–79.

Diamond, J. M. 1989. The present, past and future of human-caused extinctions. Philosphical Transaction of the Royal Society of London B 325:469–476.

Duncan, R. P., and T. M. Blackburn. 2004. Extinction and endemism in the New Zealand avifauna. Global Ecology and Biogeography 13:509–517.

Duncan, R. P., T. M. Blackburn, and T. H. Worthy. 2002. Prehistoric bird extinctions and human hunting. Proceedings of the Royal Society London B 269:517–521.

Dunning, J. B. 1992. *CRC Handbook of avian body masses*. CRC Press, Boca Raton, FL.

Ebenhard, T. 1988. Introduced birds and mammals and their ecological effects. Swedish Wildlife Research. 13:1–107.

Ehrlich, P. R. 1986. Which animals will invade? In H. A. Mooney and J. A. Drake, eds. *Ecology of biological invasions of North America and Hawaii*, pp. 79–95. Springer-Verlag, New York.

Elton, C. 1958. *The ecology of invasions by animals and plants*. Methuen, London.

Flannery, T. F. 1994. *The future eaters: an ecological history of the Australasian lands and people*. Reed, Melbourne.

Flannery, T. F. 2001. *The Eternal Frontier: an ecological history of North America and its peoples*. Heinemann, London.

Flenley, J. R., and S. M. King. 1984. Late Quaternary pollen records from Easter Island. Nature 307:47–50.

Flood, B. 2003. The New Zealand storm-petrel is not extinct. Birding World 16:479–482.

Fritts, T. H., and G. H. Rodda. 1998. The role of introduced species in the degradation of island ecosystems: a case history of Guam. Annual Review of Ecology and Systematics 29:113–140.

Gaston, K. J. 2002. Extinction. In M. Pagel, ed. *Encyclopedia of evolution*, Vol.. 1, pp. 344–349. Oxford University Press, New York.

Gaston, K. J., and T. M. Blackburn. 1995. Birds, body size and the threat of extinction. Philosphical Transaction of the Royal Society London B 347:205–212.

Gaston, K. J., and T. M. Blackburn. 2003. Macroecology and conservation biology, In T. M. Blackburn, and K. J. Gaston, eds. *Macroecology: concepts and consequences*, pp. 345–367. Blackwell Science, Oxford.

Gaston, K. J., and J. I. Spicer. 2004. *Biodiversity. an introduction*. Blackwell Science, Oxford.

Grayson, D. K. 2001. The archaeological record of human impacts on animal populations. Journal of World Prehistory 15:1–68.

Groombridge, B. 1992. *Global diversity: status of the Earth's living resources*. Chapman and Hall, London.

Hadfield, M. G. 1986. Extinction in Hawaiian achatinelline snails. Malacologia 27:67–81.

Harvey, P. H., and M. D. Pagel. 1991. *The comparative method in evolutionary biology*. Oxford University Press, Oxford.

Higham, T., A. Anderson, and C. Jacomb. 1999. Dating the first New Zealanders: the chronology of Wairau Bar. Antiquity 73:420–427.

Holdaway, R. N. 1996. Arrival of rats in New Zealand. Nature 384:225–226.

Holdaway, R. N. 1999. Introduced predators and avifaunal extinction in New Zealand. In R. D. E. MacPhee, ed. *Extinctions in near time: causes, contexts, and consequences*, pp. 189–238. Kluwer Academic/Plenum, New York.

Holdaway, R. N., and C. Jacomb. 2000. Rapid extinction of the moas (Aves: Dinornithiformes): model, tests and implications. Science 287:2250–2254.

Holdaway, R. N., T. H. Worthy, and A. J. D. Tennyson. 2001. A working list of breeding bird species of the New Zealand region at first human contact. New Zealand Journal of Zoology. 28:119–187.

Hurles, M. E., E. Matisso-Smith, R. D. Gray, and D. Penny. 2003. Untangling oceanic settlement: the edge of the knowable. Trends in Ecology and Evolution. 18:531–540.

Ihaka, R., and R. Gentleman. 1996. R: a language for data analysis and graphics. Journal of Computational and Graphical Statistics. 5:299–314.

James, H. F., and S. L. Olson. 1991. Description of thirty-two new species of birds from the Hawaiian Islands. Part II: passeriformes. Ornithology Monographs Volume 46.

Johnson, C. N. 2002. Determinants of loss of mammal species during the Late Quaternary "megafauna" extinctions: life history and ecology, but not body size. Proceedings of the Royal Society London B 269:2221–2227.

Johnson, T. H., and A. J. Stattersfield. 1990. A global review of island endemic birds. Ibis 132:167–180.

King, W. B. 1985. Island birds: will the future repeat the past? In P. J. Moors, ed. *Conservation of island birds: case studies for the management of threatened island species*, pp. 3–15. ICBP Technical Publication No. 3.

Lever, C. 1994. *Naturalized animals: The ecology of successfully introduced species*, TAD Poyser, Ltd., London.

Lockwood, J. L., T. M. Brooks, and M. L. McKinney. 2000. Taxonomic homogenization of the global avifauna. Animal Conservation. 3:27–35.

Lodge, D. M. 1993. Biological invasions: lessons for ecology. Trends in Ecology and Evolution 8:133–137.

Long, J. L. 1981. *Introduced birds of the world*. The worldwide history, distribution and influence of birds introduced to new environments. David & Charles, London.

Lyons, S. K., F. A. Smith, and J. H. Brown. 2004. Of mice, mastodons and men: human-mediated extinctions on four continents. Evolutionary Ecology Research 6:339–358.

MacArthur, R. H., and E. O. Wilson. 1967. *The theory of island biogeography*. Princeton University Press, Princeton, NJ.

Mack, R. N., D. Simberloff, W. M. Lonsdale, H. Evans, M. Clout, and F. A. Bazzaz. 2000. Biotic invasions: causes, epidemiology, global consequences, and control. Ecological Applications 10:689–710.

MacPhee, R. D. E. 1999. *Extinctions in near time: causes, contexts, and consequences*. Kluwer Academic/Plenum, New York.

Manne, L. L., T. M. Brooks, and S. L. Pimm. 1999. Relative risk of extinction of passerine birds on continents and islands. Nature 399:258–261.

Martin, P. S. 1984. Prehistoric overkill: the global model. In P. S. Martin, and R. Klein, eds. *Quaternary extinctions: a prehistoric revolution*, pp. 354–403. University of Arizona Press, Tucson.

Martin, P. S. 2001. Mammals (Late Quaternary), extinctions of. In S. A. Levin, ed. *Encyclopedia of biodiversity*, pp. 825–839. Vol. 3. Academic Press, San Diego.

Martin, P. S., and R. Klein. 1984. *Quaternary extinctions: A prehistoric revolution*. University of Arizona Press, Tucson.

Martin, P. S., and D. W. Steadman. 1999. Prehistoric extinctions on islands and continents. In R. D. E. MacPhee, ed. *Extinctions in near time: causes, contexts, and consequences*, pp. 17–56. Kluwer Academic/Plenum, New York.

May, R. M., J. H. Lawton, and N. E. Stork. 1995. Assessing extinction rates. In J. H. Lawton, and R. M. May, eds. *Extinction rates*, pp. 1–24. Oxford University Press, Oxford.

McGlone, M. S. 1978. Forest destruction by early Polynesians, Lake Poukawa, Hawkes Bay, New Zealand. Journal of the Royal Society of New Zealand. 8:275–281.

McGlone, M. S. 1983. Polynesian deforestation of New Zealand: a preliminary synthesis. Archaeology in Oceania 18:11–25.

McGlone, M. S. 1989. The Polynesian settlement of New Zealand in relation to environmental and biotic changes. New Zealand Journal of Ecology 12(suppl.):115S–129S.

McGlone, M. S., and L. R. Basher. 1995. The deforestation of the upper Awatere catchment, Inland Kaikoura Range, Marlborough, South island, New Zealand. New Zealand Journal of Ecology 19:53–66.

McNeely, J. A. 2001. *The great reshuffling: human dimensions of invasive alien species*. IUCN, Gland, Switzerland and Cambridge, U.K..

Milberg, P., and T. Tyrberg. 1993. Naïve birds and noble savages: a review of man-caused prehistoric extinctions of island birds. Ecography 16:229–250.

Miller, G. H., J. W. Magee, B. J. Johnson, M. L. Fogel, N. A. Spooner, M. T. McCulloch, and L. K. Ayliffe. 1999. Pleistocene extinction of *Genyornis newtoni*: human impact on Australian megafauna. Science 283:205–208.

Moyle, P. B., and T. Light. 1996. Biological invasions of fresh water: Empirical rules and assembly theory. Biological Conservation 78:149–161.

Olson, S. L. 1985. Pleistocene birds of Puerto Rico. National Geographic Society Research Report 18:563–566.

Olson, S. L., and H. F. James. 1982. Fossil birds from the Hawaiian Islands: evidence for wholesale extinction by man before Western contact. Science 217:633–635.

Olson, S. L. and H. F. James. 1991. Description of thirty-two new species of birds from the Hawaiian Islands. Part I: Non-passeriformes. Ornithology Monographs Volume 45.

Owens, I. P. F., and P. M. Bennett. 1995. Ancient ecological diversification explains life-history variation among living birds. Proceedings of the Royal Society London B 261:227–232.

Owens, I. P. F. and P. M. Bennett. 2000. Ecological basis of extinction risk in birds: Habitat loss versus human persecution and introduced predators. Proceedings of the National Academy of Sciences USA 97:12144–12148.

Parker, I. M., D. Simberloff, W. M. Lonsdale, K. Goodell, M. Wonham, P. M. Kareiva, M. H. Williamson et al. 1999. Impact: towards a framework for understanding the ecological effects of invaders. Biological Invasions 1:3–19.

Peters, R. H. 1983. *The ecological implications of body size*. Cambridge University Press, Cambridge.

Pimm, S. L., M. P. Moulton, and L. J. Justice. 1994. Bird extinctions in the Central Pacific. Philosophical Transaction of the Royal Society London B 343:27–33.

Pimm, S. L., M. P. Moulton, and L. J. Justice. 1995. Bird extinctions in the central Pacific. In J. H. Lawton, and R. M. May, eds. *Extinction rates*, pp. 75–87. Oxford University Press, Oxford.

Pringle, C. M., M. C. Freeman, and B. J. Freeman. 2000. Regional effects of hydrologic alterations on riverine macrobiota in the New World: tropical-temperate comparisons. BioScience 50:807–823.

Purvis, A., J. L. Gittleman, G. Cowlishaw, and G. M. Mace. 2000. Predicting extinction risk in declining species. Proceedings of the Royal Society London B 267:1947–1952.

Quintana-Murci, L., O. Semino, H.-J. Bandelt, G. Passarion, McElreavey, and A. S. Santachiara-Benerecetti. 1999. Genetic evidence of an early exit of *Homo sapiens sapiens* from Africa through eastern Africa. Nature Genetics 23:437–441.

Rahel, F. J. 2000. Homogenization of freshwater faunas. Annual Review Ecology and Systematics 33:291–315.

Rice, W. R. 1989. Analyzing tables of statistical tests. Evolution 43:223–225.

Roberts, M. 1991. Origin, dispersal routes, and geographic distribution of *Rattus exulans*, with special reference to New Zealand. Pacific Sciences 45:123–130.

Roberts, R. G., T. F. Flannery, L. K. Ayliffe, H. Yoshida, J. M. Olley, G. J. Prideaux, and G. M. Laslett. 2001. New ages for the last Australian megafauna: Continent-wide extinction about 46,000 years ago. Science 292:1888–1892.

Roff, D. A., and R. J. Roff. 2003. Of rats and Maoris: a novel method for the analysis of patterns of extinction in the New Zealand avifauna before human contact. Evolutionary Ecology Research 5:759–779.

Savidge, J. A. 1987. Extinction of an island forest avifauna by an introduced snake. Ecology 68:660–668.

Saville, S., B. Stephenson, and I. Southey. 2003. A possible sighting of an "extinct" bird, the New Zealand storm-petrel. Birding World 16:173–175.

Simberloff, D. 1995. Why do introduced species appear to devastate islands more than mainland areas? Pacific Sciences 49:87–97.

Simberloff, D. 2000. Extinction-proneness of island species: Causes and management implications. Raffles Bulletin of Zoology 48:1–9.

Simberloff, D., and B. Von Holle. 1999. Positive interactions of nonindigenous species: invasional meltdown? Biological Invasions 1:21–32.

Smith, A. P., and D. G. Quin. 1996. Patterns and causes of extinction and decline in Australian conilurine rodents. Biological Conservation 77:243–263.

Smith, F. D. M., R. M. May, R. Pellew, T. H. Johnson, and K. S. Walter. 1993a. Estimating extinction rates. Nature 364:494–496.

Smith, F. D. M., R. M. May, R. Pellew, T. H. Johnson, and K. R. Walter. 1993b. How much do we know about the current extinction rate? Trends in Ecology and Evolution 8:375–378.

Soulé, M. E. 1983. What do we really know about extinctions? In C. M. Schonewald-Cox, S. M. Chambers, B. MacBryde, and W. L. Thomas, eds. *Genetics and conservation: a reference for managing wild animal and plant populations*, pp. 111–124. Benjamin Cummings, Menlo Park, CA.

Steadman, D. W. 1995. Prehistoric extinctions of Pacific island birds: biodiversity meets zooarchaeology. Science 267:1123–1131.

Steadman, D. W. 1997. The historic biogeography and community ecology of Polynesian pigeons and doves. Journal of Biogeography 24:737–753.

Steadman, D. W., and P. V. Kirch. 1990. Prehistoric extinction of birds on Mangaia, Cook Islands, Polynesia. Proceedings of the National Academy of Sciences USA. 87:9605–9609.

van Riper, C. III. 1991. The impact of introduced vectors and avian malaria on insular passeriform bird populations in Hawaii. Bulletin of the Society for Vector Ecology 16:59–83.

Vitousek, P. M., C. M. D'Antonio, L. L. Loope, M. Rejmánek, and R. Westbrooks. 1997. Introduced species: a significant component of human-caused global change. New Zealand Journal of Ecology 21:1–16.

Walker, A. 1967. Patterns of extinction among sub-fossil Madagascan lemuroids. In P. S. Martin, and H. E. Wright, eds. *Pleistocene extinctions: the search for a cause*, pp. 425–432. Yale University Press, New Haven, CT.

Wiles, G. J., J. Bart, R. E. Beck Jr., and C. F. Aguon. 2003. Impacts of the brown tree snake: patterns of decline and species persistence in Guam's avifauna. Conservation Biology 17:1350–1360.

Williamson, M. 1996. *Biological invasions*. Chapman and Hall, London.

Williamson, M. 1999. Invasions. Ecography 22:5–12.

Wilson, E. O. 1985. The biological diversity crisis: a challenge to science. BioScience 35:700–706.

Wodzicki, K., and R. H. Taylor. 1984. Distribution and status of the Polynesian rat *Rattus exulans*. Annales Zoologici Fennici 172:99–101.

Worthy, T. H. 1997. What was on the menu? Avian extinction in New Zealand. New Zealand Journal of Archaeology 19:125–160.

Worthy, T. H. 1999. The role of climate change versus human impacts—avian extinction on South Island, New Zealand. Smithsonian Contributions to Paleobiology 89:111–123.

Wragg, G. M. 1995. The fossil birds of Henderson Island, Pitcairn group: natural turnover and human impact. Biological Journal of the Linnaean Society 56:405–414.

5

The Role of Infectious Diseases in Natural Communities

WHAT INTRODUCED SPECIES TELL US

Kevin D. Lafferty, Katherine F. Smith, Mark E. Torchin,
Andy P. Dobson, and Armand M. Kuris

Mathematical models provide many predictions about the effects of parasites on host populations, but these predictions have been challenging to test. Controlled parasite addition and removal experiments have provided some of the most valuable insights into theoretical predictions. Like these experiments, species invasions may involve the addition and removal of infectious disease agents and may therefore add to our understanding of the effects of infectious disease. For example, species that invade without their parasites give us an idea of how populations perform in the absence of parasites, whereas biological control programs and introduced diseases can provide before-and-after comparisons. Studies of introduced species indicate that parasites can reduce host performance and may dramatically reduce host abundance. These effects on host species can have indirect effects on ecosystems. Species introductions can result in novel host-parasite contacts, which provide insight into how host-parasite evolution drives host specificity and indicate that host-parasite evolution can occur in only a few generations.

Introduction

Following the introduction of livestock with rinderpest into colonial Africa, diseased carcasses, bloated vultures, and starving predators populated the plains of the Serengeti. For the ranchers and local officials, the effect of infectious disease on the savanna ecosystem was impossible to overlook. Among modern ecologists, the role of infectious disease is gaining increasing appreciation, largely through the lens of species introductions. This chapter reviews how introduced species, such as the morbillivirus that causes rinderpest, can lend considerable insight into the role of infectious diseases in host populations.

We use the term "infectious disease" to describe the pathological consequences of infection by parasites and pathogens. Disease is a pervasive element of all natural communities (Kennedy et al. 1986; Dobson et al. 1992), and the parasitic mode of life is probably the most popular of all consumer strategies (Price 1980; Toft 1986). Nearly all animal taxa have parasitic representatives, and most free-living species (and many parasite species) are hosts for infectious disease agents. At the most basic level, parasites (and pathogens) are consumers that live in physical association with their hosts. Although all parasites negatively affect the vital rates (e.g., birth, growth, death) of their hosts by consuming host energy, there is a vast diversity of parasitic strategies, including those of typical parasites (e.g., intestinal tapeworms), pathogens (e.g., many viruses, fungi, and protozoa), parasitoids (e.g., some wasps and nematodes), parasitic castrators (e.g., rhizocephalan barnacles), trophically transmitted parasites (e.g., larval acanthocephalans), and, by some considerations, micropredators (e.g., herbivorous insects) (Lafferty and Kuris 2002). Depending on the type of infectious agent, hosts can suffer a variety of consequences, including higher mortality, slower growth, lower fecundity, altered behavior, and lower social status.

Before discussing the utility of introduced species as a tool to understand host-parasite dynamics, we provide background information on existing paradigms and traditional tools for testing them. Changes in the vital rates of infected hosts can have a diverse array of effects at the population level as well as on natural selection. Here, we are specifically interested in the effect of infectious diseases on host population dynamics. An obvious prediction is that infectious disease agents with strong effects on host vital rates could noticeably reduce the mean equilibrium density of their hosts. Infectious diseases that reduce host density may have indirect effects by facilitating trophic cascades or mediating competition. Finally, due to the likelihood of density-dependent feedback between host and parasite populations, there is potential for infectious diseases to alter the stability of host populations.

In addition to valuable studies of the pathological effects of infectious diseases and the natural history of infectious disease agents, research into the population-level effects of parasites and pathogens comes from theoretical mathematical models, correlations in nature, and experiments in the laboratory and field. For example, simple mass action equations (Bernoulli 1760) used in "microparasite" disease models track *s*usceptible, *e*xposed, *i*nfectious, and *r*ecovered hosts (SEIR models), but not parasite intensities (the number of parasite individuals within an infected host) (Ross 1916; Kermack and McKendrick 1927; May 2000). Similarly, models of parasitoids used for assessing the effects of biological control agents are based on predator-prey models. Finally, intensity-dependent "macroparasite" models have proved most effective in understanding the relationship between parasite intensity and host pathology.

Perhaps the most fundamental principle of epidemiology revealed by models is that the spread of a directly transmitted infectious disease agent through a population increases with the density of susceptible and infectious hosts. Most simple epidemiological models indicate that there is a host threshold density below which a parasite cannot sustain itself within a host population (Anderson and May 1979). This feature is key to understanding the parasite release hypothesis for species invasions because introduced species usually experience a population bottleneck in the early stages of an invasion. In addition, the density-dependent nature of transmission makes infectious diseases unlikely to be agents of extinction except where alternative hosts are present (Dobson and May 1986).

The best evidence for the hypothesis that parasites can affect host populations comes from experiments in which parasites are added to an uninfected host population or removed from an infected host population (Scott and Dobson 1989). This has been done repeatedly in the laboratory, often leading to the conclusion that parasites can limit host abundance (Greenwood et al. 1936; Park 1948; Stiven 1964; Keymer 1981; Lanciani 1982; Anderson and Crombie 1984; Scott and Anderson 1984; Scott 1987). This approach is much more difficult to apply to natural populations due to logistic and ethical constraints.

One parasite species for which field experiments have been applied successfully is the parasitic nematode *Trichostrongylus tenuis* in one of its host species, the red grouse (*Lagopus lagopus scoticus*) (Hudson and Dobson 1989; Dobson and Hudson 1992; Hudson et al. 1992). The host species is a game bird restricted to the upland areas of northern Britain. The population dynamics of red grouse exhibit sustained cycles of abundance with a period of about 5 to 6 years, and variation in grouse fecundity is associated with these cycles (Jenkins et al. 1967). Empirical studies (Potts et al. 1984; Hudson et al. 1985) and mathematical models (Dobson and Hudson 1992) suggest that decreased fecundity associated with nematode infection can cause grouse populations to cycle. Hudson et al. (1998) tested this hypothesis using a parasite removal experiment. By treating a significant proportion of a grouse population with drugs (reducing the number of nematodes in infected hosts), they increased grouse fecundity and thus changed the long-term population dynamics such that grouse populations no longer suffered periodic crashes (Hudson et al. 1998).

Although such experiments are very handy for testing predictions, it is difficult to manipulate and monitor disease agents in a safe and controlled manner. Introduced species can help us to circumvent the difficulty of designing and implementing field experiments, as they present a rich set of inadvertent "natural" field experiments that we can use to expand our understanding of the role of parasites in natural communities. In this chapter, we examine species introductions with respect to parasites and disease agents in several ways. We begin by exploring what happens when introduced species leave their parasites behind. We next consider the effects of biological control. Finally, we consider, through a detailed examination of four case studies, the effects of accidentally introduced disease agents.

Unintentional Parasite Removals: Introduced Species and Escape from Natural Enemies

Species introductions can provide an opportunity to look at how host populations perform without parasites, albeit with substantially less control than field experiments. In this section, we address the evidence that introduced species can serve as parasite removal experiments and then summarize the inferences made possible by comparative approaches.

Growing evidence indicates that introduced plants and animals escape most of their native parasites and pathogens. Parasites may be absent from the host founder population, die out soon after the invasion (due to low host density), or fail to complete their life cycles in the new environment (Dobson and May 1986; Cornell and Hawkins 1993; Kennedy 1993; Torchin et al. 2001; Mitchell and Power 2003; Torchin et al. 2003; Ricklefs, this volume). Over time, and as they spread, introduced species accumulate new parasites, but these generally amount to only a fraction of those they escape, perhaps because native parasites lack a coevolutionary history with introduced species (Cornell and Hawkins 1993; Mitchell and Power 2003; Torchin et al. 2003). The resulting decrease in parasitism may explain why some introduced species proliferate in their new environment and become abundant pests (Torchin et al. 2001; Mitchell and Power 2003). Freed from the effects of old host-parasite associations and occasionally establishing new ones, introduced species provide "natural experiments" that can be used to reveal the extent to which parasites control host populations and structure ecological communities. Geographically widespread species and those that have invaded multiple regions allow particularly comprehensive analyses with replication.

Available evidence from plants and animals suggests that only a fraction of the parasite species that infect species in their native range will infect those same species' populations in their introduced range (Torchin and Mitchell 2004). Parasite species richness generally decreases by 63%–77% in introduced plant populations compared with populations in their native range, while introduced animal populations are infected with roughly half the number of para-

TABLE 5.1 *Parasite release experienced by introduced species in different taxonomic groups*

Taxonomic group	Sample size (no. of species)	Species released (%)	Mean release (SR)	S.E.	Mean release (P)	S.E.	Source
Plants	473	100	0.77	—	—	—	Mitchell and Power 2003
Insects[a]	87	67	0.63	0.04	0.6	0.1	Cornell and Hawkins 1994
Crustaceans	3	100	0.86	0.14	0.93	0.06	Torchin et al. 2003
Mollusks	7	100	0.56	0.09	0.44	0.25	Torchin et al. 2003
Fishes	6	100	0.76	0.11	0.89	0.03	Torchin et al. 2003
Amphibians and reptiles	3	100	0.57	0.14	0.38	0.07	Torchin et al. 2003
Birds	3	100	0.36	0.14	0.41	0.22	Torchin et al. 2003
Mammals	4	100	0.29	0.12	0.12	0.13	Torchin et al. 2003

Note: Parasite release is represented by the proportion $(N - I)/N$, where N is the value for the native range and I is the value for the introduced range. These values are calculated for standardized parasite species richness (SR), the proportion of parasite species found in the introduced range out of the total number found in the native range, and for average parasite prevalence (P), the proportion of the population infected.

[a]Data based on medians where the range of values was reported ($N = 22$); three of these had higher prevalence in the introduced range, and these data were included in the calculation of release.

site species found in native populations (Cornell and Hawkins 1993; Torchin et al. 2003). All things being equal, this lower diversity of parasites should add up to a reduced effect of parasitism (though this depends on which parasites are left behind and the abundance of natural enemies that remain with or colonize to the host species). An additional measure of parasite release is that parasite prevalence (percentage of individuals infected) in introduced animal populations is also typically less than half that in native populations (Cornell and Hawkins 1993; Torchin et al. 2003). Similarly, introduced plants are less frequently infected with pathogens compared with native populations (Torchin and Mitchell 2004). Introduced populations of all taxonomic groups, including plants, insects, crustaceans, mollusks, fishes, amphibians and reptiles, birds, and mammals exhibit 29%–86% parasite release compared with native populations (Table 5.1).

Species adapt to the abiotic environment in which they have evolved, and an inappropriate match of abiotic conditions probably explains why many invading species fail to establish successful populations. Those species that do become established sometimes become pests and exhibit greater densities and biomasses than in their native range (Table 5.2). Information on body size indicates that introduced populations may also exhibit a larger average body size than native populations (Torchin et al. 2001; Grosholz and Ruiz 2003). Grosholz and Ruiz (2003) examined 19 introduced invertebrate species and found that

TABLE 5.2 *Evidence for increases in demographic parameters (following parasite release) of introduced species in their naturalized range, as compared with their native range*

Taxonomic group	Species	Parameter compared	Mean % increase (response range)	Evidence for parasite release?	Source
Plants	*Prunus serotina*	Density (m²)[a]	86 (80–92)	Yes	Reinhard et al. 2003
	Lythrum salicaria	Biomass (g)	157	—	Blossey and Notzhold 1995
	Lythrum salicaria	Height (cm)	41	—	Blossey and Notzhold 1995
Crustaceans	*Carcinus maenas*	Biomass (kg)	59	Yes	Torchin et al. 2001
	Carcinus maenas	Mean size (mm)	29	Yes	Torchin et al. 2001
Mammals	6 species	Density (km²)	424 (61–735)	Yes	Freeland 1993
Amphibian	*Bufo marinus*	Density (100 m²)	7400	Yes	Lampo and Bayliss 1996a
Marine invertebrates	19 species	Max. size (mm)	13 (8–45)	—	Grosholz and Ruiz 2003

[a]Density was calculated with nearest-neighbor techniques; values show a decrease in nearest-neighbor distance and hence an increase in density.

63% had a significantly larger body size in introduced populations than in native populations. In some cases, losses of natural enemies have been implicated as a potential cause of the increased demographic performance of introduced populations. Natural enemies are an aspect of the native environment that is inherently hostile. Losses of parasites, therefore, could explain the increased demographic performance of introduced species. Still, the increased performance of introduced species and its link to natural enemies is a topic in need of substantial research.

One study specifically addressed (1) how a parasite can affect the body size and abundance of its natural host and (2) how loss of this parasite (through invasion) can result in larger sizes and greater abundances. By comparing multiple native and introduced populations of the European green crab (*Carcinus maenas*), Torchin et al. (2001) demonstrated that a particular group of parasites, parasitic castrators (a rhizocephalan barnacle and an entoniscid isopod), explained 64% of the variation in mean crab size and 36% of the variation in crab biomass in native crab populations. Parasitic castrators do not infect any of the introduced populations, which exhibit significantly higher biomass and larger body sizes compared with native populations (Torchin et al. 2001). Similarly, introduced cane toads (*Bufo marinus*) in Australia reach densities two orders of magnitude higher (1000–2000/100 m² vs. 20/100 m²) than native populations in South America (Lampo and Bayliss 1996a,b; Lampo and DeLeo

1998). Australian populations harbor fewer than 30% of the helminth parasites found in native populations (Barton 1997). In addition, they lack ectoparasites that may control toad density in South America (Lampo and Bayliss 1996a,b; Lampo and DeLeo 1998). Introduced plant populations may also experience demographic release compared with native populations. They are more likely to become noxious weeds if released from pathogens (Mitchell and Power 2003). This finding suggests that pathogens may limit native plant populations as well.

In addition to providing insight into the role of parasites in host demography, biological invasions can indicate the extent to which parasites mediate interactions among free-living species. Parasites can alter competitive dynamics among hosts (Hudson and Greenman 1998). This effect can be addressed by comparing competing native and introduced species. For example, in California's San Joaquin Valley, a suite of native parasitoids attacks the native leafhopper (*Erythroneura variabilis*). An invasive congener (*E. elegantula*) is attacked much less frequently, causing the invader to replace the native (Settle and Wilson 1990). A similar interaction has been demonstrated experimentally in microcosms (Aliabadi and Juliano 2002). Here, introduction of Asian tiger mosquitoes (*Aedes albopictus*) infected with gregarine parasites does not affect the survivorship of a native mosquito (*Ochlerotatus triseriatus*). However, addition of uninfected tiger mosquitoes causes the invader to outcompete the native by reducing its survivorship. Thus, escaping this parasite may give the invading tiger mosquito a competitive advantage and facilitate its spread (Aliabadi and Juliano 2002).

Finally, in addition to their direct effects on host survivorship, parasites may have indirect effects by facilitating changes in their host's behavior. For example, the presence of parasitoid flies alters the behavior of fire ants (*Solenopsis invicta*), reducing their competitive ability in their native Brazil (Orr et al. 1995). Release from the fly may explain the competitive dominance of the fire ant over native North American ant species. Evidence of reduced parasitism facilitating competitive interactions in invading plants is less clear. Blaney and Kotanen (2001) found no difference in the effect of fungal pathogens on seed recovery (survivorship) of native and in introduced plants. However, Klironomos (2002) demonstrated that soil pathogens significantly reduced growth of native plant species, but did not reduce growth of introduced plants (see also Callaway et al., this volume).

Escape from natural enemies could have many implications for evolution and speciation. One main benefit of colonizing a new location may be escape from coevolved natural enemies such as infectious disease agents. At geologic time scales, colonization of new locations by species is commonplace. Although the risks of failure are high, the payoff includes short-term access to abundant resources and a longer-term freedom from parasites and pathogens. These benefits have broad consequences for biodiversity. Natural invasions at remote locations establish reproductive isolation among populations, which in turn can lead to speciation. The extent to which release from parasites facilitates the success of isolated populations could conceivably influence speciation rates. On a longer time scale, taxon cycle theory predicts that eventually natural enemies will catch up with released species and erode their advantage (see Ricklefs, this volume).

The association between release from parasites and pathogens and subsequent performance in terms of body size and density provides substantial support for the hypothesis that coevolved parasites strongly affect the demography of their hosts through various pathogenic effects. Because this pattern appears in all taxa considered so far, it speaks to the evident generality of the role of infectious disease in natural populations.

Intentional Parasite Additions: Biological Control Effects on Target Host Populations

Parasite addition experiments are powerful tools for determining the effects of parasites. In this section, we argue that many classic biological control programs are essentially parasite addition experiments. Here, the host is usually an unintentionally introduced pest and the natural enemy is intentionally introduced. These large-scale field manipulations, when successful, demonstrate that infectious agents can control host populations. Although this approach provides numerous examples and powerful insights, we also consider its limitations.

The release of introduced species from natural enemies, as described above, has led some species to become pests, affecting populations of native species, altering community structure, or exerting a negative economic impact on human activities (often on managed, agricultural species). Biological control seeks to reduce the abundance of exotic species that are pests (e.g., exotic insect pests, terrestrial and aquatic plants, a few vertebrates, and some medically important mollusks) to an economically or culturally acceptable level by reconstructing a few elements of the natural control of the pest where it was native. The infectious natural enemies employed in control campaigns include many parasitoids (against insects), some microbial pathogens (against vertebrates and insects), a few parasitic castrators (against snails that can serve as vectors for human disease), and herbivorous insects (against weeds). Such natural enemies usually reduce pest density by directly influencing pest mortality or reproductive rates. A few manipulations in ponds have demonstrated that biological control of freshwater snails can be achieved by parasitic castrators (trematodes) and that these parasites can competitively exclude other trematodes that cause human disease (Lie and Ow-Yang 1973; Nassi et al. 1979; Lafferty 2002).

Although biological control agents have frequently been shown to control target host populations, there are few examples of eradication or extinction of targeted pests. The most frequently cited example of extinction of a pest by its control agent is that of the coconut moth, *Levuana iridescens*. The control agent, a tachinid, *Bessa remota*, rapidly made the coconut moth vanishingly rare (Tothill et al. 1930). However, a recent examination of this case history suggests that *L. iridescens* is probably not extinct (Kuris 2003).

Overall, only a minority of biological control efforts meet their control goals. This frequent lack of "success" is partly a result of the very high standard set for control. However, the variation in success across taxa, habitats, functional groups, and life histories underscores the concept that the effects of infectious diseases on host population dynamics, while occasionally dramatic, vary greatly, even for natural enemies that can kill or castrate a host individual. The sources of this variation are under active investigation and include the searching efficiency of the infective natural enemy, its host specificity, age structure, refugia, interactions with other mortality sources, competition among infectious agents, and the type of infectious agent (parasite, pathogen, parasitoid, parasitic castrator) (Hall et al. 1980; Murdoch et al. 1985, 1987, 2003; Kuris and Lafferty 1992; Ehler 1998; Begon et al. 1999; Bellows and Hassell 1999; Briggs et al. 1999; Shea et al. 2000).

The frequent success of biological control programs against Homoptera, such as aphids, scale insects, and whiteflies (Hall et al. 1980; Ehler 1998), emphasizes the importance of the relative scale of recruitment to host and parasite populations (Kuris and Lafferty 1992). As Murdoch et al. (1985, 2003) have shown for the red scale insect (*Aonidiella aurantii*) and the parasitoid *Aphytis melinus*, host populations are regulated because the parasitoid operates over a much larger spatial scale than does its host. Infecting many host populations sustains the parasitoid population at a sufficient density to significantly reduce most host populations at a particular moment (while preventing either the parasite or the host from going locally extinct). Consequently, this parasite does not closely track each host population and maintain stable local equilibria. We suggest that this may be a common feature of host-parasite population dynamics revealed through analysis of biological control. Infectious agents with relatively open recruitment will be most likely to control host populations (Kuris and Lafferty 1992).

Some biological control campaigns have produced unintended effects on nontarget species. This is generally undesirable because native species that are not pests may be affected. Among the best-documented studies are those of the tachinid parasitoid *Compsilura concinna*, introduced to North America to control the European gypsy moth (*Lymantria dispar*) (Boettner et al. 2000). Although the tachinid does frequently parasitize the gypsy moth to the extent that populations decline, gypsy moths remain abundant in New England forests. Thus, high gypsy moth populations sustain the abundance of the tachinid, which has broad host specificity across the lepidopterans. Consequently, this generalist biological control agent can drive other (native) hosts to very low densities. Even a host-specific biological control agent can affect native species through indirect pathways (Pearson and Callaway 2003) under a limited set of conditions (Thomas et al. 2004).

Biological control projects afford an excellent opportunity to examine host specificity. Using Combes's (2001) perspective on host specificity, a successful parasite must encounter a host and be compatible with it (Combes describes these processes as a series of filters). Encountering a host requires appropriate behavior, habitat, and temporal patterns. Compatibility requires completing development (by feeding on the host) and surmounting the immunological defenses of the host. While the question of compatibility is easily amenable to experimental investigation and is generally emphasized in analyses of host specificity, host encounter is generally only partially revealed by patterns of host use in nature. Intentional releases of infectious agents in biological control programs offer perhaps the most substantial body of evidence for the importance of host encounter because actual infection in the field can be compared with the results of compatibility studies in the laboratory. Particularly for biological control agents of weeds, but recently for other agents as well, laboratory analyses of compatibility are routinely conducted and included in risk assessment before deployment. When the encounter filter is experimentally removed in the laboratory, a specialist can often parasitize a wide range of compatible hosts. Studies following the release of biological control agents show, however, that many agents infect fewer species in nature than when placed in test arenas (Sands 1997), suggesting that limited host encounter could greatly determine host specificity.

The evolutionary interactions between vertebrate hosts and microbial pathogens have been most strikingly demonstrated through biological control campaigns. To mitigate the ecological and economic impacts of the European rabbits (*Oryctolagus cuniculus*) that became abundant exotic pests in Australia, a rabbit-specific myxoma virus was released. This virus was a relatively avirulent pathogen of South American rabbits (*Silvilagus brasiliensis*) (Hoddle 1999), but it swept through Australian rabbit populations, causing very high mortality rates. This resulted in strong selection for immune response to the virus in rabbits (Fenner and Ratcliffe 1965; Hoddle 1999). More importantly, as rabbits became rare (often decreasing below the transmission threshold), the virus experienced selection for reduced virulence, as measured by exposing laboratory rabbits to viruses taken from the wild in successive years (Fenner and Ratcliffe 1965). This case history revealed the close interplay between selective pressures and host and parasite population dynamics. The accidental release in Australia of a rabbit-specific calicivirus from Europe has begun to replay some of these dynamics as rabbit populations have experienced dramatic declines in some areas (Hoddle 1999). It will now be interesting to examine interactions in a two-pathogen one-host system.

Despite the oversimplification of agricultural systems (where many biological control experiments are performed), the overall conclusion from consideration of biological control studies is that infectious diseases can exert a significant and strong effect on host populations. However, as predicted by simple epidemiological models, biological control rarely leads to host extinction because pathogens decline in importance as hosts become rare. The ability of an infectious disease organism to persist in times or places of low host availability may determine the need for use of alternative hosts or broad dissemination. Where host specificity is low, there is the potential for broad impacts on the community of susceptible hosts (e.g., parasite-mediated changes in competitive ability). Finally, observations of novel contacts between pests and biological control agents show how virulence and host defenses can evolve.

Unintentional Parasite Additions: Effects of Introduced Pathogens on Host Populations

Perhaps the most notable insights into the role of infectious disease have come from introduced diseases. Increases in global trade have allowed diseases to cross geographic barriers and attack naïve hosts. In addition, human encroachment on wildlife habitat has led to more opportunities for transmission of disease to new hosts. Although the vast majority of disease introductions probably fail, those that succeed can have dramatic effects on new hosts, underscoring the potential for infectious diseases to influence not only host population dynamics but also entire communities. Below, we present four case studies that have shed considerable light on the effects of infectious disease on entire communities.

Introduction of avian pox and avian malaria to the endemic Hawaiian avifauna

The largest loss of the native Hawaiian avifauna occurred following the arrival and colonization of the Hawaiian Islands by Polynesians (James and Olson 1991; Olson and James 1991). Since Europeans arrived, however, the Hawaiian Islands have lost additional native species while gaining more than 125 exotic bird species, over 60 of which have become naturalized (van Riper and Scott 2001; Sax et al. 2002). This trend toward biotic homogenization of the Hawaiian avifauna can be explained in part by the post-European extinction event that peaked in the early twentieth century. At this time, the endemic Hawaiian avifauna undoubtedly experienced losses due to habitat destruction, introduced predators, and competitors. However, Warner (1969) cogently examined the evidence for the influence of habitat destruction, competition, and predation and rejected these as the primary cause of this extinction event. Warner observed that native birds of several species were common in disturbed areas in the nineteenth century, that they readily consumed a wide variety of introduced plants, and that large tracts of native vegetation remained in some lowland and most upland areas. When the die-off occurred, it was rapid and nearly total in lowland areas; by 1902, native lowland forests were largely silent. During this brief period, diseased native birds were unusually common. Their symp-

TABLE 5.3 *Prevalence and effects of avian pox and avian malaria on the native and exotic avifauna of the Hawaiian Islands*

Species	Altitude (m)	Disease[a]	Prevalence
NATIVE			
Chasiempis sandwichensis	1900	Avian pox[1]	1.0%
	1550	Avian pox[1]	40.0%
	300–2400	Avian pox[2]	19.5%
		Avian malaria[3]	6.0%
Hemignathus virens	>1600	Avian malaria[3]	Seropositive
	<1600	Avian malaria[3]	Seropositive
	300–2400	Avian malaria[3]	7.3%
		Avian pox[2]	17.6%
Vestiaria coccinea	300–2400	Avian malaria[3]	6.1%
		Avian pox[2]	16.8%
Myadestes obscurus	300–2400	Avian malaria[3]	2.1%
		Avian pox[2]	24.3%
Himatione sanguinea	300–2400	Avian malaria[3]	29.2%
		Avian pox[2]	34.9%
Telespiza cantans	Experimental	Avian malaria[3]	Seropositive
Paroreomyza montana	Experimental	Avian malaria[4]	Seropositive
EXOTIC			
Carpodacus mexicanus	300–2400	Avian malaria[3]	11.6%
		Avian pox[2]	21.5%
Passer domesticus	300–2400	Avian malaria[3]	11.4%
		Avian pox[2]	7.4%
Cardinalis cardinalis	300–2400	Avian malaria[3]	2.2%
		Avian pox[2]	2.0%
Zosterops japonicus	300–2400	Avian malaria[3]	0.9%
		Avian pox[2]	2.2%
Lonchura punctulata	300–2400	Avian malaria[3]	2.5%
		Avian pox[2]	0.0%

Note: All studies were conducted on Mauna Lao Volcano, except that of *Chasiempis sandwichensis*, which was conducted on Mauna Kea.

Sources: 1, VanderWerf 2001; 2, van Riper et al. 2002; 3, van Riper et al. 1986; 4, Jarvi et al. 2001

[a] Avian pox = poxvirus avium; avian malaria = *Plasmodium relictum capistranoae*.

[b] IUCN Red List category (www.redlist.org). NL = no listing.

toms were consistent with avian pox and possibly other diseases. These observations are all consistent with the hypothesis that a disease swept through these populations. The epidemic affected many endemic bird species, perhaps because all these species were naïve to diseases common on continents.

TABLE 5.3 *(continued)*

Species	Morbidity	Mortality	Status[b]
NATIVE			
Chasiempis sandwichensis			Vulnerable
	70% deformity		
	9.1% lesions		
Hemignathus virens		66.0%	NL
		20.0%	
	10.6% lesions		
Vestiaria coccinea			Lower risk
	10.3% lesions		
Myadestes obscurus			Vulnerable
	20.3% lesions		
Himatione sanguinea			NL
	14.1% lesions		
Telespiza cantans		100.0%	Vulnerable
Paroreomyza montana		75.0%	Vulnerable
EXOTIC			
Carpodacus mexicanus			NL
	6.3% lesions		
Passer domesticus			NL
	2.5% lesions		
Cardinalis cardinalis			NL
Zosterops japonicus			NL
	1.4% lesions		
Lonchura punctulata			NL

The avian diseases in question probably arrived with domesticated chickens imported in 1901, which probably carried avian poxvirus, and with caged passerines, brought in the early 1920s, which came with avian malaria (van Riper and Scott 2001; van Riper et al. 2002). These infectious diseases had "visited" Hawaii for millennia in migratory birds, but without a vector, were not able to infect native birds. The vector was supplied in 1826 when the ship *Wellington* drained its dregs, releasing mosquito larvae from the west coast of Mexico into a stream near Lahaina. The mosquito *Culex quinquefasciatus* was soon ubiquitous throughout the Hawaiian Islands at elevations below 1650 m.

A series of experiments with Laysan finches (*Telespiza cantans*) showed how devastating these vectored diseases were in naïve birds. Laysan (along with Nihoa) is a low, upwind island. It is still mosquito-free and retains large populations of its native birds. Warner (1969) brought Laysan finches to Honolulu and established an experiment in which some birds were exposed to mosquitoes in an unscreened cage and others protected from exposure in a screened cage. In 15 days, all the unscreened birds had died; all the screened birds were alive. Further experiments showed that an exposure as brief as 3 days was lethal. Blood analysis revealed unusually high parasitaemias for both *Plasmodium* (two species) and a species of *Haemoproteus*.

The loss of native birds was greatest in the lowlands, in the wettest areas, and during the rainy season. The effects of this extinction event are evident from IUCN's (2002) species status reports. Limitations in mosquito distributions (particularly along elevation gradients) appear to have spared the extant native avifauna from extinction due to avian pox and avian malaria (van Riper et al. 1986). It should be noted that mosquitoes are now less abundant than in the past due to partially effective mosquito abatement practices.

Pathogens infect exotic birds on Hawaii as well, but their effect on those birds is relatively minimal. This difference is a probably a result of differential exposure to vectors and resistance to disease. For example, Warner (1969) showed that the number of mosquitoes feeding on Laysan finches was 5–10 times greater than on introduced white-eyes (*Zosterops palpebrosa*). The prevalence of avian pox and avian malaria, and consequent levels of morbidity and mortality, are all significantly higher in natives than in exotics (Table 5.3). Whereas five of the seven native species listed in Table 5.3 are classified as vulnerable or at risk, none of the exotic species achieve this ranking at a global scale, and all are purportedly in stable condition on the Hawaiian Islands (IUCN 2002).

In short, the evidence strongly supports vectored blood pathogens as the primary cause of the extinctions of Hawaiian endemic avifauna, and the restriction of remnant populations to high altitudes and upwind islands, that occurred in the twentieth century. The roles of other exotic pests and habitat alterations appear to have been relatively unimportant during this time.

The effects of avian pox and avian malaria on the Hawaiian avifauna illustrate several attributes of vector-transmitted diseases that might otherwise have remained elusive had the pathogens not been introduced. The first is that host species distributions can be altered by the distributions of vectors (which, in this case, have environmental tolerances far narrower than the host species). Second, in a community in which multiple hosts share a common parasite, it is possible for the parasite to facilitate host-specific extinctions, so long as other host reservoirs persist. In the case of Hawaiian birds, it appears that the exotic hosts of avian pox and avian malaria, in part because they are relatively unaffected (and thereby remain common and infected), act as a substantial source of infection for native hosts. Finally, the differential susceptibility of exotic and naïve native hosts illustrates how hosts can evolve defenses that limit impacts at the individual, population, and species level.

Community response to the introduction and eradication of African rinderpest

In 1889, the Italian army imported cattle carrying rinderpest virus (a morbillivirus related to measles) from India to the Horn of Africa. Within a year, the pathogen infected a number of native species (Spinage 2003). Spreading at a rate of 500 km per year, rinderpest epidemics caused mass mortality in domestic and wild artiodactyls from Egypt to South Africa (Plowright 1982). By the turn of the twentieth century, the African rinderpest epidemic had claimed about 90% of the East African domestic cattle population and about 95% of the Serengeti buffalo (*Syncerus caffer*) and wildebeest (*Connochaetes taurinus*) populations (Plowright 1982; Spinage 2003). During a rinderpest outbreak between 1959 and 1961, the annual mortality of first-year wildebeest increased significantly to about 85%, with predation accounting for 45% and rinderpest accounting for 40% of all deaths. Numerous other species also succumbed to the disease (Plowright 1982). Rinderpest continued to suppress domestic and wild artiodactyls until a vaccine was introduced and the virus was largely eradicated (Spinage 2003). Although only domestic cattle were vaccinated, the disease disappeared from the wild species, implying that cattle were, in fact, the main reservoir. This outcome demonstrates the importance of reservoir hosts in maintaining and facilitating the spread of infectious diseases to other species, an issue central to how diseases can affect rare species (Lafferty and Gerber 2002). After 1961, when clinical rinderpest was absent from the Serengeti, wildlife populations experienced rapid recovery (Plowright 1982; Spinage 2003). Within a decade, the wildebeest population increased from 260,000 to 700,000 individuals, and buffalo increased from 30,000 to more than 60,000 individuals (Plowright 1982). These increases were largely attributed to the reduction in juvenile mortality following rinderpest control (Plowright 1982). This response implies that rinderpest suppressed artiodactyl populations to densities far lower than the Serengeti habitat could support (Sinclair 1979; Dobson 1995).

Given this strong suppression of artiodactyl populations, it is perhaps not surprising that rinderpest control had corresponding effects on the population dynamics and trophic structure of the Serengeti ecosystem as a whole (Sinclair 1979; Plowright 1982; Dobson 1995; Tompkins et al. 2001). Increases in the abundance of ungulate species (following rinderpest control) led to an increase in the density of carnivores, particularly lions (*Panthera leo*) and hyenas (*Crocuta crocuta*). These increases in carnivore abundance were matched by decreases in the abundance of gazelles, most likely due to increased predation pressure. The most dramatic demographic change following rinderpest control occurred in wild dogs (*Lycaon pictus*), whose numbers declined from about 500 to eventual local extinction, a likely consequence of increased competition with recovering lion and hyena populations. Furthermore, the changes in the numbers of grazing artiodactyls would certainly have had an effect on plant biomass and composition. There is even evidence to suggest that the decline in browsers, particularly impala (*Aepyceros melampus*), during the first pandemic may have allowed a large recruitment pulse in many tree species (Prins and Weyer-haeuser 1987). For example, the acacia stands in many parts of the ecosystem are remarkably even in their area and size distribution and appear to result from a narrow window of recruitment during grazer population minima. Collectively, the trophic cascades and other regulatory effects initiated by rinderpest virus on the East African ecosystem suggest that certain viruses can play keystone roles in ecosystem functioning and structure.

Evolution of host specificity and host switching in distemper viruses

The devastating epidemics of another morbillivirus, canine distemper virus (CDV), in Serengeti predators (Table 5.4) provide additional insight into the effects of disease on population dynamics. Plowright (1982) recorded that when he developed the rinderpest vaccine in Nairobi, dead cattle were disposed of by supplying them to the local dog owners. Distemper effectively disappeared from the domestic dog population at this time! This observation implies that exposure to rinderpest in infected carcasses may have caused cross-immunity to distemper in canids. Therefore, it may be that loss of rinderpest from wildebeest subsequently increased the susceptibility of carnivores to distemper.

Between 1984 and 1988, CDV prevalence in the Serengeti lion population declined to zero from about 75%, a value believed to reflect a previously unde-

TABLE 5.4   *Emergent epidemics of distemper viruses in novel species*

Species	Location	Prevalence (%)	Morbidity	Mortality[a]	Source
Baikal seal	Lake Baikal	71.4	100% lesions	18,000 total	Grachev et al. 1989; Ohashi et al. 2001
Caspian Sea	Caspian Sea	75.0–100.0	100% lesions	>10,000 total	Kennedy et al. 2000
Harbor seal[b]	Europe	47.0–100.0	Acute CDV symptoms	60% regional, >17,000 total	Osterhaus and Vedder 1988; Heide-Jorgensen et al. 1992
African wild dog	Masai Mara, Kenya	4.0–75.0	Decreased appetite, diarrhea, listlessness	21%–50% regional	Alexander and Appel 1994
Serengeti lion	Serengeti, Tanzania	85.5–99.0	> 94% lesions	30% total	Roelke-Parker et al. 1996; Haas et al. 1996
Spotted hyena	Serengeti, Tanzania	99.0	100% lesions	—	Haas et al. 1996

[a]Percentage or absolute mortality resulting from epidemics.
[b]Values reported for phocine distemper virus (PDV).

tected epidemic (Packer et al. 1999). In 1994, another epidemic occurred, infecting the majority of the population in less than 3 months and reducing fecundity and survival in all age groups (Table 5.4; Cleaveland et al. 2000). Interestingly, evidence suggests that the 1980s epidemic corresponded with only 70 susceptible individuals, whereas the 1994 epidemic did not occur until 100% of the population (250 individuals) was susceptible (Packer et al. 1999). This discrepancy implies that CDV existed in other species in the early 1990s, so that the pool of susceptible hosts in the ecosystem was far larger than it had been in the 1980s (Packer et al. 1999). Indeed, other Serengeti species, including spotted hyenas and wild dogs, also experienced high morbidity and mortality from CDV in 1994 (Table 5.4; Cleaveland et al. 2000). This cross-species transmission may have been facilitated by the severe drought conditions that plagued the Serengeti throughout 1993 and increased the probability of contact between domestic dogs (the purported source of the disease), lions, and spotted hyenas at waterholes and carcasses (Cleaveland et al. 2000).

Since 1988, CDV and related morbilliviruses, such as phocine distemper virus (PDV), have also emerged in a number of marine hosts (see Table 5.4). The majority of recent CDV and PDV epidemics are attributed to interspecific virus transfer between species that do not typically experience close contact. As in the Serengeti epidemic, there is speculation that domestic dogs are the source of morbillivirus in marine mammals. For example, the introduction of sled dogs to Antarctica in 1955 is believed to be responsible for a mass mortality event that was induced by CDV and caused a 97% decline in the local crabeater seal (*Lobodon carcinophagus*) population (Bengston and Boveng 1991).

In 1987–1988, an epidemic of PDV occurred for the first time in European harbor seals (*Phoca vitulina vitulina*) (Osterhaus and Vedder 1988). Several hypotheses have been proposed to explain the origin of the virus, including incipient contact with harp seals (*Phoca groenlandica*) and terrestrial canids (Heide-Jorgensen and Harkonen 1992). The virus infected more than 90% of the total European harbor seal population, causing about 60% mortality in most regions (see Table 5.4; Osterhaus and Vedder 1988; Heide-Jorgensen and Harkonen 1992). Population recovery was rapid following 1988, with a 6%–12% increase in population size per year at the epicenter in Danish waters and more than a fourfold increase between 1989 and 2000 in the Wadden Sea (Jensen et al. 2002). In 2002, when the harbor seal population exceeded pre-epidemic levels and the majority of individuals were again susceptible to PDV, another outbreak occurred. The 2002 epidemic mirrored the 1988 epidemic in timing, geography, rate of spread, morbidity, and mortality (Jensen et al. 2002). Recent theory, although controversial, predicts a 1%–18% risk that recurrent outbreaks will reduce the European harbor seal population by 90% (Harding et al. 2002; Lonergan and Harwood 2003).

It is evident from the emergence of distemper virus in European harbor seals and Serengeti carnivores that unusual contact events between typically segregated species have the potential to spread disease to novel species groups, even across the land-sea interface. Such host switching can occur rapidly and with devastating impacts on the novel host.

Indirect effects of disease on competitive interactions in red squirrels and gray partridges

At the turn of the twentieth century, the gray squirrel (*Sciurus carolinensis*), a native of North America, was introduced to the United Kingdom as a wildlife novelty and, like many exotics, experienced a rapid range expansion (Middleton 1930). Since its introduction, the gray squirrel has "replaced" the native red squirrel (*S. vulgaris*) throughout much of its range, primarily through competition for food resources (MacKinnon 1978; Bryce 1997; O'Teangane et al. 2000). Evidence supporting this replacement is apparent in Table 5.5, which shows that carrying capacity, growth rate, reproductive rate, and competitive effect are all greater for the gray squirrel. However, recent theory demonstrates that parapoxvirus, which was introduced by the gray squirrel, may also be contributing to the decline in red squirrel abundance (Tompkins et al. 2003). Whereas gray squirrels are resistant to the effects of parapoxvirus, red squirrels experience considerable virus-induced mortality. Tompkins et al. (2003) theorized that competition-mediated replacement of local red squirrel populations by gray squirrels could occur within 15 years, but when the effects of parapoxvirus are incorporated, replacement time drops to only 6 years. Furthermore, without the effects of parapoxvirus, natural rates of competition alone cannot explain the decline in red squirrel populations or gray squirrel range expansion (Tompkins et al. 2003).

A similar effect is apparent in the wild gray partridge (*Perdix perdix*) in the United Kingdom. Although declines in gray partridge populations over the last 40 years have been largely attributed to changes in agricultural regimes (Potts 1986), there is evidence that released pheasants (*Phasianus colchicus*) may

TABLE 5.5 *Variation in life history parameters for red and gray squirrels*

Parameter	Red squirrel	Gray squirrel
Carrying capacity	12/km^2	16/km^2
Net growth rate	0.61/year	0.82/year
Maximum reproductive rate	1.0/year	1.2/year
Competitive effect[a]	0.61	1.65
Natural mortality rate	0.40/year	0.40/year
Mortality rate due to virus	0.26/year	Resistant

Source: From Tompkins et al. 2003.

Note: Red squirrels are native to the UK. Gray squirrels were introduced together with parapoxvirus, which has a negative impact on naïve red squirrel populations.

[a] Competitive effect of one squirrel species on the other.

also be a contributing factor. Pheasants appear to be the driving force behind the spread of a cecal nematode (*Heterakis gallinarum*) that induces morbidity in gray partridge populations, but not in the pheasants themselves (Tompkins et al. 2000). Theoretical research by Tompkins et al. (2000) predicts parasite-mediated competition between the species, whereby partridges are excluded from regions where they overlap with pheasants due to the negative effects of *H. gallinarum*.

These examples illustrate how infectious disease can mediate competitive interactions, usually to the disadvantage of the host that is naïve to the pathogen. Although these two examples suggest an advantage for introduced species, it is also reasonable to expect that introduced species will face challenges from the new diseases they encounter in the areas they invade (though, except for agricultural introductions, it is difficult to observe cases in which diseases prevent establishment of an invader).

Conclusions

Species introductions can provide considerable insight into the role of infectious diseases in nature. We can use host species introductions to understand the dynamics of populations and communities in the absence of disease (disease removal experiments). In this case, the introduction may be replicated spatially or temporally, and the control can be the introduced species in its native range (with its native parasites). Similarly, we can use pathogen introductions to understand the dynamics of populations and communities in the presence of disease (disease addition experiments). Here, the control can be the status of the host population before the disease invades or after the disease is eradicated (as in the case of rinderpest) or other host populations that have not been exposed to the pathogen. Despite the large number of possible comparisons to be made, most of our knowledge is anecdotal and nonsystematic. An exception is the biological control literature, in which careful comparisons are often made and conclusions are easier to reach. Unfortunately, biological control is typically carried out in simplified agricultural settings, making it difficult to extrapolate what the effects of natural enemies would be in natural settings (Hawkins et al. 1999).

In spite of the limitations that exist, disease additions and removals lead to several ecological and evolutionary generalizations. The first ecological generalization is that not all infectious disease organisms have dramatic effects on host populations. This is most obvious from studies of biological control programs, in which even parasites that are chosen for their high potential to affect the host population often fail at regulating the host to the extent that economic damage becomes insignificant. Second, some infectious diseases do have appreciable effects on hosts. By affecting host vital rates, they can cause host performance (body size, density, and biomass) to decline. Third, the densities of populations subject to disease can fall, sometimes to low levels, but typically not to extinction. This pattern may be a result of reduced parasite transmission at low host densities in the absence of alternative hosts or of restricted environmental tolerances of disease vectors. Fourth, effects on hosts can be broad (and potentially lead to extinction) if the disease is able to infect several different species, which may occur if the presence of a novel host creates opportunities for host switching. When more than one host species is affected, rare hosts can be differentially affected because disease transmission does not decline as the density of the rare species drops. Fifth, if the host plays a keystone role in the community, the disease may have considerable indirect effects. These effects may take the form of trophic cascades, in which hosts unaffected by the disease (competitors or prey) may gain a competitive advantage. Sixth, if a pathogenic disease has a different geographic distribution from the host (due to the distribution of a vector, for example), the host distribution may shift away from areas where the risk of infection is high. Finally, release from infectious diseases (via long-distance dispersal and species introduction) may have substantial benefits for the host species with respect to population abundance, growth rates, and so forth. Several evolutionary insights arise as well. For instance, novel encounters between hosts and infectious disease agents probably rarely result in disease. When they do, naïve hosts may suffer substantially, but natural selection can rapidly select for host resistance. In addition, release from parasites may have played a role in speciation events by aiding the performance of recently isolated host species.

Although many of these insights are speculative and based on only one or a few examples, more systematic study of introduced species and infectious disease may help us ascertain the generalities of these insights. In particular, it remains to be determined (1) how frequently disease effects are demographically or ecologically important, (2) how important the effects of disease are relative to those of other factors, and (3) what types of hosts and ecosystems are most affected. Researchers have been able to uncover valuable insights into infectious disease processes through their study of introduced species thus far. In our opinion, however, these investigators have just scratched the surface of what is possible.

Acknowledgments

This work was conducted as a part of the "Exotic Species: A Source of Insight into Ecology, Evolution, and Biogeography" Working Group supported by the National Center for Ecological Analysis and Synthesis, a center funded by the National Science Foundation (grant DEB-0072909), the University of California, and the Santa Barbara campus. In addition, this chapter has benefited from support received from the National Science Foundation through the NIH/NSF Ecology of Infectious Disease Program (DEB-0224565).

Literature Cited

Alexander, K. A., and M. J. G. Appel. 1994. African wild dogs (*Lycaon pictus*) endangered by a canine distemper epizootic among domestic dogs near the Masai Mara National Reserve, Kenya. Journal of Wildlife Diseases 30:481–485.

Aliabadi, B. W., and S. A. Juliano. 2002. Escape from gregarine parasites affects the competitive interactions of an invasive mosquito. Biological Invasions 4:283–297.

Anderson, R. M., and J. A. Crombie. 1984. Experimental studies of age-prevalence curves for *Schistosoma mansoni* infections in populations of *Biomphalaria glabrata*. Parasitology 89:79–105.

Anderson, R. M., and R. M. May. 1979. Population biology of infectious diseases. Part 1. Nature 280:361–367.

Barton, D. P. 1997. Introduced animals and their parasites: the cane toad, *Bufo marinus*, in Australia. Australian Journal of Ecology 22:316–324.

Begon, M., S. Sait, and D. Thompson. 1999. Host-pathogen-parasitoid systems. In B. A. Hawkins, and H. V. Cornell, eds. *Theoretical approaches to biological control*, pp. 327–348. Cambridge University Press, Cambridge.

Bellows, T. S., and M. P. Hassell. 1999. Theories and mechanisms of natural population regulation, In T. S. Bellows, and T. W. Fisher, eds. *Handbook of biological control*, pp. 17–44. Academic Press, San Diego.

Bengston, J. L., and P. Boveng. 1991. Antibodies to canine distemper virus in Antarctic seals. Marine Mammal Science 7:85–87.

Bernoulli, D. 1760. Essai d'une nouvelle analyse de la mortalité causée par la petite variole et des avantages de l'inoculation pour la prevenir. Memoires Mathematiques et Physiques Tires Registres de l'Academie Royale des Sciences:1–45.

Blaney, C. S., and P. M. Kotanen. 2001. Effects of fungal pathogens on seeds of native and exotic plants: a test using congeneric pairs. Journal of Applied Ecology 38:1104–1113.

Blossey, B., and R. Notzhold. 1995. Evolution of increased competitive ability in invasive nonindigenous plants: a hypothesis. Journal of Ecology 83:887–889.

Boettner, G. H., J. S. Elkington, and C. J. Boettner. 2000. Effects of a biological control introduction on three nontarget native species of saturniid moths. Conservation Biology 14:1798–1806.

Briggs, C. J., W. W. Murdoch, and R. M. Nisbet. 1999. Recent developments in theory for biological control of insect pests by parasitoids, In B. A. Hawkins, and H. V. Cornell, eds. *Theoretical approaches to biological control*, pp. 22–42. Cambridge University Press, Cambridge.

Bryce, J. 1997. Changes in the distribution of red and grey squirrels in Scotland. Mammal Review 27:171–176.

Cleaveland, S., M. G. J. Appel, W. S. K. Chalmers, C. Chillingworth, M. Kaare, and C. Dye. 2000. Serological and demographic evidence for domestic dogs as a source of canine distemper virus infection for Serengeti wildlife. Veterinary Microbiology 72:217–227.

Combes, C. 2001. *Parasitism: the ecology and evolution of intimate interactions*. University of Chicago Press, Chicago.

Cornell, H. V., and B. A. Hawkins. 1993. Accumulation of native parasitoid species on introduced herbivores: a comparison of hosts as natives and hosts as invaders. American Naturalist 141:847–865.

Cornell, H. V., and B. A. Hawkins. 1994. Patterns of parasitoid accumulation on introduced herbivores. In B. A. Hawkins, and W. Sheehan, eds. *Parasitoid community ecology*, pp. 77–89. Oxford University Press, New York.

Dobson, A. P. 1995. The ecology and epidemiology of rinderpest virus in Serengeti and Ngorongoro crater conservation area, In A. R. E. Sinclair, and P. Arcese, eds. *Serengeti II: Research, management and conservation of an ecosystem*, pp. 485–505. University of Chicago Press, Chicago

Dobson, A. P., and P. J. Hudson. 1992. Regulation and stability of a free-living host-parasite system, *Trichostrongylus tenuis* in red grouse. II. Population models. Journal of Animal Ecology 61:487–500.

Dobson, A. P., and R. M. May. 1986. Patterns of invasions by pathogens and parasites, In H. A. Mooney, and J. A. Drake, eds. *Ecology of biological invasions of North America and Hawaii*, pp. 58–76. Springer-Verlag, New York.

Dobson, A. P., P. J. Hudson, and A. M. Lyles. 1992. Macroparasites: worms and others. In M. J. Crawley, ed. *Natural enemies: the population biology of predators, parasites and diseases*, pp. 329–348. Blackwell Scientific, Oxford.

Ehler, L. E. 1998. Invasion biology and biological control. Biological Control 13:127–133.

Fenner, F., and F. N. Ratcliffe. 1965. *Myxomatosis*. Cambridge University Press, Cambridge.

Freeland, W. J. 1993. Parasites, pathogens and the impacts of introduced organisms on the balance of nature in Australia. In C. Moritz and J. Kikkawa, eds. *Conservation biology in Australia and Oceania*, pp. 171–180. Surrey Beatty and Sons, Chipping Norton.

Grachev, M. A., V. P. Kumarev, and L. V. Mamaev. 1989. Distemper virus in Baikal seals. Nature 338:209–210.

Greenwood, M., A. Bradford-Hill, W. W. C. Topely, and J. Wilson. 1936. Experimental epidemiology. Medical Research Council Special Report 209:204.

Grosholz, E. D., and G. M. Ruiz. 2003. Biological invasions drive size increases in marine invertebrates. Ecology Letters 6:700–705.

Haas, L., H. Hofer, M. East, P. Wohlsein, B. Liess, T. and Barrett. 1996. Canine distemper virus infection in Serengeti spotted hyaenas. Veterinary Microbiology 49:147–152.

Hall, R. W., L. E. Ehler, and B. Bisabri-Ershadi. 1980. Rate of success in classical biological control of arthropods. Bulletin of the Entomological Society of America 26:111–114.

Harding, K. C., T. Harkonen, and H. Caswell. 2002. The 2002 European seal plague: epidemiology and population consequences. Ecology Letters 5:727–732.

Hawkins, B. A., N. J. Mills, M. A. Jervis, and P. W. Price. 1999. Is the biological control of insects a natural phenomenon? Oikos 86:493–506.

Heide-Jorgensen, M. P., and T. Harkonen. 1992. Epizootiology of the seal disease in the Eastern North Sea. Journal of Applied Ecology 29:99–107.

Heide-Jorgensen, M. P., T. Harkonen, R. Dietz, and P. M. Thompson. 1992. Retrospective of the 1988 European seal epizootic. Diseases of Aquatic Organisms 13:37–62.

Hoddle, M. S. 1999. Biological control of vertebrate pests, In T. S. Bellows, and T. W. Fisher, eds. Handbook of biological control, pp. 955–974. Academic Press, San Diego.

Hudson, P. J., and A. P. Dobson. 1989. Population biology of Trichostrongylus tenuis, a parasite of economic importance for red grouse management. Parasitology Today 5:283–291.

Hudson, P. J., and J. Greenman. 1998. Competition mediated by parasites: biological and theoretical progress. Trends in Ecology and Evolution 13:387–390.

Hudson, P. J., A. P. Dobson, and D. Newborn. 1985. Cyclic and non-cyclic populations of red grouse: a role for parasitism? In D. Rollinson, and A. M. Anderson, eds. Ecology and genetics of host-parasite interactions, pp. 77–89. Academic Press, London.

Hudson, P. J., A. P. Dobson, and D. Newborn. 1992. Do parasites make prey vulnerable to predation? Red grouse and parasites. Journal of Animal Ecology 61:681–692.

Hudson, P. J., A. P. Dobson, and D. Newborn. 1998. Prevention of population cycles by parasite removal. Science 282:2256–2258.

IUCN. 2002. Red List of Threatened Species. IUCN Species Survival Commission. IUCN, Gland, Switzerland.

James, H. F., and S. L. Olson. 1991. Descriptions of thirty-two new species of birds from the Hawai'ian Islands. Part II: Passeriformes. Ornithological Monographs Volume 46.

Jarvi, S. I., C. T. Atkinson, and R. C. Fleischer. 2001. Immunogenetics and resistance to avian malaria in Hawaiian honeycreepers. Studies in Avian Biology 22:254–263.

Jenkins, D., A. Watson, and G. R. Miller. 1967. Population fluctuations in the red grouse Lagopus lagopus scoticus. Journal of Animal Ecology 36:97–122.

Jensen, T., M. van de Bildt, H. H. Dietz, H. Anderson, A. S. Hammer, T. Kuiken, and A. Osterhaus. 2002. Another phocine distemper outbreak in Europe. Science 297:209.

Kennedy, C. R. 1993. Introductions spread and colonization of new localities by fish helminth and crustacean parasites in the British Isles: a perspective and appraisal. Journal of Fish Biology 43.

Kennedy, C. R., A. O. Bush, and J. M. Aho. 1986. Patterns in helminth communities: why are birds and fish different? Parasitology 93:205–215.

Kennedy, S., T. Kuiken, P. D. Jepson, R. Deaville, M. Forsyth, T. Barrett, M. W. G. van de Bildt, A. D. M. E. Osterhaus, T. Eybatov, C. Duck, A. Kydyrmanov, I. Mitrofanov, and S. Wilson. 2000. Mass die-off of Caspian seals caused by canine distemper virus. Emerging Infectious Diseases 6:637–639.

Kermack, W. O., and A. G. McKendrick. 1927. Contributions to the mathematical theory of epidemics. Proceedings of the Royal Society of London B 115:700–721.

Keymer, A. E. 1981. Population dynamics of Hymenolepis diminuta: the influence of infective-stage density and spatial distribution. Parasitology 79:195–207.

Klironomos, J. N. 2002. Feedback with soil biota contributes to plant rarity and invasiveness in communities. Nature 417:67–70.

Kuris, A. M. 2003. Did biological control cause extinction of the coconut moth, Levuana iridescens, in Fiji? Biological Invasions 5:133–141.

Kuris, A. K., and K. D. Lafferty. 1992. Modelling crustacean fisheries: effects of parasites on management strategies. Canadian Journal of Fisheries and Aquatic Sciences 49:327–336.

Lafferty, K. D. 2002. Interspecific interactions in trematode communities. In E. E. Lewis, M. V. K. Sukhdeo, and J. F. Campbell, eds. The Behavioral Ecology of Parasites, pp. 153–169. CAB International, Oxford.

Lafferty, K. D., and L. Gerber. 2002. Good medicine for conservation biology: the intersection of epidemiology and conservation theory. Conservation Biology 16:593–604.

Lafferty, K. D., and A. M. Kuris. 2002. Trophic strategies, animal diversity, and body size. Trends in Ecology and Evolution 17:507–513.

Lampo, M., and P. Bayliss. 1996a. Density estimates of cane toads from native populations based on mark-recapture data. Wildlife Research 23:305–315.

Lampo, M., and P. Bayliss. 1996b. The impact of ticks on *Bufo marinus* from native habitats. Parasitology 113:199–206.

Lampo, M., and G. A. DeLeo. 1998. The invasion ecology of the toad *Bufo marinus* from South America to Australia. Ecological Applications 8:388–396.

Lanciani, C. A. 1982. Parasite-mediated reductions in the survival and reproduction of the backswimmer *Buenoa scimitra* (Hemiptera: Notonectidae). Parasitology 85:593–603.

Lie, K. J., and C. K. Ow-Yang. 1973. A field trial to control *Trichobilharzia brevis* by dispersing eggs of *Echinostoma audyi*. Southeast Asian Journal of Tropical Medicine and Public Health 4:208–217.

Lonergan, M., and J. Harwood. 2003. The potential effects of repeated outbreaks of phocine distemper among harbor seals: a response to Harding et al. (2002). Ecology Letters 6:889–893.

MacKinnon, K. 1978. Competition between red and grey squirrels. Mammal Review 8:185–190.

May, R. 2000. Simple rules with complex dynamics. Science 287:601–602.

Middleton, A. D. 1930. Ecology of the American gray squirrel in the British Isles. Proceedings of the Zoological Society of London 2:809–843.

Mitchell, C. E., and A. G. Power. 2003. Release of invasive plants from fungal and viral pathogens. Nature 421:625–627.

Murdoch, W. W., J. Chesson, and P. L. Chesson. 1985. Biological control in theory and practice. American Naturalist 125:344–366.

Murdoch, W. W., R. M. Nisbet, S. P. Blythe, and W. S. C. Gurney. 1987. An invulnerable age class and stability in delay-differential parasitoid-host models. American Naturalist 129:263–282.

Murdoch, W. W., C. J. Briggs, and R. M. Nisbet. 2003. *Consumer-resource dynamics*. Princeton University Press. Princeton, NJ.

Nassi, H., J. P. Pointier, and Y. J. Golvan. 1979. Bilan d'un essai de contrôle de *Biomphalaria glabrata* en Guadalupe à l' aide d'un Trématode stérilisant. Annales de Parasitologie 52:277–323.

Ohashi, K., N. Miyazaki, S. Tanabe, H. Nakta, R. Miura, K. Fujita, C. Wakasa, M. Uema, M. Shiotani, E. Takahashi, C. Kai. 2001. Seroepidemiological survey of distemper virus infection in the Caspian Sea and in Lake Baikal. Veterinary Microbiology 82:203–210.

Olson, S. L., and H. F. James. 1991. Descriptions of thirty-two new species of birds from the Hawai'ian islands. Part I: Non-passeriformes. Ornithological Monographs Volume 45.

Orr, M. R., S. H. Seike, W. W. Benson, and L. E. Gilbert. 1995. Flies suppress fire ants. Nature 373:292–293.

Osterhaus, A. D. M. E., and E. J. Vedder. 1988. Identification of virus causing recent seal deaths. Nature 335:20.

O'Teangane, D., S. Reilly, W. I. Montgomery, and J. Rochford. 2000. Distribution and status of the red squirrel and grey squirrel in Ireland. Mammal Review 30:45–56.

Packer, C., S. Altizer, M. Appel, E. Brown, J. Martenson, S. J. O. O'Brien, M. Roelke-Parker et al. 1999. Viruses of the Serengeti: patterns of infection and mortality in African lions. Journal of Animal Ecology 68:1161–1178.

Park, T. 1948. Experimental studies of interspecies competition 1. Competition between populations of the flour beetles, *Tribolium confusum* Duval and *Tribolium castaneum* Herbst. Ecological Monographs 18:267–307.

Pearson, D. E., and R. M. Callaway. 2003. Indirect effects of host-specific biological control agents. Trends in Ecology and Evolution 18:456–461.

Plowright, W. 1982. The effects of rinderpest and rinderpest control on wildlife in Africa. Symposia of the Zoological Society of London 50:1–28.

Potts, G. R. 1986. *The partridge: pesticides, predation and conservation*. Collins, London.

Potts, G. R., S. C. Tapper, and P. J. Hudson. 1984. Population fluctuations in red grouse: analysis of bag records and a simulation model. Journal of Animal Ecology 53:21–36.

Price, P. W. 1980. *Evolutionary biology of parasites*. Princeton University Press, Princeton, NJ.

Prins, H. H. T., and F. J. Weyerhaeuser. 1987. Epidemics in populations of wild ruminants: anthrax and impala, rinderpest and buffalo in Lake Manyara National Park, Tanzania. Oikos 49:28–38.

Reinhart, K. O., A. Packer, W. H. van der Putten, and K. Clay. 2003. Plant-soil biota interactions and spatial distribution of black cherry in its native and invasive ranges. Ecology Letters 6:1046–1050.

Roelke-Parker, M. E., L. Munson, C. Packer, R. Kock, S. Cleaveland, M. Carpenter, O'Brien, S. J., A. Pospischill, R. Hofmann-Lehmann, H. Luts, G. L. M. Mwamengele, B. A. Summers, and M. J. G. Appel. 1996. A canine distemper virus epidemic in Serengeti lions (*Panthera leo*). Nature 379:441–445.

Ross, R. 1916. An application of the theory of probabilities to the study of a priori pathometry. Part I. Proceedings of the Royal Society of London A 92:212–240.

Sands, D. P. A. 1997. The "safety" of biological control agents: assessing their impact on beneficial and other non-target hosts. Memoirs of the Museum of Victoria 56:611–616.

Sax, D. F., S. D. Gaines, and J. H. Brown. 2002. Species invasions exceed extinctions on islands worldwide: a comparative study of plants and birds. American Naturalist 160:766–783.

Scott, M. E. 1987. Regulation of mouse colony abundance by *Heligmosomoides polygyrus* (Nematoda). Parasitology 95:111–129.

Scott, M. E., and A. M. Anderson. 1984. The population dynamics of *Gyrodactylus bullatarudis* (Mongenea) on guppies (*Poecilia reticulata*). Parasitology 89:159–194.

Scott, M. E., and A. P. Dobson. 1989. The role of parasites in regulating host abundance. Parasitology Today 5:176–183.

Settle , W. H., and L. T. Wilson. 1990. Invasion by the variegated leafhopper and biotic interactions: parasitism, competition, and apparent competition. Ecology 71:1461–1470.

Shea, K., P. H. Thrall, and J. J. Burdon. 2000. An integrated approach to management in epidemiology and pest control. Ecology Letters 3:150–158.

Sinclair, A. R. E. 1979. The eruption of the ruminants. In A. R. E. Sinclair, and M. Norton-Griffiths, eds. *Serengeti: Dynamics of an ecosystem*, pp. 82–103. University of Chicago Press, Chicago.

Spinage, C. A. 2003. *Cattle plague: a history*. Kluwer/Plenum, New York.

Stiven, A. E. 1964. Experimental studies on the host parasite system hydra and *Hydramoeba hydroxena* (Entz.). II. The components of a single epidemic. Ecological Monographs 34:119–142.

Thomas, M. B., P. Casula, and A. Wilby. 2004. Biological control and indirect effects. Trends in Ecology and Evolution 19:61.

Toft, C. A. 1986. Communities of parasites with parasitic life-styles. In J. M. Diamond, and T. J. Case, eds. *Community ecology*. pp. 445–463. Harper and Row, New York.

Tompkins, D. M., J. V. Greenman, P. A. Robertson, and P. J. Hudson. 2000. The role of shared parasites in the exclusion of wildlife hosts: *Heterakis gallinarum* in the ring-necked pheasant and the grey partridge. Journal of Animal Ecology 69:829–840.

Tompkins, D. M., A. P. Dobson, P. Arneberg, M. E. Begon, I. M. Cattadori, J. V. Greenman, J. A. P. Heesterbeek et al. 2001. Parasites and host population dynamics. In P. J. Hudson, A. Rizzoli, B. T. Grenfell, H. Heesterbeek, and A. Dobson, eds. *The ecology of wildlife diseases*. Oxford University Press, Oxford.

Tompkins, D. M., A. R. White, and M. Boots. 2003. Ecological replacement of native red squirrels by invasive greys driven by disease. Ecology Letters 6:189–196.

Torchin, M. E., and A. J. Mitchell. 2004. Parasites, pathogens, and invasions by plants and animals. Frontiers in Ecology and the Environment 2:183–190.

Torchin, M. E., K. D. Lafferty, and A. M. Kuris. 2001. Release from parasites as natural enemies: increased performance of a globally introduced marine crab. Biological Invasions 3:333–345.

Torchin, M. E., K. D. Lafferty, A. P. Dobson, V. J. McKenzie, and A. M. Kuris. 2003. Introduced species and their missing parasites. Nature 421:628–630.

Tothill, J. D., T. H. C. Taylor, and R. W. Paine. 1930. *The Coconut Moth in Fiji: a History of its Control by Means of Parasites*. Imperial Bureau of Entomology, London.

VanderWerf, E. 2001. Distribution and potential impacts of avian poxlike lesions in 'Elepaio at Hakalau Forest National Wildlife Refuge. Studies in Avian Biology: Evolution, Ecology, Conservation, and Management of Hawaiian Birds: a Vanishing Avifauna 22:247–253.

van Riper, C., and M. J. Scott. 2001. Limiting factors affecting Hawaiian native birds. In M. J. Scott, S. Conant, and C. van Riper, eds. *Studies in Avian Biology: Evolution, Ecology, Conservation, and Management of Hawaiian Birds: a Vanishing Avifauna*, pp. 221–233. Cooper Ornithological Society, Camarillo.

van Riper, C., S. G. van Riper, M. L. Goff, and M. Laird. 1986. The epizootiology and ecological significance of malaria in Hawaiian land birds. Ecological Monographs 56:327–344.

van Riper, C., S. G. van Riper, and W. R. Hansen. 2002. Epizootiology and effect of avian pox on Hawaiian forest birds. The Auk 119:929–942.

Warner, R. E. 1969. The role of introduced diseases in the extinction of the endemic Hawaiian avifauna. Condor 70:101–120.

PART II

Insights into

EVOLUTION

Mark Vellend, A. Randall Hughes, Richard K. Grosberg, and Robert D. Holt

The most dramatic examples of evolutionary change often occur when a lineage is faced with a novel set of biotic or abiotic conditions. The driving force may be local or widespread environmental change, such as when pollution altered the selection regime on melanism in peppered moths in England (Kettlewell 1955), or it may be a species' arrival in a novel geographic locale, such as the adaptive radiation of Darwin's finches following arrival on the Galápagos islands (Grant and Grant 2002). Our ability to study and understand the evolutionary consequences of species introductions in new regions is often limited because we must use indirect evidence to reconstruct the sequence of events connecting an initial introduction to the current distribution of phenotypes and genotypes in space. Thus, in order to evaluate evolutionary processes directly, we often resort to laboratory experiments in which conditions are greatly simplified. This simplification can create doubt as to the general applicability of the results to natural settings.

Ecological geneticists occupy a middle ground, measuring selection in natural populations in contemporary time (Endler 1986). Exotic species provide exemplary fodder for such studies. As pointed out throughout this book, exotic species introductions present an unprecedented number of natural experiments replicated across space (the same species is often introduced into multiple places) and across taxa (multiple species are often introduced into the same place). What can we learn about the evolutionary process from studying exotic species?

It is always humbling to revisit classic texts in evolutionary biology. The present volume finds several remarkable parallels in the 1965 symposium volume *The Genetics of Colonizing Species*, edited by Herbert Baker and G. Ledyard Stebbins. As described in the Preface to that volume, the symposium "had as its object the bringing together of geneticists, ecologists, taxonomists, and scientists working in some of the more applied phases of ecology … to present facts and exchange ideas about the kinds of evolutionary change which take place when organisms are introduced into new territories." Forty years later, we have assumed much the same objective. Many of the themes raised in the Baker and

Stebbins volume are echoed in the chapters of this and other sections of the book: changing evolutionary pressures during different stages of invasion (Huey et al., Chapter 6; Ricklefs, Chapter 7); genetic bottlenecks and the maintenance of genetic variation (Novak and Mack, Chapter 8; Wares et al., Chapter 9); the development of local adaptation and spatial genetic structure (Holt et al., Chapter 10; Huey et al.; Wares et al.); the elucidation of traits important for successful invasion via cross-species comparisons (Rice and Sax, Chapter 11); the role of hybridization and introgression in adaptive evolution (Rice and Sax; Wares et al.); and the evolutionary consequences of novel biotic interactions (Callaway et al., Chapter 13).

Despite foreshadowing many of the themes of the present volume, the participants in Baker and Stebbins' symposium largely had their hands tied by a lack of direct evidence. As Ernst Mayr put it in the volume's summary, "When it comes to actual observations of genetic changes in colonizing species, the number of clear-cut cases is disappointingly small." The past 40 years have seen tremendous progress in elucidating evolutionary changes following the introduction of exotic species, in many instances providing case studies destined to be textbook examples of fundamental evolutionary processes (Cox 2004). The following chapters summarize much of this progress and point to a range of new directions.

The Baker and Stebbins symposium occurred on the eve of the "molecular revolution," sparked in the late 1960s by the widespread application of protein electrophoresis to the study of natural populations (Hubby and Lewontin 1966). Molecular markers have thoroughly infiltrated a wide range of subdisciplines in evolution and ecology (Avise 2004), and their application to most of the evolutionary phenomena discussed here provides perhaps the most striking contrast with

the contents of Baker and Stebbins' volume. Molecular markers have allowed enormous advances to be made in extracting evolutionary lessons from exotic species invasions. For example, John Harper suggested in 1965 that knowing the source of the propagules that start an exotic invasion was the key to unraveling subsequent evolutionary trajectories, but admitted that possession of such knowledge was "rarely the case." Molecular studies have helped pinpoint the source (or, just as often, the sources) of exotic introductions, knowledge that has indeed proven vital to studies of evolutionary change. In addition, hybridization and introgression, which frequently accompany exotic invasions, have been much more thoroughly characterized using molecular markers, providing a springboard for examining the evolution of particular traits as they enter novel genetic backgrounds. Putatively neutral molecular markers allow strong tests for genetic bottlenecks following invasion, and also provide a critical basis of comparison when testing for adaptive evolution in traits of presumed ecological significance. Finally, DNA sequence data add a temporal element when used to reconstruct phylogenies, allowing the study of the timing of particular colonization events. Advances on these fronts feature prominently in the following chapters, building on the foundation laid by the Baker and Stebbins volume.

This book also highlights research whose conceptual basis has emerged more recently. While there has long been interest in the manifold consequences of exotic species arriving in new territories, where they are free of many natural enemies, the coevolution of competitors *within* trophic levels has developed only recently as a topic of potential importance in invasion biology. Callaway et al. review compelling evidence that allelopathic chemicals in some exotic plants make a major contribution to their invasive success, providing novel insights into the importance of competitor coevolution in terms of evolved tolerance to allelochemicals. Huey et al. also present some compelling case studies of exotic species revealing the simultaneous predictability of evolution in some respects (e.g., comparable latitudinal clines have arisen repeatedly for independent introductions into different biogeographic regions) and its unpredictability in others (e.g., the developmental basis for a given phenotype may vary among regions).

Interestingly, a number of issues raised in 1965 remain major unsolved puzzles today, representing important future directions in evolutionary studies of invasive species. First, if populations of exotic species in their introduced ranges are genetically isolated from founder populations, can we expect a spike in the global rate of speciation? Polyploidy and hybridization in plants has provided some striking examples of novel species creation resulting in part from exotic introduction (Cox 2004), but in how many other cases is speciation underway? Second, invasive species often remain quite rare for a long time following their introduction before experiencing explosive population growth and expansion. How often is this "lag phase" a result of the time it takes for adaptive evolution to produce invasive phenotypes? Given that adaptation via natural selection is involved, what is the relative importance of adaptation to abiotic versus biotic factors in the environment? Finally, perhaps the greatest limitation to research on the evolution of exotic species is that we can study only those species that have successfully established. How often is the *failure* of exotic species to become invasive due to a lack of appropriate genetic variation and therefore evolutionary potential? At the moment, we simply don't know. Tremendous progress has been made in the last 40 years, but in many ways we have only scratched the surface of the potential for exotic species to provide fundamental lessons in evolutionary biology.

Literature Cited

Avise, J. C. 2004. *Molecular markers, natural history, and evolution*, 2nd Ed. Sinauer Associates, Sunderland, MA.

Baker, H. G. and G. L. Stebbins, eds. 1965. *The genetics of colonizing species*. Academic Press, New York.

Cox, G. W. 2004. *Alien species and evolution*. Island Press, Washington DC.

Endler, J. A. 1986. *Natural selection in the wild*. Princeton University Press, Princeton, NJ.

Grant, P. R. and B. R. Grant. 2002. Unpredictable evolution in a 30-year study of Darwin's Finches. Science 296:707–711

Hubby, J. L., and R. C. Lewontin. 1966. A molecular approach to the study of genic heterozygosity in natural populations. I. the number of alleles at different loci in *Drosophila pseudoobscura*. Genetics 54:577–594.

Kettlewell, H. B. D. 1955. Selection experiments on industrial melanism in the Lepidoptera. Heredity 10:287–301.

6

Using Invasive Species to Study Evolution

CASE STUDIES WITH *DROSOPHILA* AND SALMON

Raymond B. Huey, George W. Gilchrist, and Andrew P. Hendry

As invasive species spread through a new environment, they encounter novel selection pressures and challenges. Invasives thus offer rich opportunities to monitor the rate and predictability of evolution in the wild. Moreover, their evolutionary responses can alter their rate of spread as well as their interactions with native species; thus understanding whether invasive species evolve quickly or not is directly relevant to evolutionary biologists, ecologists, and conservation biologists. Here we review empirical studies of invasive species of fruit flies (Drosophila, focusing primarily on D. subobscura) and salmon (mainly Oncorhynchus). Both taxa have been introduced multiple times (the former by accident, the latter intentionally), offering replicated "experiments" of evolution in action. D. subobscura is an Old World fly that was introduced into both South and North America in the late 1970s. Studies pioneered by Catalonian and Chilean scientists document not only that some traits (e.g., chromosome inversions, wing size) evolved with extraordinary rapidity, but also that some (though not all) evolved predictably. Studies of introduced salmon have shown that most introductions fail; but when they succeed, many life history and morphological traits evolve quickly and (often) predictably. These and related studies show that invasive species can evolve remarkably quickly; therefore, evolutionary processes probably affect ecological ones. Future studies need to address how adaptive changes alter the spread of biological invaders and their interactions with native species, as well as how native species evolve in response to invaders.

Introduction

… for in all countries, the natives have been so far conquered by naturalised productions, that they have allowed foreigners to take firm possession of the land. And as foreigners have thus everywhere beaten some of the natives, we may safely conclude that the natives might have been modified with advantage, so as to have better resisted such intruders. (Darwin 1859, p. 83)

Invasive species pose multiple threats to native species (Ruesink et al. 1995; Vitousek et al. 1996). An invader can add to the competitive load pressuring native species (Callaway and Ridenour 2004), disrupt the physical structure of natural habitats (Singer et al. 1984; Pollock et al. 1995), and sometimes introduce parasites or diseases that decimate natives (Daszak et al. 2000). Not surprisingly, invaders sometimes overwhelm native species, often with disastrous ecological and economic consequences (Wilcove et al. 1998). Because invasive species are growing in number around the globe, they are increasingly a major ecological concern.

Ecologists and conservationists have long drawn attention to the negative effects of introduced species (e.g., Elton 1958), but they have tended to focus on the ecological effects of invaders on the community dynamics of native species. Thus they have generally considered evolution to be largely irrelevant to the dynamics and consequences of invasion (see Thompson 1998). As a first approximation, this assumption is quite reasonable. To be sure, the initial damage caused by invasive species is often so rapid as to preclude any significant role for evolution. Moreover, invasive species might have a limited potential to evolve in general, simply because they *may* experience severe genetic bottlenecks during their introduction (e.g., Franklin and Frankham 1998; but see Goodnight 2000; Novak and Mack, this volume; Wares et al., this volume). Most importantly, however, evolution has until recently (Hendry and Kinnison 1999) generally been considered too slow to play a role in the dynamics of invasive species.

Despite such considerations, we argue here that invasive species are not just an ecological problem, but also an evolutionary one. As evidence, we review selected examples of invasive species that are evolving with extraordinary rapidity in their new environments. Moreover, parallel examples are accumulating (Diniz-Filho et al. 1999; Mooney and Cleland 2001; Palumbi 2001; Blair and Wolfe 2004; see also Drummond et al. 2003). Such rapid evolution by invaders is not just of academic interest, for it may have serious ecological and conservation consequences for native species (Stockwell et al. 2003). Consider a novel invader that harms native species. As the invader begins to adapt to the local environment, its effect on native species is likely to be exacerbated (Mooney and Cleland 2001; Lee et al. 2003). Furthermore, an invader's spread can be accelerated by adaptation to its new environment (García-Ramos and Rodríguez 2002; Holt et al., this volume). Consequently, even though ecological interactions will dominate the initial effects of invaders on native species, local adaptation will eventually modify those interactions. Of course, native species could evolve in response to invaders (see the Darwin quote at the beginning of this chapter), so the long-term dynamics could be complex.

Invasive species cause many problems, but they nonetheless offer superb research opportunities to evolutionary biologists (Baker and Stebbins 1965). Joseph Grinnell (1919) was probably the first to suggest using invasive species to observe the dynamics of adaptation to new environments. Johnston and Selander (1964) implemented Grinnell's suggestion in their classic studies on the evolution of house sparrows (*Passer domesticus*) introduced to North America. Recently, the use of invasive species to study evolution has accelerated dramatically (Hendry and Berg 1999; Kinnison et al. 2001; Reznick and Travis 2001; Lee 2002; Parker et al. 2003; Blair and Wolfe 2004). Furthermore, the relevance of the evolution of invasive species to conservation issues is increasingly appreciated (Vermeij 1996; Mooney and Cleland 2001; Allendorf and Lundquist 2003; Stockwell et al. 2003).

Here we extend the Grinnellian perspective by turning an ecological problem into an evolutionary opportunity. Specifically, we show how introduced species can be used to address several issues of direct relevance to evolutionary biologists. We conclude by arguing that evolution feeds back on the ecology of interactions between native and invasive species and thus presents research opportunities to ecologists as well.

Invasive species can be used to address several general problems in evolution. We focus here on two:

1. How fast does evolution occur *in nature* (Darwin 1859; Simpson 1944; Hendry and Kinnison 1999)? In simple laboratory experiments, evolution is often extremely rapid (Rose et al. 1987; Lenski et al. 1991; Partridge et al. 1995). To determine whether evolution can also be fast in nature, one can monitor invasive species—or the native species with which they interact—and quantify changes in ecologically relevant traits over time scales of a few years, decades, or centuries.

2. Are patterns of evolution predictable (Gould 1989)? Laboratory or field experiments evaluate the predictability of evolutionary trajectories by setting up and monitoring replicate populations that are subject to some common selective factor (Reznick et al. 1990b; Travisano et al. 1995; Losos et al. 1998). Similarly, one can monitor the independent evolutionary responses of "replicate" introduced populations. If evolution is predictable, then the evolutionary responses of these replicate populations to specific environmental gradients (e.g., climate) should converge on those seen among ancestral populations. On the other hand, if evolution is unpredictable or highly sensitive to local conditions, then the responses of the replicate introduced populations may diverge from one another and from those of their ancestors.

To exemplify these issues, we focus on empirical studies of flies (*Drosophila*) and salmon (primarily *Oncorhynchus*). Both taxa have been repeatedly introduced around the world—the former by accident, the latter usually by intention. The repeated introductions of each taxon conveniently serve as replicated natural experiments (Ayala et al. 1989) that allow us to determine the rate, pattern, and predictability of evolutionary change in different venues and with biologically different players. Consequently, they offer an opportunity to explore the evolutionary roles of adaptation, chance, and history (Travisano et al. 1995).

Drosophila are of interest in part because numerous laboratory studies have consistently shown them capable of evolving very rapidly under controlled conditions (Harshman and Hoffmann 2000; but see Hoffmann et al. 2003). Although *Drosophila* clearly have the genetic potential to evolve quickly, laboratory studies do not imply that flies will necessarily evolve as quickly in nature (Huey et al. 1991). After all, flies in nature face uncontrolled temporal variation in environments as well as selection pressures (e.g., predators, parasites, competitors) that are traditionally excluded from the benign and constant environments of laboratory cages. Moreover, gene flow among natural populations may constrain local adaptation (Lenormand 2002), and behavioral adjustments may buffer selection on physiological and morphological traits (Huey et al. 2003). In any case, established geographic patterns for natural populations provide an evolutionary baseline (Figure 6.1A) that can be used to predict the evolutionary trajectories of introduced species.

Salmon are of interest primarily because their strong philopatry results in thousands of isolated and locally adapted populations. By studying associations between traits and environments in native salmon populations (Figure 6.1B,C), biologists can predict how salmon will evolve when introduced into new locations. Introduced salmon will face many of the same challenges as introduced *Drosophila*, including uncontrolled temporal variation, multiple selective pressures, and ongoing gene flow. Nonetheless, the two taxa differ in at least one critical way: salmon have much longer generation times (2 to 7 years) than do *Drosophila* (a few weeks in warm seasons). As a result, salmon probably face greater challenges in adapting to local environments and interacting with native species. An additional advantage of studying salmon is that precise records of sources, numbers, and times of introductions are often available.

Terminology Issues

The terms "introduced," "invasive," and "colonizing" are often used somewhat interchangeably in the literature. Here we use "introduced" to imply an intentional introduction by humans (as with salmon), but otherwise we use "invasive." We do not use "colonizing" here, as this term should apply only to natural range expansions.

We use the term "rapid evolution" to imply observable genetic changes over a short time frame (a few decades), but caution that actual rates for such "rapid"

(A)

(B)

(C)

Figure 6.1 Geographic patterns in native populations provide predictive scenarios for how introduced species will evolve in new environments. (A) An example of latitudinal variation in size among Old World *D. subobscura*. As in many *Drosophila* species, high-latitude females (right) are substantially larger than low-latitude females. These flies are from Valencia, Spain (39°) and Aarhus, Denmark (56°). (B, C) Examples of morphological variation among populations of sockeye salmon in their native range. (B) A typical mature male from a medium-sized creek in Alaska (Lynx Creek). (C) A typical mature male from a beach site in Iliamna Lake, Alaska. Note the substantially deeper body of the beach male.

evolution may vary by orders of magnitude. For this reason, "contemporary evolution" has been suggested as the general term, with "rapid evolution" being reserved only for truly exceptional rates (Hendry and Berg 1999; Kinnison and Hendry 2001; Stockwell et al. 2003).

Case Study 1: Evolution of *Drosophila subobscura* on a Continental Scale

Many species of *Drosophila* are invasive (Parsons 1983), with *Drosophila melanogaster* at the forefront. This species is thought to be native to Africa, but (thanks to inadvertent human help) has successfully invaded broad ranges of latitude on all nonpolar continents. Many studies have shown that these invasive populations have evolved conspicuous and repeatable clines with latitude in diverse traits (David and Bocquet 1975; Cohan and Graf 1985; Boussy 1987; Simmons et al. 1989; James et al. 1995; Gilchrist and Partridge 1999; Hoffmann and Harshman 1999; van't Land et al. 1999). Nevertheless, *D. melanogaster* is not ideal for addressing the rate and predictability of evolution. For one thing, the initial introductions probably took place hundreds of years ago, such that we can now observe only the outcome of many generations of accumulated evolution. Moreover, repeated introductions to each continent have almost certainly taken place, contaminating and confounding evolutionary trajectories.

An alternative system for studying the rate and predictability of evolution involves *Drosophila subobscura*. The history of invasions of these flies has been well chronicled (Ayala et al. 1989), so we give only an outline here. The species is native to the Old World, where it is widespread from North Africa to Scandinavia and shows marked latitudinal clines in genetic markers and in many other traits (Krimbas 1993).

In the late 1970s, Chilean biologists discovered *D. subobscura* in the coastal city of Puerto Montt, Chile (Brncic and Budnik 1980). Very likely the flies had recently arrived by ship from the Mediterranean (Brncic et al. 1981; Ayala et al. 1989). The invaders spread rapidly; in less than a year they had colonized much of the habitable coast of Chile and had become the dominant *Drosophila* in many localities. Soon thereafter they crossed the Andes and are currently spreading in Argentina and Uruguay (Prevosti et al. 1983; Goñi et al. 1998).

In June 1982, Andrew Beckenbach discovered *D. subobscura* in central Oregon and on the Olympic Peninsula of Washington State. Intensive collecting soon thereafter revealed that the flies had already spread north into British Columbia and south to central California (Beckenbach and Prevosti 1986). As in Chile, mid- to high-latitude populations were (and still are) very abundant, and native *obscura*-group flies simultaneously became hard to collect (A. Beckenbach, personal communication). Abundant genetic evidence establishes that the North and South American populations were founded from a single stock (Ayala et al. 1989; Mestres and Serra 1995; Pascual et al. 2001).

Evolutionary geneticists immediately seized this opportunity (Brncic et al. 1981; Ayala et al. 1989), which they recognized as a "grand experiment in evolution" (Ayala et al. 1989). They recognized that the Old World populations, which probably have been evolving in place since the last glaciation, serve as an evolutionary baseline that establishes the long-term patterns of clinal evolution with latitude (see Figure 6.1A). Moreover, they recognized that the North and South American populations serve as independent evolutionary replicates. Finally, because the invasions probably took place in the late 1970s, they realized that the time frame for evolution in the New World was very short—just a few decades. Thus the invasions of *D. subobscura* offered special opportunities for monitoring rates and patterns of clinal evolution, essentially in real time (Prevosti et al. 1988).

Drosophila subobscura is a continuing focus of diverse evolutionary studies. For two traits (chromosomal inversions and wing length), several different sets of latitudinal samples have been made over time, generating a rare "time series" of evolution in action. Here we review the key points emerging from these studies, then discuss their implications for the themes of this chapter.

The bottleneck

Whether *D. subobscura* would show rapid evolution in the New World was initially debatable, simply because the genetic diversity of New World populations is greatly reduced in comparison with native European populations (Balanyà et al. 1994, 2003; Rozas et al. 1990; Mestres et al. 1992, 2004; Pascual et al. 2001). In fact, recent studies suggest that fewer than 15 individuals (probably from the Mediterranean: see Mestres et al. 2004) founded the New World populations (Pascual et al. 2001; Mestres et al. 2004). As shown below, the North and South American flies have rapidly evolved clinal patterns despite these major bottlenecks (Prevosti et al. 1985; Ayala et al. 1989; Huey et al. 2000; Pascual et al. 2001; Gilchrist et al. 2004). Nevertheless, those bottlenecks may be serving as a brake on the evolution of some traits (Balanyà et al. 2003) and may have influenced the evolutionary particulars of others (Gilchrist et al. 2001b).

Chromosomal inversions

The chromosomal arrangements that crossed to the New World are generally among the most common ones in Europe, occurring in 79% to 95% of all Old World samples (Balanyà et al. 2003). One interesting exception is the O_5 inversion, which is rare in Europe (~0.5%) but more common (6% to 7%) in the New World, and which is linked with a lethal allele in the colonizing, but not the ancestral, populations (Mestres et al. 1995). All 18 of the chromosomal arrangements that came over from the Old World were present in the first samples collected in South America in 1981 (Prevosti et al. 1985) and also in the first samples from North America collected in 1982–1983 (Beckenbach and Prevosti 1986). Furthermore, no new European chromosomal arrangements have appeared in subsequent New World collections (Balanyà et al. 2003). These facts, combined with the reduced diversity of microsatellites (Pascual et al. 2001) and DNA sequences (F. Mestres and M. Pascual, personal communication) observed in the New World, suggest that no further introductions are likely to have occurred since the initial colonization.

Most of the common (i.e., overall frequency greater than 0.01) chromosomal arrangements show significant latitudinal clines in the Old World (Menozzi and Krimbas 1992). Two alternative hypotheses have been suggested to explain these clines. First, latitudinal gradients in climate may favor different inversions at different latitudes (Prevosti et al. 1988). Second, the clines could reflect the chance colonization of northern Europe following the last glaciation by flies carrying certain inversions (Krimbas and Loukas 1980). Thus the hypotheses invoke either selection or historical contingency (e.g., phylogenetic history, founder effects, and genetic drift).

The invasion and rapid spread of *D. subobscura* in the New World provided a dramatic opportunity to test these competing hypotheses (Prevosti et al. 1988). Only 3 years after the discovery of *D. subobscura* in Puerto Montt, evolutionary geneticists collected flies from seven Chilean sites spanning 12° of latitude. They discovered that latitudinal clines in inversion frequencies were beginning to evolve (Brncic et al. 1981; Prevosti et al. 1985). Remarkably, clines for 17 of the 18 chromosomal arrangements were in the same direction with latitude as those in Europe! Collections in North America just a few years later (eight sites spanning 13°, 1985–1986) revealed clines that were again in the same direction with latitude as in Europe for 14 of 18 arrangements. Such rapidly evolving and concordant patterns on all three continents (Prevosti et al. 1988; Menozzi and Krimbas 1992) unambiguously suggest that the clines are driven by potent natural selection (Endler 1986), not by historical contingency (Krimbas and Loukas 1980).

Although the latitudinal patterns of inversion frequency are similar *in sign* on all three continents, slopes in the New World (regressing inversion frequency on latitude for Chile in 1999 and for North America in 1994) are generally far less steep than those in Europe (Balanyà et al. 2003). Climatic differences among the three continents may provide an explanation. Old World populations span more than 30° of latitude, whereas New World populations span only about 15° on each continent, suggesting that selection might be acting over a larger range of climates in the Old World. Moreover, seasonal variation in temperature at a given latitude is greater in Europe than in either New World continent, particularly South America (Addo-Bediako et al. 2000; Gilchrist et al. 2004). Thus climate-based selection might be stronger in the Old World, resulting in steeper slopes there.

Because flies had evolved shallow inversion clines within only 3 years of arriving in Chile, one might expect that the steepness of those clines would continue to converge on that of Old World clines. Surprisingly, a recent analysis of data from Chile spanning nearly two decades revealed no continued directional evolution of the clines since the early 1980s (Balanyà et al. 2003), contrary to an earlier suggestion (Prevosti et al. 1990). Apparently, the evolution of inversion clines was initially explosive, but stalled soon thereafter (Balanyà et al. 2003). This finding argues against the alternative explanation that clines might still be shallow simply because of insufficient time for divergence.

What drives latitudinal clines in inversion frequencies? Several lines of evidence suggest that temperature may be important. First, frequencies of inversions on the O chromosome of Spanish *D. subobscura* fluctuate seasonally in ways consistent with expectations based on their latitudinal patterns (Rodríguez-Trelles et al. 1996). Second, long-term shifts in frequencies have been detected within sites in Spain (Orengo and Prevosti 1996; Rodríguez-Trelles and Rodríguez 1998; Solé et al. 2002). Specifically, "southern" inversions are increasing in frequency, suggesting a response to documented climatic warming at these sites. Surprisingly, however, expected shifts in inversion frequencies have not been detected in laboratory stocks of *D. subobscura* currently evolving at three different temperatures (Santos et al. 2004), even though temperature has been shown to influence frequencies in *D. pseudoobscura* (Dobzhansky 1943; Wright and Dobzhansky 1946; Dobzhansky 1948).

In conclusion, the observed chromosomal inversion patterns in the New World conclusively resolve debates over the importance of selection versus historical contingency in this system (Prevosti et al. 1988; Ayala et al. 1989; Menozzi and Krimbas 1992). Moreover, the New World data provide a clear testament to the efficacy, speed, and repeatability of natural selection. They also show that the pace of clinal evolutionary change can be quite episodic, even over the span of a few decades. However, the hunt for unambiguous selective factors promoting latitudinal clines in inversion frequencies in *D. subobscura* (and in other *Drosophila*) continues to the present (Santos et al. 2004).

Evolutionary changes in wing size

Wing size varies across latitudinal clines in most species of *Drosophila*: flies from low latitudes have wings that are genetically smaller than flies from high latitudes (*D. melanogaster*: David and Bocquet 1975; Coyne and Beecham 1987; James et al. 1995; van't Land et al. 1999; Gilchrist and Partridge 1999; *D. obscura*: Pegueroles et al. 1995; *D. pseudoobscura*: Sokoloff 1966; *D. robusta*: Stalker and Carson 1947; *D. simulans*: David and Bocquet 1975; *Zaprionus indianus*: Karan et al. 2000). Prominent latitudinal size clines are also present in the ancestral European population of *D. subobscura* (Prevosti 1955; Misra and Reeve 1964; Pfriem 1983). Studies with *D. melanogaster* (cited above) have shown that size clines evolve predictably, at least over the course of several centuries, and that large size may be adaptive at low temperatures (Reeve et al. 2000). The invasion of the Americas by *D. subobscura* provided an opportunity to discover whether these clinal patterns could also evolve over a few decades.

Pegueroles and colleagues (1995) made the first comparative analysis of clinal variation in morphology across latitudes in populations of the ancestral and invading flies. They reared flies collected between 1986 and 1988 from seven sites in South America, six in North America, and four in Europe. They found ample clinal variation in wing size and tarsus length among the ancestral European flies, but no significant clines in either North or South America. Furthermore, wing and tarsus size were uncorrelated with environmental variables such as temperature and altitude. Thus, less than a decade after the discovery of *D. subobscura* in Chile, clines in morphology had not formed in the invading populations, even though clines in inversion frequencies had been detected earlier (as described above).

Limited evidence links wing size and chromosomal clines. Pfriem (1983) found that wing length was correlated with the frequencies of two classes of gene arrangements on the O chromosome, and Santos and colleagues (2004) recently found associations between gene arrangements and wing shape in laboratory stocks evolving at different temperatures. Nevertheless, the rapid evolution of chromosome clines in the New World (Prevosti et al. 1988) without a corresponding shift in wing length (Pegueroles et al. 1995) suggests that wing length is not inevitably influenced by gene arrangements.

About two decades after the introduction, Huey and colleagues reexamined variation in wing size along the New World and Old World clines (Huey et al. 2000; Gilchrist et al. 2001b; Gilchrist et al. 2004). Our motivation was to deter-

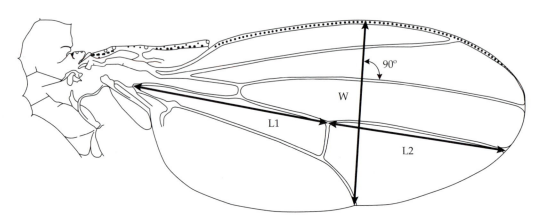

Figure 6.2   Wing of *Drosophila subobscura*, indicating the size dimensions used in our analyses.

mine whether size clines were finally detectable in the Americas. From 1997 to 1999, flies were collected from 10 sites in North America, 11 sites in Europe, and 10 sites in Chile. The sites spanned approximately 13° of latitude on each continent, were relatively near the west coast of each continent, and were below 525 m altitude. All flies were reared in a common garden in the laboratory for several generations before being measured. Three wing dimensions have been used in studies published to date: L1 is the length of the basal portion of the wing along vein IV, L2 is the length of the distal portion of the wing along vein IV, and W is the width of the wing, measured from the intersection of vein V and the trailing edge to the leading edge along a path perpendicular to vein III (Figure 6.2). These data were combined using principal components analysis. The first axis (PC1) explains about 70% of the variation in overall wing size. We also examined the component measurements independently to assess allometric change among the wing regions.

The repeatability of morphological evolution was assessed by comparing across continents the regression slopes of a given trait on latitude or on a local temperature index. Figure 6.3 shows the clinal pattern of PC1 and of the component wing traits as a function of latitude on all three continents. Slopes for females are statistically indistinguishable among the continents, although the latitudinal range in Chile is shifted approximately 6° toward the equator (probably reflecting the effect of the cold Humboldt Current: see Gilchrist et al. 2004). Thus, approximately two decades after colonization of the New World, natural selection has independently created similar patterns of geographic variation in both invading populations. But not everything is so elegantly repeated. North American males, for example, show less steep clines than do their European counterparts. The genetic correlation between male and female wing size has not been estimated in *D. subobscura*, but is essentially unity in *D. melanogaster* (Cowley et al. 1986; Reeve and Fairbairn 1996). If the genetic basis

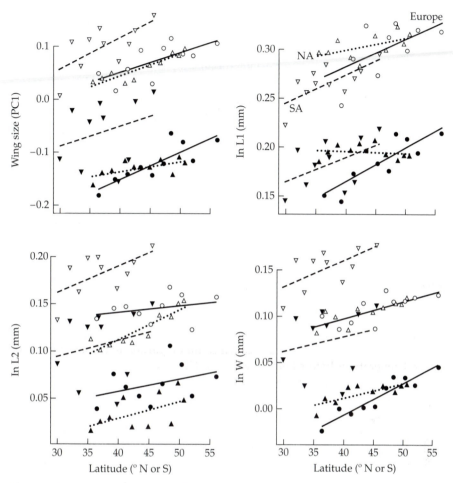

Figure 6.3   Wing size changes with latitude in Old World (circles), North American (tri-angles), and South American (inverted triangles) populations. Solid and open symbols represent females and males, respectively. PC1 refers to the first principal component of the three wing dimensions, L1, L2, and W (illustrated in Figure 6.2). (After Figure 2 in Gilchrist et al. 2004.)

of sexual dimorphism is similar in these two species, then the bottleneck in the founding population may have had a dramatic effect on the genetic architecture of sexual dimorphism. Additionally, males might face a different range of selection pressures in North America.

Morphological details of wing dimensions differ among continents (see Figure 6.3). The basal portion of the wing (L1) shows little clinal pattern for North American females or males, whereas it shows significant slopes in both the European ancestors and the South American invaders. South American females, but not males, tend to have shorter L1s than do European or North

American lines. In contrast, the length of the distal portion of the wing (L2) increases with latitude in both sexes in both New World populations, but shows little significant clinal pattern in Europe. Wing width (W) increases with latitude on all three continents, especially for females. South American flies of both sexes have longer and wider distal wings than their European and North American counterparts.

Concluding remarks on D. subobscura

Studies to date demonstrate that *D. subobscura* is undergoing extraordinarily rapid evolution on a continental scale. In less than 25 years, many traits have evolved—some strikingly so (see Figure 6.3). In fact, measured rates of evolution in this species are among the fastest ever documented in nature (Hendry and Kinnison 1999; Kinnison and Hendry 2001; Gilchrist et al. 2001b).

The observed patterns certainly highlight the predictability of evolutionary trajectories, but they simultaneously highlight some unpredictable aspects of evolution. Consider latitudinal clines in overall wing size (see Figure 6.3). Latitudinal patterns for females are predictable and always converge on the Old World pattern. However, those of males are much less predictable. Why patterns should differ between the sexes is unclear. Furthermore, the actual portion of the wing involved in size clines differs among the three continents (Gilchrist et al. 2004)! So even though the cline in total wing length is largely predictable, how that cline is achieved developmentally is decidedly not (Huey et al. 2000; Santos et al. 2004). Similarly, latitudinal shifts in inversion frequencies are predictable and usually converge on Old World patterns in sign, but not in magnitude (Balanyà et al. 2003).

We do not yet know whether these rapid changes will affect the competitive relations (Blossey and Nötzold 1995; Weber and Schmid 1998; Siemann and Rogers 2001) of *D. subobscura* with native species. To be sure, native *obscura*-group species, which were once abundant in the Pacific Northwest, are now hard to collect there. All in all, evolution is likely to have exacerbated the effect of this invader, but direct studies will be required to test this assumption. Surprisingly, *D. subobscura* fares poorly in competition with native species in the laboratory (Pascual et al. 1998; Pascual et al. 2000), in stark contrast to its apparent superiority in the field (Pascual et al. 1993). Nevertheless, field or outdoor-enclosure experiments remain to be done.

Case Study 2: Evolution in Introduced Salmon

At the outset of this chapter, we outlined two problems that can be addressed by studying introduced organisms: the rate and the predictability of evolution. Research on introduced salmonids (salmon, trout, char, and whitefish) readily informs both of these problems, but we concentrate here on the latter. Detailed reviews of salmonid evolutionary rates can be found elsewhere (e.g., Haugen and Vøllestad 2001; Hendry 2001; Quinn et al. 2001a; Koskinen et al. 2002; Kinnison and Hendry 2004). In brief, introduced salmonids evolve at rates typical of other introduced organisms (i.e., neither exceptionally fast nor exceptionally slow), which is itself surprising given the substantial ongoing gene flow among diverging populations.

Salmonids are well suited to a consideration of evolutionary predictability. First, they form a multitude of breeding populations that are reproductively isolated owing to strong philopatry (reviewed in Hendry et al. 2004a). Second, these populations typically adapt to their local environments (Taylor 1991; Quinn et al. 2001b). Third, groups of populations in different watersheds often have independent evolutionary origins, providing convenient replication of adaptive patterns (Wood 1995; Taylor et al. 1996; Waples et al. 2004). Fourth, salmonids have been introduced throughout the world and are now found on all major continents except Antarctica (Lever 1996). In several cases, natural dispersal after the initial introduction has generated multiple new populations, which now occupy environments that closely mirror those occupied by native (non-introduced) populations. These properties facilitate informed predictions as to how salmonids should evolve when introduced to new locations.

Research on introduced salmonids necessarily focuses on the successful introductions, but most introduction attempts have actually failed (Withler 1982; Wood 1995; Altukhov et al. 2000; Utter 2001). The record is particularly poor for anadromous salmon, which breed in fresh water but spend part of their lives in the ocean (Altukhov et al. 2000). In many cases, attempts to introduce anadromous salmonids have failed utterly despite massive and repeated efforts. Where such introductions have been successful, the new populations often forgo the anadromous life history, remaining in fresh water for their entire lives. The difficulty of establishing new populations, particularly anadromous ones, implies that introduced organisms (at least those with complex life cycles) often fail to adapt to novel environments. The successful introductions should therefore be viewed as exceptional, presumably succeeding either by chance or because environmental conditions were particularly favorable.

Several successful salmonid introductions deserve special mention. Most striking among these has been the establishment of several Pacific salmon species in the North American Great Lakes (e.g., pink salmon, *Oncorhynchus gorbuscha*: Gharrett and Thomason 1987). Interestingly, these introduced fish have adopted a quasi-anadromous life history in which the Great Lakes substitute for the ocean. Successful introductions where the true anadromous life history has been retained include sockeye salmon (*O. nerka*) in Frazer Lake, Alaska (Burger et al. 2000), and Atlantic salmon (*Salmo salar*) on the Kerguelen Islands in the southern Indian Ocean (Ayllon et al. 2004). Successful introductions are much more common for nonanadromous salmonids (Lever 1996), but these have rarely been used to examine the rate or predictability of evolution. One exemplary exception is work on European grayling (*Tymallus thymallus*) introduced to Norwegian lakes (Haugen and Vøllestad 2001; Koskinen et al. 2002). In the following sections, we describe research on two successful introductions of anadromous salmon: Lake Washington sockeye salmon (*O. nerka*) and New Zealand chinook salmon (*O. tshawytscha*).

Lake Washington sockeye salmon

In the 1930s and 1940s, more than 3 million juvenile sockeye salmon from Baker Lake (in northwestern Washington State) were introduced into Lake Washington (near Seattle, Washington) (Hendry et al. 1996; Hendry and Quinn 1997). These fish soon established new anadromous populations in several different ecological environments, with the most striking contrast being that between a large river (Cedar River) and a lake beach (Pleasure Point). Strong divergent selection is expected between these environments over the approximately 13 subsequent generations (the typical life cycle is 4 years for this species). First, breeding adults experience strong water flows in the river, but not at the beach. Second, embryos incubating in gravel nests experience floods that cause gravel "scour" in the river, but not at the beach. Third, incubating embryos experience variable and cold temperatures in the river, but constant and warm temperatures at the beach (Hendry and Quinn 1997; Hendry et al. 1998). These and other putative selective agents, coupled with observed phenotypic variation among native (non-introduced) populations found in other watersheds, allow robust predictions of how deterministic evolution should proceed within the Lake Washington watershed. Deviations from these predictions would suggest that contingency may have played a role. Unfortunately, the ancestral Baker Lake population has not bred in its natural environment for more than a century (Hendry 2001), which precludes direct comparisons between Lake Washington populations and their ancestral source.

Perhaps the most obvious deterministic prediction relates to male body depth, a secondary sexual trait that increases dramatically with maturation (Hendry and Berg 1999). Deep-bodied males are favored by sexual selection because they are dominant during male-male competition for breeding females (Quinn and Foote 1994). In fast-flowing rivers, however, a deep body is hydrodynamically inefficient (Kinnison et al. 2003; Crossin et al. 2004), particularly because males often orient perpendicular to the flow of water during breeding competition. Accordingly, males from native beach populations consistently have deeper bodies at a common length than do males from native river populations (Blair et al. 1993; Quinn et al. 2001b) (see Figure 6.1B and C and Figure 6.4).

Has the same pattern evolved within Lake Washington? Males in this watershed generally have shallower bodies than males in the native populations surveyed thus far, but Pleasure Point (beach) males nevertheless have deeper bodies than Cedar River males at a common body length (Figure 6.4; Hendry and Quinn 1997; Hendry et al. 2000). Moreover, the relative difference in standardized body depth between beach and river males in Lake Washington (13.0%–13.8%) is similar to the relative difference between beach and river males in native populations (Hendry et al. 2000). For example, the mean difference in standardized body depth between beach and river males in Iliamna

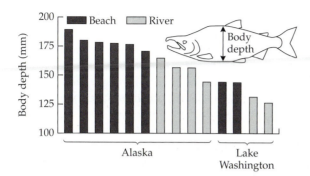

Figure 6.4 Average adult male body depths in beach and river populations from Alaska (native) and Lake Washington (introduced). (Data for Alaska from Quinn et al. 2001b; data for Lake Washington represent samples from two different years and are from Hendry and Quinn 1997 after standardization to the same body length [450 mm] used by Quinn et al. 2001b.)

Lake, Alaska, is 12.8% (Blair et al. 1993; Hendry 2001). Thus the evolution of male body depth seems to have been rapid and predictable, with a caveat that the relative contributions of genetic variation versus phenotypic plasticity are not yet known. Detailed considerations, however, suggest that adaptive divergence in male body depth within Lake Washington is at least partially genetically based (Hendry 2001), just as it is for populations of chinook salmon introduced to New Zealand (Kinnison et al. 2003).

Another deterministic prediction relates to female body length. Females oviposit into gravel nests, where their eggs then incubate for several months. High water flows during this period can mobilize the gravel and "scour" nests, causing high embryo mortality. Larger females can bury their eggs deeper in the gravel (Steen and Quinn 1999), which reduces the risk that their eggs will be lost to scour. Thus selection should favor larger females in high-scour environments. Within Lake Washington, the Cedar River is subject to strong gravel scour that causes high embryo mortality (Thorne and Ames 1987), whereas the Pleasure Point beach is largely devoid of scour. Matching this selective difference, Cedar River females are 5.5%–7.1% longer than Pleasure Point females (Hendry and Quinn 1997; Hendry et al. 2000; Hendry 2001). However, as with male body depth, a genetic basis for divergence in female length has not been unequivocally documented (Hendry 2001).

A third set of deterministic predictions relates to the effects of water temperature on the development of salmonid embryos incubating in the gravel. Conveniently, Pleasure Point beach and Cedar River embryos are exposed to very different temperature regimes. The former incubate in upwelling groundwater with a constant temperature of 10°C, whereas the latter incubate in temperatures that range from 8°C in mid-November to 4°C in the middle of winter and back up to 9°C by mid-April. Relying on observed patterns for native populations, we predicted that these temperature differences would cause local adaptation of survival, development rate, and developmental efficiency to river and beach sites within Lake Washington (see Hendry et al. 1998).

To test these predictions, we captured adults from both populations, used artificial fertilization to generate full-sib families, and then raised siblings from each family at 5°C, 9°C, and 12.5°C. Contradicting our prediction, river and beach embryos did not differ in survival or in development rates (Hendry et al. 1998). Matching our prediction, however, embryos from each population attained their largest size when incubating at the temperature closest to that which they would experience in nature (Hendry et al. 1998). This last result suggests the presence of genetic divergence in temperature-specific developmental properties that maximize embryo size, a critical trait in wild salmon (reviewed in Einum et al. 2004).

Did any aspects of the Lake Washington study system facilitate evolutionary divergence? Historical records are incomplete, but the original Baker Lake population seemingly contained both beach and river fish, with hatcheries then mixing both into a single panmictic group for about 10 generations preceding the Lake Washington introductions (Hendry 2001). This mixing of beach and river gene pools may have generated a highly variable group of introduced fish, thus facilitating evolutionary divergence following the colonization of beach and river environments in Lake Washington (for a detailed discussion, see Hendry 2001).

In summary, results for Lake Washington suggest that when evolutionary divergence occurs, it does so predictably. Each of the traits that *did* differ between the introduced populations did so in accordance with the expected role of divergent selection and with patterns previously documented in native populations. Moreover, the degree of differentiation between river and beach populations is similar for introduced and native populations. Thus divergent natural selection appears to generate predictable adaptive divergence in very short order—here, in fewer than 13 generations! Furthermore, the role of contingency seems limited to determining which traits evolve and which do not. As a caveat, however, trait divergence in Lake Washington has not been studied at the level of detail that was necessary to reveal a role for contingency in introduced *Drosophila* (Gilchrist et al. 2001a).

New Zealand chinook salmon

In 1901 and 1904–1907, juvenile chinook salmon from the Sacramento River, California, were introduced into the Hakataramea River on South Island, New Zealand (McDowall 1994; Quinn et al. 1996; Kinnison et al. 2002). Descendants of these fish then dispersed and established self-sustaining populations in other rivers on South Island. Environmental characteristics differ among these rivers, suggesting that selection might have promoted evolutionary diversification over the 26 subsequent generations (generation length for these populations is estimated at 3.2 years). Indeed, genetic differences among these populations have been confirmed for many phenotypic traits (reviewed in Quinn et al. 2001a). Here we focus on a single selective factor—migratory distance—that allows particularly clear interpretations.

Maturing salmon cease to feed when they enter fresh water. Their upstream migration, as well as all subsequent breeding activity, must then be fueled with stored energy (Hendry and Berg 1999). This strict energy budget leads to trade-offs among various aspects of reproductive investment and generates strong selection to maximize energy use efficiency. One axis of this trade-off is the amount of energy required for migration to the breeding grounds, which depends critically on the difficulty and length of the upstream migration. Accordingly, native populations that migrate longer distances typically (1) store more energy before entering fresh water, (2) use more energy during upstream migration, and (3) invest less energy in ovaries and have smaller eggs and smaller secondary sexual characteristics (e.g., Beacham and Murray 1993; Hendry and Berg 1999; Healey 2001; Kinnison et al. 2001; Crossin et al. 2004). Some of this variation probably reflects the proximate costs of migration (i.e., plasticity), but some of it probably also has a genetic basis (reviewed in Hendry et al. 2004b).

Research on chinook salmon introduced to New Zealand confirmed that upstream migrations are energetically costly. Juveniles were produced by artificial fertilization from two populations that differ in migration difficulty: Glenariffe (100 km and 430 m elevation) and Hakataramea (60 km and 200 m elevation). Representatives from each family were then released at two locations, one requiring a more difficult migration (Glenariffe) than the other (Silverstream, 17 km and 17 m elevation). The juveniles "imprinted" on the release sites, migrated to the ocean, and returned with strong fidelity to the release sites as adults. Sampling of the returning adults revealed that fish migrating the longer distance had substantially smaller ovaries, smaller eggs, smaller secondary sexual characteristics (hump size and snout length), and lower energy stores, but not fewer eggs, compared with their siblings migrating the shorter distance (Figure 6.5; Kinnison et al. 2001; Kinnison et al. 2003). Thus, migration imposed a substantial proximate cost, manifested as a phenotypically plastic reduction in the size of several traits.

Representatives from these same experimental families were also raised for their entire lives in a hatchery (i.e., a common garden), allowing a test for evolutionary divergence after 26 generations of potential adaptation to migration difficulty (Hakataramea vs. Glenariffe). In this common hatchery environment, the population adapting to the longer migration had relatively larger ovaries, more eggs, and smaller humps; however, the two populations had similar egg sizes, snout lengths, and energy stores (Kinnison et al. 2001; Kinnison et al. 2003) (see Figure 6.5). Thus adaptation to different migration distances led to substantial genetic changes in some traits, but not others.

How might these evolutionary changes (or the lack thereof) be interpreted in the context of natural selection? One predictable result was the genetically smaller hump size for the population migrating the longer distance (see also Crossin et al. 2004). Individuals with smaller humps should have enhanced migratory ability because they are more hydrodynamically efficient. Moreover, developing larger humps expends energy that might otherwise be used for migration or breeding. Supporting these ideas, hump size and somatic energy stores were negatively correlated for adults that migrated 100 km to Glenariffe, but not for those that migrated only 17 km to Silverstream (Kinnison et al. 2003).

Females

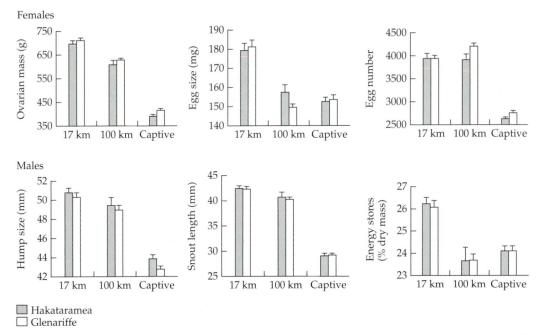

Males

☐ Hakataramea
☐ Glenariffe

Figure 6.5 Reproductive traits for females of all ages and secondary sexual traits and energy stores for 2-year-old males from two New Zealand chinook salmon populations. The Hakataramea population naturally migrates 60 km before breeding, while the Glenariffe population naturally migrates 100 km. Juveniles were released at two river sites (100 km and 17 km from the ocean), where they imprinted, migrated to the ocean, and subsequently returned as adults. Representatives of the same families were also reared in a common hatchery environment (Captive). Mean trait values (with standard errors) are shown after standardization to a common body size. The difference between the 17 km and the 100 km migration treatments is significant ($P < 0.05$) for all traits. Comparisons between the migration treatments and the captive treatment were not performed because of entirely different diets and conditions. The difference between the two populations in the captive environment is significant ($P < 0.05$) for ovarian mass and hump size, and marginally significant ($P = 0.07$) for egg number. Comparisons between the two populations within each migration treatment were not performed because we cannot be sure they shared a common environment after release. (From Kinnison et al. 2001, 2003; and M. T. Kinnison, unpublished data.)

In addition, males experienced greater somatic energy losses than did females, presumably because males have larger humps (Kinnison et al. 2003).

Another predictable result was the genetically larger ovarian mass in the longer-migrating population, a difference that partially offsets the proximate cost of migration on this trait (see Figure 6.5). This result might seem to be in conflict with data from native populations, wherein ovary size and migration distance are negatively correlated. The work in New Zealand, however, shows that such phenotypic trends in wild populations are probably caused by proximate effects of migration that obscure an opposite trend in the genetic contribution to ovarian investment (i.e., countergradient variation). That is, longer-migrating populations have a genetic tendency to invest more energy in ovaries, but this tendency is not reflected in a higher ovarian mass at maturity because they also expend more energy during the migration itself.

Not all of the results for New Zealand chinook salmon were immediately predictable. Neither egg size, snout length, nor energy stores differed between the populations after rearing in a common hatchery environment, whereas egg number did. These results seem puzzling because each trait, *except* for egg number, was influenced by the proximate cost of migration. Interestingly, the genetic architecture of these traits may help explain the results for egg size and number. Specifically, the genetic correlation between ovary size and egg number appears greater than that between ovary size and egg size (Kinnison et al. 2001). As a result, selection to increase ovarian mass (to compensate for the proximate cost of migration) should lead to a greater initial increase in egg number than in egg size. This argument assumes little or no variation among populations in direct selection on egg size or number. Alternatively, smaller eggs may be favored in populations that migrate longer distances (Healey 2001; Kinnison et al. 2001). The lack of evolutionary divergence for snout length and somatic energy stores, however, currently remains unexplained.

Closing remarks on salmon

Studies of introduced salmon suggest that evolution is driven primarily by deterministic (predictable) processes such as natural selection. The role of contingency (e.g., phylogenetic history, founder effects, and genetic drift)at this taxonomic level appears related to which traits evolve and which do not. The importance of deterministic processes has also been confirmed for introduced European grayling (Koskinen et al. 2002). This is not to say that contingency does not have important effects. For example, it probably plays a substantial role in determining which populations survive the initial introduction. Moreover, patterns of life history variation show a signature of both deterministic and contingent events even among native salmon populations (Kinnison and Hendry 2004; Waples et al. 2004).

Parallel Opportunities That Can Be Exploited

We have highlighted some examples from studies of *Drosophila* and salmon, but we would be remiss not to mention that other introduced species offer similar opportunities. Indeed, diverse studies have recently documented rapid evolution of introduced and invasive species (Diniz-Filho et al. 1999; Losos et al. 2001; Maron et al. 2004), sometimes even involving reaction norms (Lee et al. 2003).

Many species of animals and plants are (like salmon) being intentionally introduced around the globe for agriculture or sport. For example, honey bees, trout, chickens, cattle, and sheep have been introduced into most continents;

these introductions provide biologists with many opportunities to study (often in replicate) adaptation to local environments. Of course, concomitant selective breeding may enhance (or sometimes confound) patterns of local adaptation of these species.

Similarly, although our focus has been on invasive species themselves, one could just as well look at the evolution of native species responding to an exotic. For example, the intentional introduction of predators into streams has had dramatic effects on the life history, morphology, behavior, and physiology of native guppies (Reznick et al. 1990a). Similarly, the introduction of exotic plants has led to the evolution of soapberry bugs (Carroll and Boyd 1992; Carroll et al. 2001) and of apple maggot flies (Filchak et al. 2000). The introduction of the European periwinkle *Littorina* into Connecticut prompted evolutionary shifts in the shell preferences of native hermit crabs (Blackstone and Joslyn 1984). Interestingly, plastic and genetic changes in the shell shape of *Littorina* have since been changed by an invasive crab (Trussell and Etter 2001).

Ecological Implications of Rapid Evolution

Our review of a few selected studies shows that the evolution of invasive and introduced species is often—though not always—rapid, predictable, and dramatic. These observations are directly relevant to classic debates in evolution (Darwin 1859; Simpson 1944; Gould and Eldredge 1977). In addition, they have profound significance for those attempting to monitor (as well as to blunt) the negative effects of introduced species on native species. Specifically, as introduced species rapidly adapt to local physical and biotic environments, their ecological effects are likely to grow. If they are "bad" just after arriving, they may well become even worse (Maron et al. 2004).

Even so, we can try to turn this problem to our advantage, at least from an academic perspective. Consider *D. subobscura*, which was probably introduced to the Americas from the Mediterranean (roughly 40° latitude) and which has now colonized much colder sites in both North (51°) and South America (46°). As these "southern" flies adapt to cold environments at high latitudes, their competitive effects on native, cold-adapted species might well increase. But does this actually occur? Currently no one knows. However, this possibility can be tested by setting up competition experiments, either in the laboratory (Pascual et al. 1998; Pascual et al. 2000) or in seminatural enclosures. At a high-latitude site in Washington State, for example, one might set up competition experiments involving native *Drosophila* versus local *subobscura*, versus *subobscura* from southern California, or perhaps even versus the presumed source Mediterranean population. If local adaptation has enhanced competitive ability since the introductions (see Bossdorf et al. 2004), then high-latitude *subobscura* should fare better in competition with native high-latitude congeners than should *subobscura* either from California or from the Mediterranean. Should that prove to be the case, it will further validate our central thesis; namely, that invasion is an evolutionary as well as an ecological problem.

Acknowledgments

We thank Dov Sax, Jay Stachowicz, and Steve Gaines for inviting us to participate in this volume. RBH's interest in introduced species was initially inspired by Johnston and Selander's pioneering work on the house sparrow, and later by Prevosti and colleagues' brilliant studies of introduced *D. subobscura*. We express deep appreciation to our friends and colleagues at the University of Barcelona (A. Prevosti, L. Serra, M. Pascual, J. Balanyà, F. Mestres, and E. Solé), who not only have shared this opportunity to study this species, but are wonderful collaborators. We also thank D. Sax, J. A. F. Diniz-Filho, and anonymous reviewers for constructive comments on the manuscript. RBH and GWG's research on *Drosophila* has been generously supported by the Royalty Research Fund (University of Washington), by the National Science Foundation (NSF grants DEB 9629822 to RBH and GWG, DEB 9981598 to RBH, and DEB 9981555 to GWG), by a grant from the Jeffress Foundation to GWG, and by an international collaborative grant from the Comisión Conjunta Hispano-Norteamericana de Cooperación Científica y Tecnológica to L. Serra and RBH. APH thanks the many people who have contributed to the work on introduced salmon, particularly M. Kinnison and T. Quinn. APH is supported by a Natural Sciences and Engineering Research Council of Canada Discovery Grant.

Literature Cited

Addo-Bediako, A., S. L. Chown, and K. J. Gaston. 2000. Thermal tolerance, climatic variability and latitude. Proceedings of the Royal Society of London B 267:739–745.

Allendorf, F. W., and L. L. Lundquist. 2003. Introduction: Population biology, evolution, and control of invasive species. Conservation Biology 17:24–30.

Altukhov, Y. P., E. A. Salmenkova, and V. T. Omelchenko. 2000. Salmonid fishes: population biology, genetics and management. Blackwell Science, Oxford.

Ayala, F. J., L. Serra, and A. Prevosti. 1989. A grand experiment in evolution: the *Drosophila subobscura* colonization of the Americas. Genome 31:246–255.

Ayllon, F., J. L. Martinez, P. Davaine, E. Beall, and E. Garcia-Vazquez. 2004. Interspecific hybridization between Atlantic salmon and brown trout introduced in the subantarctic Kerguelen Islands. Aquaculture 230:81–88.

Baker, H. G., and G. L. Stebbins. 1965. *The genetics of colonizing species.* Academic Press, New York.

Balanyà, J., C. Segarra, A. Prevosti, and L. Serra. 1994. Colonization of America by *Drosophila subobscura*: the founder event and a rapid expansion. Journal of Heredity 85:427–432.

Balanyà, J., L. Serra, G. W. Gilchrist, R. B. Huey, M. Pascual, F. Mestres, and E. Solé. 2003. Evolutionary pace of chromosomal polymorphism in colonizing populations of *Drosophila subobscura*: an evolutionary time series. Evolution 57:1837–1845.

Beacham, T. D., and C. B. Murray. 1993. Fecundity and egg size variation in North American Pacific salmon (*Oncorhynchus*). Journal of Fish Biology 42:485–508.

Beckenbach, A. T., and A. Prevosti. 1986. Colonization of North America by the European species *Drosophila subobscura* and *D. ambigua*. American Midlands Naturalist 115:10–18.

Blackstone, N. W., and A. R. Joslyn. 1984. Utilization and preference for the introduced gastropod *Littorina littorea* (L.) by the hermit crab *Pagurus longicarpus* (Say) at Guilford, Connecticut. Journal of Experimental Marine Biology and Ecology 80:1–9.

Blair, A. C., and L. M. Wolfe. 2004. The evolution of an invasive plant: an experimental study with *Silene latifolia*. Ecology 85: 3035–3042.

Blair, G. R., D. E. Rogers, and T. P. Quinn. 1993. Variation in life history characteristics and morphology of sockeye salmon in the Kvichak River system, Bristol Bay, Alaska. Transactions of the American Fisheries Society 122:550–559.

Blossey, B., and R. Nötzold. 1995. Evolution of increased competitive ability in invasive nonindigenous plants: a hypothesis. Journal of Ecology 83:877–889.

Bossdorf, O., D. Prati, H. Auge, and B. Schmid. 2004. Reduced competitive ability in an invasive plant. Ecology Letters 7:346–353.

Boussy, I. A. 1987. A latitudinal cline in P-M gonadal dysgenesis potential in Australian *Drosophila melanogaster* populations. Genetical Research, Cambridge 49:11–18.

Brncic, D., and M. Budnik. 1980. Colonization of *Drosophila subobscura* Collin in Chile. Drosophila Information Service 55:20.

Brncic, D., A. Prevosti, M. Budnik, M. Monclús, and J. Ocaña. 1981. Colonization of *Drosophila subobscura* in Chile I. First population and cytogenetic studies. Genetica 56:3–9.

Burger, C. V., K. T. Scribner, W. J. Spearman, C. O. Swanton, and D. E. Campton. 2000. Genetic contribution of three introduced life history forms of sockeye salmon to colonization of Frazer Lake, Alaska. Canadian Journal of Fisheries and Aquatic Sciences 57:2096–2111.

Callaway, R. M., and W. M. Ridenour. 2004. Novel weapons: invasive success and the evolution of increased competitive ability. Frontiers in Ecology and the Environment 8:436–443.

Carroll, S. P., and C. Boyd. 1992. Host race radiation in the soapberry bug: natural history with the history. Evolution 46:1052–1069.

Carroll, S. P., H. Dingle, T. R. Famula, and C. W. Fox. 2001. Genetic architecture of adaptive differentiation in evolving hot races of the soapberry bug, *Jadera haematoloma*. Genetica 112/113:273–286.

Cohan, F. M., and J.-D. Graf. 1985. Latitudinal cline in *Drosophila melanogaster* for knockdown resistance to ethanol fumes and for rates of response to selection for further resistance. Evolution 39:278–293.

Cowley, D. E., W. R. Atchley, and J. J. Rutledge. 1986. Quantitative genetics of *Drosophila melanogaster*. Genetics 114:549–566.

Coyne, J. A., and E. Beecham. 1987. Heritability of two morphological characters within and among natural populations of *Drosophila*. Genetics 117:727–737.

Crossin, G. T., S. G. Hinch, A. P. Farrell, D. A. Higgs, A. G. Lotto, J. D. Oakes, and M. C. Healey. 2004. Energetics and morphology of sockeye salmon: effects of upriver migratory distance and elevation. Journal of Fish Biology 65:1–23.

Darwin, C. 1859. *The origin of species by means of natural selection, or the preservation of favoured races in the struggle for life.* J. Murray, London.

Daszak, P., A. A. Cunningham, and A. D. Hyatt. 2000. Emerging infectious diseases of wildlife—threats to biodiversity and human health. Science 287:443–449.

David, J. R., and C. Bocquet. 1975. Similarities and differences in latitudinal adaptation of two *Drosophila* sibling species. Nature 257:588–590.

Diniz-Filho, J. A. F., S. Fuchs, and M. C. Arias. 1999. Phylogeographical autocorrelation of phenotypic evolution in honey bees (*Apis mellifera* L.). Heredity 83:671–680.

Dobzhansky, T. 1943. Genetics of natural populations. IX. Temporal changes in the composition of populations of *Drosophila pseudoobscura*. Genetics 28:162–186.

Dobzhansky, T. 1948. Genetics of natural populations. XVI. Altitudinal and seasonal changes produced by natural selection in certain populations of *Drosophila pseudoobscura* and *Drosophila persimilis*. Genetics 33:158–176.

Drummond, A. J., O. G. Pybus, A. Rambaut, R. Forsberg, and A. G. Rodrigo. 2003. Measurably evolving populations. Trends in Ecology and Evolution 18:481–488.

Einum, S., M. T. Kinnison, and A. P. Hendry. 2004. The evolution of egg size and number. In A. P. Hendry, and S. C. Stearns, eds. *Evolution illuminated: salmon and their relatives*, pp. 126–153. Oxford University Press, Oxford.

Elton, C. S. 1958. *The ecology of invasions by animals and plants.* Methuen, London.

Endler, J. A. 1986. *Natural selection in the wild.* Princeton University Press, Princeton, NJ.

Filchak, K. E., J. B. Roethele, and J. L. Feder. 2000. Natural selection and sympatric divergence in the apple maggot *Rhagoletis pomenella*. Nature 407:739–742.

Franklin, I. R., and R. Frankham. 1998. How large must populations be to retain evolutionary potential. Animal Conservation 1:699–673.

García-Ramos, G., and D. Rodríguez. 2002. Evolutionary speed of species invasion. Evolution 56:661–668.

Gharrett, A. J., and M. A. Thomason. 1987. Genetic changes in pink salmon (*Oncorhynchus gorbuscha*) following their introduction into the Great Lakes. Canadian Journal of Fisheries and Aquatic Sciences 44:787–792.

Gilchrist, A. S., and L. Partridge. 1999. A comparison of the genetic basis of wing size divergence in three parallel body size clines of *Drosophila melanogaster*. Genetics 153:1775–1787.

Gilchrist, G. W., R. B. Huey, and L. Serra. 2001. Rapid evolution of wing size clines in *Drosophila subobscura*. Genetica 112/113:273–286.

Gilchrist, G. W., R. B. Huey, J. Balanyà, M. Pascual, and L. Serra. 2004. A time series of evolution in action: latitudinal cline in wing size in South American *Drosophila subobscura*. Evolution 58:768–780.

Goñi, B., M. E. Martinez, V. L. S. Valente, and C. R. Vilela. 1998. Preliminary data on the *Drosophila* species (Diptera, Drosophilidae) from Uruguay. Revista Brasileira de Entomologia 42:131–140.

Goodnight, C. J. 2000. Modeling gene interaction in structured populations. In J. B. Wolf, E. D. Brodie, III, and M. J. Wade, eds. *Epistasis and the evolutionary process*. Oxford University Press, Oxford.

Gould, S. J. 1989. *Wonderful life: the Burgess Shale and the nature of history*. W. W. Norton, New York.

Gould, S. J., and N. Eldredge. 1977. Punctuated equilibria: the tempo and mode of evolution reconsidered. Paleobiology 3:115–151.

Grinnell, J. 1919. The English sparrow has arrived in Death Valley: an experiment in nature. The American Naturalist 43:468–473.

Harshman, L. G., and A. A. Hoffmann. 2000. Laboratory selection experiments using *Drosophila*: what do they really tell us? Trends in Ecology and Evolution 15:32–36.

Haugen, T. O., and L. A. Vøllestad. 2001. A century of life-history evolution in grayling. Genetica 112/113:475–491.

Healey, M. C. 2001. Patterns of gametic investment by female stream- and ocean-type chinook salmon. Journal of Fish Biology 58:1545–1556.

Hendry, A. P. 2001. Adaptive divergence and the evolution of reproductive isolation in the wild: an empirical demonstration using introduced sockeye salmon. Genetica 112/113:515–534.

Hendry, A. P., and O. K. Berg. 1999. Secondary sexual characters, energy use, senescence, and the cost of reproduction in sockeye salmon. Canadian Journal of Zoology 77:1663–1675.

Hendry, A. P., and M. T. Kinnison. 1999. The pace of modern life: measuring rates of micro-evolution. Evolution 53:1667–1653.

Hendry, A. P., and T. P. Quinn. 1997. Variation in adult life history and morphology among populations of sockeye salmon (*Oncorhynchus nerka*) within Lake Washington, WA, in relation to habitat features and ancestral affinities. Canadian Journal of Fisheries and Aquatic Sciences 54:74–84.

Hendry, A. P., T. P. Quinn, and F. M. Utter. 1996. Genetic evidence for the persistence and divergence of native and introduced sockeye salmon (*Oncorhynchus nerka*) within Lake Washington, WA. Canadian Journal of Fisheries and Aquatic Science 53:823–832.

Hendry, A. P., J. E. Hensleigh, and R. R. Reisenbichler. 1998. Incubation temperature, developmental biology, and the divergence of sockeye salmon (*Oncorhynchus nerka*) within Lake Washington. Canadian Journal of Fisheries and Aquatic Sciences 53:1387–1394.

Hendry, A. P., J. K. Wenburg, P. Bentzen, E. C. Volk, and T. P. Quinn. 2000. Rapid evolution of reproductive isolation in the wild: evidence from introduced salmon. Science 290:516–5 18.

Hendry, A. P., V. Castric, M. T. Kinnison, and T. P. Quinn. 2004a. The evolution of philopatry and dispersal: homing vs. straying in salmonids, In A. P. Hendry, and S. C. Stearns, eds. *Evolution illuminated: salmon and their relatives*, pp. 52–91. Oxford University Press, Oxford.

Hendry, A. P., J. E. Hensleigh, M. T. Kinnison, and T. P. Quinn. 2004b. To sea or not to sea? Anadromy vs. non-anadromy in salmonids. In A. P. Hendry, and S. C. Stearns, eds. *Evolution illuminated: salmon and their relatives*, pp. 95–125. Oxford University Press, Oxford.

Hoffmann, A. A., and L. G. Harshman. 1999. Desiccation and starvation resistance in *Drosophila*: patterns of variation at the species, population and intrapopulation levels. Heredity 83:637–643.

Hoffmann, A. A., R. J. Hallas, J. A. Dean, and M. Schiffer. 2003. Low potential for climatic stress adaptation in a rainforest *Drosophila* species. Science 301:100–102.

Huey, R. B., L. Partridge, and K. Fowler. 1991. Thermal sensitivity of *Drosophila melanogaster* responds rapidly to laboratory natural selection. Evolution 45:751–756.

Huey, R. B., G. W. Gilchrist, M. L. Carlson, D. Berrigan, and L. Serra. 2000. Rapid evolution of a geographic cline in an introduced species of fly. Science 287:308–309.

Huey, R. B., P. E. Hertz, and B. Sinervo. 2003. Behavioral drive versus behavioral inertia: a null model approach. American Naturalist 161:357–366.

James, A. C., R. B. R. Azevedo, and L. Partridge. 1995. Cellular basis and developmental timing in a size cline of *Drosophila melanogaster*. Genetics 140:659–666.

Johnston, R. F., and R. K. Selander. 1964. House sparrows: rapid evolution of races in North America. Science 144:548–550.

Karan, D., S. Dubey, B. Moreteau, R. Parkash, and J. R. David. 2000. Geographical clines for quantitative traits in natural populations of a tropical drosophilid: *Zaprionus indianus*. Genetica 108:91–100.

Kinnison, M. T., and A. P. Hendry. 2001. The pace of modern life II: from rates of contemporary microevolution to pattern and process. Genetica 112/113:145–164.

Kinnison, M. T., and A. P. Hendry. 2004. Tempo and mode of evolution in salmonids. In A. P. Hendry, and S. C. Stearns, eds. *Evolution illuminated: salmon and their relatives*, pp. 208–231. Oxford University Press, Oxford.

Kinnison, M. T., M. J. Unwin, A. P. Hendry, and T. P. Quinn. 2001. Migratory costs and the evolution of egg size and number in introduced and indigenous salmon populations. Evolution 55:1656–1667.

Kinnison, M. T., P. Bentzen, M. J. Unwin, and T. P. Quinn. 2002. Reconstructing recent divergence: evaluating nonequilibrium population structure in New Zealand chinook salmon. Molecular Evolution 11:739–754.

Kinnison, M. T., M. J. Unwin, and T. P. Quinn. 2003. Migratory costs and contemporary evolution of reproductive allocation in male chinook salmon. Journal of Evolutionary Biology 16:1257–1269.

Koskinen, M. T., T. O. Haugen, and C. R. Primmer. 2002. Contemporary Fisherian life-history evolution in small salmonid populations. Nature 419:826–830.

Krimbas, C. B. 1993. *Drosophila subobscura: Biology, genetics and inversion polymorphism*. Verlag Dr. Kovac, Hamburg.

Krimbas, C. B., and M. Loukas. 1980. The inversion polymorphism of *Drosophila subobscura*. Evolutionary Biology 12:163–234.

Lee, C. E. 2002. Evolutionary genetics of invasive species. Trends in Ecology and Evolution 17:386–391.

Lee, C. E., J. L. Remfert, and G. W. Gelembiuk. 2003. Evolution of physiological tolerance and performance during freshwater invasions. Integrative and Comparative Biology 43:439–449.

Lenormand, T. 2002. Gene flow and the limits to natural selection. Trends in Ecology and Evolution 17:183–189.

Lenski, R. E., M. R. Rose, S. C. Simpson, and S. C. Tadler. 1991. Long-term experimental evolution in *Escherichia coli*. I. Adaptation and divergence during 2000 generations. American Naturalist. 138:1315–1341.

Lever, C. 1996. *Naturalized fishes of the world*. Academic Press, New York.

Losos, J. B., A. Larson, T. R. Jackman, K. de Queiroz, and L. Rodíguez-Schettino. 1998. Contingency and determinism in replicated adaptive radiations of island lizards. Science 279:2115–2118.

Losos, J. B., T. W. Schoener, K. I. Warheit, and D. Creer. 2001. Experimental studies of adaptive differentiation in Bahamian *Anolis* lizards. Genetica 112/113:399–415.

Maron, J. L., M. Vila, R. Bommarco, S. Elmendorf, and P. Beardsley. 2004. Rapid evolution of an invasive species. Ecological Monographs 74:261–280.

McDowall, R. M. 1994. The origins of New Zealand's chinook salmon, *Oncorhynchus tshawytscha*. Marine Fisheries Review 56:1–7.

Menozzi, P., and C. B. Krimbas. 1992. The inversion polymorphism of *D. subobscura* revisited: Synthetic maps of gene arrangement frequencies and their interpretation. Journal of Evolutionary Biology 5:625–641.

Mestres, F., and L. Serra. 1995. On the origin of the O_5 chromosomal inversion in American populations of *Drosophila subobscura*. Hereditas 123:39–46.

Mestres, F., J. Balañá, C. Segarra, A. Prevosti, and L. Serra. 1992. Colonization of America by *Drosophila subobscura*: Analysis of the O_5 inversions from Europe and America and their implications for the colonizing process. Evolution 46:1564–1568.

Mestres, F., L. Serra, and F. J. Ayala. 1995. Colonization of the Americas by *D. subobscura*: lethal-gene allelism and association with chromosomal arrangements. Genetics 140:1297–1305.

Mestres, F., L. Abad, B. Sabater-Muñoz, A. Latorre, and L. Serra. 2004. Colonization of America by *Drosophila subobscura*: association between *Odh* gene haplotypes, lethal genes, and chromosomal arrangements. Genes Genetics and Systematics 79:233–244.

Misra, R. K., and E. C. R. Reeve. 1964. Clines in body dimensions in populations of *Drosophila subobscura*. Genetical Research 5:240–256.

Mooney, H. A., and E. E. Cleland. 2001. The evolutionary impact of invasive species. Proceedings of the National Academy of Sciences USA 98:5446–5451.

Orengo, D. J., and A. Prevosti. 1996. Temporal changes in chromosomal polymorphism of *Drosophila subobscura* related to climatic changes. Evolution 50:1346–1350.

Palumbi, S. R. 2001. Humans as the world's greatest evolutionary force. Science 293:1786–1790.

Parker, I. M., J. Rodriguez, and M. E. Loik. 2003. An evolutionary approach to understanding the biology of invasions: local adaptation and general-purpose genotypes in the weed *Verbascum thapsus*. Conservation Biology 17:59–72.

Parsons, P. A. 1983. *The evolutionary biology of colonizing species*. Cambridge University Press, Cambridge.

Partridge, L., B. Barrie, N. H. Barton, K. Fowler, and V. French. 1995. Rapid laboratory evolution of adult life history traits in *Drosophila melanogaster* in response to temperature. Evolution 49:538–544.

Pascual, M., F. J. Ayala, A. Prevosti, and L. Serra. 1993. Colonization of North America by *Drosophila subobscura*. Ecological analysis of three communities of drosophilids in California. Zeitschrift für zoologische Systematik und Evolutionsforschung 31:216–226.

Pascual, M., L. Serra, and F. J. Ayala. 1998. Interspecific laboratory competition on the recently sympatric species *Drosophila subobscura* and *Drosophila pseudoobscura*. Evolution 52:269–274.

Pascual, M., E. Sagarra, and L. Serra. 2000. Interspecific competition in the laboratory between *Drosophila subobscura* and *D. azteca*. American Midland Naturalist 144:19–27.

Pascual, M., C. F. Aquadro, V. Soto, and L. Serra. 2001. Microsatellite variation in colonizing and Palearctic populations of *Drosophila subobscura*. Molecular Biology and Evolution 18:731–740.

Pegueroles, G., M. Papaceit, A. Quintana, A. Guillén, A. Prevosti, and L. Serra. 1995. An experimental study of evolution in progress: clines for quantitative traits in colonizing and Palearctic populations of *Drosophila*. Evolutionary Ecology 9:453–465.

Pfriem, P. 1983. Latitudinal variation in wing size in *Drosophila subobscura* and its dependence on polygenes of chromosome O. Genetica 61:221–232.

Pollock, M. M., R. J. Naiman, H. E. Erickson, C. A. Johnston, J. Pastor, and G. Pinay. 1995. Beaver as engineers: influences on biotic and abiotic characteristics of drainage basins. In C. G. Jones, and J. H. Lawton, eds. *Linking species and ecosystems*, pp. 117–126. Chapman and Hall, New York.

Prevosti, A. 1955. Geographical variability in quantitative traits in populations of *Drosophila subobscura*. Cold Spring Harbor Symposium on Quantitative Biology 20:294–298.

Prevosti, A., L. Serra, and M. Monclús. 1983. *Drosophila subobscura* has been found in Argentina. *Drosophila* Information Service 59:103.

Prevosti, A., L. Serra, G. Ribó, M. Aguadé, E. Sagarra, M. Monclús, and M. P. García. 1985. The colonization of *Drosophila subobscura* in Chile. II. Clines in the chromosomal arrangements. Evolution 39:838–844.

Prevosti, A., G. Ribó, L. Serra, M. Aguadé, J. Balaña, M. Monclús, and F. Mestres. 1988. Colonization of America by *Drosophila subobscura*: Experiment in natural populations that supports the adaptive role of chromosomal-inversion polymorphism. Proceedings of the National Academy of Sciences USA 85:5597–5600.

Prevosti, A., L. Serra, C. Segarra, M. Aguadé, G. Ribó, and M. Monclús. 1990. Clines of chromosomal arrangements of *Drosophila subobscura* in South America evolve closer to Old World patterns. Evolution 44:218–221.

Quinn, T. P., and C. J. Foote. 1994. The effects of body size and sexual dimorphism on the reproductive behaviour of sockeye salmon, *Oncorhynchus nerka*. Animal Behaviour 48:751–761.

Quinn, T. P., J. L. Nielsen, C. Gan, M. J. Unwin, R. Wilmot, C. Guthrie, and F. M. Utter. 1996. Origin and genetic structure of chinook salmon, *Oncorhynchus tshawytscha*, transplanted from California to New Zealand: allozyme and mtDNA evidence. Fishery Bulletin 94:506–521.

Quinn, T. P., M. T. Kinnison, and M. J. Unwin. 2001a. Evolution of chinook salmon (*Oncorhynchus tshawytscha*) populations in New Zealand: pattern, rate, and process. Genetica 112/113:493–513.

Quinn, T. P., L. Wetzel, S. Bishop, K. Overberg, and D. E. Rogers. 2001b. Influence of breeding habitat on bear predation and age at maturity and sexual di-

morphism of sockeye salmon populations. Canadian Journal of Zoology 79:1782–1793.

Reeve, J. P., and D. J. Fairbairn. 1996. Sexual size dimorphism as a correlated response to selection on body size: an empirical test of the quantitative genetic model. Evolution 50:1927–1938.

Reeve, M. W., K. Fowler, and L. Partridge. 2000. Increased body size confers greater fitness at lower experimental temperature in male *Drosophila melanogaster*. Journal of Evolutionary Biology 13:836–844.

Reznick, D., and J. Travis. 2001. Adaptation, In C. Fox, D. Roff, and D. Fairbairn, eds. *Evolutionary ecology*, pp. 44–57. Oxford University Press, New York.

Reznick, D. A., H. Bryga, and J. A. Endler. 1990. Experimentally induced life-history evolution in a natural population. Nature 346:357–359.

Rodríguez-Trelles, F., and M. A. Rodríguez. 1998. Rapid microevolution and loss of chromosomal diversity in *Drosophila* in response to climate-warming. Evolutionary Ecology 12:829–838.

Rodríguez-Trelles, F., G. Alvarez, and C. Zapata. 1996. Time-series analysis of seasonal changes of the O inversion polymorphism of *Drosophila subobscura*. Genetics 142:179–187.

Rose, M. R., P. M. Service, and E. W. Hutchinson. 1987. Three approaches to trade-offs in life-history evolution. In V. Loeschcke, ed. *Genetic constraints on adaptive evolution*, pp. 91–105. Springer-Verlag, Berlin.

Rozas, J., M. Hernandez, V. M. Cabrera, and A. Prevosti. 1990. Colonization of America by *Drosophila subobscura*: effect of the founder event on the mitochondrial DNA polymorphism. Molecular Biology and Evolution 7:103–109.

Ruesink, J. L., I. M. Parker, M. J. Groom, and P. M. Kareiva. 1995. Reducing the risks of nonindigenous species introductions. BioScience 45:465–477.

Santos, M., P. J. F. Iriarte, W. Céspedes, J. Balanyà, A. Fontdevila, and L. Serra. 2004. Swift laboratory thermal evolution of wing shape (but not size) in *Drosophila subobscura* and its relationship with chromosomal inversion polymorphism. Journal of Evolutionary Biology 17:841–855.

Siemann, E., and W. E. Rogers. 2001. Genetic differences in growth of an invasive tree species. Ecology Letters. 4:514–518.

Simmons, G. M., M. E. Kreitman, W. F. Quattlebaum, and N. Miyashita. 1989. Molecular analysis of the alleles of alcohol dehydrogenase along a cline in *Drosophila melanogaster*. I. Maine, North Carolina, and Florida. Evolution 43:393–409.

Simpson, G. G. 1944. *Tempo and mode in evolution* (1984 reprint). Columbia University Press, NewYork.

Singer, F. J., W. T. Swank, and E. E. C. Clebsch. 1984. Effects of wild pig rooting in a deciduous forest. Journal of Wildlife Management 48:464–473.

Sokoloff, A. 1966. Morphological variation in natural and experimental populations of *Drosophila pseudoobscura* and *Drosophila persimilis*. Evolution 20:49–71.

Solé, E., J. Balanyà, D. Sperlich, and L. Serra. 2002. Long-term changes of the chromosomal inversion polymorphism of *Drosophila subobscura*. I. Mediterranean populations from South-western Europe. Evolution 56:830–835.

Stalker, H. D., and H. L. Carson. 1947. Morphological variation in natural populations of *Drosophila robusta* Sturtevant. Evolution 1:237–248.

Steen, R. P., and T. P. Quinn. 1999. Egg burial depth by sockeye salmon (*Oncorhynchus nerka*): implications for survival of embryos and natural selection on female body size. Canadian Journal of Zoology 77:836–841.

Stockwell, C. A., A. P. Hendry, and M. T. Kinnison. 2003. Contemporary evolution meets conservation biology. Trends in Ecology and Evolution 18:99–101.

Taylor, E. B. 1991. A review of local adaptation in Salmonidae, with particular reference to Pacific and Atlantic salmon. Aquaculture 98:185–207.

Taylor, E. B., C. J. Foote, and C. C. Wood. 1996. Molecular genetic evidence for parallel life-history evolution within a Pacific salmon (sockeye salmon and kokanee, *Oncorhynchus nerka*). Evolution 50:401–416.

Thompson, J. N. 1998. Rapid evolution as an ecological process. Trends in Ecology and Evolution 13:329–332.

Thorne, R. E., and J. J. Ames. 1987. A note on variability of marine survival of sockeye salmon (*Oncorhynchus nerka*) and effects of flooding on spawning success. Canadian Journal of Fisheries and Aquatic Sciences 44:1791–1795.

Travisano, M., J. A. Mongold, A. F. Bennett, and R. E. Lenski. 1995. Experimental tests of the roles of adaptation, chance, and history in evolution. Science 267:87–90.

Trussell, G. C., and R. J. Etter. 2001. Integrating genetic and environmental forces that shape the evolution of geographic variation in a marine snail. Genetica 112/113:321–337.

Utter, F. M. 2001. Patterns of subspecific anthropogenic introgression in two salmonid genera. Reviews in Fish Biology and Fisheries 10:265–279.

van't Land, J., P. van Putten, B. Zwaan, A. Kamping, and W. van Delden. 1999. Latitudinal variation in wild populations of *Drosophila melanogaster*: heritabilities and reaction norms. Journal of Evolutionary Biology 12:222–232.

Vermeij, G. J. 1996. An agenda for invasion biology. Biological Conservation 78:3–9.

Vitousek, P. M., C. M. D'Antonio, L. L. Loope, and R. Westbrooks. 1996. Biological invasions as global environmental change. American Scientist 84:468–478.

Waples, R. S., D. J. Teel, J. M. Myers, and A. R. Marshall. 2004. Life-history divergence in chinook salmon: historical contingency and parallel evolution. Evolution 58:386–403.

Weber, E., and B. Schmid. 1998. Latitudinal population differentiation in two species of *Solidago* (Asteraceae) introduced into Europe. American Journal of Botany 85:1110–1121.

Wilcove, D. S., D. Rothstein, J. Dubow, A. Phillips, and E. Losos. 1998. Quantifying threats to imperiled species in the United States. Bioscience 48:607–615.

Withler, F. C. 1982. *Transplanting Pacific salmon*. Canadian Technical Report of Fisheries and Aquatic Sciences No. 1079, 21 pp.

Wood, C. C. 1995. Life history variation and population structure in sockeye salmon. American Fisheries Society Symposium 17:195–216.

Wright, S., and T. Dobzhansky. 1946. Genetics of natural populations. XII. Experimental reproduction of some of the changes caused by natural selection in certain populations of *Drosophila pseudoobscura*. Genetics 31:125–136.

7

Taxon Cycles

INSIGHTS FROM INVASIVE SPECIES

Robert E. Ricklefs

The taxon cycle consists of phases of expansion and contraction in the geographic and ecological ranges of populations, easily visualized in island archipelagoes but also a feature of continental regions. Taxon cycles and human-aided introductions of invasive species share phases of colonization, establishment, range expansion, and eventual contraction. The primary difference between the expansion phase of a natural taxon cycle and the invasive phase of an introduced species is that human-assisted introductions do not require adaptations for long-distance dispersal. Thus, trade-offs between dispersal ability and competitiveness may be circumvented by introduced species. Because the eventual decline of naturally expanded populations probably requires evolutionary time spans, the commonalities between species invasions and taxon cycles speak to the qualities that lead to population expansion. Most of the variance in range size and ecological distribution of native species resides at a low taxonomic level, suggesting that the traits that influence the "success" of a population are evolutionarily labile and probably do not include conservative adaptations for tolerance of physical conditions or use of particular habitats. Recent analyses have failed to identify attributes of receiving communities, such as native diversity, that make them more or less resistant to invasion. More likely, invasion success hinges on coevolutionary relation-

ships with predators and pathogens, which might depend on a small number of gene mutations in native populations. Phylogenetic analysis of character evolution among closely related species with varied geographic and ecological distributions might help to identify traits associated with invasiveness and to understand the forces that drive taxon cycles.

Introduction

The term "taxon cycle" was coined 45 years ago by E. O. Wilson (1959, 1961), but its basic premise has been a common theme among evolutionary biologists for over a century (Willis 1922; Darlington 1943; Brown 1957; Brown and Lomolino 1998; Ricklefs and Bermingham 2002). Wilson's insight, based on detailed studies of the evolution and distribution of ants in Melanesia, was that taxa at all levels, but most noticeably and accessibly at the species or population level, go through phases of geographic expansion and contraction associated with similar changes in ecological distribution. These cycles tend to degrade toward eventual extinction, moving from larger to smaller landmasses and islands. However, populations at any point are capable of initiating a new cycle. During expansion phases, populations can be considered invasive species— that is, species that become established and spread in non-native regions.

Expansion presumably follows release from factors that control populations, or at least the relaxation of these constraints. Limiting factors potentially include extrinsic properties of the environment, such as climate and other physical conditions, and intrinsic features of biological systems, as in the case of control by predators or pathogens.

Examining species that have become invasive in association with human activities—for example, species transported between regions or gaining footholds in altered landscapes—can provide insights into taxon cycles in natural systems, and vice versa. It is reasonable to ask whether human-assisted invaders share attributes with natural invaders in the early, expanding stages of the taxon cycle. Of particular interest are features of invasive species that enhance their competitive success and extrinsic factors that influence colonization and the spread of populations in new areas.

Four stages of invasion can be recognized: colonization, establishment, spread, and decline. When comparing the establishment and invasion of non-native species with natural invasions, the most obvious differences are seen in the first two stages. Invasive species often gain transport from their native area to other regions with the intentional or unintentional help of humans, and they often become established in habitats that have been modified by humans. In contrast, natural invaders must traverse barriers under their own steam. This requirement sets constraints beyond the ability shared by all invasive species to spread through biological communities in their new homelands.

Natural invasions are often considered primarily in the context of island colonization, whereas human-aided invasions include introductions to islands and continents alike. The key point is that one can identify the invasive spread of a population most readily when it colonizes an area away from its native range. Nevertheless, an equally important aspect of natural range expansion is the spread of populations into new habitats and areas within their native regions. This phase of spreading by a native population may parallel the expansion of established non-native species, and this is where the most useful parallels between native and non-native invasive species might be developed.

This chapter considers the qualities of invasive species and factors contributing to invasion success in the context of the taxon cycle concept. Special attention is paid to the idea that invasive species, non-native and native alike, are released from herbivores or predators and pathogens, which contributes to their productivity and competitive ability (Keene and Crawley 2002; Torchin et al. 2003). Even in the absence of change in the physical environment, a shift in the balance between consumer and resource populations can shift a species from an endemic to an invasive (expanding) phase.

Although non-native invasive species often bypass long-distance dispersal, natural colonists of islands and both native and non-native invasive species may share attributes if the demographic basis for habitat expansion and long-distance dispersal have common roots in increased population productivity. This possibility can be explored by comparing the distributions within continental regions of populations that have or have not been sources of colonists to islands. Particular consideration will be given to the land birds of the Lesser Antilles, whose evolutionary and geographic history I have investigated for several years with my colleague Eldredge Bermingham of the Smithsonian Tropical Research Institute (Ricklefs and Bermingham 1999, 2001, 2002). The methods of molecular phylogenetics allow one to distinguish species in early stages of the taxon cycle from older colonists in the declining phase of the taxon cycle. If young colonists were to occupy particular habitats or could otherwise be linked by common ecological associations, one might identify extrinsic drivers of expansion and invasion. In contrast, idiosyncratic emergence of natural colonists from a potential source biota would strengthen the case for coevolutionary dynamics having a role, perhaps based on the unpredictable generation of genetic variation.

One may ask whether it is useful to seek insights from invasive species if one is uncertain about the application of such insights to natural populations. Two answers come to mind. First, insight is often complementary, and we may be able to learn something of invasive species from natural colonizers, as well as vice versa, by considering the two together. Second, invasions depending on human assistance by means of transportation or habitat alteration are much more frequent and better characterized than natural range expansions. We know so much more about non-native invasive species than native invasive species that those of us interested in natural processes cannot afford to ignore them.

To provide a proper context for comparing native and non-native invasive species, it will be useful to briefly describe the taxon cycle and related ideas about natural invasions, and to address the more general issue of equilibrium and change in biological systems as it relates to expansion and contraction of the ranges of individual species. This discussion is followed by a consideration of key issues in the taxon cycle and invasion biology literature and an explicit comparison of the taxon cycle with species invasion. The chapter ends with some concluding thoughts and suggestions for future work.

What Is the Taxon Cycle?

The idea that the distributions of species go through phases of expansion and contraction has been a persistent theme in evolutionary biology and biogeography (Ricklefs and Cox 1972; Ricklefs and Bermingham 2002). The idea has taken many forms, from comparison with the life cycles of individuals through growth, maturity, and senescence (Cain 1944; Simpson 1949) to extrinsically driven cycles paralleling change in the physical environment (Pregill and Olson 1981; Dynesius and Jansson 2000) and cycles expressing the changing balance of host-pathogen coevolution (Ricklefs and Cox 1972). Interest in such cyclic dynamics extends far beyond individual populations, even to the rise and fall of states (Turchin 2003).

The evidence for historical expansion and contraction of populations is most apparent in archipelagoes because island systems have a discrete geography that clearly reveals variation in distribution and where telltale signs of extinction appear as gaps in distributions. Thus, it is no accident that the concept of the taxon cycle was formalized by E. O. Wilson based on his studies of the systematics, geographic distribution, and ecology of ants on Melanesian islands (Wilson 1959, 1961). Wilson distinguished expanding and contracting species by a combination of geographic and habitat distributions. Expanding species had identifiable continental sources and occupied marginal habitats, through which they spread to other islands by rafting. Colony sizes were generally large, and expanding species were considered good competitors. After invading a new island, colonists extended their distributions into the larger area of less densely occupied forested habitat in the island interior, eventually evolving smaller colonies in more cryptic microhabitats (Wilson 1959). Competition from new invaders at the margins of islands helped to push this process along. Once a species had become restricted to interior habitats, it lost much of its potential colonizing ability, and contraction and fragmentation of the range began to follow upon loss of individual island populations. The eventual fate of ant species in the archipelago was extinction. Wilson's concept of the taxon cycle included evolution, competition, and habitat shifts, and therefore invoked the interplay of many processes.

Wilson presumed that ant species in marginal habitats in continental source areas initiated taxon cycles, but he did not discuss the origins of further population expansions, except that these involved a reinvasion of marginal habitats. Ricklefs and Cox (1972), in applying Wilson's taxon cycle concept to the birds of the West Indies, proposed that the expanding and contracting phases of the cycle reflected the balance between the adaptations of populations and those of their antagonists, whether predators, pathogens, competitors, or food resources. They referred to this process as "counteradaptation" and, following upon the experimental work of David Pimentel and his colleagues on flies (Pimentel 1961; Pimentel and Al-Hafidh 1963; Pimentel et al. 1965), they suggested that the relative success of antagonist populations might shift owing to frequency dependence coupled with evolutionary lag times.

Several studies of the distributions of birds and insects within archipelagoes uncovered patterns similar to those described by Wilson (Greenslade 1968; Greenslade 1969; Ricklefs 1970; Ricklefs and Cox 1972). Since 1961, several modifications of the taxon cycle have been published, notably Erwin's (1981) taxon pulse and Roughgarden and Pacala's (1989) variant of the taxon cycle for Lesser Antillean *Anolis*, and the concept has been subjected to further analysis and criticism (see Ricklefs and Bermingham 2002). Nonetheless, recent molecular phylogeographic studies have confirmed the existence of expansion and contraction phases previously inferred from distribution and taxonomic differentiation (Ricklefs and Bermingham 2001). To fully understand both the potential and the difficulties of the taxon cycle concept, it is useful to consider perceptions of ecological systems more generally.

Static versus Dynamic Views of Ecological and Geographic Distributions

Ecology has for the most part been an equilibrium science. Population and community processes have been perceived as leading toward equilibrium states. Populations grow toward a carrying capacity, and species coexist stably within a community having a particular matrix of interaction coefficients (Vandermeer 1972; May 1975). This perspective led to the idea of community saturation, whereby ecological space became filled with species and further invasion of a community was prevented (MacArthur and Levins 1967; Terborgh and Faaborg 1980; Alroy 1998; McKinney 1998; Kelt and Brown 1999). Accordingly, differences between communities in numbers of species reflected the influence of the physical environment on either the ecological space available in a community or the average ecological space occupied by each species (MacArthur 1965; Cody 1975; Cody and Diamond 1975). MacArthur and Wilson's (1963, 1967) equilibrium theory of island biogeography envisioned a balance between the rate of colonization of an island by new species and the rate of extinction of island populations. Although their model of island diversity was dynamic, they did not discuss in detail the origin of new colonists or the factors resulting in extinction of established populations, topics that have remained relatively unexplored in the ensuing decades (Whittaker 1998). In particular, whether new colonists hasten the extinction of established island populations is obscure. David Lack (1976) retreated further from a dynamic concept of island biogeography when he asserted that few island populations went extinct; he believed that colonization stopped when islands became sat-urated with species. For Lack, diversity on islands was correlated with the eco-logical variety of habitats and the structure of those habitats—that is, the total ecological space available to be occupied (Ricklefs and Lovette 1999; Ricklefs and Bermingham 2004). Several new analyses suggest that establishment of exotic plants on islands has not caused extinction in the native floras (Sax et al. 2002; Davis 2003; Sax and Gaines 2003).

According to this equilibrium view of ecology, changes in communities are driven by external factors, primarily changes in the environment, including its physical conditions and biological resources. For example, warm-cold and wet-dry cycles associated with the advance and retreat of continental glaciers are seen as external forces that shifted ecological conditions and the distributions of species, changing the composition of communities locally because of the dif-ferent responses of their component species to changes in different aspects of the environment. In many cases, this process must have led to disruption of communities as well as their shifting in space, although even this point is con-tentious (Davis 1986; Jacobson et al. 1987; Webb 1987; Jackson et al. 2000). A view of ecology in which species are conservative entities whose distributions are pushed about by the physical environment is essentially static or, at least, passive (Janzen 1985). Colonization is perceived as haphazard, caused by storms or currents that transport animals or seeds from one place to another (Lack 1976; Pielou 1979), or serendipitously resulting from the dispersal of occa-sional individuals into the void. The tropical ecologist Daniel Janzen (1985) has said that if a population arrives at a place and it "fits" ecologically, it will per-sist. Sometimes the fit is aided by individual phenotypic flexibility (Callaway et al. 2003) or by the absence of coevolved antagonistic species (Callaway and Aschehoug 2000; Callaway et al. 2004).

Many ecologists believe that ecological systems are intrinsically more dynamic than this view allows because of the evolutionary responses of species to one another and to changing environments. The question for ecologists has been whether evolution occurs rapidly enough compared with the approach of systems to ecological equilibrium to influence the composition of commu-nities (Ricklefs 1989). In the present context, are invasive species afforded enough time to evolve an improved fit to available ecological space? More gen-erally, is evolution relevant to ecology?

Ehrlich and Raven (1964) made ecologists aware of the importance of coevo-lution, but the outcome they envisioned was a conservative association between particular groups of herbivores and particular groups of plants, punc-tuated at times by shifts onto new host plants (Farrell et al. 1992; Farrell and Mitter 1994). Robert H. Whittaker (1972, 1977) and Leigh Van Valen (1973) adopted a more dynamic view focused on populations of predators and prey, herbivores and plants, pathogens and hosts, and so on, each engaged in a con-tinuing coevolutionary struggle to improve its position with respect to mutu-ally antagonistic interactions and changing environments. Presumably, coevo-lution maintains the relationships between species in a community in a dynamic state. However, in developing the idea of coevolution, ecologists did not explicitly address community relationships or changes in the composition of communities. Indeed, most believed that coevolution leads to a dynamic equilibrium, at least over ecological time spans (Rosenzweig 1973; Ricklefs 1990). That is, although particular adaptations for attack and defense come and go, the outcome of the interaction between antagonists remains more or less a stalemate. Indeed, coevolved relationships between antagonists can become entrenched and static, as in the case of many insects specialized to feed on plants with particularly toxic compounds (Berenbaum 1983).

In this context, several ecologists have emphasized the long evolutionary history of herbivore-plant and predator-prey relationships, which fossils suggest have persisted more or less unchanged for millions of years (e.g., Opler 1973). The literature on the evolution of virulence in pathogens emphasizes conditions that can shift the equilibrium level of virulence achieved (Ewald 1983; Bull 1994; Day 2001). Thus, in spite of the evolutionary changes occurring within communities, these changes are not seen as challenging the outcomes of species interactions or the stability of community composition. A static, equilibrium view does not readily allow for the reshuffling of ecological relationships, which would otherwise boost some populations into phases of expansion or provide openings for the invasion of new species from outside. This view is consistent with the prevalent ideas that intact communities are resistant to invasion, and that invasive species do best where human activities have badly torn the fabric of the community (Elton 1958; Williamson 1996; Hector et al. 2001).

The most important sources of transformation in dynamic models of ecological systems are lag times in the responses of biological systems to change. Delays in the population responses of predators to changes in the numbers of their prey, and vice versa, destabilize population interactions and are responsible for the dramatic cycling of some predator-prey systems (Nisbet and Gurney 1982; Ricklefs 1990). The time required in consumer and host populations for shifts in the frequencies of virulence and resistance genes can establish population limit cycles with dynamics on the order of tens to hundreds of generations (Lenski and May 1994; Frank 1996; Roy and Kirchner 2000). When coupled to other populations in a system, these cycles could lead to changes in community composition and provide openings for new "invasive" species (e.g., Shea and Chesson 2002). Longer-term dynamics in ecological systems might reflect the origin of new variation in species interactions that depends on functionally important mutations arising by chance and becoming established at long intervals.

An Empirical Look at Natural Range Expansions

Natural invasions involving colonization

Islands have provided important settings for examining natural invasions and the development of ecological communities (Carlquist 1974; Wagner and Funk 1995; Losos 1996; Williamson 1996; Whittaker 1998; Ricklefs and Bermingham

2001). Islands are discrete geographic entities with well-defined biotas. Multiple islands in archipelagoes provide an opportunity to examine generalized (i.e., consistent across islands) aspects of invasion and community development (Losos et al. 1998). Individual large islands and groups of small islands also illustrate the buildup of communities internally through evolutionary radiations of endemic lineages (Givnish and Sytsma 1997; Schluter 2000). Where these radiations involve allopatric formation of new species, they can lead to repeated reinvasion of areas (secondary sympatry), which increases diversity locally (Lack 1947; Grant 1986). Thus, the dynamics of this community-building process might provide insight into the likelihood of a species becoming established in an intact natural community, especially one that includes a sister species that, because of common ancestry, is ecologically similar.

Adaptive radiation within isolated islands and island groups can build diversity more rapidly than repeated colonization of an island from outside sources (Losos and Schluter 2000). For example, the contemporary passerine (songbird) avifauna of the Hawaiian Islands represents the establishment of only six non-native lineages. Yet, one of these, the Drepanidinae (honeycreepers), derived from a cardueline finch colonist, has diversified into 30 or so named species and others known only from the bones of extinct forms (James and Olson 1991). At least 15 species co-occur on the largest (and youngest) of the islands, Hawaii, all of them having colonized the island from elsewhere. The morphological and ecological diversity of the drepanid clade, which has diversified over perhaps 5 million years (Fleischer et al. 1998; Fleischer and McIntosh 2001), exceeds that of all other passerine birds in several respects, particularly the morphology of the beak (Lovette et al. 2002). Similarly, a single lineage of *Drosophila* fruit flies that reached the Hawaiian Islands has diversified into hundreds of species that collectively exceed both the non-Hawaiian species richness and the morphological diversity of the genus worldwide (Carson and Kaneshiro 1976). Although one can argue that these radiations build diversity within relatively empty ecological space, there is no evidence that the accumulation of diversity is slowing or that it will reach a saturation level (compare Terborgh and Faaborg 1980; Ricklefs 1987; Ricklefs 2000; Ricklefs 2004a).

Ricklefs and Bermingham (2001, 2004) used a molecular phylogenetic approach to examine in more detail the dynamics of bird species in the Lesser Antilles. These islands are relatively close to sources of colonists in northern South America and the Greater Antilles. The contemporary small land bird fauna of the archipelago includes 57 lineages, the oldest of which may be approximately 10 million years old. Based on the distribution of colonization times, there appears to be little or no background extinction of lineages within the archipelago as a whole, but gaps in the distributions of some old taxa indicate that individual island populations have disappeared. Established populations on the larger core islands of the Lesser Antilles (Guadeloupe, Dominica, Martinique, St. Lucia) appear to have life spans averaging about 2 million years. The molecular phylogenetic data also indicate that endemic lineages occasionally undergo phases of reexpansion within the Lesser Antilles, which undoubtedly minimizes archipelago-wide extinction. Phases of contraction and expansion apparent in the molecular phylogenetic record occur at intervals of hundreds of thousands to millions of years, making their connection with currently invasive species tenuous and reinforcing the appearance of stability in natural systems. Three "natural" invasions of islands by land birds within the Lesser Antilles have occurred during the past century: the shiny cowbird (*Molothrus bonairiensis*), Carib grackle (*Quiscalus lugubris*), and bareeyed thrush (*Turdus nudigenis*) (Raffaele et al. 1998). All these species are closely associated with landscapes altered by human activities, particularly agricultural and residential areas.

Only two small evolutionary radiations have occurred within the archipelago, one producing two species of *Eulampis* hummingbirds (E. Bermingham and R. E. Ricklefs, unpublished data) and the other producing four species of endemic thrashers (Mimidae) in three or four genera, depending on the taxonomy adopted (Hunt et al. 2001). The absence of marked adaptive radiation of land birds in the Lesser Antilles as compared with Hawaii supports the idea that invasion and secondary sympatry are hindered when ecological space is densely packed. Indeed, the endemic thrashers were one of the earliest of the currently extant lineages to invade the islands, and they might initially have diversified within relatively open ecological space. Examination of island biotas thus presents conflicting views. On one hand, existing species and packed ecological space appear to constrain evolutionary radiation within island groups, while on the other hand, colonization from outside appears to face little resistance.

Natural range expansion within continental regions

Invasion dynamics are more difficult to perceive within large continental areas than on islands. The biota of a region can be considered as a single large community within which the distribution of species among areas and habitats reflects the evolutionary history of their ecological relationships (Ricklefs 2004b). These relationships are established partly through adaptations for tolerance of physical conditions of the environment and ability to utilize ecological substrates unique to particular habitats. Thus, the ecological (and geographic) distributions of species within a region expand and contract because of changes in the physical environment (e.g., climate), but also in response to the relative balance of adaptations for defense against consumers and exploitation of resources (Ricklefs 2004b). In the first case, species with similar tolerances would exhibit parallel changes; in the second case, phases of expansion and contraction would probably be more idiosyncratic.

The relative importance of general and particular factors is apparent in the distribution of variation among levels of the taxonomic hierarchy. If variance in range size or breadth of ecological distribution, for example, is concentrated at a high taxonomic level, then differences between taxa most likely reflect conserved adaptations to stable attributes of the environment, including physical conditions and general habitat preferences. If most of the variance resides at a lower taxonomic level, then differences are more likely reflect special relationships between organisms and their environments, including specialized herbivores, predators, parasites, and pathogens, as well as physiological adaptations to particular physical conditions. Ricklefs and Cox (1972) argued that the lack of association between expansion or contraction phases of West Indian birds and taxonomic affiliation reflected special relationships of this kind with predators or pathogens, although no data were (or are) available to assess this directly.

More recently, Gaston (1998, 2003) conducted nested analyses of variance on range size in a variety of organisms, finding that most of the variance resided on the taxonomic level of species within genera. In a similar analysis, Webb and Gaston (2003) found no correlation of range size in 103 pairs of avian sister species; if range size were conservative, the correlation should have been strong and positive. Thus, most of the variation in range size occurs between species in the same genus. The implication of these findings is that the factors that determine range size are unrelated to the general adaptations of organisms to climate or the relative availability of different habitat types, which one would expect to be more conservative.

Scheuerlein and Ricklefs (2004) conducted a similar partitioning of variance in a large sample of European passerine birds. They found that variation in most measures of distribution, including overall population size and the north-south location of the midpoint of a species range, resided at the level of species within genera (85% and 87% respectively). In contrast, 80% of the variance in distribution along a habitat gradient from open habitats, including wetlands, to forests was at the level of genera within families, reflecting more conserved adaptations for using different vegetation strata and habitat types. Variation in several life history traits, such as body mass, clutch size, and the duration of incubation, resided primarily at the level of genera within families or families within passerine birds. The implication of this finding for invasion biology is that habitat structure is probably more critical to establishment success than climate. As Janzen (1985) pointed out, the ranges of many organisms extend from temperate regions right through the tropics.

Stotz et al. (1996) compiled several measures of abundance and breadth of ecological and geographic distributions for all Neotropical birds. As found by Gaston and by Scheuerlein and Ricklefs, between 74% and 87% of the variance in relative abundance, number of habitats occupied, and number of zoogeographic regions occupied in species of non-raptorial land birds resided at the level of species within genera (Table 7.1). Only the stratum of vegetation (i.e., terrestrial, understory, mid-canopy, canopy, and aerial) varied substantially among genera within families (45%) and families within orders (43%), again reflecting more conserved morphological adaptations for performing well in different parts of the habitat (R. E. Ricklefs, unpublished data). At a more local scale, 71% of the variance in abundance of trees on the 50-ha Forest Dynamics Plot on Barro Colorado Island, Panama, resides at the level of species within genera, with an additional 22% among genera within families

TABLE 7.1   *Distribution of variance in measures of ecological and geographic distribution of Neotropical nonraptorial land birds within a nested taxonomic hierarchy*

	df	Stratum[a]	Relative abundance[b]	Sqrt (min)[c]	Sqrt (max)[c]	Nhab[d]	Logzoo[e]
Order	8	1.3	1.3	3.8	0.0	0.0	0.7
Family	48	42.7	1.8	1.5	5.7	8.8	6.3
Genus	684	45.4	20.8	40.5	24.1	17.3	6.5
Species	2292	10.5	76.2	54.3	70.1	73.9	86.5

Source: Based on data in Stotz et al. 1996.
[a] Stratum: terrestrial (1), understory (2), midstory (3), canopy (4), aerial (5).
[b] Relative abundance: rare (1), uncommon (2), fairly common (3), common (4).
[c] Sqrt, square root of the minimum (min) elevation and maximum (max) elevations (meters).
[d] Nhab, number of habitats (out of 15 forest and 14 open scrub/grassland habitat types).
[e] Logzoo, logarithm of the number of zoogeographic regions (out of 22).

(Ricklefs 2005). These analyses imply that local communities must be dynamic on time scales that correspond to the formation of new species within genera, and conceivably on shorter time scales. Thus, species continually establish themselves in and disappear from habitats and zoogeographic regions, creating continental-scale expansions and contractions of ecological extent and geographic range. Relative abundance, number of habitats, and number of zoogeographic regions are also significantly correlated with one another at the level of species within genera, suggesting that geographic extent and habitat breadth share a common, local, demographic basis (Ricklefs and Cox 1978). This pattern could reflect the relative balance of adaptations governing relationships with other, relatively specialized organisms.

The connection between natural colonization and ecological spread

If a population's distribution reflected specific attributes or interactions, natural colonization might be relatively deterministic; for example, following release from consumer pressure resulting in high population density and an increase in dispersal. Although it has been difficult to predict which species should make good colonists at any particular time (e.g., Ricklefs and Cox 1972; Terborgh et al. 1978), one can test whether colonists to one destination have independently colonized other destinations accessible from the source area. If so, this would indicate that some species have attributes that generally enhance their capacities as colonists (Simberloff and Boecklen 1991; Duncan 1997). It is probably more than a coincidence that many resident birds of the Galápagos Islands are derived from taxa that have colonized other islands off the west coast of the Americas or in the West Indies [for example, mockingbirds (*Mimus*), yellow warblers (*Dendroica petechia*), *Myiarchus* flycatchers, *Tiaris*-like emberizid finches, *Zenaida* doves].

Figure 7.1   Number of zoogeographic regions throughout the Neotropics occupied by birds that occur on the island of Trinidad and which have, or have not, colonized the Lesser Antilles. (Zoogeographic data from Stotz et al. 1996: $G = 8.7$, df = 3, $P = 0.034$; analysis from R. E. Ricklefs and E. Bermingham, unpublished data).

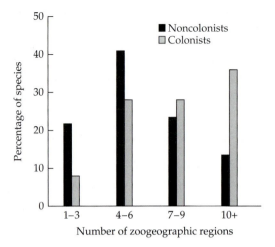

Of the small, non-raptorial land birds present on Trinidad, 25 species have invaded the Lesser Antilles, separated from that island by a 150-km water gap to the north, whereas 169 have no extant descendants in that archipelago. The colonists have a wider distribution among zoogeographic regions in the Neotropics than the noncolonists (Figure 7.1; see also Blackburn and Duncan 2001).

The determinism of colonization among species that have invaded the Lesser Antilles, as well as its transient nature, can also be seen by examining the occurrence of the source populations on small islands close to their distributions in northern South America (islands off the northern coast of Venezuela) or the Greater Antilles (small islands close to Puerto Rico). Colonizing species are distinguished as being young colonists (less than 2% mtDNA sequence divergence [ca. 1 million years] and taxonomically undifferentiated) or old colonists. Twenty-four of 33 young colonists to the Lesser Antilles, but only 7 of 24 colonists that have given rise to older, generally endemic species, occur on these peripheral islands ($G = 10.6$, df = 1, $P = 0.001$) (R. E. Ricklefs and E. Bermingham, unpublished data). Evidently, the colonizing ability of the source populations of the older Antillean taxa has waned, and extinction has claimed former peripheral island populations.

The picture of natural invasiveness that emerges from such analyses of birds is one of substantial dynamism of ecological and geographic distributions of species within regions and of turnover of species within local communities following disappearance of established species and establishment of new species from outside. The time scale of these processes appears to be less than the durations of species, but how much so remains unclear. Our work on Lesser Antillean birds suggests that 10^5 to more than 10^6 years separate phases of expansion (including initial colonization). The lability of geographic and ecological distributions indicates that the dynamics of natural invasions are largely independent of evolutionary change in relation to the physical environment, which has gone through dramatic glacial cycles over this period. Rather, this lability appears to reflect changing demography influenced by the balance of evolutionary relationships of populations with various types of consumers. These processes are difficult to observe on the time scales of natural expansions and contractions of ranges and ecological extents. Thus, an examination of invasions assisted by human interventions might provide insight into natural processes.

Key Issues in the Taxon Cycle and Invasion Biology Literature

Recent invasions and the properties of invasive species

The study of invasive species has led to a number of generalizations about the qualities of good invaders and the susceptibility of environments to invasion (Mack et al. 2000). Conventional wisdom states, first, that species invasions are limited to the range of conditions to which the colonists are adapted in their native areas (Peterson 2003), and second, that intact natural habitats resist invasion because ecological space is saturated and invading species cannot get a foothold (Williamson 1996). This is equivalent to saying that non-native species are not good competitors because they are not adapted to local environments. The first statement surely must apply to invasive species over the short term, which does not provide time for tolerance ranges to evolve. Each species occurs within what is called an environmental envelope circumscribing the range of environmental conditions over which the population extends in its native area (Cumming 2002). This range of conditions, when overlaid on an area of introduction, frequently determines the range within which a species can become established (Peterson 2003). In this context, Blackburn and Duncan (2001) determined that the invasion success of birds depended on how well the abiotic environment at the introduction site corresponded to conditions within the native range. In their analysis, the relative paucity of successful avian introductions in tropical regions was statistically unrelated to the resistance of these regions to invasions because of their high species richness, but rather resulted from poor matches, or "ecological fits" (Janzen 1985), with the native ranges of most introduced species.

The generalization that intact, diverse communities resist establishment of non-native species derives from the susceptibility of islands to introductions and the common occurrence of invasive species as weeds in disturbed habitats. It seems possible to me that this pattern is seen because both intentionally and unintentionally introduced species most often originate from human-influenced environments in their native ranges. Few truly forest-inhabiting species of birds have been introduced to Hawaii, for example, and the failure of introduced species to penetrate native forests in the Hawaiian Islands (Scott 1986) could be attributed in part to this selectivity. Where suitable intact environments are available, introduced species sometimes do quite well. For example, in an extensive survey of alien species in mixed-grass prairie in North Dakota, Larson et al. (2001) determined that despite the influence of anthropogenic disturbance, habitat type, and alien plant invasion, five of the six most abundant alien species had distributions unrelated to disturbance.

WHAT MAKES A SPECIES A GOOD INVADER?   There have been many attempts to link introduction success and invasiveness to intrinsic properties of species (e.g., Rejmánek and Richardson 1996). Characteristics proposed to enhance the invasion success of plants have included various life history traits associated with weediness, particularly those contributing to a high reproductive rate and the ability to become established on disturbed or degraded areas (Daehler 2003). Among birds, invasion success has been linked to body size (Cassey 2001), behavioral flexibility (Sol et al. 2002), habitat generalism, lack of migratory tendency, and sexual monochromatism (Cassey 2002).

If invasive species have particular attributes that make them good invaders, then independent introductions to different areas should produce a strong correlation in invasion success, as is evident in natural colonizations. Such a pattern was found in an analysis of the introduced birds of Hawaii by Simberloff and Boeklin (1991). With respect to birds introduced to New Zealand, Duncan (1997) reported a high correlation in success or failure between attempts when species were introduced to more than one area. Duncan also showed that this correlation was probably an artifact of propagule pressure (introduction effort), which was highly correlated between different introduction locations for the same species. In another analysis that considered a broader geographic region, Duncan found no correlation in the invasion success of the same bird species in different places. However, the eventual sizes of the ranges of introduced species in New Zealand were correlated with their native ranges in Great Britain (Duncan et al. 1999).

Ricklefs and Cox (1972) determined that recent immigrants to the Lesser Antilles from South America tend to inhabit open country, including grasslands, agricultural habitats, and gardens, and are relatively abundant in those habitats within their source areas. P. Cassey et al. (unpublished data) recently examined avian introductions on a global scale, analyzing the influence of introduction effort and several attributes of species (body mass, native geographic range size, annual fecundity, dietary generalism, habitat generalism, sexual dichromatism, and migratory tendency) on invasion success. In this analysis, introduction effort was the most important factor, and among the species attributes, only habitat generalism (the number of seven major habitat types included in a species' range: mixed lowland forest, alpine scrub and forest, grassland, mixed scrub, marsh and wetland, cultivated and farmlands, urban environs) had a significant effect. Diamond (1975) emphasized that not all successful colonists are good competitors. Indeed, many species of birds that are widely distributed throughout Melanesia and Micronesia, called "supertramps," are restricted to small islands with few other species. Diamond inferred that while these species readily colonize larger islands with well-developed avifaunas, they do not become established there.

PROPAGULE PRESSURE   The importance of propagule pressure suggests that the best intrinsic predictor of invasion success is the desirability of a species to humans—its sporting, aesthetic, biological control, and other values—which determines introduction effort. Of course, naturally invasive species colonize new areas under their own steam, and determinants of propagule pressure in such cases are more likely to include the intrinsic properties of species that make them capable of crossing barriers to dispersal or the demographic pressures of high population density that cause individuals to seek living space away from their natal areas. Psychological and physiological factors might also be important in the sense that few inner forest species—for example, antbirds, ovenbirds, woodcreepers, and manakins from South America—colonize islands across water (Ricklefs and Cox 1972), or even venture across narrow roadways through Amazonian forests (Laurance et al. 2004).

Colonization, which is aided by human intercession in the case of most present-day range expansions, is only the first step to becoming invasive. It must be followed by establishment and then spread throughout the new area. In the case of natural invasions, we have little evidence of failed colonization events, which, because they are transient, leave little or no trace. The stochastic element of invasiveness owing to small propagule size must cause many colonists to fail to become established (Williamson 1996).

In the Lesser Antilles, two major colonizations of land birds in the last century, the glossy cowbird and the bare-eyed thrush, have moved northward from island to island in a remarkably deterministic manner (Bond 1956). Thus, whatever qualities have made these species effective colonizers from northern South America to Grenada, the first island in the chain, have led to repeated success in crossing further gaps between islands. The intervals between colonizations were about 20 years, which would allow time for populations to fill available habitat on one island before moving on to the next. It is clear that these species have generated their own propagule pressure, which has depended on the intrinsic qualities of the colonizing species and the availability of favorable agricultural and garden habitats.

The distributions of older colonists of the Lesser Antilles suggest a difference between those with northern and southern origins (Ricklefs and Bermingham 2004). Many invasions from the south stop somewhere short of the northern Lesser Antilles, whereas most from the north extend all the way to Grenada (but rarely invade South America) (Figure 7.2). The cause of the truncated invasions of the southern species, of which approximately 20% with established populations have not colonized the next island, could be either a stochastic failure to become established and spread or a reduction of propagule pressure, caused by changes in the ecological relationships of a species on an island, as the invasion proceeds. As far as I can surmise, successive island-hopping along this linear chain depends on the establishment and growth of populations on each island independently. The more consistent penetration of the archipelago by species colonizing from the north may be related to the fact that these pop-

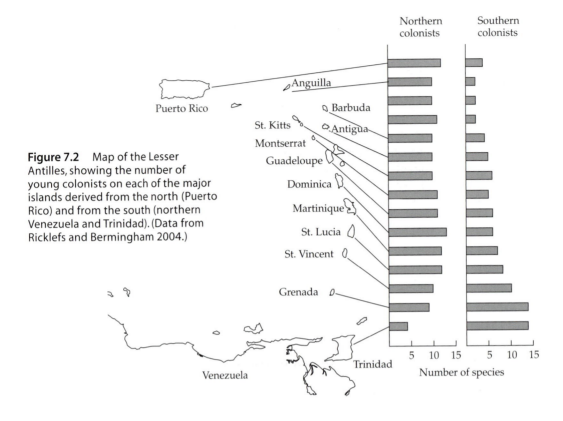

Figure 7.2 Map of the Lesser Antilles, showing the number of young colonists on each of the major islands derived from the north (Puerto Rico) and from the south (northern Venezuela and Trinidad). (Data from Ricklefs and Bermingham 2004.)

ulations started from island sources in the Greater Antilles (presumably Puerto Rico). They had to cross a larger water gap against prevailing winds to colonize smaller islands with less favorable habitats before arriving at the larger core islands of the Lesser Antilles.

Lessons from biotic interchanges

Occasionally, changes in climate or the bridging of land or water gaps by sea level change and tectonic movements lead to major exchanges of species between regions, commonly referred to as biotic interchanges. One of the best documented of these cases resulted from completion of the land bridge between North and South America during the Pliocene, creating what Simpson (1980) and Webb (1991) have called "the Great American Interchange." Many lineages of mammals crossed the Isthmus of Panama in both directions. In South America, several of these lineages underwent rapid adaptive radiations, apparently at the expense of the native mammalian fauna. Other exchanges have been equally spectacular, including the invasion and eventual dominance of the Holarctic by passerine birds with evolutionary roots in Australasia (Ericson et al. 2002) at the expense of many diverse non-passerine clades

(Mayr and Manegold 2004); broad exchanges between the Nearctic and Palearctic across Greenland and the Bering Strait at various times during the Tertiary (Tiffney 1985; Donoghue et al. 2001; Sanmartin et al. 2001); and exchanges across a Miocene mesic land connection between Africa and Eurasia (Cooke 1972; Kappelman et al. 2003). Vermeij (1991), who reviewed cases of marine biotal exchange during the Neogene, pointed out that invaders usually constitute a small percentage—typically less than 10%—of the available species pool, and that invaders constitute a variable percentage of the recipient species pool depending on the relative diversity of the regions and the asymmetry of movement during the interchange (also see Vermeij, this volume). Nonetheless, these exchanges can have dramatic and long-lasting effects on species pools and local communities.

Biotic interchanges could teach us useful lessons about biotic invasions if we had information about the attributes of the species taking part in the interchanges and the fates of those species and the of regional species pools following their establishment in a non-native area. Specifically, one would like to know whether exchanges of species between regions increased their species pools. One explanation for the greater plant species diversity in temperate eastern Asia compared with similar environments in eastern North America is the broad boundary over which species can colonize temperate Asia from more diverse tropical areas to the south (Latham and Ricklefs 1993a,b; Qian and Ricklefs 2000). As in the case of species diversity on islands (MacArthur and Wilson 1967), colonization pressure from outside tends to increase native species pools. When that pressure is released, do native species pools return to pre-exchange levels? In the case of North American mammals, pulses of immigration of taxa from Asia in the Miocene and Pliocene did not boost species richness regionally or locally (Van Valkenburgh and Janis 1993). Indeed, Webb (1989) argued that immigration from Asia was possible only after extrinsic climatic factors had reduced local species richness and opened ecological space for new colonists. Webb has not, however, applied this argument to the Great American interchange.

Range expansion

DIVERSITY AND RESISTANCE TO INVASION   In a given environment, increasing numbers of species should pack ecological space more tightly and offer increasing resistance to the establishment of new invaders. This reasoning has led to the idea that diverse communities resist invasion by non-native species (Tilman 1997; Wiser et al. 1998; Knops et al. 1999; Levine and D'Antonio 1999; Levine 2000; Naeem et al. 2000; Sax and Brown 2000; Hector et al. 2001; Shea and Chesson 2002), although not all empirical studies support that idea (Hawkins and Marino 1997; Wiser et al. 1998). A corollary is that colonists should become established in island environments more readily than in similar continental environments. In apparent confirmation, few cases of island forms invading continents are known. This observation should be tempered, however, by the possibility that colonizers from islands might be difficult to identify as such. A second caveat is that the odds of invasion from continent to island overwhelm the odds of events in the reverse direction because of the difference in relative diversities and population sizes between islands and continents. This difference makes cases of avian invasion of South America from the Lesser Antilles—for example, the bananaquit (*Coereba flaveola*: Seutin et al. 1994), *Icterus* orioles (Omland et al. 1999), and possibly *Tiaris* finches (Burns et al. 2002)—all the more remarkable.

With respect to human-assisted introductions of birds, Blackburn and Duncan (2001) determined that the diversity of species in the non-native area was not an important consideration in invasion success. However, this test is meaningful only when diversity is scaled with respect to the capacity of an environment to support coexisting species. When natural biotas have achieved ecological equilibrium diversity, resistance to invasion should be uniformly high and not strongly dependent on local species richness. Sol (2000) came to the related conclusion for birds that invasion success on islands was no greater than within continental areas.

Evidence that competition or other interactions restrict the establishment of species comes from studies relating the composition of communities on islands to that of communities drawn at random from potential colonists. Such tests have been conducted for a number of island systems, including the native avifauna of the West Indies, based on species-area relationships for members of different feeding guilds on islands (Terborgh 1973; Faaborg 1982), saturation curves relating local diversity within selected habitats to total island (regional) diversity (Terborgh and Faaborg 1980; but see Ricklefs 2000 for conflicting data on this point), size distributions within feeding guilds on islands compared with null distributions (Faaborg 1982; Case et al. 1983), and distributions of species among feeding guilds on oceanic and land-bridge islands (Faaborg 1985). These studies provide evidence for ecological sorting and species interactions on islands, but they do not directly address whether colonization is depressed by high species diversity on islands.

Working with the introduced Hawaiian avifauna, Moulton and Pimm (1983) showed that failures of introduced island populations increased dramatically with increasing diversity of introduced species, implying that the intensity of competition set upper limits to the size of the avian community. Alternatively, because introductions to Hawaii came to a halt by 1940, the pattern observed by Moulton and Pimm might have resulted simply from stochastic extinctions with time lags of a few years to a few decades combined with reduced introduction pressure. Furthermore, extinctions were strongly concentrated among doves and pigeons on some islands, suggesting that these results might not be generalizable.

In another analysis of the introduced Hawaiian avifauna, Moulton and Pimm (1987) demonstrated that average morphological minimum spanning tree segments constructed from successfully introduced species that have invaded native forest were significantly greater than those calculated by random draws from all introduced species, successful or not. The implication of this result is that extinctions were more frequent among groups of similar species (see also Moulton 1985). However, the morphological positions of the native Hawaiian avifauna were not included in this analysis, and it is unclear how the introduced species fit into native communities. Moulton and Lockwood (1992) obtained a similar result showing overdispersion of surviving introduced populations of finches on Oahu. These finches inhabit primarily open, disturbed habitats that harbor few species of native Hawaiian birds, so these results pertain more to the de novo construction of avian communities than to the establishment of new species in endemic communities. Viewed as a whole, the evidence concerning the role of competition in organizing the species that constitute a community is more convincing than that for the role of competition in preventing invasion.

Although species interactions structure relationships in established ecological communities, they do not necessarily limit membership in those communities. Nor does the establishment of new colonists necessarily lead to extinction of natives (Sax et al. 2002; Davis 2003). One strong lesson from MacArthur and Wilson's equilibrium theory is that island biotas are shaped by extrinsic factors and that the pressure of colonization influences diversity, both for entire islands and within local habitats. Terborgh and Faaborg (1980) found evidence for saturating species richness in island habitats, but only on the largest islands in the Greater Antilles. Cox and Ricklefs (1977) and Ricklefs (2000) found, instead, that diversity within individual habitats increased continuously as approximately the square root of island diversity, indicating graded resistance to establishment in new habitats by colonists, but not saturation of local communities. Based on a morphological analysis of land bird communities in the Caribbean basin, Travis and Ricklefs (1983) concluded that assemblies of species within individual habitats on small islands in the Lesser Antilles (St. Lucia and St. Kitts) were less densely packed than on the mainland (Panama and Trinidad) and Greater Antillean islands (Jamaica). The latter had nearest-neighbor distances in morphological space typical of continental assemblages. Ricklefs and Bermingham's (2001) analysis of the distribution of colonization times of Lesser Antillean birds suggested that the avifauna was not close to saturation.

RELEASE OF INVASIVE SPECIES FROM PATHOGENS   An important point that has been made with respect to introduced species is that invasion success is associated with a release from herbivores, parasites, and pathogens (Maron and Vila 2001; Keene and Crawley 2002; Wolfe 2002; Mitchell and Power 2003; Torchin et al. 2003; see also Lafferty et al., this volume), although factors such as range sizes of native and non-native species might confound this relationship (Clay 1995). Small propagules probably sample only a small part of the total pathogen community in the native range, and additional pathogens can be lost after introduction owing to stochastic effects in initially small populations and absence of suitable vectors or environmental conditions for free-living stages of pathogens (Dobson and May 1986). Provided that introduced species do not pick up new pathogens, they may experience rapid population growth and superior competitive ability in the absence of suppression by parasites, herbivores, and disease. This is the rationale behind biological control programs that seek to reverse this situation by restoring consumers of invasive pest species from their native areas wherever possible (Charudattan 2001).

Ricklefs and Cox (1972) also suggested that release from predators and pathogens could lead to the initiation of an expansion phase of the taxon cycle, although there may be essential differences between natural and human-aided invaders. While species that naturally colonize new areas might escape pathogens in a manner analogous to anthropogenic introductions, this release cannot explain increases in propagule pressure that might lead to colonization success in the first place. In the framework of coevolution, loss of pathogens depends on the acquisition of genetic factors through mutation and selection that effectively defend the individual against infection or predation or control their harmful effects at a low level. Simple models of the cycling of virulence and resistance genes in pathogen and host populations have dynamics that are too fast to explain the long intervals between expansion phases in natural populations (Lenski and May 1994; Frank 1996; Roy and Kirchner 2000). If host resistance depended on the fixation of rare mutations that could not be readily countered by pathogen evolution, then phases of expansion might arise abruptly, but at long intervals determined by the appearance of suitable genetic variability.

If coevolution were important, colonizing and noncolonizing species would not necessarily differ ecologically, and the composition of invaded communities would be relatively unimportant. Rather, invasion success would be determined by demographic rather than habitat factors, the dispersal of individuals from crowded populations would be the major driver of range and habitat expansion, and invasiveness would decline over time owing to pathogen coevolution independently of change in the physical environment. Presumably, any advantage that a host population enjoys owing to escape from pathogens would be transient because the diverse array of pathogens in the new environment would be selected to utilize the newly abundant resource. Thus, one might expect the colonization phase of the taxon cycle to be relatively short-lived. Rapid coevolution of pathogens and hosts is well known from studies of human diseases, including syphilis (Knell 2003) and falciparum malaria (Hartl et al. 2002), and one would not expect a host species to maintain an advantage over a pathogen population for long.

Torchin et al. (2003) and others have demonstrated that introduced individuals of non-native species typically harbor fewer parasites than individuals in their native range. Relatively few data are available to address this point in the case of natural introductions. Most examples come from islands, which typically lack diverse communities of parasites and pathogens. A classic case concerns birds in the Hawaiian Islands (Van Riper et al. 1986). Prior to colonization by Europeans, the islands lacked suitable mosquito vectors of such avian diseases as poxvirus and malaria (*Plasmodium / Haemoproteus*). As a result, the introduced species (that is, the now native honeycreepers and other lineages that colonized the islands on their own) had lost most of their genetic defenses against these pathogens. When *Culex* mosquitoes, and then both poxvirus and *Plasmodium relictum*, were introduced to the Hawaiian Islands during the nineteenth and twentieth centuries, these diseases decimated many of the native populations. Bird species introduced more recently from continental areas are infected by the malaria pathogen, but host individuals control infections at low levels (Van Riper et al. 1986; Jarvi et al. 2001).

Loss of parasite diversity is evident in bird communities on small islands in the West Indies, but the patterns of parasite distribution are also idiosyncratic. For example, populations of bananaquits and of Lesser Antillean bullfinches (*Loxigilla noctis*) on Barbados harbor only the commonest one of the variety of malaria parasite lineages found in these species on St. Lucia and St. Vincent (Fallon et al. 2005), which were the source areas for the Barbados populations (Lovette et al. 1999). Yet, on Barbados, that single parasite is now found in other species of avian hosts that are not infected by the lineage elsewhere. Clearly, the colonization of Barbados acted as a filter for avian malaria parasites, but the one parasite lineage that became established there was also able to switch to new hosts.

Variation in the relative prevalence of particular lineages of malaria parasites in different island populations of the same host species suggests independent host-parasite coevolution in each of the island populations (Fallon et al. 2003). The further observation that the total prevalence of all parasite lineages in a particular host is relatively constant suggests that these pathogens interact with one another, potentially indirectly through the host immune system.

Another aspect of invasion is that colonizing species might bring endemic pathogen populations with them, which might spread to the detriment of native species in the colonized areas, increasing the competitive ability of the invaders. This has certainly been the case with the European colonization of the New World (Diamond 1997). This model has many interesting implications for natural invasions and the structuring of natural communities. For example, invasion might be easier in remote islands because the few pathogens there cannot prevent the establishment of invading populations. Beyond this, the absence of pathogens might remove a block to the establishment of secondary sympatry within archipelagoes, facilitating evolutionary radiation of species (R. E. Ricklefs and E. Bermingham, unpublished data). Pathogens might also control the direction of invasion between continental regions and islands where islands are fairly close to the mainland. Invaders from more diverse communities might harbor more varied pathogens and present a formidable defense against new pathogens.

Range contraction: Decline and extinction of populations

The eventual fate of a species is to decline in numbers and distribution and eventually to disappear. Although extinction is accepted as a fact of life, just as mortality is the accepted fate of individuals, its causes are not well understood and may be quite diverse. In the case of non-native species, notable declines have been brought about by the importation of natural enemies. Examples include myxoma virus in the case of the European rabbit introduced to Australia (Fenner and Ratcliffe 1965; Kerr and Best 1998), the cactus moth for the prickly pear cactus introduced to Australia (Dodd 1959), and chrysomelid beetles for Klamath weed introduced to western North America (Huffaker and Kennett 1959; Harris et al. 1969). In most cases, control agents have been effective in controlling populations at endemic levels, but rarely in causing extinction. Indeed, most invasive populations coexist with these control agents in their native ranges.

In recent times, humans have been the most effective agents of extinction in many regions. Relatively few cases of introduced species causing extinction of native island populations through competitive interactions have been reported (Sax et al. 2002; Davis 2003; Sax and Gaines 2003). However, many introduced species have greatly expanded their distributions at the expense of native populations, particularly in continental regions. Examples include the shrub *Miconia calvescens* on Pacific islands (Meyer and Florence 1996), the American crayfish in Europe (Lodge 2001), zebra mussels in North America (Martel et al. 2001; Nalepa et al. 2001), various fishes in North America (Miller et al. 1989), and gray squirrels in Great Britain (Usher et al. 1992). Most extinctions on islands are caused by human activity, either directly by hunting or indirectly by introduction of predators and pathogens. The lack of strong evidence for competition in such cases further undermines the idea that higher diversity or greater community "intactness" helps communities resist invasion. If this were so, then one would expect cases in which excess diversity and community saturation drive extinction to be more apparent.

In a detailed examination of extinction in the New Zealand avifauna, Duncan and Blackburn (2004) distinguished a "prehistoric" period of Maori colonization of the islands and a "historic" period of European colonization. During the first period, hunting of large species by native islanders and egg predation of small species by introduced Polynesian rats appear to have been the major factors in the disappearance of bird species. Later, after Europeans had introduced Norway rats, cats, and other highly effective mammalian predators, direct predation on adult birds, particularly flightless species, wiped out a different set of the native avifauna. Introduced birds were not a factor in prehistoric extinctions, and there is little evidence that they played a direct role in later extinctions through competition. Few introduced birds have penetrated native habitats. Extinction was also well under way in the Hawaiian avifauna before Europeans reached the islands (James and Olson 1991) and before the introduction of potentially competing species of birds. Historic population declines and extinctions in Hawaii can be related to the combined effects of hunting, introduced predators, introduced diseases, and habitat destruction, rather than competition from non-native species.

What characteristics make a species vulnerable or, alternatively, resistant to extinction? Are the same factors relevant in both natural and human-caused extinctions? All species present today are the living representatives of lineages that have not gone extinct. And because these species represent the totality of contemporary diversity, it follows that all have features that resist extinction. Any morphological or physiological attribute that is shared among close relatives cannot be the cause of extinction of any species independently of the particular environment of that species. Thus, natural extinctions in the contemporary biota of the earth must be independent of most characteristics that we associate with their life histories and general ecological relationships.

Human-caused extinctions, and other mass extinction events, appear to be selective (e.g., Jablonski 1989, 1991; McKinney 1997; von Euler 2001), resulting from specific mortality agents whose effects are generally beyond the capacity of evolutionary response. Bolide impacts causing mass extinctions are the extreme example. Unless the normal fate of all species is not to become extinct, and extinction results primarily from catastrophic agents, human-caused extinctions can provide little insight into the natural decline of species. Clearly, however, the distributions and abundances of species vary tremendously, notably among closely related species, and this variation suggests a more dynamic aspect to the normal condition of life on earth. Many observers have commented on the vulnerability of large ground-dwelling animals to hunting by humans. Other attributes make certain island populations particularly vulnerable to introduced predators. Small population size is also seen as an important factor, although this might be like saying that old age is a factor in human mortality; that is, small populations may simply be the penultimate stage of a long decline.

The concept of the taxon cycle encompasses both expansion and decline toward extinction. Before species disappear, they are likely to exist as small populations with limited distributions. How they become so is the important issue. Ricklefs and Cox (1972) recognized stages of the taxon cycle in West Indian birds and devised a logical analysis for inferring both the relative ages of species in different stages, taking into account the fact that expansion phases could be initiated by declining taxa. The temporal sequence of the stages was confirmed by molecular phylogenetic analyses (Ricklefs and Bermingham 1999; Ricklefs and Bermingham 2002). Ricklefs and Cox (1978) had already shown that late-stage species, regardless of their taxonomic relationships or general ecological position, exhibited restricted habitat distributions centered on forested and montane habitats and were less abundant within occupied habitats than were early-stage species. Furthermore, individual island populations of species in later stages of the taxon cycle were more likely to have gone extinct during historic times or to be in a precarious position owing to hunting, introduced predators, habitat destruction, and other human-related causes (Figure 7.3).

This picture, which has been developed for a very limited sample of the earth's biota, suggests that the normal progression for invasive species is to decline and, in the extreme, to disappear. About the causes of this decline, and its reversal when new phases of expansion occur, I can do little more than speculate. I am reasonably certain that phases of expansion are short-lived and reflect transient periods of release from factors that limit populations under

Figure 7.3 Proportion of individual island populations that either became extinct since their scientific discovery or are in critical danger of extinction at present, as a function of their stage of the taxon cycle, in three archipelagoes. Taxon cycle stages, which represent increasing age in the West Indies (Ricklefs and Bermingham 1999), as defined by Ricklefs and Cox (1972), are (1) widespread and undifferentiated taxonomically; (2) widespread and differentiated; (3) differentiated with distributional gaps; (4) single-island endemics. (Data for the West Indies are from Ricklefs and Cox 1972; others are unpublished data from R. E. Ricklefs.)

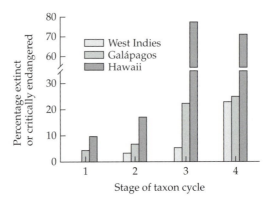

normal circumstances. I am even more confident that subsequent decline takes place over evolutionary time spans—in the case of birds on islands, perhaps hundreds of thousands of generations. The lack of coincidence in taxon cycle stages among closely related species suggests that expansion and decline depend on the special relationship of each species to factors in its environment. These factors are most likely to be other specialized resources, consumers, or pathogens. The fact that small island populations persisted through major cycles of climate change in the Pleistocene suggests that such changes are not important and that populations adjust easily. Following a decline, extinction may depend on the colonization of an area by new predators, diseases, or competitors. Thus, species in late stages of the taxon cycle might persist for very long periods on isolated islands. The implication of this pattern for non-native species is that, once invasive species have become firmly established, it is unlikely that they will decline and disappear rapidly through restorative community processes. The responsibility for controlling invasive species would appear to lie with their initial benefactors—namely, us (Simberloff 2003).

Explicit Comparisons of the Taxon Cycle and Species Invasions: Reciprocal Insights and Predictions

The general attributes of human-assisted and natural invasions are compared in Table 7.2. All invasions are destined to pass through four distinct phases: (1) introduction or colonization, (2) establishment or naturalization, (3) geographic range expansion within the non-native area, and (4) eventual decline. Observations on natural range expansions and their eventual decline—a complete taxon cycle—suggest that phases 1 through 3 are relatively brief, at least in geologic and molecular-clock time, and that, once established, the decline stage may last for time spans recognizable only in the fossil record or through inference of evolutionary diversification within the expanded range. The complete revolution of the taxon cycle is most readily recognizable in island archipela-

TABLE 7.2 *General attributes of invasions aided by human activities compared to natural invasions*

Invasion stage	Human-assisted invasion	Natural invasions	
		Colonization-based	Expansion-based
Introduction or colonization	Intentional or accidental human-assisted introductions; highly dependent on introduction effort	Depends on ability to disperse across barriers and on population pressures that promote dispersal; haphazard introductions	Dispersal ability generally not a factor, as this type of invasion involves expansion into adjacent habitats
Establishment or naturalization	Requires a suitable environment; environment often anthropogenic, generally within the range of conditions in the native area	Less of a consideration because these invasions usually involve nearby areas having similar conditions	Not a factor, as the initial stage is an established population
Geographic range expansion	Must be able to fit into the competitive and pathogenic environment; evolutionary adjustment may play a role	Presumably the same criteria of ecological compatibility apply to both colonization-based and expansion-based natural invasions; evolutionary adjustment to new environments may be important in both cases (see Cox 2004)	
Eventual decline	Rarely a consideration; generally not observed without substantial human intervention	Over evolutionary time, taxa eventually decline ecologically and withdraw from invaded habitats or become extinct in colonized areas; this phase most apparent in island systems	

goes, in which distribution is partitioned into discrete island units, history reveals itself through genetic divergence between island populations, and decline is apparent as gaps in the distribution of a species across islands.

Typically, studies of human-assisted species invasions address the first three stages of this process, particularly establishment and geographic range expansion. Nevertheless, relatively little is known about colonization—the rain of propagules that arrive in areas and soon disappear without becoming established. The transience of this stage of the taxon cycle makes it difficult to observe, except in the case of intentional human introductions (e.g., Duncan 1997). At the other end of the cycle, except in the case of obvious habitat change or human control programs, declining phases are rarely observed in human-assisted invasions. Once established, populations are difficult to get rid of, and left on their own they tend to persist for evolutionary time spans.

Early stages of colonization and establishment often include periods of rapid evolutionary adaptation to local conditions and in response to interactions with local species (Cox 2004). This adaptation process may account for a lag time between the first establishment of a population and its subsequent expansion to occupy a substantial part of its adopted region. Why some species expand in their non-native ranges and others do not is an open question. Ecologists have not reached a consensus concerning the general attributes that make a species a good invader. Those species that eventually do expand their ranges following establishment are thought to benefit from leaving behind in their native ranges consumers and pathogens that could otherwise keep their populations in check. This demographic release is thought to be a factor in natural invasions as well. Indeed, it is my belief that the acquisition of natural genetic resistance to one or more significant predators, parasites, or pathogens provides the initiation for new phases of expansion.

The decline phase of the taxon cycle under natural conditions may require substantial time for evolutionary adjustment of native populations to the new invader or for the arrival of new competitors as colonists from elsewhere (Wilson 1961; Ricklefs and Cox 1972). Thus, it is unlikely that recently established non-native species will provide insights into the historical decline of natural populations, or that inferences about these declines from the fossil record or molecular phylogeographic analysis will enlighten us about the fates of recently established non-natives or the control of those non-native populations that we would like to be rid of. Nevertheless, natural and human-assisted invasions can inform each other in a number of ways.

Perhaps the most important insights to be gained will come from continuing studies of the attributes of species that determine naturalization success and expansion within non-native areas. Considering that closely related species vary tremendously in this regard, as well as in range size and ecological distribution within their native areas, success may depend on subtle genetic factors, possibly associated with herbivore or disease resistance or with direct competitive abilities (e.g., Callaway and Aschehoug 2000; Callaway et al. 2001, 2004).

In many cases of human-assisted invasions, particularly intentional introductions, dispersal ability is not a critical factor for initial establishment, and colonization can occur over global distances to regions greatly isolated in evolutionary time. With intentional introductions, population bottlenecks are less frequent, and genetic diversity is maintained in the colonizing population (see Wares et al., this volume). Indeed, multiple introductions from different parts of a species' natural range might lead to increased genetic variation in local non-native populations, as in the case of the brown anole (*Anolis sagrei*), introduced to Florida from several locations in Cuba and the Bahamas (Kolbe et al. 2004; see also Novak and Mack, this volume). In bypassing adaptations required for long-distance dispersal, such colonists might be in a position to take advantage of traits that enhance their competitive ability in diverse communities. Cody and Overton (1996) demonstrated rapid loss of dispersal ability in several weedy plants that colonized islands in Barkley Sound, British Columbia, suggesting that light, wind-dispersed seeds, which are essential for dispersing to disturbed habitats, are impediments to persistence on small offshore islands.

Whether a general trade-off between dispersal and competitive ability is a general feature of invasion biology, as it seems to be for "supertramp" species of birds in Melanesia (Diamond 1975), may depend on the remoteness of the destination. However, for the rare colonists to Hawaii and other similarly isolated spots, competitive ability might not be a critical attribute early in the development of the island fauna. In general, however, one expects a decline in long-distance dispersal ability, including the loss of flight in insects and birds, following colonization of remote sites (Darlington 1943; Carlquist 1974). To the extent that humans introduce species that otherwise lack the capacity for long-distance dispersal, they short-circuit this natural phase of the taxon cycle and potentially increase the probability of establishment and expansion.

The contributions of human-assisted introductions to our understanding of the initial phases of the taxon cycle lie in the large sample sizes available for comparative analyses, the experimental manipulations of introduction effort and the conditions in receiving environments that can be performed, and the detailed historical records of the establishment and spread of introduced populations. Studying the fates of naturally expanding and contracting populations emphasizes the importance of evolved interactions of species with other populations, including competitors, predators, and pathogens, in determining distribution and abundance. The long persistence times of established natural populations suggest that individuals can tolerate broad ranges of physical (climatic) conditions and that coevolutionary relationships tend to strike long-term balances. The implication for the control of invasive species is that their populations will not tend to dwindle of their own accord; rather, their control will often require substantial human intervention.

Conclusions and Suggestions for Future Work

Easily measured attributes of species are unlikely to predict relative invasion success beyond broad generalizations. Success or failure is more likely to depend on finer adjustments to environmental conditions or with respect to relations with predators, pathogens, and perhaps mutualists and competitors. These relationships reveal themselves in the ecological and geographic distributions of species within their native areas, and possibly also in their relative abundances within their natural ranges. The two are often correlated (Bock and Ricklefs 1983; Brown 1984; Brown 1995; Blackburn and Gaston 2001). Most introductions involve species that are common and widespread in their native ranges, and thus the correlation of invasion success with native distribution might not be adequately quantified. However, although many naturally rare and localized species have been successful in horticulture or captive propagation (for example, *Metasequoia, Gingko, Franklinia*), such species do not often become invasive. Nonetheless, a more systematic examination of the relationship between invasion success and distribution within the native range would provide useful indications of the types of factors that are likely to influence invasion success (e.g., Duncan et al. 1999). Factors responsible for variation in native ranges are also an important open question for ecological research.

Certainly, physiological tolerance ranges must have some bearing on the location of a population's center of abundance and in some cases the conditions within which a population can maintain itself. However, because common and rare species within the same lineages often live side by side in some environments, physiological tolerance is not likely to provide the entire answer to distribution.

To the extent that variations in range size and ecological distribution reflect coevolved relationships with predators, pathogens, and perhaps resource populations, ecologists will have to undertake detailed experimental studies on, for example, host-pathogen and host-parasite relationships to evaluate the effects of these factors. In addition, these studies will have to be carried out within the context of community ecology, rather than the more typical perspective of population interactions in narrowly circumscribed systems. One can imagine broadly comparative studies of experimental infection or parasitism involving both common and rare (hence logistically challenging) host species. Considering the nature of this work, invertebrates will perhaps be more felicitous subjects than vertebrates, and plants more than animals. The simple prediction is that common and widespread species not only should exhibit lower intensities of infections and infestations, but also should resist infection better than narrowly distributed host species. Common species might also be expected to harbor fewer specialized parasites and pathogens that can circumvent the main host defenses. Understanding the mechanisms by which resistance is achieved would indicate the kinds of evolutionary adjustments that are responsible for variation in distribution and abundance.

The history of these evolutionary adjustments, or at least their effects on abundance and distribution, might be appreciated in some cases by tracing character change over phylogenetic trees. Of course, extremely labile characters responsible for variation in distribution would not exhibit a recognizable history of change on the scale of phylogenetic relationships of species or higher taxa, because such characters would change within terminal branches. Nonetheless, to the extent that the distributions of smaller, population-level units could be characterized within species, phylogenetic analysis might provide a picture of infraspecific evolution. Within the Lesser Antilles, for example, populations of some species on different islands vary widely in both habitat distribution and abundance.

At a larger scale, ecologists could use historical-phylogenetic analyses of the development of island and continental biotas to gain insights into community assembly (Ricklefs 2002, 2004c; Webb et al. 2002; Ackerly 2003; Cavender-Bares and Wilczek 2003). Such longer-term processes, which must involve evolutionary adjustments of ecological distributions that govern competition and other relationships, would not be closely related to expansion and contraction phases in the context of taxon cycles. They might, however, indicate prerequisites for species to fit into established ecological assemblages. Discrete islands or habitat continua would provide suitable contexts for investigating these processes. Contemporary assemblages of species on an island or at a point along a continuum have a history of assembly (Ricklefs and Bermingham 2001; Ricklefs 2004b) that might involve the nonrandom accumulation of species over time or the nonrandom selection of colonists from a potential pool (Terborgh et al. 1978; Weiher et al. 1998; Weiher and Keddy 1999). The autochthonous buildup of diversity within a region through adaptive radiation might also require predictable adjustments of morphology and behavior before reproductively isolated sister populations can tolerate secondary sympatry. Aspects of genetic incompatibility have been discussed (e.g., Coyne and Orr 1997), but ecological compatibility has received less attention (see, however, Schluter 2000).

Taxon cycles represent one point along a continuum of phenomena that involve evolutionary adjustment of ecological relationships, and they reflect the balance of these relationships in the demography of populations. The success or failure of non-native species introduced to new areas presumably also depends on the kinds of attributes that control distribution and abundance in native species. Thus, it is likely that studies of each kind of phenomenon will benefit from keeping the other in mind.

Acknowledgments

I am grateful to the editors of this book and to two anonymous reviewers for insightful and constructive comments on an earlier draft of this chapter. This work was conducted as part of the "Exoctic Species: A Source of Insight into Ecology, Evolution, and Biogeography" working group supported by the National Center for Ecological Analysis and Synthesis, a Center funded by the National Science Foundation (Grand DEB-0072909), the University of California, and the Santa Barbara Campus.

Literature Cited

Ackerly, D. D. 2003. Community assembly, niche conservatism, and adaptive evolution in changing environments. International Journal of Plant Sciences 164:S165–S184.

Alroy, J. 1998. Equilibrial diversity dynamics in North American mammals. In M. L. McKinney, ed. *Biodiversity dynamic: turnover of populations, taxa, and communities*, pp. 232–287. Columbia University Press, New York.

Berenbaum, M. R. 1983. Coumarins and caterpillars: a case for coevolution. Evolution 37:163–179.

Blackburn, T. M., and R. P. Duncan. 2001. Determinants of establishment success in introduced birds. Nature 414:195–197.

Blackburn, T. M., and K. J. Gaston. 2001. Linking patterns in macroecology. Journal of Animal Ecology 70:338–352.

Bock, C. E., and R. E. Ricklefs. 1983. Range size and local abundance of some North American songbirds: a positive correlation. American Naturalist 122:295–299.

Bond, J. 1956. *Checklist of birds of the West Indies*. Academy of Natural Sciences, Philadelphia.

Brown, J. H. 1984. On the relationship between the abundance and distribution of species. American Naturalist 124:255–279.

Brown, J. H. 1995. *Macroecology*. University of Chicago Press, Chicago.

Brown, J. H., and M. V. Lomolino. 1998. *Biogeography*, 2nd Ed. Sinauer Associates, Sunderland, MA.

Brown, W. L., Jr. 1957. Centrifugal speciation. Quarterly Review of Biology 32:247–277.

Bull, J. J. 1994. Virulence. Evolution 48:1423–1437.

Burns, K. J., S. J. Hackett, and N. K. Klein. 2002. Phylogenetic relationships and morphological diversity in Darwin's finches and their relatives. Evolution 56:1240–1252.

Cain, S. A. 1944. *Foundations of plant geography*. Harper and Brothers, New York.

Callaway, R. M., and E. T. Aschehoug. 2000. Invasive plants versus their new and old neighbors: a mechanism for exotic invasion. Science 290:521–523.

Callaway, R. M., B. Newingham, C. A. Zabinski, and B. E. Mahall. 2001. Compensatory growth and competitive ability of an invasive weed are enhanced by soil fungi and native neighbours. Ecology Letters 4:429–433.

Callaway, R. M., S. C. Pennings, and C. L. Richards. 2003. Phenotypic plasticity and interactions among plants. Ecology 84:1115–1128.

Callaway, R. M., G. C. Thelen, A. Rodriguez, and W. E. Holben. 2004. Soil biota and exotic plant invasion. Nature 427:731–733.

Carlquist, S. J. 1974. *Island biology*. Columbia University Press, New York.

Carson, H. L., and K. Y. Kaneshiro. 1976. *Drosophila* of Hawaii: systematics and ecological genetics. Annual Review of Ecology and Systematics 7:311–345.

Case, T. J., J. Faaborg, and R. Sidell. 1983. The role of body size in the assembly of West Indian bird communities. Evolution 37:1062–1074.

Cassey, P. 2001. Are there body size implications for the success of globally introduced land birds? Ecography 24:413–420.

Cassey, P. 2002. Life history and ecology influences establishment success of introduced land birds. Biological Journal of the Linnaean Society 76:465–480.

Cassey, P., T. M. Blackburn, D. Sol, R. P. Duncan, and J. L. Lockwood. 2004. Global patterns of introduction effort and establishment success in birds. Proceedings of the Royal Society of London B 271 (Suppl. 6):S405–S408.

Cavender-Bares, J., and A. Wilczek. 2003. Integrating micro- and macroevolutionary processes in community ecology. Ecology 84:592–597.

Charudattan, R. 2001. Biological control of weeds by means of plant pathogens: significance for integrated weed management in modern agro-ecology. BioControl 46:229–260.

Clay, K. 1995. Correlates of pathogen species richness in the grass family. Canadian Journal of Botany / Revue Canadienne de Botanique 73:S42-S49.

Cody, M. L. 1975. Towards a theory of continental species diversities. In M. L. Cody and J. M. Diamond, eds. *Ecology and evolution of communities*, pp. 214–257. Harvard University Press, Cambridge, MA.

Cody, M. L., and J. M. Diamond. 1975. *Ecology and evolution of communities*. Harvard University Press, Cambridge, MA.

Cody, M. L., and J. M. Overton. 1996. Short-term evolution of reduced dispersal in island plant populations. Journal of Ecology 84:53–61.

Cooke, H. B. S. 1972. The fossil mammal fauna of Africa. In A. Keast, F. C. Erk and B. Glass, eds. *Evolution, mammals, and southern continents*, pp. 89–139. State University of New York Press, Albany, NY.

Cox, G. W. 2004. *Alien species and evolution: the evolutionary ecology of exotic plants, animals, microbes, and interacting native species*. Island Press, Washington, DC.

Cox, G. W., and R. E. Ricklefs. 1977. Species diversity, ecological release, and community structuring in Caribbean land bird faunas. Oikos 29:60–66.

Coyne, J. A., and H. A. Orr. 1997. Patterns of speciation in *Drosophila* revisited. Evolution 51:295–303.

Cumming, G. S. 2002. Comparing climate and vegetation as limiting factors for species ranges of African ticks. Ecology 83:255–268.

Daehler, C. C. 2003. Performance comparisons of co-occurring native and alien invasive plants: implications for conservation and restoration. Annual Review of Ecology and Systematics 34:183–211.

Darlington, P. J., Jr. 1943. Carabidae of mountains and islands: data on the evolution of isolated faunas, and on atrophy of wings. Ecological Monographs 13:37–61.

Davis, M. A. 2003. Biotic globalization: does competition from introduced species threaten biodiversity? BioScience 53:481–489.

Davis, M. B. 1986. Climatic instability, time lags, and community disequilibrium. In J. M. Diamond and T. J. Case, eds. *Community ecology*, pp. 269–284. Harper and Row, New York.

Day, T. 2001. Parasite transmission modes and the evolution of virulence. Evolution 55:2389–2400.

Diamond, J. M. 1975. Assembly of species communities. In M. L. Cody and J. M. Diamond, eds. *Ecology and evolution of communities*, pp. 342–444. Harvard University Press, Cambridge, MA.

Diamond, J. M. 1997. *Guns, germs, and steel. The fates of human societies*. W. W. Norton, New York.

Dobson, A., and R. M. May. 1986. Patterns of invasions by pathogens and parasites. In H. A. Mooney and J. A. Drake, eds. *Ecology of biological invasions of North America and Hawaii*, pp. 58–75. Springer-Verlag, New York.

Dodd, A. P. 1959. The biological control of prickly pear in Australia. In A. Keast, R. L. Crocker and C. S. Christian, eds. *Biogeography and ecology in Australia*, pp. 565–577. Dr. W. Junk BV, The Hague.

Donoghue, M. J., C. D. Bell, and J. H. Li. 2001. Phylogenetic patterns in Northern Hemisphere plant geography. International Journal of Plant Sciences 162:S41–S52.

Duncan, R. P. 1997. The role of competition and introduction effort in the success of passeriform birds introduced to New Zealand. American Naturalist 149:903–915.

Duncan, R. P., and T. M. Blackburn. 2004. Extinction and endemism in the New Zealand avifauna. Global Ecology and Biogeography 13:509–517.

Duncan, R. P., T. M. Blackburn, and C. J. Veltman. 1999. Determinants of geographical range sizes: a test using introduced New Zealand birds. Journal of Animal Ecology 68:963–975.

Dynesius, M., and R. Jansson. 2000. Evolutionary consequences of changes in species' geographical distributions driven by Milankovitch climate oscillations. Proceedings of the National Academy of Sciences USA 97:9115–9120.

Ehrlich, P. R., and P. H. Raven. 1964. Butterflies and plants: a study in coevolution. Evolution 18:586–608.

Elton, C. S. 1958. *The ecology of invasions by animals and plants*. Methuen, London.

Ericson, P. G. P., L. Christidis, A. Cooper, M. Irestedt, J. Jackson, U. S. Johansson, and J. A. Norman. 2002. A Gondwanan origin of passerine birds supported by DNA sequences of the endemic New Zealand wrens. Proceedings of the Royal Society of London B 269:235–241.

Erwin, T. C. 1981. Taxon pulses, vicariance, and dispersal: an evolutionary synthesis illustrated by carabid beetles. In G. Nelson and D. E. Rosen, eds. *Vicariance biogeography: a critique*, pp. 159–196. Columbia University Press, New York.

Ewald, P. W. 1983. Host-parasite relations, vectors, and the evolution of disease severity. Annual Review of Ecology and Systematics 14:465–485.

Faaborg, J. 1982. Trophic and size structure of West Indian bird communities. Proceedings of the National Academy of Sciences USA 79:1563–1567.

Faaborg, J. 1985. Ecological constraints on West Indian bird distributions. Ornithological Monographs 36:621–653.

Fallon, S. M., E. Bermingham, and R. E. Ricklefs. 2003. Island and taxon effects in parasitism revisited: avian malaria in the Lesser Antilles. Evolution 57:606–615.

Fallon, S. M., E. Bermingham, and R. E. Ricklefs. 2005. Host specialization and geographic localization of avian malaria parasites: A regional analysis in the Lesser Antilles. American Naturalist 165:466–480.

Farrell, B., C. Mitter, and D. J. Futuyma. 1992. Diversification at the insect-plant interface. BioScience 42:34–42.

Farrell, B. D., and C. Mitter. 1994. Adaptive radiation in insects and plants: time and opportunity. American Zoologist 34:57–69.

Fenner, F., and F. N. Ratcliffe. 1965. *Myxomatosis*. Cambridge University Press, London.

Fleischer, R. C., and C. E. McIntosh. 2001. Molecular systematics and biogeography of the Hawaiian avifauna. Studies in Avian Biology 22:51–60.

Fleischer, R. C., C. E. McIntosh, and C. L. Tarr. 1998. Evolution on a volcanic conveyor belt: using phylogeographic reconstructions and K-Ar-based ages of the Hawaiian Islands to estimate molecular evolutionary rates. Molecular Ecology 7:533–545.

Frank, S. 1996. Models of parasite virulence. Quarterly Review of Biology 71:37–78.

Gaston, K. J. 1998. Species-range size distributions: products of speciation, extinction and transformation. Philosophical Transactions of the Royal Society of London B 353:219–230.

Gaston, K. J. 2003. *The structure and dynamics of geographic ranges*. Oxford University Press, Oxford.

Givnish, T. J., and K. J. Sytsma. 1997. *Molecular evolution and adaptive radiation*. Cambridge University Press, Cambridge.

Grant, P. R. 1986. *Ecology and evolution of Darwin's finches*. Princeton University Press, Princeton, NJ

Greenslade, P. J. M. 1968. Island patterns in the Solomon Islands bird fauna. Evolution 22:751–761.

Greenslade, P. J. M. 1969. Insect distribution patterns in the Solomon Islands. Philosophical Transactions of the Royal Society of London B 255:271–284.

Harris, P., D. Peschken, and J. Milroy. 1969. The status of biological control of the weed *Hypericum perforatum* in British Columbia. Canadian Entomologist 101:1–15.

Hartl, D. L., S. K. Volkman, K. M. Nielsen, A. E. Barry, K. P. Day, D. F. Wirth, and E. A. Winzeler. 2002. The paradoxical population genetics of *Plasmodium falciparum*. Trends in Parasitology 18:266–272.

Hawkins, B. A., and P. C. Marino. 1997. The colonization of native phytophagous insects in North America by exotic parasitoids. Oecologia 112:566–571.

Hector, A., K. Dobson, A. Minns, E. Bazeley-White, and J. H. Lawton. 2001. Community diversity and invasion resistance: an experimental test in a grassland ecosystem and a review of comparable studies. Ecological Research 16:819–831.

Huffaker, C. B., and C. E. Kennett. 1959. A ten-year study of vegetational changes associated with biological control of Klamath weed. Journal of Range Management 12:69–82.

Hunt, J. S., E. Bermingham, and R. E. Ricklefs. 2001. Molecular systematics and biogeography of Antillean thrashers, tremblers, and mockingbirds (Aves: Mimidae). Auk 118:35–55.

Jablonski, D. 1989. The biology of mass extinction: a palaeontological view. Philosophical Transactions of the Royal Society of London B 325:357–368.

Jablonski, D. 1991. Extinctions: a paleontological perspective. Science 253:754–757.

Jackson, S. T., R. S. Webb, K. H. Anderson, J. T. Overpeck, T. Webb, J. W. Williams, and B. C. S. Hansen. 2000. Vegetation and environment in eastern North America during the last glacial maximum. Quaternary Science Reviews 19:489–508.

Jacobson, G. L., Jr., T. Webb, III, and E. C. Grimm. 1987. Patterns and rates of vegetation change during the deglaciation of eastern North America. In W. F. Ruddiman, and H. E. Wright, Jr., eds. *North American and adjacent oceans during the last deglaciation*, Geological Society of America, Boulder, CO.

James, H. F., and S. L. Olson. 1991. Descriptions of thirty-two new species of birds from the Hawaiian Islands. Part II: Passeriformes. Ornithological Monographs Volume 46.

Janzen, D. H. 1985. On ecological fitting. Oikos 45:308–310.

Jarvi, S. I., C. T. Atkinson, and R. C. Fleischer. 2001. Immunogenetics and resistance to avian malaria in Hawaiian honeycreepers (Drepanidinae). Studies in Avian Biology 22:254–263.

Kappelman, J., D. T. Rasmussen, W. J. Sanders, M. Feseha, T. Bown, P. Copeland, J. Crabaugh, J. Fleagle, M. Glantz, A. Gordon, B. Jacobs, M. Maga, K. Muldoon, A. Pan, L. Pyne, B. Richmond, T. Ryan, E. R. Seiffert, S. Sen, L. Todd, M. C. Wiemann, and A. Winkler. 2003. Oligocene mammals from Ethiopia and faunal exchange between Afro-Arabia and Eurasia. Nature 426:549–552.

Keene, R. M., and M. J. Crawley. 2002. Exotic plant invasions and the enemy release hypothesis. Trends in Ecology and Evolution 17:164–170.

Kelt, D. A., and J. H. Brown. 1999. Community structure and assembly rules: confronting conceptual and statistical issues with data on desert rodents. In E. Weiher and P. A. Keddy, eds. *The search for assembly rules in ecological communities*, pp. 75–107. Cambridge University Press, Cambridge.

Kerr, P. J., and S. M. Best. 1998. Myxoma virus in rabbits. Revue Scientifique et Technique 17:256–268.

Knell, R. J. 2003. Syphilis in Renaissance Europe: rapid evolution of an introduced sexually transmitted disease? Proceedings of the Royal Society of London B 271:S174–S176.

Knops, J. M. H., D. Tilman, N. M. Haddad, S. Naeem, C. E. Mitchell, J. Haarstad, M. E. Ritchie, K. M. Howe, P. B. Reich, E. Siemann, and J. Groth. 1999. Effects of plant species richness on invasion dynamics, disease outbreaks, insect abundances and diversity. Ecology Letters 2:286–293.

Kolbe, J. J., R. E. Glor, L. R. G. Schettino, A. C. Lara, A. Larson, and J. B. Losos. 2004. Genetic variation increases during biological invasion by a Cuban lizard. Nature 431:177–181.

Lack, D. 1947. *Darwin's finches*. Cambridge University Press, Cambridge.

Lack, D. 1976. *Island biology illustrated by the land birds of Jamaica*. University of California Press, Berkeley, CA.

Larson, D. L., P. J. Anderson, and W. Newton. 2001. Alien plant invasion in mixed-grass prairie: effects of vegetation type and anthropogenic disturbance. Ecological Applications 11:128–141.

Latham, R. E., and R. E. Ricklefs. 1993a. Continental comparisons of temperate-zone tree species diversity. In R. E. Ricklefs and D. Schluter, eds. *Species diversity in ecological communities*, pp. 294–314. University of Chicago Press, Chicago.

Latham, R. E., and R. E. Ricklefs. 1993b. Global patterns of tree species richness in moist forests: energy-diversity theory does not account for variation in species richness. Oikos 67:325–333.

Laurance, S. G. W., P. C. Stouffer, and W. E. Laurance. 2004. Effects of road clearings on movement patterns of understory rainforest birds in central Amazonia. Conservation Biology 18:1099–1109.

Lenski, R. E., and R. M. May. 1994. The evolution of virulence in parasites and pathogens: reconciliation between two competing hypotheses. Journal of Theoretical Biology 169:253–265.

Levine, J. M. 2000. Species diversity and biological invasions: relating local process to community pattern. Science 288:852–854.

Levine, J. M., and C. M. D'Antonio. 1999. Elton revisited: a review of evidence linking diversity and invasibility. Oikos 87:15–26.

Lodge, D. M. 2001. Lakes. In F. S. Chapin, III, O. E. Sala and E. Huber-Sannwald, eds. *Future scenarios of global biodiversity*, pp. 277–312. Springer-Verlag, New York.

Losos, J. B. 1996. Community evolution in Greater Antillean *Anolis* lizards: phylogenetic patterns and experimental tests. In P. H. Harvey, A. J. L. Brown, J. M. Smith and S. Nee, eds. *New uses for new phylogenies*, pp. 308–321. Oxford University Press, Oxford.

Losos, J. B., T. R. Jackman, A. Larson, K. De Queiroz, and L. Rodriguez-Schettino. 1998. Contingency and determinism in replicated adaptive radiations of island lizards. Science 279:2115–2118.

Losos, J. B., and D. Schluter. 2000. Analysis of an evolutionary species-area relationship. Nature 408:847–850.

Lovette, I. J., E. Bermingham, and R. E. Ricklefs. 2002. Clade-specific morphological diversification and adaptive radiation in Hawaiian songbirds. Proceedings of the Royal Society of London B 269:37–42.

Lovette, I. J., E. Bermingham, G. Seutin, and R. E. Ricklefs. 1999. The origins of an island fauna: a genetic assessment of sources and temporal patterns in the avian colonization of Barbados. Biological Invasions 1:33–41.

MacArthur, R. H. 1965. Patterns of species diversity. Biological Reviews 40:510–533.

MacArthur, R. H., and R. Levins. 1967. The limiting similarity, convergence, and divergence of coexisting species. American Naturalist 101:377–385.

MacArthur, R. H., and E. O. Wilson. 1963. An equilibrium theory of insular zoogeography. Evolution 17:373–387.

MacArthur, R. H., and E. O. Wilson. 1967. *The theory of island biogeography*. Princeton University Press, Princeton, NJ.

Mack, R. N., D. Simberloff, W. M. Lonsdale, H. Evans, M. Clout, and F. A. Bazzaz. 2000. Biotic invasions: causes, epidemiology, global consequences, and control. Ecological Applications 10:689–710.

Maron, J. L., and M. Vila. 2001. When do herbivores affect plant invasion? Evidence for the natural enemies and biotic resistance hypotheses. Oikos 95:361–373.

Martel, A. L., D. A. Pathy, J. B. Madill, C. B. Renaud, S. L. Dean, and S. J. Kerr. 2001. Decline and regional extirpation of freshwater mussels (Unionidae) in a small river system invaded by *Dreissena polymorpha*: the Rideau River, 1993–2000. Canadian Journal of Zoology / Revue Canadienne de Zoologie 79:2181–2191.

May, R. M. 1975. *Stability and complexity in model ecosystems*. Princeton University Press, Princeton, NJ.

Mayr, G., and A. Manegold. 2004. The oldest European fossil songbird from the early Oligocene of Germany. Naturwissenschaften 91:173–177.

McKinney, M. L. 1997. Extinction vulnerability and selectivity: combining ecological and paleontological views. Annual Review of Ecology and Systematics 28:495–516.

McKinney, M. L. 1998. Biodiversity dynamics: niche preemption and saturation in diversity equilibria. In M. L. McKinney, ed. *Biodiversity dynamics. Turnover of populations, taxa and communities*, pp. 1–16. Columbia University Press, New York.

Meyer, J. Y., and J. Florence. 1996. Tahiti's native flora endangered by the invasion of *Miconia calvescens* DC (Melastomataceae). Journal of Biogeography 23:775–781.

Miller, R. R., J. D. Williams, and J. E. Williams. 1989. Extinctions of North American fishes during the past century. Fisheries 14:22–38.

Mitchell, C. E., and A. G. Power. 2003. Release of invasive plants from fungal and viral pathogens. Nature 421:625–627.

Moulton, M. P. 1985. Morphological similarity and coexistence of congeners: an experimental test with introduced Hawaiian birds. Oikos 44:301–305.

Moulton, M. P., and J. L. Lockwood. 1992. Morphological dispersion of introduced Hawaiian finches: evidence for competition and a Narcissus effect. Evolutionary Ecology 6:45–55.

Moulton, M. P., and S. L. Pimm. 1983. The introduced Hawaiian avifauna: biogeographical evidence for competition. American Naturalist 121:669–690.

Moulton, M. P., and S. L. Pimm. 1987. Morphological assortment in introduced Hawaiian passerines. Evolutionary Ecology 1:113–124.

Naeem, S., J. M. H. Knops, D. Tilman, K. M. Howe, T. Kennedy, and S. Gale. 2000. Plant diversity increases resistance to invasion in the absence of covarying extrinsic factors. Oikos 91:97–108.

Nalepa, T. F., D. J. Hartson, D. L. Fanslow, and G. A. Lang. 2001. Recent population changes in freshwater mussels (Bivalvia : Unionidae) and zebra mussels (*Dreissena polymorpha*) in Lake St. Clair, USA. American Malacological Bulletin 16:141–145.

Nisbet, R. M., and W. S. C. Gurney. 1982. *Modelling fluctuating populations*. Wiley, New York.

Omland, K. E., S. M. Lanyon, and S. J. Fritz. 1999. A molecular phylogeny of the new world orioles (*Icterus*): the importance of dense taxon sampling. Molecular Phylogenetics and Evolution 12:224–239.

Opler, P. A. 1973. Fossil lepidopterous leaf-mines demonstrate the age of some insect-plant relationships. Science 179:1321–1323.

Peterson, A. T. 2003. Predicting the geography of species' invasions via ecological niche modeling. Quarterly Review of Biology 78:419–433.

Pielou, E. C. 1979. *Biogeography*. Wiley, New York.

Pimentel, D. 1961. Animal population regulation by the genetic feedback mechanism. American Naturalist 95:65–79.

Pimentel, D., and R. Al-Hafidh. 1963. The coexistence of insect parasites and hosts in laboratory populations. Annals of the Entomological Society of America 56:676–678.

Pimentel, D., E. H. Feinberg, P. W. Wood, and J. T. Hayes. 1965. Selection, spatial distribution, and the coexistence of competing fly species. American Naturalist 99:97–109.

Pregill, G. K., and S. L. Olson. 1981. Zoogeography of West Indian vertebrates in relation to Pleistocene climatic cycles. Annual Review of Ecology and Systematics 12:75–98.

Qian, H., and R. E. Ricklefs. 2000. Large-scale processes and the Asian bias in species diversity of temperate plants. Nature 407:180–182.

Raffaele, H., J. Wiley, O. Garrido, A. Keith, and J. Raffaele. 1998. *A guide to the birds of the West Indies*. Princeton University Press, Princeton, NJ.

Rejmánek, M., and D. M. Richardson. 1996. What attributes make some plant species more invasive? Ecology 77:1655–1661.

Ricklefs, R. E. 1970. Stage of taxon cycle and distribution of birds on Jamaica, Greater Antilles. Evolution 24:475–477.

Ricklefs, R. E. 1987. Community diversity: relative roles of local and regional processes. Science 235:167–171.

Ricklefs, R. E. 1989. Speciation and diversity: integration of local and regional processes. In D. Otte and J. A. Endler, eds. *Speciation and its consequences*, pp. 599–622. Sinauer Associates, Sunderland, MA.

Ricklefs, R. E. 1990. *Ecology*. W. H. Freeman, New York.

Ricklefs, R. E. 2000. The relationship between local and regional species richness in birds of the Caribbean Basin. Journal of Animal Ecology 69:1111–1116.

Ricklefs, R. E. 2002. Splendid isolation: historical ecology of the South American passerine fauna. Journal of Avian Biology 33:207–211.

Ricklefs, R. E. 2004a. Cladogenesis and morphological diversification in passerine birds. Nature 430:338–341.

Ricklefs, R. E. 2004b. A comprehensive framework for global patterns in biodiversity. Ecology Letters 7:1–15.

Ricklefs, R. E. 2004c. Phylogenetic perspectives on patterns of regional and local species richness. In E. Bermingham, C. Dick and C. Moritz, eds. *Tropical rainforests: past, present, and future*, pp. 16–40. University of Chicago Press, Chicago.

Ricklefs, R. E. 2005. Historical and ecological dimensions of global patterns in plant diversity. Biologiske Skrifter (Royal Danish Academy of Sciences and Letters) 55:583–603.

Ricklefs, R. E., and E. Bermingham. 1999. Taxon cycles in the Lesser Antillean avifauna. Ostrich 70:49–59.

Ricklefs, R. E., and E. Bermingham. 2001. Nonequilibrium diversity dynamics of the Lesser Antillean avifauna. Science 294:1522–1525.

Ricklefs, R. E., and E. Bermingham. 2002. The concept of the taxon cycle in biogeography. Global Ecology and Biogeography 11:353–361.

Ricklefs, R. E., and E. Bermingham. 2004. History and the species-area relationship in Lesser Antillean birds. American Naturalist 163:227–239.

Ricklefs, R. E., and G. W. Cox. 1972. Taxon cycles in the West Indian avifauna. American Naturalist 106:195–219.

Ricklefs, R. E., and G. W. Cox. 1978. Stage of taxon cycle, habitat distribution, and population density in the avifauna of the West Indies. American Naturalist 112:875–895.

Ricklefs, R. E., and I. J. Lovette. 1999. The roles of island area *per se* and habitat diversity in the species-area relationships of four Lesser Antillean faunal groups. Journal of Animal Ecology 68:1142–1160.

Rosenzweig, M. L. 1973. Evolution of the predator isocline. Evolution 27:84–94.

Roughgarden, J., and S. Pacala. 1989. Taxon cycles among *Anolis* lizard populations: review of the evidence. In D. Otte and J. A. Endler, eds. *Speciation and its consequences* pp. 403–432. Sinauer Associates, Sunderland, MA.

Roy, B. A., and J. W. Kirchner. 2000. Evolutionary dynamics of pathogen resistance and tolerance. Evolution 54:51–63.

Sanmartin, I., H. Enghoff, and F. Ronquist. 2001. Patterns of animal dispersal, vicariance and diversification in the Holarctic. Biological Journal of the Linnaean Society 73:345–390.

Sax, D. F., and J. H. Brown. 2000. The paradox of invasion. Global Ecology and Biogeography Letters 9:363–371.

Sax, D. F., and S. D. Gaines. 2003. Species diversity: from global decreases to local increases. Trends in Ecology and Evolution 18:561–566.

Sax, D. F., S. D. Gaines, and J. H. Brown. 2002. Species invasions exceed extinctions on islands worldwide: a comparative study of plants and birds. American Naturalist 160:766–783.

Scheuerlein, A., and R. E. Ricklefs. 2004. Prevalence of blood parasites in European passeriformes. Proceedings of the Royal Society of London B 271:1363–1370.

Schluter, D. 2000. *The ecology of adaptive radiation*. Oxford University Press, Oxford.

Scott, J. M. 1986. Forest Bird Communities of the Hawaiian Islands. Their Dynamics, Ecology, and Conservation. Cooper Ornithological Society, Santa, Barbara, CA.

Seutin, G., N. K. Klein, R. E. Ricklefs, and E. Bermingham. 1994. Historical biogeography of the bananaquit (*Coereba flaveola*) in the Caribbean region: a mitochondrial DNA assessment. Evolution 48:1041–1061.

Shea, K., and P. Chesson. 2002. Community ecology theory as a framework for biological invasions. Trends in Ecology and Evolution 17:170–176.

Simberloff, D. 2003. Eradication-preventing invasions at the outset. Weed Science 51:247–253.

Simberloff, D., and W. Boecklen. 1991. Patterns of extinction in the introduced Hawaiian avifauna: a re-examination of the role of competition. American Naturalist 138:300–327.

Simpson, G. G. 1949. *The meaning of evolution*. Yale University Press, New Haven, CT.

Simpson, G. G. 1980. *Splendid isolation. The curious history of South American mammals*. Yale University Press, New Haven, CT.

Sol, D., R. Jovani, and J. Torres. 2000. Geographical variation in blood parasites in feral pigeons: the role of vectors. Ecography 23:307–314.

Sol, D., S. Timmermans, and L. Lefebvre. 2002. Behavioural flexibility and invasion success in birds. Animal Behaviour 63:495–502.

Stotz, D. F., J. W. Fitzpatrick, T. A. Parker, III, and D. K. Moskovits. 1996. *Neotropical birds: ecology and conservation*. With Ecological and Distributional Databases by Theodore A. Parker III, Douglas F. Stotz, and John W. Fitzpatrick. University of Chicago Press, Chicago and London.

Terborgh, J. 1973. Chance, habitat and dispersal in the distribution of birds in the West Indies. Evolution 27:338–349.

Terborgh, J., J. Faaborg, and H. J. Brockman. 1978. Island colonization by Lesser Antillean birds. Auk 95:59–72.

Terborgh, J. W., and J. Faaborg. 1980. Saturation of bird communities in the West Indies. American Naturalist 116:178–195.

Tiffney, B. H. 1985. The Eocene North Atlantic land bridge: its importance in Tertiary and modern phyogeography of the Northern Hemisphere. Journal of the Arnold Arboretum 66:243–273.

Tilman, D. 1997. Community invasibility, recruitment limitation, and grassland biodiversity. Ecology 78:81–92.

Torchin, M. E., K. D. Lafferty, A. P. Dobson, V. J. McKenzie, and A. M. Kuris. 2003. Introduced species and their missing parasites. Nature 421:628–630.

Travis, J., and R. E. Ricklefs. 1983. A morphological comparison of island and mainland assemblages of Neotropical birds. Oikos 41:434–441.

Turchin, P. 2003. Historical Dynamics: *Why states rise and fall*. Princeton University Press, Princeton, NJ.

Usher, M. B., T. J. Crawford, and J. L. Banwell. 1992. An American invasion of Great Britain: the case of the native and alien squirrel (*Sciurus*) species. Conservation Biology 6:108–115.

Van Riper, C., III, S. G. Van Riper, M. L. Goff, and M. Laird. 1986. The epizootiology and ecological significance of malaria in Hawaiian land birds. Ecological Monographs 56:327–344.

Van Valen, L. 1973. A new evolutionary law. Evolutionary Theory 1:1–30.

Van Valkenburgh, B., and C. M. Janis. 1993. Historical diversity patterns in North American large herbivores and carnivores. In R. E. Ricklefs and D. Schluter, eds. *Species diversity in ecological communities. Historical and geographical perspectives*, pp. 330–340. University of Chicago Press, Chicago.

Vandermeer, J. H. 1972. Niche theory. Annual Review of Ecology and Systematics 3:107–132.

Vermeij, G. J. 1991. When biotas meet: understanding biotic interchange. Science 253:1099–1104.

von Euler, F. 2001. Selective extinction and rapid loss of evolutionary history in the bird fauna. Proceedings of the Royal Society of London B 268:127–130.

Wagner, W. L., and V. A. Funk. 1995. *Hawaiian biogeography: evolution on a hot spot archipelago*. Smithsonian Institution Press, Washington, DC.

Webb, C. O., D. D. Ackerly, M. A. McPeek, and M. J. Donoghue. 2002. Phylogenies and community ecology. Annual Review of Ecology and Systematics 33:475–505.

Webb, S. D. 1989. The fourth dimension in North American terrestrial mammal communities. In D. W. Morris, Z. Abramsky, B. J. Fox and M. Willig, eds. *Patterns in the structure of mammalian communities*, pp. 181–203. Texas Tech University Press, Lubbock.

Webb, S. D. 1991. Ecogeography and the Great American Interchange. Paleobiology 17:266–280.

Webb, T., III. 1987. The appearance and disappearance of major vegetational assemblages: long-term vegetational dynamics in eastern North America. Vegetatio 69:177–187.

Webb, T. J., and K. J. Gaston. 2003. On the heritability of geographic range sizes. American Naturalist 161:553–566.

Weiher, E., G. D. P. Clarke, and P. A. Keddy. 1998. Community assembly rules, morphological dispersion, and the coexistence of plant species. Oikos 81:309–322.

Weiher, E., and P. Keddy. 1999. *Ecological assembly rule: perspectives, advances, retreats*. Cambridge University Press, Cambridge and New York.

Whittaker, R. H. 1972. Evolution and measurement of diversity. Taxon 21:213–251.

Whittaker, R. H. 1977. Evolution of species diversity in land communities. Evolutionary Biology 10:1–67.

Whittaker, R. J. 1998. *Island biogeography*. Oxford University Press, Oxford.

Williamson, M. 1996. *Biological invasions*. Chapman and Hall, London.

Willis, J. C. 1922. *Age and area: a study in geographical distribution and origin in species*. Cambridge University Press, Cambridge.

Wilson, E. O. 1959. Adaptive shift and dispersal in a tropical ant fauna. Evolution 13:122–144.

Wilson, E. O. 1961. The nature of the taxon cycle in the Melanesian ant fauna. American Naturalist 95:169–193.

Wiser, S. K., R. B. Allen, P. W. Clinton, and K. H. Platt. 1998. Community structure and forest invasion by an exotic herb over 23 years. Ecology 79:2071–2081.

Wolfe, L. M. 2002. Why alien invaders succeed: support for the escape-from-enemy hypothesis. American Naturalist 160:705–711.

8

Genetic Bottlenecks in Alien Plant Species

INFLUENCE OF MATING SYSTEMS AND INTRODUCTION DYNAMICS

Stephen J. Novak and Richard N. Mack

Small populations are prone to alterations or reductions, or both, of genetic diversity through genetic drift, founder effects, and genetic bottlenecks. Such events can reduce a population's evolutionary potential and increase the risk of extinction. In this chapter, we use plant immigrants as a model system to examine the circumstances in which genetic bottlenecks do (and do not) occur and where in the hierarchical partitioning of genetic diversity such bottlenecks take place. We assess the amount and distribution of genetic diversity among plant species in their native (or donor) and introduced ranges. Two factors are important in influencing genetic bottlenecks for alien plants: their mating system, and the circumstances of their introduction. Among outcrossing alien species (e.g., Echium plantagineum, Epipactis helleborine, Trifolium hirtum*), the likelihood of a genetic bottleneck can be low because even a few immigrants may represent much of the species' genetic variation. Similarly, multiple introductions can reduce the likelihood and severity of bottlenecks among selfing alien species (e.g.,* Bromus tectorum *and* Avena barbata*). The loss of genetic diversity most often takes place across populations and not at the within-population level. Within-population genetic diversity can be higher in naturalized than in native populations if the naturalized populations are a by-product of genetic admixtures from multiple sources. As a result, multiple in-*

troductions of different genotypes from diverse donor ranges can trigger novel outcrossing events, lead to adaptive radiation, and even initiate an invasion. Accumulating evidence suggests that many plant invasions are sparked by multiple immigrations, a large founder population, or both. Consequently, the role of putative genetic bottlenecks in affecting the fate of alien species must be evaluated in the light of the species' immigration history.

Modern evolutionary biology (referred to as the modern evolutionary synthesis) merges population genetic theory with Darwinian natural selection (Fisher 1930; Wright 1931; Haldane 1932; Dobzhansky 1937; Huxley 1942; Mayr 1942; Simpson 1944; Stebbins 1950). One goal of this modern synthesis has been to understand the factors that influence or determine the type, amount, and distribution of genetic diversity within and among populations. Genetic diversity arises through mutation and recombination and can be altered across generations through natural selection, gene flow, and random genetic drift during periods when a population is small.

Small populations can experience an alteration or a reduction, or both, in genetic diversity through genetic drift, founder effects, and genetic bottlenecks (Lande 1988). As specific examples of genetic drift, founder effects are defined as genetic changes in populations that occur when few founders establish a new colony. In contrast, genetic bottlenecks are defined as reductions in genetic diversity in populations experiencing rapid, severe reductions in the number of individuals for one to a few generations (Lande 1988; Ridley 1993).

Reduction of genetic diversity during founder events and population bottlenecks is important because it can reduce the evolutionary potential of populations (Fisher 1930; Nei et al. 1975; Falconer and Mackay 1996) and increase the risk of extinction (Frankel and Soulé 1981; Lande 1988, 1993; Frankham 1995; Saccheri et al. 1998). Loss of heterozygosity may limit a population's ability to respond rapidly to selection after a bottleneck (Maruyama and Fuerst 1985), whereas reductions in allelic richness may limit a population's long-term capacity to respond to a changing environment (Allendorf 1986). When founders establish a colony or population bottlenecks occur, genetic drift becomes a major evolutionary force. Random genetic drift, in combination with demographic stochasticity (Shaffer 1981; Menges 1991, 1998), could exert potentially powerful effects within small populations.

Our goal here is to provide an understanding of the genetic and evolutionary consequences of population bottlenecks. Specifically, we examine the circumstances in which genetic bottlenecks do (and do not) occur and where in the hierarchical partitioning of genetic diversity bottlenecks take place. In this context, plant immigrants serve as excellent model systems because reductions in population size appear as recurring aspects of their population biology (Baker 1974). Despite the negative consequences associated with these reductions, plants do not always appear to be adversely affected. How often do such species experience genetic bottlenecks? Do they experience multiple bottlenecks during their introduction into a new territory? What are the genetic and evolutionary consequences (e.g., for hybridization, for adaptive radiation) of these events?

In exploring these questions, we first describe the results of theoretical models that predict genetic changes and responses for populations undergoing reductions in size. Next, we describe the characteristics of alien plants that make them an appropriate model system for evaluating the genetic and evolutionary consequences of population bottlenecks. We then assess the amount and distribution of genetic diversity in an array of plant species in their native and introduced ranges. These assessments provide insights into the occurrence and severity of genetic bottlenecks in alien plants and highlight the role of two factors that influence genetic bottlenecks in these plants: their mating system and specific details associated with their introduction (i.e., single compared with multiple introductions and small compared with large founder populations). Finally, in counterpoint, we provide historical evidence that many alien plant species have arrived so frequently and in such numbers as to add a caveat to explanations of the genetics of these species in their new ranges as the products of bottlenecks.

Theoretical Models

Genetic diversity can be partitioned at three hierarchical levels: the DNA sequence, Mendelian traits, and quantitative genetic traits (Hartl and Clark 1997; Futuyma 1998). Genetic diversity at the DNA sequence level is not our focus here. Rather, we examine Mendelian and quantitative genetic trait variation. A Mendelian (or discrete) trait is determined by the alleles at a single gene, or locus. In contrast, several to many genes or loci may contribute to the expression of a quantitative (or continuous) trait (Hartl and Clark 1997).

Theory that assesses change(s) in the genetic diversity of populations following a founder event or genetic bottleneck is invaluable for making formal predictions that can then be evaluated in nature with alien plant species. These models generally emphasize several factors: the effective population size of the founder or bottlenecked population (N_e), the duration of the population bottleneck (i.e., the number of generations), the number of population bottlenecks, the level of inbreeding in the post-bottleneck population, and life history traits. All these parameters can vary independently in affecting genetic diversity. In addition, Mendelian and quantitative traits would not be expected to respond to founder effects and genetic bottlenecks in the same manner, based on differences in the number of loci involved in their expression (Lewontin 1974). The circumstances associated with reductions in population size vary among species (and even among populations of the same species); consequently, these models make predictions about general, rather than specific, outcomes.

Mendelian traits

Models predicting genetic change in single-locus traits in populations experiencing founder effects and genetic bottlenecks have usually employed two parameters: average heterozygosity per locus and the mean number of alleles per locus (also referred to as allelic richness). Wright (1931) predicted that populations experiencing a sudden reduction in size would experience a reduction in average heterozygosity per locus. More recent theoretical results indicate, however, that average heterozygosity per locus may not be adversely affected by a reduction in population size (Nei et al. 1975) and so may not be as useful as allelic richness in detecting a genetic bottleneck (Allendorf 1986).

During founder events and population bottlenecks, random sampling can alter the frequency of alleles at polymorphic loci among individuals and may result in the fixation of alleles. This process decreases allelic richness within populations and may contribute to increased genetic differentiation among populations (Brown and Marshall 1981; Barrett and Husband 1990). Alleles with the highest frequency (common alleles) have the highest probability of being sampled during a founder event or population bottleneck; rare alleles have the lowest probability of being sampled (Nei et al. 1975; Chakraborty and Nei 1977; Watterson 1984; Maruyama and Fuerst 1985). Thus, allelic richness in a population can decrease rapidly following a bottleneck through the elimination of rare alleles, unless by chance alone rare alleles are also sampled (Barrett and Husband 1990). Conversely, the frequency of heterozygotes following a bottleneck would be maintained through the union of gametes bearing the common alleles that persist in a population. The relationship between the size of a population bottleneck (N_e) and the number of alleles remaining in the population after one generation (Table 8.1) suggests that allelic richness can serve as a sensitive indicator of genetic bottlenecks (Nei at al. 1975; Allendorf 1986).

Studies of the genetic signature of population bottlenecks have contributed to conservation biology and the study of invasions (Leberg 2002). This work clearly indicates that allelic richness is more sensitive to the effects of short,

TABLE 8.1   *Relationship between effective population size (N_e) and the number of alleles remaining in a population after one generation*

N_e	Number of alleles[a]
1000	8.00
100	7.81
10	3.86
5	2.69
1	1.35

[a] In this hypothetical example, the population contained 8 alleles before a bottleneck, 7 of which were rare. For more details, see Meffe and Carroll 1997.

severe bottlenecks than is heterozygosity (Leberg 1992, 2002; Luikart et al. 1998b; Spencer et al. 2000). Approaches used to detect recent genetic bottlenecks include testing for the distortion of allele frequency distributions (Luikart et al. 1998a), assessing temporal changes in allele frequencies with repeated sampling (Waples 1989; Richards and Leberg 1996; Luikart et al. 1999), and the "variance test," based on the standardized variance in allele frequencies (Luikart et al. 1998b). In turn, different techniques can assess genetic diversity following a bottleneck, including analyses of allozymes (Leberg 1992; Richards and Leberg 1996) and microsatellite DNA (Luikart et al. 1998b, 1999; Spencer et al. 2000). Regardless of the method used, employing loci with many alleles enhances the detection of bottlenecks. Moreover, loci with many alleles are particularly helpful for detecting cryptic genetic bottlenecks that occur in the absence of, or before the detection of, a demographic bottleneck (e.g., when there are few breeders of one sex due to a skewed sex ratio) (Luikart et al. 1998b). Recovery of both allelic richness and heterozygosity, but particularly heterozygosity, following a bottleneck will be greatest when population size increases rapidly (Nei et al. 1975). The specific circumstances associated with a founder (or colonization) event may also directly influence the amount and distribution of genetic diversity in subsequent populations.

Repeated recolonizations have genetic consequences that depend on the manner in which the emigrants are assembled and on the levels of gene flow among the resultant colonies (Slatkin 1977, 1985; Hamrick 1987; Barrett and Husband 1990; McCauley 1991). Slatkin's (1977) "propagule pool" and "migrant pool" models are two extreme cases of colony formation. In the propagule pool model, all individuals in a colony are sampled from only one source population and thus exhibit little genetic diversity. In the migrant pool model, colonies exhibit greater genetic diversity (compared with the propagule pool model) because migrants are derived from multiple populations across the entire range of a species (Slatkin 1977; Wade and McCauley 1988; Whitlock and McCauley 1990). Under the migrant pool model, moderate to high levels of gene flow among populations would result in more genetic diversity within colonizing populations and less genetic differentiation among populations, whereas, under the propagule pool model, little gene flow would yield the opposite outcome. These two scenarios are simplifications of colonization and gene flow (Whitlock and McCauley 1990); they do, however, provide a framework for assessing the consequences of founder effects and genetic bottlenecks arising from small population size during colonization.

Quantitative genetic traits

The metric value of a quantitative (or phenotypic) trait is the result of genetic factors, environmental factors, and their interaction (Falconer and Mackay 1996). The genetic variation that contributes to the total phenotypic variation can be subdivided into additive genetic variance, dominance genetic variance, and interaction genetic variance. Additive genetic variance is determined by the fixed value that each allele at each locus contributes to the expression of a quantitative trait; dominance genetic variance occurs when a dominant allele at a locus masks the contribution of a recessive allele. Epistasis, the interactions between genes, can influence the expression of quantitative traits. The variance that results from epistatic interactions is interaction genetic variance (Falconer and Mackay 1996).

The effects of founder events and population bottlenecks on the expression of quantitative traits depend on the underlying genetic architecture associated with each trait. For instance, if the genetic variation component of a quantitative trait is purely additive, a bottleneck will result in a decrease in additive genetic variance and decrease the total phenotypic variation for the quantitative trait (Wright 1931; Lande 1980; Carson and Templeton 1984). Theory also suggests that additive genetic variance may increase following a bottleneck. Models by Robertson (1952), Rose (1982), Willis and Orr (1993), Wang et al. (1998), and Lopez-Fanjul et al. (2002) show an increase in additive genetic variance following a bottleneck that is attributable to dominance genetic variance. Epistasis may also contribute to an increase in additive genetic variance (Goodnight 1987, 1988; Whitlock et al. 1993; Lopez-Fanjul et al. 1999, 2000, 2002), but these models of epistatic interactions are based on only two interacting loci. More recent models find that as the number of loci increases from two to three or four, additive genetic variance also increases (Naciri-Graven and Goudet 2003). This result is significant because the number of loci contributing to a quantitative trait associated with fitness is likely to be greater than two. Regardless of the underlying mechanism (dominance or epistasis), this process suggests that additive genetic variance, and therefore quantitative trait variation, can actually increase following a founder event or population bottleneck.

Increases in additive genetic variance following population bottlenecks have been demonstrated for morphometric and behavioral traits in the housefly (*Musca domestica*) (Bryant et al. 1986a,b; Bryant and Meffert 1996) and for components of fitness in *Drosophila melanogaster* (Lopez-Fanjul and Villaverde 1989; Garcia et al. 1994), *Tribolium castaneum* (Fernandez et al. 1995), *Bicyclus anynana* (Saccheri et al. 1996), and *Mus musculus* (Cheverud et al. 1999). Unfortunately, we know of no experimental studies of this type for plants (however, see Polans and Allard 1989), yet such studies are needed to determine whether the conversion of nonadditive variance to additive variance also occurs in plant populations experiencing population bottlenecks. Clearly, alien plant immigrations present ample opportunities and appropriate model systems for such investigations.

Quantitative traits are often ecologically important and can contribute directly to the components of fitness. A reduction in the additive genetic variance for a quantitative trait may lead to a reduction in fitness. In addition to reductions in genetic variation, founder effects and genetic bottlenecks can result in the fixation of deleterious alleles and contribute to inbreeding depression, thus further reducing the likelihood that a small population will persist. These issues are examined here by assessing the response of alien plant species to these two categories of genetic drift that could arise with immigration. (For a discussion of these issues with animal invasions, see Wares et al., this volume.)

Alien Plant Species as Model Systems

Humans have become increasingly adept at transporting plants into new ranges, often far removed from the species' native range (Mack et al. 2000). Although plants are routinely dispersed accidentally (as contaminants in seed lots or as hitchhikers in or on cargo), a huge number of species have been introduced deliberately to new ranges (Bryant 1998). The fates of such deliberately introduced species vary widely (Ridley 1930). These species' seeds or other dispersal units are not only protected in transit from environmental hazards, but are also usually cultivated upon entry into the new range. Cultivation of deliberately introduced aliens is key to their survival and eventual persistence, as it effectively buffers the immigrants and their descendants from many forms of environmental stochasticity (Mack 2000). Equally important is the opportunity for repeated deliberate introductions of these alien species, regardless of the fate of the initial founders. Not surprisingly, then, the largest fraction of the naturalized flora in the United States and elsewhere has arisen from deliberately introduced species (Mack and Erneberg 2002, and see references therein). Cultivation and the likelihood of repeated introduction enhance the opportunity for a deliberately introduced alien species not only to become naturalized (i.e., persistent in a new range), but also to proliferate, spread, and cause damage (Mack 2000). At this point in its history in the new range, the species is termed an invader.

The characteristics that correlate with invasions vary substantially, and the epidemiology of biological invasions is not easily categorized. For example, naturalized plant species often exhibit uniparental reproduction, either self-pollination (selfing) or some form of asexual (or clonal) reproduction (Mulligan and Findlay 1970; Baker 1974; Price and Jain 1981). As a result, Baker (1967, 1974) suggested that uniparental reproduction would be an advantage during colonization. Furthermore, there may be selection for selfing during colonization, even in alien species that display an outcrossing mating system in their native range. Primarily selfing alien species typically exhibit low amounts of genetic diversity across and within populations and a high level of genetic differentiation among populations (Barrett and Richardson 1986; Barrett and Husband 1990; Hamrick and Godt 1990; Schoen and Brown 1991). Uniparental reproductive systems do, however, free a plant from dependence on the proximity of mates (Baker 1955), can confer reproductive success even under unfavorable environmental conditions (Stebbins 1957), and preserve adaptive gene complexes (Antonovics 1968; Brown 1979)—all potential advantages for an immigrant population. But many invasive species are not selfers. *Apera spicaventi* (Warwick et al. 1987), *Echium plantagineum* (Brown and Burden 1983; Burdon and Brown 1986), *Epipactis helleborine* (Squirrell et al. 2001), and *Trifolium hirtum* (Molina-Freaner and Jain 1992) all have become invasive, yet these species exhibit outcrossing mating systems.

The range of characteristics that invasive plant species exhibit probably reflects the range of characteristics exhibited by vascular plants in general. Variation in the invasion process and in the characteristics of invasive species has resulted in our inability to predict accurately which species will become invasive. As a result, a comprehensive list of the characteristics of invasive species remains elusive (National Research Council 2002) and may not be feasible (Williamson 1998).

Alien plants have served as model systems for studying the genetic and evolutionary consequences of dispersal and colonization (Baker and Stebbins 1965; Brown and Marshall 1981; Clegg and Brown 1983; Barrett and Husband 1990). In the remainder of this chapter, we discuss specific examples of genetic diversity in alien plant species. To this end, we attempt to provide a thorough assessment of the amount of genetic diversity reported for these species and how it is partitioned within and among populations. Such detail allows us to determine the presence of genetic bottlenecks and their extent (i.e., where in the hierarchical partitioning of genetic diversity they take place). Finally, this examination provides insights into the factors that influence the genetic and evolutionary consequences of founder events and population bottlenecks for plants in new ranges (Box 8.1).

Case Studies

Assessments of the genetic and evolutionary consequences of founder events and population bottlenecks for alien plants should include a comparison of populations from introduced and native ranges, thereby providing a relevant gauge of changes in genetic diversity among the immigrants (Barrett and Husband 1990; Novak et al. 1991, 1993; Novak and Mack 1993, 2001). Unfortunately, few studies include this comparison. Additionally, such assessments should document the size of the species' native and introduced ranges and its introduction history. Knowing the ranges for alien species can guide population sampling so that the breadth of the species' genetic diversity can be determined. Dates and locations of early collection sites and the circumstances of introduction (e.g., accidental or deliberate) provide evidence on which introduction scenarios can be based and by which the genetic consequences of founder effects and genetic bottlenecks can be evaluated (Bartlett et al. 2002).

The studies we review here collectively meet several criteria: (1) they include several plant families (to minimize taxonomic bias), (2) they involve plants introduced in widely separated parts of the world (to minimize geographic bias), (3) they include analysis of both native and introduced populations within the same design, (4) they employ species that vary from highly outcrossing to highly selfing, (5) they assess both the amount and the distribution

---

BOX 8.1 *Insights Alien Plants Yield about Genetic Bottlenecks*

1. For outcrossing alien plants, a few immigrants may represent a significant amount of the genetic variation in the species, reducing the likelihood of genetic bottlenecks following introduction.

2. Even for selfing alien plants, multiple introductions can reduce the expected loss of genetic variation with immigration, reducing the likelihood that genetic bottlenecks will occur.

3. Loss of genetic variation with introduction appears to occur most often across populations and not at the within-population level.

4. Within-population genetic diversity is sometimes higher in naturalized than in native populations, particularly when naturalized populations are formed as a by-product of genetic admixtures from multiple native populations.

5. Within-population genetic diversity may contribute directly to initiating an invasion and may also set the stage for evolutionary change.

6. Multiple introductions of different genotypes allow outcrossing events, even if rare, to give rise to adaptive radiations that have not occurred in the native range (e.g., *Avena barbata* in different climatic regimes in California).

7. Many plant invasions, including those of ruderals and deliberately introduced species, are sparked by multiple immigrations (which can span centuries) or large founder populations, or both.

8. Consequently, the results of putative bottleneck effects in an alien species must be evaluated in the light of the species' immigration history.

9. Both current theory and empiricism fail to adequately examine the level of genetic variation associated with failed (extirpated) populations in new ranges. As a result, the extent to which bottlenecks characterize the obvious alternative outcome to immigration is unknown.

---

of the species' genetic diversity, and (6) they provide at least some historical context for the species' entry and spread in the new range.

Outcrossing mating system

Despite the prediction that plants exhibiting uniparental reproduction, including a selfing mating system, would be at an advantage during colonization, a number of plants with an outcrossing mating system have become invasive (Brown and Burdon 1987, see also Rice and Sax, this volume). Here we compare the results of studies of genetic diversity in native and introduced populations of *Echium plantagineum* (Burdon and Brown 1986), *Epipactis helleborine* (Squirrell et al. 2001), and *Trifolium hirtum* (Molina-Freaner and Jain 1992). In general, these studies reveal that alien plant species with an outcrossing mating system do not always experience the reduction in genetic diversity anticipated with founder events or population bottlenecks. If a reduction does occur, it is not severe. Additionally, these results suggest that the circumstances associated with the introduction of alien plant species can also mitigate the effects of population bottlenecks.

Echium plantagineum (Patterson's curse) is native to the western Mediterranean region, and it has become invasive in regions with Mediterranean-like climates. *Echium plantagineum* was initially introduced into Australia in the mid-nineteenth century both accidentally, as a contaminant of seed and livestock feed, and deliberately, as an ornamental (Piggin 1982). Burdon and Brown (1986) analyzed allozyme diversity in ten populations of *E. plantagineum*: two from the native European range and eight from the introduced range in Australia. Both across and within populations, the populations of *E. plantagineum* from Australia exhibit more alleles per locus and a higher percentage of polymorphic loci than European populations (Table 8.2). In addition, expected heterozygosity (i.e., the heterozygosity expected under Hardy-Weinberg equilibrium) levels are virtually the same among introduced and native populations (0.34 and 0.35, respectively); introduced populations possess a slightly higher value for observed heterozygosity than those sampled in the native range (0.32 and 0.29, respectively) (Table 8.2). Values for total allelic diversity (H_T) for introduced and native populations of *E. plantagineum* are similar, as is the manner in which genetic variation is partitioned within and among populations (Table 8.3). For both introduced and native populations, most of the allozyme diversity is partitioned within rather than among populations (G_{ST} values were 0.12 and 0.08, respectively).

Australian populations of *E. plantagineum* do not appear to exhibit a reduction in genetic diversity associated with long-distance dispersal and subsequent spread (e.g., in the number of alleles across and within populations). Burdon and Brown (1986) suggested that this lack of reduction in genetic diversity is due to high levels of outcrossing and genetic recombination in *E. plantagineum*. Additionally, introductions of this species probably occurred over many years (Piggin 1982; Burdon and Brown 1986). Furthermore, either the size of founder populations or the number of introductions, or both, must have been sufficiently large to prevent any reduction in genetic diversity with introduction. The reported population sample sizes of *E. plantagineum* are small, especially the number of native populations analyzed ($N = 2$) (Burdon and Brown 1986), and may bias the comparison of genetic diversity between native and introduced populations. Fortunately, sample sizes are larger in other studies assessing genetic diversity in primarily outcrossing alien plants.

The pattern of genetic diversity in *Epipactis helleborine* (broad-leaved helleborine) reflects less outcrossing than in *E. plantagineum*. *Epipactis helleborine* is a multiflowered, wasp-pollinated, self-compatible terrestrial orchid native to Europe and Asia that has become invasive in North America. It was first recorded near Syracuse, New York, in 1879 and soon thereafter near Toronto, Ontario (1890), and Montreal, Quebec (1892). The plant spread rapidly across North America (Squirrell et al. 2001).

TABLE 8.2   *Amount of genetic diversity across and within native and introduced populations of alien plant species*

Species/Region	N	A_a	P_a	A_w	P_w	H_o	H_e	Source
Echium plantagineum								
Europe (native)	2	2.86	86%	2.61	82%	0.29	0.35	Burdon and Brown 1986
Australia (alien)	8	3.56	100%	2.72	94%	0.32	0.34	Burdon and Brown 1986
Epipactis helleborine								
Europe (native)	35	2.33	78%	1.77	55%	−0.019*	0.230	Squirrel et al. 2001
Canada (alien)	12	2.22	78%	1.90	58%	0.068*	0.232	Squirrel et al. 2001
Trifolium hirtum								
Eurasia (native)	22	1.62	38%	1.07	5%	—	0.01	Molina-Freaner and Jain 1992
California (alien)	22	1.29	19%	1.18	14%	—	0.05	Molina-Freaner and Jain 1992
Bromus tectorum								
Eurasia (native)	51	1.64	52%	1.01	2%	0.0001	0.005	Novak and Mack 1993
North America (alien)	60	1.32	28%	1.05	5%	0.0000	0.012	Novak et al. 1991
Eastern North America (alien)	38	1.08	16%	1.01	1%	0.0000	0.002	Bartlett et al. 2002
Other ranges (alien)	19	1.28	24%	1.03	4%	0.0002	0.008	Novak and Mack 2001
Avena barbata								
Mediterranean (native)	51	2.40	100%	1.04	4%	—	—	Clegg and Allard 1972
California (alien)	16	2.20	100%	1.24	23%	—	—	Clegg and Allard 1972

Note: N refers to the number of populations sampled, A_a and P_a refer to the number of alleles per locus and the percentage of polymorphic loci across populations, A_w and P_w refer to the average number of alleles per locus and percentage of polymorphic loci within populations, and H_o and H_e refer to the observed and expected heterozygosity. An asterisk indicates values for the fixation index, which was reported in this study rather than observed heterozygosity.

Squirrell et al. (2001) used allozyme and chloroplast DNA markers to assess the amount and distribution of genetic diversity in native and European populations of *E. helleborine*. They analyzed the allozyme diversity in 35 European populations and 12 introduced populations from Ontario and Quebec. Allelic composition and the level of allelic diversity across native and introduced populations of *E. helleborine* are similar, although the value for the latter is slightly higher in the native range: European and Canadian populations average 2.33

TABLE 8.3 *Distribution of genetic diversity*

Species/Region	N	H_T	H_S	D_{ST}	G_{ST}	Source
Echium plantagineum						
Europe	2	0.381	0.351	0.030	0.080	Burdon and Brown 1986
Australia	8	0.388	0.341	0.047	0.120	Burdon and Brown 1986
Epipactis helleborine						
Europe	35	—	—	—	0.200*	Squirrel et al. 2001
Canada	12	—	—	—	0.090*	Squirrel et al. 2001
Trifolium hirtum						
Eurasia	22	0.082	0.014	0.068	0.824	Molina-Freaner and Jain 1992
California	22	0.078	0.055	0.024	0.300	Molina-Freaner and Jain 1992
Bromus tectorum						
Eurasia	51	0.087	0.009	0.077	0.754	Novak and Mack 1993
North America	60	0.115	0.046	0.069	0.478	Novak et al. 1991
Eastern North America	38	0.075	0.014	0.061	0.560	Bartlett et al. 2002
Other ranges	19	0.117	0.033	0.084	0.321	Novak and Mack 2001

Note: Distribution of genetic diversity using Nei's gene diversity statistics (Nei 1977), where H_T is the total gene diversity, H_S in the portion of the total diversity distributed within populations, D_{ST} is the portion of the total diversity distributed among populations, and G_{ST} is the proportion of the total diversity distributed among populations. An asterisk indicates values for F_{ST}, which was reported in this study rather than G_{ST}.

and 2.22 alleles per locus, respectively. The percentage of polymorphic loci (78%) is the same across both native and introduced populations (see Table 8.2). Within populations, this pattern is reversed, as plants from Canada have more alleles (1.90 vs. 1.77), slightly more polymorphic loci (58% vs. 55%), and a slightly higher value of expected heterozygosity (0.232 vs. 0.230) than plants from Europe. Observed heterozygosity was not reported, but the fixation index (*f*) for each population was provided (see Table 8.2). Fixation index values for both native and introduced populations are not significantly different from zero and indicate random mating in both regions. The level of genetic differentiation among introduced populations of *E. helleborine* is much less ($F_{ST} = 0.090$) than the level reported for native populations ($F_{ST} = 0.200$) (see Table 8.3).

Twenty-nine populations (17 European and 12 Canadian populations) of *E. helleborine* were analyzed for their chloroplast DNA haplotypes through the presence (or absence) of a 10-bp duplication in the *trnL* intron. Only 4 of 17 populations from Europe exhibited this polymorphism, whereas 11 of 12 populations from North America were polymorphic. Similar to the results reported for the allozyme data, values of F_{ST} for chloroplast DNA polymorphism indicate less genetic differentiation among introduced populations than among those from the native range ($F_{ST} = 0.367$ and 0.506, respectively).

Thus, allozyme and chloroplast DNA genetic markers both reveal that *E. helleborine* has apparently not undergone a reduction in genetic diversity upon introduction into North America (Squirrell et al. 2001). In fact, the average level of genetic diversity within introduced populations is higher, and the level of genetic differentiation is lower, than in native populations. These data did not allow Squirrell et al. (2001) to determine whether single or multiple introductions had occurred, but they do suggest that the introduction was sufficiently large to include much of the genetic diversity in the native range. Given the outcrossing mating system of *E. helleborine*, its level of heterozygosity, and its genetic structure, a founder population with as few as five individuals would have been sufficient to maintain this genetic diversity, provided that large population sizes were rapidly restored following introduction (Squirrell et al. 2001).

Unlike *E. plantagineum* and *E. helleborine*, *Trifolium hirtum* (rose clover) does appear to have experienced a reduction in genetic diversity upon its introduction into California. However, this reduction occurs at the among-population level and not within introduced populations. *Trifolium hirtum* is an annual legume native to the Mediterranean region. It was deliberately introduced into California in 1944 as livestock forage (Love and Sumner 1952) by seeds collected in 1936 in the Adana Province, Turkey. By the late 1960s, *T. hirtum* had invaded highway roadside habitats (Jain and Martins 1979).

Genetic and demographic studies assessed the clover's spread (Jain and Martins 1979; Martins and Jain 1979, 1980), and more recently Molina-Freaner and Jain (1992) compared allozyme diversity in native and introduced populations of *T. hirtum*. Twenty-two populations from Eurasia and the same number from California were included; 14 of the Eurasian populations were collected in Turkey. Across populations, Eurasian populations exhibit higher levels of genetic diversity than Californian populations (1.62 and 1.29 alleles per locus, and 38% and 19% polymorphic loci, respectively) (see Table 8.2). In contrast, genetic diversity at the within-population level is, on average, higher for Californian populations (1.18 alleles per locus and 14% polymorphic loci) than that reported within Eurasian populations (1.07 alleles per locus and 5% polymorphic loci). In addition, expected heterozygosity for introduced populations is five times greater than the value for native populations. Values for total allelic diversity (H_T) for introduced and native populations of *T. hirtum* are similar; however, the manner in which genetic variation is partitioned within and among populations is very different (see Table 8.3). Genetic differentiation among native populations is nearly three times greater than that among introduced populations (G_{ST} values were 0.824 and 0.300, respectively). Thus, *T. hirtum* appears to have experienced a reduction in genetic diversity across introduced populations in California, but within populations the Californian plants exhibit higher levels of genetic diversity, and have much less genetic structure, than populations in Eurasia.

Molina-Freaner and Jain (1992) contend that these results are a product of the outcrossing mating system of *T. hirtum*; that is, that even a few founding individuals would possess a high proportion of the species' genetic diversity. Alternatively, given that the species was deliberately introduced for forage, the founding population may have been large and may even have been composed of admixtures of genotypes derived from multiple source populations in Turkey. This introduction scenario is similar to the migrant pool model described by Slatkin (1977). Moreover, Turkey includes the most polymorphic portion of the clover's native range: 72% of all polymorphic loci detected across native populations were found in the Turkish samples. Differences in the distribution of genetic diversity between European and Californian populations of *T. hirtum* may also be due to a shift toward a mixed mating system with colonization of roadside habitats in California. This shift could have resulted in higher outcrossing rates and an intermixing of genetic diversity (Molina-Freaner and Jain 1992).

Assessments of the amount and distribution of genetic diversity in other primarily outcrossing alien plant species are generally in keeping with the results for these species (e.g., Warwick et al. 1987; Balfourier and Charmet 1994; Meekins et al. 2001); that is, evidence for genetic bottlenecks is not apparent. Genetic bottlenecks were observed, however, in a comparison of continental and island populations of *Turnera ulmifolia*, a herbaceous perennial native to the New World tropics, including the Caribbean region (Barrett and Shore 1989; Barrett and Husband 1990). The species' continental and island populations are both self-incompatible and outcrossing. Despite its outcrossing mating system, *T. ulmifolia* exhibits a marked reduction in genetic diversity within Caribbean populations. For example, the value for the percentage of polymorphic loci across Caribbean populations (20%) is less than half the value (46%) in South American populations (Barrett and Shore 1989; Barrett and Husband 1990). Additionally, average genetic diversity within South American populations is threefold greater than genetic diversity within island populations (0.12 and 0.04, respectively). Colonization of islands can impose demographic constraints on populations that often lead to genetic bottlenecks (Barrett and Husband 1990). However, evidence from *Rubus alceifolius* suggests that these demographic constraints can be overcome by multiple introductions.

Rubus alceifolius (blackberry) is native to Southeast Asia and has been introduced onto Madagascar and other Indian Ocean islands, including Mayotte, Réunion Island, and Mauritius. Amsellem et al. (2000) determined the genetic diversity in native and introduced populations of *R. alceifolius* using amplified fragment length polymorphic (AFLP) markers. Genetic diversity within native populations is high, while the diversity in Madagascar is somewhat lower. In contrast, populations from Mayotte, Réunion, and Mauritius are each composed of a single genotype, all of which are similar to genotypes on Madagascar. Amsellem et al. (2000) proffer that *R. alceifolius* was initially introduced onto Madagascar multiple times, and then spread to the other Indian Ocean islands from Madagascar (rather than directly from the native range). Thus, the high level of diversity within Madagascan populations probably stems from multiple introductions, whereas subsequent founder events resulted in a reduction in genetic diversity on each of the other islands.

Another consequence of the introduction of *R. alceifolius* onto Madagascar and Réunion Island is an apparent shift from sexual reproduction in its native range to predominantly asexual reproduction (apomixis) (Amsellem et al. 2001). Although plant species exhibiting apomixis are typically expected to possess extremely low genetic diversity, recent findings reveal that such species vary widely in the amount and distribution of diversity within and among populations (Ellstrand and Roose 1987; Hamrick and Godt 1990). In fact, occasional sexual reproduction (especially outcrossing) in a species that reproduces predominantly through asexual means may generate unique, highly adapted, or aggressive genotypes that are subsequently conserved through apomixis (Novak and Mack 2000). Amsellem et al. (2001) suggest that just such a mechanism may have contributed to the invasion of *R. alceifolius* on Madagascar and the other Indian Ocean islands.

These studies collectively provide insight into the fates of alien plants with an outcrossing mating system: such species may undergo little or no reduction in genetic diversity with founder events or population bottlenecks. Furthermore, if a genetic bottleneck does occur in these alien plants, it is most likely to occur at the across-population level (as seen with *T. hirtum*, and to a lesser degree with *E. helleborine*). These conclusions stem directly from an additional insight provided by these studies: several of the outcrossing species described above were deliberately transported to their new range(s), and such deliberate introductions often involve multiple introductions and founder populations with many individuals. This observation runs contrary to conventional wisdom and suggests that multiple introductions may be more common among plants than previously thought. Moreover, in at least one instance (*T. hirtum*), immigrants appear to have been drawn from multiple source populations, suggesting that these results are in keeping with the migrant pool model of Slatkin (1977). These alien plant species possess an outcrossing mating system in their native range, so in most cases, the majority of their genetic diversity is partitioned within rather than across populations. These studies also reveal that even a small number of founding individuals or founder events will probably sample much of an outcrossing species' genetic diversity (Warwick 1990). Thus, a reduction in genetic diversity with founder events or population bottlenecks is not a certainty for naturalized species with an outcrossing mating system. Any reduction depends on the degree to which genetic diversity is partitioned within native range populations and on the specific details associated with a species' introduction (the number of introduction or colonization events and the size of the founder populations).

Selfing mating system

We next examine the influence of primarily selfing mating systems on the amount and distribution of genetic diversity following immigration. Results from these studies also illustrate the post-immigration genetic consequences of multiple introductions.

Bromus tectorum (cheatgrass) is a predominantly selfing annual grass with wide distribution in temperate grasslands (Upadhyaya et al. 1986). Its native range includes most of Europe, the northern rim of Africa, and southwestern Asia (Pierson and Mack 1990). The grass has been accidentally introduced into many temperate environments worldwide (Upadhyaya et al. 1986; Stace 1997).

The amount and distribution of genetic diversity within and among native and introduced populations of *B. tectorum* have been compared, and the introduction and spread of the plant in North America and around the world evaluated, using genetic markers. The dynamics of introduction and spread have been explored by combining historical information with the detection of the same geographically restricted multilocus genotypes in populations from both the native and introduced ranges. This protocol has allowed assessment of the genetic and evolutionary consequences of the founder effects and genetic bottlenecks associated with the grass's introduction and invasion.

In total, 51 Eurasian populations, 94 North American populations, and 19 populations from other naturalized ranges of this species (Canary Islands, Argentina, Chile, Hawaii, and New Zealand) have been analyzed (Novak et al. 1991, 1993; Novak and Mack 1993, 2001; Bartlett et al. 2002). Based on multilocus genotype distributions, source populations for the introduction of *B. tectorum* into North America and its other naturalized ranges appear to have been drawn exclusively from populations in either central Europe or the western Mediterranean region. For instance, the multilocus genotype characterized by the allele *Got-3c* has been detected in only two central European populations (Bayreuth, Germany, and Libochovice, Czech Republic), yet this genotype is now widespread throughout the grass's range in western North America, Argentina, Hawaii, and New Zealand. In contrast, the *Pgi-2b* multilocus genotype is the most widespread of any genotype detected so far in the native range (e.g., France, Spain, and Morocco); it is, however, restricted to the western Mediterranean region. This genotype has been detected in widely separated populations across the introduced range of *B. tectorum*, including three from Nevada, one from California (Truckee, CA), four from the Canary Islands, and two from Chile, confirming that emigrants of *B. tectorum* departed from at least two European regions.

Genetic markers also prove their efficiency in determining the dynamics of introduction and spread. The pattern of genetic markers for *B. tectorum* in its introduced range supports historical evidence that its invasion in western North America stems from multiple introductions (Novak et al. 1993; Novak and Mack 1993, 2001). Diagnostic genetic markers were detected for populations in six locales in western North America, indicating a minimum of six independent founder events. Conversely, only two genetic markers were detected among the 38 eastern North American populations of *B. tectorum* that have been analyzed; 33 of these 38 populations exhibit the most common genotype (Bartlett et al. 2002). Thus, populations from eastern North America do not appear to have had the same history of introductions as seen in the West. Furthermore, the detection of more diagnostic genetic markers in western

North America suggests that its populations were derived directly via introductions from the native range (Novak and Mack 2001). However, the distribution of one multilocus genotype (*Pgm-1a* and *Pgm-2a*) in several populations across the United States suggests that the descendants of some plants introduced into eastern North America were subsequently transported westward.

Multiple introductions may have also occurred in the other naturalized ranges of *B. tectorum*. For instance, the occurrence of different marker genotypes in Argentinean (*Pgm-1a* and *Pgm-2a*) and in Chilean (*Pgi-2b*) populations suggests that separate introductions occurred in each country (Novak and Mack 2001). Likewise, the distribution of multilocus genotypes in populations in the Canary Islands suggests that multiple introductions from populations in Europe and North Africa led to the grass's establishment there (Novak and Mack 2001). Despite cheatgrass having been introduced almost exclusively by accident, multiple introductions occurred repeatedly.

Populations of *B. tectorum* from Eurasia exhibit more alleles and polymorphic loci than populations from any of the species' introduced ranges (see Table 8.2). *Bromus tectorum* has apparently undergone a reduction in genetic diversity with introduction, but this reduction has not been detected within all introduced populations. Based on genetic markers, populations of *B. tectorum* in western North America and other naturalized ranges appear to have been established through a more complicated introduction history than has been the case for populations in eastern North America. On average, populations from western North America and other naturalized ranges contain more alleles and polymorphic loci than populations from eastern North America and even more than native Eurasian populations (see Table 8.2). *Bromus tectorum* is reportedly a highly selfing species (Upadhyaya et al. 1986). Heterozygosity levels in this grass are indeed exceedingly low: only a handful of heterozygotes have been detected in the plant's native and introduced ranges (see Table 8.2). As with alien plants with an outcrossing mating system, the genetic signature of founder effects and population bottlenecks may vary among the hierarchical levels at which genetic variation can be partitioned.

The loss of genetic diversity across introduced populations of *B. tectorum* is, in part, a consequence of the manner in which genetic diversity is partitioned within and across its native populations. Most of the genetic diversity in its native range is partitioned across rather than within populations ($G_{ST} = 0.754$) (see Table 8.3). Thus, during the "random sampling" of individuals upon emigration, many alleles or genotypes were not represented, especially rare ones. This situation is exacerbated for those selfing species for which the emigrants are drawn from only a small portion of an enormous native range. In the case of *B. tectorum*, levels of genetic differentiation across populations are lower in the introduced ranges than in the native range (see Table 8.3). Even the few introductions that occurred in eastern North America reduced the level of genetic differentiation among populations compared with the level in the native range.

Identification of source populations for the introduction of *B. tectorum* worldwide reveals that immigrants were drawn from at least two regions in the native range: Central Europe and the western Mediterranean. Similarly, the geographic distribution of marker genotypes suggests that multiple introductions have occurred across much of the species' naturalized range. Furthermore, the genetic diversity within introduced populations of *B. tectorum* is, on average, higher, and the level of genetic differentiation among introduced populations lower, than in native populations. As in our examination of alien plants with an outcrossing mating system, these findings are consistent with the migrant pool model of Slatkin (1977). These findings for *B. tectorum* are revealing: multiple introductions, even in a highly selfing plant species, appear to partially offset the reduction in genetic diversity associated with founder events and population bottlenecks.

Avena barbata (slender wild oat), which is naturalized in California, presents an alternative genetic outcome of immigration. *Avena barbata* is an annual, predominantly selfing, diploidized tetraploid grass (i.e., a tetraploid plant with gene expression similar to that of a diploid) whose native range extends from the Mediterranean basin to southwestern Asia and eastward to Nepal (Marshall and Allard 1970; Allard et al. 1993). In Eurasia, *A. barbata* persists in a wide range of environments (Allard et al. 1993). The grass has become a successful colonizer of Mediterranean-like climates worldwide and was accidentally introduced to California from the western Mediterranean basin in a large number of independent introductions extending over more than a century (Kahler et al. 1980). *Avena barbata* spread rapidly upon introduction and is now a major component of grasslands and grass-oak savanna habitats (Marshall and Allard 1970; Clegg and Allard 1972).

Clegg and Allard (1972) described the allozyme diversity in 16 Californian populations of *A. barbata*. Of these 16 populations, 9 were sampled from the Central Valley and adjacent foothills (region I), and 7 were sampled from the intermontane regions of the coastal strip and higher-elevation foothills of the Sierra Nevada (region II). The diversity of *A. barbata* in California was compared with that of 51 populations in 9 areas in the Mediterranean region. Only a portion of the data can be summarized in the format we use above. However, one goal of the investigators was to compare the allelic compositions of native and introduced populations. Consequently, their data do reveal genetic consequences of this species' introduction into California.

The extent of genetic diversity across native and introduced populations of *A. barbata* is similar. The number of alleles per locus and the percentage of polymorphic loci across populations are 2.40 and 100% in native populations and 2.20 and 100% in introduced populations for five allozyme loci (see Table 8.2). In contrast, genetic diversity within Californian populations (1.24 alleles per locus and 23% polymorphic loci) is, on average, higher than the values within the Mediterranean populations (1.04 alleles per locus and 4% polymorphic loci) (see Table 8.2). Thus, Californian populations do not appear to have experienced a severe genetic bottleneck, either at across-population or within-population levels (Clegg and Allard 1972). Similar results were reported for populations of *A. barbata* from the Mediterranean region and central California (see Singh and Jain 1971). This close correspondence in the amount of genetic diversity and in allelic composition between populations from the western Mediterranean region and California is probably a result of multiple introductions from Spain.

A more comprehensive analysis of Spanish populations of *A. barbata* subsequently compared the allelic and genotypic composition of native and introduced gene pools (Garcia et al. 1989). Plants from 42 populations across southern and western Spain were analyzed for genetic diversity at 15 loci. Once again, Spanish and Californian gene pools were found to be similar. The similarity is especially high for populations sampled from southwestern Spain, a likely point of embarkation for ships to the New World, including California (Garcia et al. 1989). However, the level of allelic diversity in Spain is lower than the level observed elsewhere in the native range, such as Israel (Kahler et al. 1980). Similar to many other species in Europe with cryptogenic origin (Kornas 1990), *A. barbata* may have evolved in the eastern Mediterranean region and southwestern Asia and subsequently spread across the Mediterranean basin. As a consequence, neither Spanish nor Californian populations are likely to reflect all the genetic diversity of this species across its native range (Kahler et al. 1980; Garcia et al. 1989).

The multilocus structure of Californian populations of *A. barbata* is quite different from that of populations in the native range, despite the similarity of allelic composition in native and introduced populations (Clegg and Allard 1972; Kahler et al. 1980; Garcia et al. 1989; Perez de la Vega 1991). For instance, 80% of populations in region I of California are fixed or nearly fixed for the "xeric" multilocus genotype, while 10% are fixed or nearly fixed for the "mesic" genotype (found in region II); few populations are polymorphic. Spanish multilocus genotypes are intermediate between the xeric and mesic Californian genotypes (Garcia et al. 1989; Perez de la Vega 1991). As a result, overall genetic diversity is generally higher within Spanish than within Californian populations, and the genetic diversity of Californian populations is distributed mainly among populations—the opposite of the patterns seen in the species' native and introduced populations discussed thus far.

Such a pattern of genetic diversity could be attributed to founder effects or genetic bottlenecks associated with introduction. Alternatively, these patterns could result from natural selection for locally adapted genotypes within xeric and mesic habitats (Clegg and Allard 1972)—selection that could spark an "adaptive radiation" in the introduced range. The implication of this scenario is potentially far-reaching: it suggests that a combination of sufficient genetic variation in introduced populations together with novel selection regimes can set the stage for adaptive evolution, as suggested by Ellstrand and Schierenbeck (2000). Invasions, then, need not proceed only through the introduction of preadapted genotypes (*sensu* Futuyma 1998); the genetic raw material originally introduced can be reorganized into more invasive genotypes. Consequently, a species' performance in other ranges may not serve as the sole indicator of its invasiveness, if novel genetic recombination can occur anywhere in a new range (National Research Council 2002). How these events may relate to the potential for speciation is unresolved, but they do appear to set the stage for the diversification of genotypes between native and introduced populations. These issues can be resolved thorough detailed genetic analysis of both native and introduced populations, followed by comparisons of the performance of these novel genotypes in their new ranges.

A variety of other genetic diversity patterns are observed in predominately selfing plant invaders, particularly when (in contrast to *B. tectorum* and *A. barbata*) introduced populations are a consequence of fewer introduction events. For example, *Capsella bursa-pastoris* (shepherd's purse) has experienced an overall reduction in genetic diversity in its introduced range, presumably as a consequence of few introduction events (Neuffer and Hurka 1999). In other cases, some highly selfing alien species that occupy large naturalized ranges are apparently represented by only one or a few genotypes: *Xanthium strumarium* (Moran and Marshall 1978), *Emex spinosa* (Marshall and Weiss 1982), and *Emex australis* (Panetta 1990) in Australia and *Sorghum halepense* (Warwick et al. 1984) in Canada. Similar findings have been reported for alien species that reproduce clonally, such as the highly apomictic *Chondrilla juncea* (Chaboudez 1994). In these cases, naturalization stems from few introductions. As a consequence, genetic diversity is sharply reduced in the introduced range.

Do Deliberate Introductions Mitigate Genetic Drift?

We opened this chapter with questions about the consequences of founder events and genetic bottlenecks for alien plants as well as the larger question of the extent to which immigrant plants face these potentially severe constraints. While the examples cited above provide a mixed answer, the general conclusion would be "yes, sometimes plant immigrants do experience a reduction in genetic variation." Left largely unaddressed, however, is whether most other naturalized and invasive species stem from large, not small, immigrant populations and from multiple (even many) introductions, not a single introduction. Unlike other large taxonomic groups of immigrants, such as insects (Simberloff 1989), plants may not routinely experience these hazards because opportunities for founder events and population bottlenecks have been rendered moot by immigrant plants' prevalence as seed contaminants in imported seeds and as deliberate commodities in international trade. As a consequence, founder and bottleneck events may be infrequent. Hundreds, and possibly thousands, of plant species have importation histories that exceed 100 years (Rehder 1940; Mack 1991); that is, they were not simply imported once, nor were they always imported in small numbers (Box 8.2; Mack 1991). Furthermore, as the international exchange in all manner of horticultural species has grown since 1800 (McCracken 1997), the potential donor ranges have also grown, providing an increase in the opportunity for selection in another naturalized range before entry into the United States. Future comprehensive examinations of the genetic

BOX 8.2   *A History of Seed Contamination in the United States*

Long after U.S. agriculture became based largely on crops of foreign origin, much, if not all, of the seed sown for these crops was annually imported from Europe. Hicks (1895) provided a detailed account of the total dependence on imported seeds for crops that included alfalfa, beet, broccoli, cress, endive, radish, spinach, and turnip, as well as many forage grasses (e.g., *Alopecurus pratensis*, *Anthoxanthum odoratum*, *Festuca ovina*), and substantial importation for many others (cabbage, carrot, leek, onion, pepper, parsley). Such massive and prolonged importation of these species created opportunities for two categories of naturalized species. Some of these imports have become naturalized themselves (e.g., *A. odoratum*, *F. ovina*), but more likely their importation enormously facilitated the entry and cultivation of other alien species that were seed contaminants.

Extraneous, potentially invasive seeds were a routine component of these seed imports. The inclusion of the seeds of ruderal or otherwise unwanted species would have been a common by-product of harvesting seed crops before the development of effective seed sieving procedures in the late nineteenth century. Consequently, many of the ruderal species now naturalized in the United States had probably arrived there repeatedly before 1800

(Mack 2003). But the genetic diversity of these species' gene pools in the U.S. would have been bolstered by the huge volume and extent of European seed sources that entered the trade post-1865. Unfortunately, the U.S. government placed few, if any, safeguards on seed purity in that era, and some unscrupulous European seed merchants took full advantage of this regulatory gap by shipping extensively contaminated seeds to the U.S., even while their own nations were mandating the purity of seeds for domestic trade (Hicks 1895).

Newly created seed testing laboratories in the United States chronicled the extent and diversity of seed contamination in the late nineteenth century. In one extreme case, 90% of the seed sold as clover was extraneous and nonindigenous (Hicks 1895). Moreover, the sources of imported seeds spanned Western Europe, thereby ensuring that a wide breadth of the genotypic diversity among European ruderals was annually sampled and transported to the U.S. For some of the most widespread naturalized species in the U.S. (e.g., *Capsella bursa-pastoris*, *Nepeta cataria*, *Plantago lanceolata*, *Rumex acetosella*, *Rumex crispus*, *Verbascum thapsus*), any constraints imposed by genetic drift would have been eliminated via these massive immigrations (Mack and Erneberg 2002).

diversity and origins of naturalized and invasive horticultural species in the United States may well reveal that any latent constraints arising from founder effects and population bottlenecks are the exception, not the rule.

Conclusions

An abundance of theory predicts reductions in genetic diversity following founder events or population bottlenecks (e.g., Wright 1931; Nei et al. 1975; Chakraborty and Nei 1977; Watterson 1984; Maruyama and Fuerst 1985; Lande

1993). Although such reductions have indeed occurred among some alien plants, the extent of reduction and its location can vary substantially within the hierarchical partitioning of genetic diversity. Additionally, the mating system and history of introduction appear to influence the likelihood of any reduction in diversity (see Box 8.1). For instance, outcrossing species partition most of their diversity within rather than among populations. Consequently, only a few immigrants could represent a significant amount of the genetic diversity of native populations. With multiple introductions, the probability of a reduction in genetic diversity among alien outcrossing species appears to decline further.

Genetic diversity in highly selfing (and primarily clonal) plant species is mostly partitioned among rather than within populations, and a few immigrants would probably incorporate little of the diversity in the species' native range. However, multiple introductions of even selfing aliens may include much genetic diversity, especially if individuals are drawn from across a species' native range. This scenario closely resembles the "migrant pool" model of Slatkin (1977). As reflected in a growing body of examples for both outcrossing and selfing alien plant species, multiple introductions appear to be the rule rather than the exception (Novak et al. 1993; Mack and Erneberg 2002). Either as deliberate introductions or as seed contaminants, large numbers of propagules can be drawn repeatedly from much of the donor range (Mack and Erneberg 2002). Consequently, the prevalence of genetic bottlenecks following a founder event or a population bottleneck, at least for plants (for data on animals see Wares et al., this volume), may be much less than once envisioned and should be evaluated in light of each species' immigration history.

When genetic diversity is lost upon introduction, the loss most often occurs across populations and not at the within-population level. This observation is important because the maintenance of genetic diversity within populations may directly foster an invasion and also sets the stage for evolutionary change. In fact, multiple introductions can result in the establishment of populations that contain admixtures of genotypes from the native range; these naturalized populations may contain far greater within-population diversity than occurs in the native range. Furthermore, outcrossing events in the naturalized range, even if rare, would lead to recombination and the creation of novel genotypes. The emergence of such genotypes would provide opportunities for adaptive radiation that could not occur in the native range (as seen with *A. barbata* in the different climatic regimes of California).

The discrepancy between the predictions of theory and the empirical evidence needs to be resolved. Future investigations should include direct side-by-side comparisons of the performance of native and introduced populations, wide sampling across a species' native and introduced range, and identification of as many genetic markers as possible, and should incorporate historical information for evaluation of introduction dynamics (Novak and Mack 2001; Bartlett et al. 2002). Based on recent models for detecting genetic bottlenecks (Luikart et al. 1998b, 1999), highly polymorphic genetic markers should be employed in which individual alleles (and heterozygous individuals) can be identified. Microsatellite DNA analysis is often well suited for this task (Parker et al. 1998; Sunnucks 2000). Investigations that assess the effects of population bottlenecks on the additive genetic variance in alien plants are especially needed, in contrast to the bulk of investigations that have assessed the effects of genetic bottlenecks with single-locus genetic markers.

Many aspects of the population biology of plants, especially the production of many progeny, make plants particularly suitable as models in quantitative genetics. Identifying the underlying genetic controls of ecologically important traits will be key; these identifications will contribute substantially to understanding the consequences of introduction events and to predicting invasive potential. Finally, we need to apply the same diligence to deciphering explanations for plant introductions that fail that we now do to those that persist (Harper 1982): for example, do failures stem more from genetic, demographic, or environmental limitations, or from some combination of these factors? The study of alien species has already fulfilled much of their promise to serve as natural experiments in evolution (Waddington 1965) and population biology (Mack 1985). Much more has yet to be discovered.

Acknowledgments

We sincerely thank Dov Sax, Steve Gaines, and Jay Stachowicz for their diligence and industry in all aspects of proposing and organizing the meeting from which this book grew. Without their efforts, that meeting would not have taken place. We also thank the other authors of this volume and several reviewers for their constructive comments and interaction. This work was conducted as part of the "Exotic Species: A Source of Insight into Ecology, Evolution, and Biogeography" working group supported by the National Center for Ecological Analysis and Synthesis, a Center funded by the National Science Foundation (Grant DEB-0072909), the University of California, and the Santa Barbara campus.

An earlier version of this chapter was completed while SJN was a visiting professor in the University Studies Abroad Consortium (USAC) program at the Université de Pau, France. He is grateful to the USAC staff, Robina Muller, Ryan Findley, and Aurelie Escada, for their assistance and patience while this work was undertaken. SJN also thanks Magdalen Kadlecsovics and Stephane Defraine for being welcoming and generous hosts during visits to Chateau de Fontenille, and M. and Mme. Mialou and M. and Mme. Guilhemet for their hospitality and generosity during his family's stay in Laroin, France.

Literature Cited

Allard, R. W., P. Garcia, L. E. Saenzdemiera, and M. P. Perez de la Vega. 1993. Evolution of multilocus geneticstructure in *Avena hirtula* and *Avena barbata*. Genetics 135:1125–1139.

Allendorf, F. W. 1986. Genetic drift and loss of alleles versus heterozygosity. Zoo Biology 5:181–190.

Amsellem, L., J. L. Noyer, T. Le Bourgeois, and M. Hossaert-McKey. 2000. Comparison of genetic diversity of the invasive weed *Rubus alceifolius* Poir.

(Rosaceae) in its native range and in areas of intro-duction, using amplified fragment length polymor-phism (AFLP) markers. Molecular Ecology 9:443–455.

Amsellem, L., J. L. Noyer, and M. Hossaert-McKey. 2001. Evidence for a switch in reproductive biology of *Rubus alceifolius* (Rosaceae) towards apomixis, between its native range and its area of introduc-tion. American Journal of Botany 88:2243–2251.

Antonovics, J. 1968. Evolution in closely adjacent pop-ulations. V. Evolution of self fertility. Heredity 23:219–238.

Baker, H. G. 1955. Self-compatibility and establishment after "long-distance" dispersal. Evolution 9:347–348.

Baker, H. G. 1967. Support for Baker's Law—as a rule. Evolution 21:853–856.

Baker, H. G. 1974. The evolution of weeds. Annual Review of Ecology and Systematics 5:1–25.

Baker, H. G., and G. L. Stebbins. 1965. *The genetics of colonizing species.* Academic Press, New York.

Balfourier, F., and G. Charmet. 1994. Geographic pat-terns of isozyme variation in Mediterranean popu-lations of perennial ryegrass. Heredity 72:55–63.

Barrett, S. C. H., and B. C. Husband. 1990. The genetics of plant migration and colonization. In A. H. D. Brown, M. T. Clegg, A. L. Kahler and B. S. Weir, eds. *Plant population genetics, breeding, and germplasm resources*, pp. 254–277. Sinauer Associates, Sunderland, MA.

Barrett, S. C. H., and B. J. Richardson. 1986. Genetic at-tributes of invading species. In R. H. Groves and J. J. Burdon, eds. *Ecology of biological invasions: an Australian perspective*, pp. 21–33. Cambridge University Press, Cambridge.

Barrett, S. C. H., and J. S. Shore. 1989. Isozyme varia-tion in colonizing plants. In D. E. Soltis and P. S. Soltis, eds. *Isozymes in plant biology*, pp. 106–126. Dioscorides, Portland, OR.

Bartlett, E. A., S. J. Novak, and R. N. Mack. 2002. Genetic variation in *Bromus tectorum* (Poaceae): dif-ferentiation in eastern United States. American Journal of Botany 89:626–636.

Brown, A. H. D. 1979. Enzyme polymorphisms in plant populations. Theoretical Population Biology 15:1–42.

Brown, A. H. D., and J. J. Burdon. 1983. Multilocus di-versity in an outbreeding weed, *Echium plan-tagineum*. Australian Journal of Biological Science 36:503–509.

Brown, A. H. D., and J. J. Burdon. 1987. Mating system and colonization success in plants. In A. J. Gray, M. J. Crawley, and P. J. Edwards, eds. *Colonization, suc-cession and stability*, pp. 115–131. Blackwell Scientific, Oxford.

Brown, A. H. D., and D. R. Marshall. 1981. Evolutionary changes accompanying colonization in plants. In G. C. E. Scudder and J. L. Reveal, eds. *Evolution today*, pp. 351–363. Hunt Institute of Botanical Documents, Carnegie Mellon University Press, Pittsburgh.

Bryant, E. H., S. A. McCommas, and L. M. Combs. 1986a. The effect of an experimental bottleneck upon quantitative genetic variation in the housefly. Genetics 114:1191–1211.

Bryant, E. H., L. M. Combs, and S. A. McCommas. 1986b. Morphometric differentiation among experi-mental lines of the housefly in relation to the bottle-neck. Genetics 114:1213–1223.

Bryant, E. H., and L. M. Meffert. 1996. Nonadditive ge-netic structuring of morphometric variation in rela-tion to a population bottleneck. Heredity 77:168–176.

Bryant, G. 1998. *Botanica*, 2nd Ed. Bateman, North Shore City, New Zealand.

Burdon, J. J., and A. H. D. Brown. 1986. Population ge-netics of *Echium plantagineum* L., target weed for bi-ological control. Australian Journal of Biological Sciences 30:369–378.

Carson, H. L., and A. R. Templeton. 1984. Genetic revo-lutions in relation to speciation phenomena: the founding of new populations. Annual Review of Ecology and Systematics 15:97–131.

Chaboudez, P. 1994. Patterns of clonal variation in skeleton weed (*Chondrilla juncea*), an apomictic species. Australian Journal of Botany 42:283–295.

Chakraborty, R., and M. Nei. 1977. Bottleneck effects on average heterozygosity and genetic distance with the stepwise mutation model. Evolution 31:347–356.

Cheverud, J. M., T. T. Vaughn, L. S. Pletscher, K. King-Ellison, J. Bailiff, E. Adams, C. Erickson, and A. Bonislawski. 1999. Epistasis and the evolution of additive genetic variance in populations that pass through a bottleneck. Evolution 53:1009–1018.

Clegg, M. T., and R. W. Allard. 1972. Patterns of genetic diversity in the slender wild oat species, *Avena bar-bata*. Proceedings of the National Academy of Science USA 69:1820–1824.

Clegg, M. T., and A. H. D. Brown. 1983. The founding of plant populations. In C. M. Schonewald-Cox, S. M. Chambers, B. Macbryde, and W. L. Thomas, eds. *Genetics and conservation*, pp. 216–228. Benjamin Cummings, Menlo Park, CA.

Dobzhansky, T. 1937. *Genetics and the origin of species.* Columbia University Press, New York.

Ellstrand, N. C., and M. L. Roose. 1987. Patterns of genotypic diversity in clonal plant species. American Journal of Botany 74:123–131.

Ellstrand, N. C., and K. A. Schierenbeck. 2000. Hybridization as a stimulus for the evolution of in-vasiveness in plants. Proceedings of the National Academy of Science USA 97:7043–7050.

Falconer, D. S. and T. F. C. Mackay. 1996 *Introduction to quantitative genetics*, 4th Ed. Longmans, Essex, U.K.

Fernandez, A., M. A. Toro, and C. Lopez-Fanjul. 1995. The effect of inbreeding on the redistribution of ge-netic variance of fecundity and viability in *Tribolium castaneum*. Heredity 75:376–381.

Fisher, R. A. 1930. *The general theory of natural selection*. Oxford University Press, London.

Frankel, O. H., and M. E. Soulé. 1981. *Conservation and evolution*. Cambridge University Press, Cambridge.

Frankham, R. 1995. Inbreeding and extinction—a threshold effect. Conservation Biology 9:792–799.

Futuyma, D. J. 1998. *Evolutionary biology*, 3rd Ed. Sinauer Associates, Sunderland, MA.

Garcia, N., C. Lopez-Fanjul, and A. Garcia-Dorado. 1994. The genetics of viability in *Drosophila melanogaster*: effects of inbreeding and artificial selection. Evolution 48:1277–1285.

Garcia, P., F. J. Vences, M. Perez de la Vega, and R. W. Allard. 1989. Allelic and genetic composition of ancestral Spanish and colonial Californian gene pools of *Avena barbata*. Genetics 122:687–694.

Goodnight, C. J. 1987. On the effect of founder events on epistatic genetic variance. Evolution 41: 80–91.

Goodnight, C. J. 1988. Epistasis and the effect of founder principle on additive genetic variance. Evolution 42:441–454.

Haldane, J. B. S. 1932. *The causes of evolution*. Longmans, Green, New York.

Hamrick, J. L. 1987. Gene flow and the distribution of genetic variation in plant populations. In K. Urbanska, ed. *Differentiation in higher plants*, pp. 3–67. Academic Press, New York.

Hamrick, J. L., and M. J. W. Godt. 1990. Allozyme diversity in plant species. In A. D. H. Brown, M. T. Clegg, A. L. Kahler, and B. S. Weir, eds. *Plant population genetics, breeding and germplasm resources*, pp. 43–63. Sinauer Associates, Sunderland, MA.

Harper, J. L. 1982. After description. In E. I. Newman, ed. *The plant community as a working mechanism*, pp. 11–25. British Ecological Society Special Publications, No. 1. Blackwell Scientific, Oxford.

Hartl, D. L., and A. G. Clark. 1997. *Principles of population genetics*, 3rd Ed. Sinauer Associates, Sunderland, MA.

Hicks, G. H. 1895. Pure seed investigations. In *Yearbook of the United States Department of Agriculture*, 1894, pp. 389–408. Washington, DC.

Huxley, J. S. 1942. *Evolution: the modern synthesis*. Allen and Unwin, London.

Jain, S. K., and P. S. Martins. 1979. Ecological genetics of the colonizing ability of rose clover (*Trifolium hirtum* All.). American Journal of Botany 66:361–366.

Kahler, A. L., R. W. Allard, M. Krzakowa, C. F. Wehrhahn, and E. Nevo. 1980. Associations between isozyme phenotypes and environment in the slender wild oat (*Avena barbata*) in Israel. Theoretical Applied Genetics 56:31–47.

Kornas, J. 1990. Plant invasions in Central Europe: historical and ecological aspects. In F. di Castri, A. J. Hansen, and M. Debussche, eds. *Biological invasions in Europe and the Mediterranean Basin*, pp. 19–36. Kluwer, Dordrecht.

Lande, R. 1980. Genetic variation and phenotypic evolution during allopatric speciation. American Naturalist 116:463–479.

Lande, R. 1988. Genetics and demography in biological conservation. Science 241:1455–1460.

Lande, R. 1993. Risks of population extinction from demographic and environmental stochasticity and random catastrophes. American Naturalist 142:911–927.

Leberg, P. L. 1992. Effects of population bottlenecks on genetic diversity as measured by allozyme electrophoresis. Evolution 46:477–494.

Leberg, P. L. 2002. Estimating allelic richness: Effects of sample size and bottlenecks. Molecular Ecology 11:2445–2449.

Lewontin, R. C. 1974. Detecting population differences in quantitative characters as opposed to gene frequencies. American Naturalist 123:115–124.

Lopez-Fanjul, C., and A. Villaverde. 1989. Inbreeding increases genetic variance for viability in *Drosophila melanogaster*. Evolution 43:1800–1804.

Lopez-Fanjul, C., A. Fernandez, and M. A. Toro. 1999. The role of epistasis in the increase in the additive genetic variance after population bottlenecks. Genetical Research 73:45–59.

Lopez-Fanjul, C., A. Fernandez, and M. A. Toro. 2000. Epistasis and the conversion of non-additive to additive genetic variance at population bottlenecks. Theoretical Population Biology 58:49–59.

Lopez-Fanjul, C., A. Fernandez, and M. A. Toro. 2002. The effect of epistasis on the excess of the additive and nonadditive variances after population bottlenecks. Evolution 56:865–876.

Love, R. M., and D. C. Sumner. 1952. Rose clover, a new winter legume. California Agricultural Experiment Station Circular 407.

Luikart, G., F. W. Allendorf, J. M. Cornuet, and W. B. Sherwin. 1998a. Distortion of allele frequency distributions provides a test for recent population bottlenecks. Journal of Heredity 89:238–247.

Luikart, G., W. B. Sherwin, B. M. Steele, and F. W. Allendorf. 1998b. Usefulness of molecular markers for detecting population bottlenecks via monitoring genetic change. Molecular Ecology 7:963–974.

Luikart, G., J. M. Cornuet, and F. W. Allendorf. 1999. Temporal changes in allele frequencies provide estimates of population bottleneck size. Conservation Biology 13:523–530.

Mack, R. N. 1985. Invading plants: Their potential contribution to population biology. In J. White, ed. *Studies on plant demography: John L. Harper Festschrift*, pp. 127–142. Academic Press, New York.

Mack, R. N. 1991. The commercial seed trade: an early disperser of weeds. Economic Botany 45:257–273.

Mack, R. N. 2000. Cultivation fosters plant naturalization by reducing environmental stochasticity. Biological Invasions 2:111–122.

Mack, R. N. 2003. Plant naturalizations and invasions in the eastern United States: 1634–1860. Annals of the Missouri Botanical Garden 90:77–90.

Mack, R. N., and M. Erneberg. 2002. The United States naturalized flora: largely the product of deliberate introductions. Annals of the Missouri Botanical Garden 89:176–189.

Mack, R. N., D. Simberloff, W. M. Lonsdale, H. Evans, M. Clout, and F. A. Bazzaz. 2000. Biotic invasions: causes, epidemiology, global consequences and control. Ecological Applications 10:689–710.

Mayr, E. 1942. *Systematics and the origin of species.* Columbia University Press, New York.

Marshall, D. R., and R. W. Allard. 1970. Isozyme polymorphisms in natural populations of *Avena fatua* and *Avena barbata*. Heredity 25:373–382.

Marshall, D. R., and P. W. Weiss. 1982. Isozyme variation within and among Australian population of *Emex spinosa* (L.) Campd. Australian Journal of Biological Science 35:327–332.

Martins, P. S., and S. K. Jain. 1979. Role of genetic variation in the colonizing ability of rose clover (*Trifolium hirtum* All.) American Naturalist 114:591–595.

Martins, P. S., and S. K. Jain. 1980. Interpopulation variation in rose clover—a recently introduced species in California rangelands. Journal of Heredity 71:29–32.

Maruyama, T., and P. A. Fuerst. 1985. Population bottlenecks and nonequilibrium models in population genetics. II. Number of alleles in a small population that was formed by a recent bottleneck. Genetics 111:675–689.

McCauley, D. E. 1991. Genetic consequences of local-population extinction and recolonization. Trends in Ecology and Evolution 6:5–8.

McCracken, D. P. 1997. *Gardens of empire: botanical institutions of the Victorian British empire.* Leicester University Press, London.

Meekins, J. F., H. E. Ballard, and B. C. McCarthy. 2001. Genetic variation and molecular biogeography of a North American invasive plant species (*Allaria petiolata*, Brassicaceae). International Journal of Plant Science 162:161–169.

Meffe, G. K., and C. R. Carroll. 1997. *Principles of conservation biology.* 2nd Edition. Sinauer Associates, Sunderland, MA.

Menges, E. S. 1991. The application of minimum viable population theory to plants. In D. A. Falk and K. E. Holsinger, eds. *Genetics and conservation of rare plants*, pp. 45–61. Oxford University Press, New York.

Menges, E. S. 1998. Population viability analyses in plants: challenges and opportunities. Trends in Ecology and Evolution 15:51–56.

Molina-Freaner, F., and S. K. Jain. 1992. Isozyme variation in Californian and Turkish populations of the colonizing species *Trifolium hirtum*. Journal of Heredity 83:423–430.

Moran, G. F., and D. R. Marshall. 1978. Allozyme uniformity within and variation between races of the colonizing species *Xanthium strumarium* L. (Noogoora Burr). Australian Journal of Biological Science 31:283–291.

Mulligan, G. A., and J. N. Findlay. 1970. Reproductive systems and colonization in Canadian weeds. Canadian Journal of Botany 48:859–860.

Naciri-Graven, Y., and J. Goudet. 2003. The additive genetic variance after bottlenecks is affected by the number of loci involved in epistatic interactions. Evolution 57:706–716.

National Research Council. 2002. *Predicting invasions of nonindigenous plants and plant pests.* National Academy Press, Washington, DC.

Nei, M. 1977. *F*-statistics and analysis of gene diversity in subdivided populations. Annals of Human Genetics 41:225–233.

Nei, M., T. Maruyama, and R. Chakraborty. 1975. The bottleneck effect and genetic variability in populations. Evolution 29:1–10.

Neuffer, B. and H. Hurka. 1999. Colonization history and introduction dynamics of *Capsella bursa-partoris* (Brassicaceae) in North America: isozyme and quantitative traits. Molecular Ecology 8:1667–1681.

Novak, S. J., and R. N. Mack. 1993. Genetic variation in *Bromus tectorum* (Poaceae): comparison between native and introduced populations. Heredity 71:167–176.

Novak, S. J., and R. N. Mack. 2000. Clonal diversity within and among introduced populations of the apomictic vine *Bryonia alba* (Cucurbitaceae). Canadian Journal of Botany 78:1469–1481.

Novak, S. J., and R. N. Mack. 2001. Tracing plant introduction and spread into naturalized ranges: genetic evidence from *Bromus tectorum* (cheatgrass). BioScience 51:114–122.

Novak, S. J., R. N. Mack, and D. E. Soltis. 1991. Genetic variation in *Bromus tectorum* L.: population differentiation in its North American range. American Journal of Botany 78: 1150–1161.

Novak, S. J., R. N. Mack, and P. S. Soltis. 1993. Genetic variation in *Bromus tectorum* (Poaceae): introduction dynamics in North America. Canadian Journal of Botany 71:1441–1448.

Panetta, F. D. 1990. Isozyme variation in Australian and South-African populations of *Emex australis* Steinh. Australian Journal of Botany 38:161–167.

Parker, P. G., A. A. Snow, M. D. Schug, G. C. Booton, and P. A. Fuerst. 1998. What molecules can tell us about populations: Choosing and using a molecular marker. Ecology 79: 361–382.

Perez de la Vega, M., P. Garcia, and R. W. Allard. 1991. Multilocus genetic structure of ancestral Spanish and colonial Californian populations of *Avena barbata*. Proceedings of the National Academy of Sciences USA 88:1202–1206.

Pierson, E. A., and R. N. Mack. 1990. The population biology of *Bromus tectorum* in forests: distinguishing the opportunity for dispersal from environmental restriction. Oecologia 84:519–525.

Piggin, C. M. 1982. The biology of Australian weeds 8. *Echium plantagineum* L. Journal of the Australian Institute of Agricultural Science 48:3–16.

Polans, N. O., and R. W. Allard. 1989. An experimental evaluation of the recovery potential of ryegrass populations from genetic stress resulting from restriction of population size. Evolution 43:1320–1324.

Price, S. C., and S. K. Jain. 1981. Are inbreeders better colonizers? Oecologia 49:283–286.

Rehder, A. 1940. *Manual of cultivated trees and shrubs hardy in North America, exclusive of the subtropical and warmer temperate regions*, 2nd Ed. Macmillan, New York.

Richards, C., and P. L. Leberg. 1996. Temporal changes in allele frequencies and a population's history of severe bottlenecks. Conservation Biology 10:832–839.

Ridley, H. N. 1930. *The dispersal of plants throughout the world*. L. Reeve, Kent, U.K.

Ridley, M. 1993. *Evolution*. Blackwell Scientific, London.

Robertson, A. 1952. The effect of inbreeding on the variation due to recessive genes. Genetics 37:189–207.

Rose, M. R. 1982. Antagonistic pleiotropy, dominance, and genetic variance. Heredity 48:63–78.

Saccheri, I. J., P. M. Brakefield, and R. A. Nichols. 1996. Severe inbreeding depression and rapid fitness rebound in the butterfly *Bicyclus anynana* (Satyridae). Evolution 50:2000–2013.

Saccheri, I., M. Kuussaari, M. Kankare, P. Vikman, W. Fortelius, and I. Hanski. 1998. Inbreeding and extinction in a butterfly metapopulation. Nature 392:491–494.

Schoen, D., and A. D. H. Brown. 1991. Intraspecific variation in population gene diversity and effective population size correlates with the mating system of plants. Proceedings of the National Academy of Science USA 88:4494–4497.

Shaffer, M. L. 1981. Minimum population sizes for species conservation. BioScience 31:131–134.

Simberloff, D. 1989. Which insect introductions succeed and which fail? In J. Drake, H. A. Mooney, F. di Castri, R. H. Groves, F. J. Kruger, M. Rejmanek and M. Williamson, eds. *Biological invasions: a global perspective*, pp. 61–76. Wiley, New York.

Simpson, G. G. 1944. *Tempo and mode in evolution*. Columbia University Press, New York.

Singh, R. S., and S. K. Jain. 1971. Population biology of *Avena*. II. Isozyme polymorphisms in populations of the Mediterranean region and central California. Theoretical Applied Genetics 41:79–84.

Slatkin, M. 1977. Gene flow and genetic drift in a species subject to frequent local extinctions. Theoretical Population Biology 12:253–262.

Slatkin, M. 1985. Rare alleles as indicators of gene flow. Evolution 39:53–65.

Spencer, C. C., J. E. Neigel, and P. L. Leberg. 2000. Experimental evaluation of the usefulness of microsatellite DNA for detecting demographic bottlenecks. Molecular Ecology 9: 1517–1528.

Squirrell, J., P. M. Hollingsworth, R. M. Bateman, J. H. Dickson, M. H. S. Light, M. MacConaill, and M. C. Tebbitt. 2001. Partitioning and diversity of nuclear and organelle markers in native and introduced populations of *Epipactis helleborine* (Orchidaceae). American Journal of Botany 88:1409–1418.

Stace, C. A. 1997. *New flora of the British Isles*. Cambridge University Press, Cambridge.

Stebbins, G. L. 1950. *Variation and evolution in plants*. Columbia University Press, New York.

Stebbins, G. L. 1957. Self-fertilization and population viability in the higher plants. American Naturalist 91:337–354.

Sunnucks, P. 2000. Efficient genetic markers for population biology. Trends in Ecology and Evolution 15:199–203.

Upadhyaya, M. K., R. Turkington, and D. McIlvride. 1986. The biology of Canadian weeds. 75. *Bromus tectorum* L. Canadian Journal of Plant Science 66:689–709.

Waddington, C. H. 1965. Introduction to the symposium. In H. G. Baker, and G. L. Stebbins, eds. *The genetics of colonizing species*, pp. 1–6. Academic Press, New York.

Wade, M. J., and D. E. McCauley. 1988. Extinction and recolonization: their effects on the genetic differentiation of local populations. Evolution 42:995–1005.

Wang, J., A. Caballero, P. D. Knightley, and W. G. Hill. 1998. Bottleneck effect on genetic variance: a theoretical investigation of the role of dominance. Genetics 150:435–447.

Waples, R. S. 1989. A generalized approach for estimating effective population size from temporal changes in allele frequency. Genetics 121:379–391.

Warwick, S. I. 1990. Genetic variation in weeds—with particular reference to Canadian agricultural weeds. In S. Kawano, ed. *Biological approaches and evolutionary trends in plants*, pp. 3–18. Academic Press, London.

Warwick, S. I., B. K. Thompson, and L. D. Black. 1984. Population variation in *Sorghum halepense*, Johnson grass, at the northern limits of its range. Canadian Journal of Botany 62:1781–1790.

Warwick, S. I., B. K. Thompson, and L. D. Black. 1987. Genetic variation in Canadian and European populations of the colonizing weed species *Apera spica-venti*. New Phytologist 106: 301–317.

Watterson, G. A. 1984. Allele frequencies after a bottle-neck. Theoretical Population Biology 26:387–407.

Whitlock, M. C., and McCauley, D. E. 1990. Some population genetic consequences of colony formation and extinction—genetic correlations within founding groups. Evolution 44:1717–1724.

Whitlock, M. C., P. C. Phillips, and M. J. Wade. 1993. Gene interaction affects the additive genetic variance in subdivided populations with migration and extinction. Evolution 47:1758–1769.

Williamson, M. 1998. Measuring the impact of plant invaders in Britain. In U. Starfinger, K. Edwards, I. Kowarik, and M. Williamson, eds. *Ecological mechanisms and human resources*, Pp. 57–68. Backhuys, Leiden, The Netherlands.

Willis, J., and H. A. Orr. 1993. Increased heritable variation following population bottleneck: the role of dominance Evolution 47:949–957.

Wright, S. 1931. Evolution in Mendelian populations. Genetics 16:97–159.

9

Mechanisms that Drive Evolutionary Change

INSIGHTS FROM SPECIES INTRODUCTIONS AND INVASIONS

John P. Wares, A. Randall Hughes, and Richard K. Grosberg

One way to summarize the evolutionary dynamics of species introductions is to estimate how levels of genetic diversity in non-native populations are different from those of their source populations. While it is typically assumed that a significant loss of diversity will be associated with species introductions, the actual effect may be more complex, depending on propagule pressure and patterns of diversity and population structure in the native range of a species. We review a number of studies of animal species introductions in which allelic diversity and heterozygosity in the non-native and source ranges of each species can be compared, and find that the typical loss of diversity is minimal. The generality of this pattern may provide new insight into debates over the prevalence of stochastic processes in generating novel phenotypes or coadapted gene complexes in founder populations. These results suggest that the response of founder populations to natural selection in a novel environment is generally more important than the stochastic effects of the founder event itself in determining the evolutionary trajectory of a population.

Introduction

In 1880, local fishermen introduced grayling (*Thymallus thymallus*) into Lake Lesjaskogsvatn in Norway. Subsequent introductions and dispersal of this fish into nearby lakes provided an exceptional window into the phenotypic and genetic changes that occurred as grayling populations became established in novel environments. After fewer than 25 generations of isolation, levels of among-population variance for seven heritable life history traits (such as yolk sac volume and growth rate) were consistently higher than levels at microsatellite (presumably neutral) loci (Koskinen et al. 2002). Despite the fact that a small number of fish initiated each population, the loss of genetic variation due to founder effects, population bottlenecks, and random genetic drift (see Box 9.1 for definitions of key terms and concepts used in this chapter) apparently had little effect on the overall phenotypic divergence of these populations. Instead, the observed shifts in life history traits were largely attributable to natural selection (Koskinen et al. 2002). This example is consistent, therefore, with the view that introductions and invasions frequently involve episodes of rapid evolution (reviewed in Reznick and Ghalambor 2001; Sakai et al. 2001; Lee 2002). It also embodies one of the most fundamental debates in evolutionary biology; namely, the relative importance of deterministic processes such as natural selection versus stochastic processes such as genetic drift as causes of evolutionary change (reviewed in Reznick and Ghalambor 2001; Dupont et al. 2003).

Ever since the classic debate between Fisher (1930) and Wright (1932), one of the major challenges facing evolutionary biologists has been to understand how selection and drift, along with additive and higher-order interactions (e.g., dominance and epistasis) among genes, contribute to evolution in natural populations (Coyne et al. 1997; Wade and Goodnight 1998; reviewed in Brodie 2000). The Fisherian paradigm holds that natural selection generally drives evolutionary change within populations, giving primacy to the role of mass selection acting on additive genetic variance produced by the small, independent contributions of numerous loci to fitness. In contrast, Sewall Wright emphasized the importance of random genetic drift in population differentiation. Specifically, his shifting balance theory conceived of an "adaptive landscape" that described the relationship between the summed effects of interactions between the environment and additive genetic variance, as well as higher-order epistatic interactions among genes, on fitness (Wright 1932). Whereas selection acted to drive a population toward any of the fitness peaks on the landscape, drift explained how populations crossed valleys of low fitness separating adaptive peaks (Figure 9.1). Thus, Wright (1932) emphasized that stochastic processes, in concert with higher-order epistatic interactions among alleles at different loci, could constrain mass selection, especially in small, subdivided populations.

To what extent do founder effects or population bottlenecks, genetic drift, and higher-order genetic interactions influence the outcomes of Fisherian mass selection, particularly when populations experience novel selective regimes? Progress toward answering this fundamental question requires, first, docu-

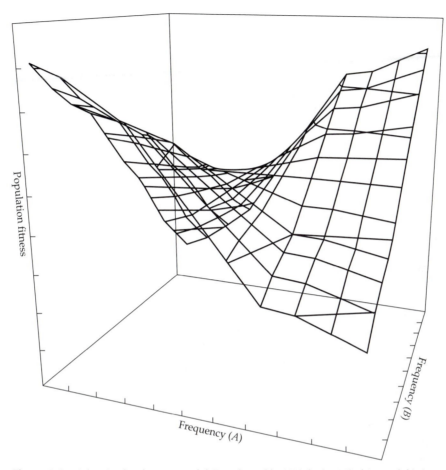

Figure 9.1   Adaptive landscape model. Developed by Wright (1932), this model is intended to portray the interactions of genes and their effect on fitness. While it is typically represented as a simple model balancing the effects of two loci (here the frequency of an allele at locus *A* and an allele at locus *B*), the interactions across an entire genome are of course more complex. The question is how populations move from one phenotypic optimum to another, given that this requires passing through genotypic combinations that are maladaptive (the "valleys"). While proponents of mass selection (*sensu* Fisher 1931) argue that most evolutionary adaptation is the result of natural selection acting on mutations, recombination events, and gene interactions, Wright (1932) proposed that genetic drift in smaller populations—which may stochastically push populations toward suboptimal genotypic combinations—aids the process of moving from one peak to another. The debate over the generality of these models for evolution is a fundamental and persistent issue (Skipper 2002).

menting genetic and phenotypic responses of natural populations to changes in selective regimes, and second, determining the contributions of additive, dominance, and epistatic variance to fitness (see papers in Wolf et al. 2000). Because it is often possible to infer divergence times from source populations,

BOX 9.1 *Key Terms and Concepts*

Here we discuss a few terms that are necessary for biologists to critically evaluate genetic characterizations of species introductions and the potential for evolution following these events. Demonstrating whether populations have evolved following a species introduction typically involves characterizing whether the mean phenotype of individuals has shifted, and whether it has shifted in a heritable way. A **trait** or set of traits may be measured to determine how much variance there is for a phenotypic character, whether the trait is structural, behavioral, or ecological. The variation in this phenotypic trait is determined by the level of **genetic variation** (V_G) at related loci, the plastic response of an individual to the environment, and the interaction between particular genotypes and environmental variables.

The genetic variation that contributes to a heritable trait may be separated into three primary components. **Additive genetic variance** (V_A) involves the whole-genome allelic diversity that contributes to phenotypic variation and is considered the most direct target for natural selection. If the contribution of a locus is additive, then each allele contributes a fixed value to the measure of a quantitative trait; higher-order interactions modify this contribution, and thus do not contribute additively to the phenotype. For example, genetic variance due to **dominance** (V_{Aa}) arises because of interactions among alleles at the same locus, such that the expression of one may mask the expression of the other, and may be important following a **population bottleneck** (the short-term restriction in population size associated with species introductions) because the frequency of particular dominant/recessive allelic combinations may change dramatically. Variance due to **epistasis** (V_{AB}) is caused by the interactions among

genes. It is thought by some to be particularly important during and after genetic bottlenecks because particular gene complexes may be altered dramatically by the loss of alleles at particular loci.

A bottleneck in terms of allelic diversity and **heterozygosity** (a measure of allelic diversity within as well as between individuals in a population) arises because a species introduction involves taking a sample from the original source population. If this sample is large, it is likely to represent the whole of source population diversity, but as the sample size of the introduced population shrinks, it is more likely that rare alleles will be omitted. Further, because of strong inbreeding among individuals in the founder population, there will be transient linkage disequilibrium between loci (genes that are not physically linked in the genome may be associated with each other because of the small number of gametes that are involved in the founding of a population) that may temporarily change the evolutionary dynamics of the population.

What is frequently debated by evolutionary biologists is the importance of these transient **founder effects**. For example, rare alleles may be lost or overrepresented in the introduced population. If these alleles are **selectively neutral**, it is of no consequence for the fitness or phenotype of the individuals, but if the alleles are rare in the source population, either because they are deleterious (that is, they potentially reduce the fitness of individuals) or because they are maintained by frequency-dependent selection, then the resulting evolutionary change could be more dramatic and long-lasting. For the same reason, the contribution of dominance interactions to genetic variance following a bottleneck is extremely variable (Barton and Turelli

BOX 9.1 *(continued)*

2004), outweighing much of the contribution of interactions among loci.

The often-discussed **"conversion"** of epistatic genetic variance to additive genetic variance involves changes in the background of allelic diversity at interacting loci; the loss of diversity at one locus may allow greater expression of the allelic diversity at another locus, contributing to additive variation and exposing this variation to natural selection (Brodie 2000). For example, gene interactions may include variation at distinct loci with balancing effects on fitness (e.g., for two loci A and B, each with two alleles, where selection favors the A_1B_2 and A_2B_1 genotypes over the A_1B_1 and A_2B_2 genotypes), and the expression of additive genetic variation may increase following a population bottleneck (e.g., the loss of A_1 due to a founder effect will result in the expression of the B alleles with no compensatory interactions—the A_2B_1 genotype should still be favored over the A_2B_2 genotype in this case, and the epistatic variance caused by interactions between these two loci is then converted into the additive effect on fitness of the B locus alone).

The strength of a bottleneck may be measured by the proportional loss (F) of **heterozygosity** at a number of presumably **neutral markers** (those that are believed to be unimportant for fitness, such as silent mutations that do not change the amino acid sequence of a protein) or by the loss of allelic diversity itself. The proportion of additive genetic variance that may be gained through epistatic interactions, following a bottleneck, is on the order of F^k (where k is the number of interacting loci; Barton and Turelli 2004). Because the contribution of epistatic interactions to phenotypic variance only declines as a function of $(1 - F^k)$, and the amount of additive genetic vari-

ance that makes it through a bottleneck is $(1 - F)$, epistatic and dominance interactions can contribute relatively more to phenotypic variance after a severe bottleneck.

The effects of population bottlenecks are inherently linked to the concept of **effective population size (N_e)**, a value that suggests the average number of individuals in a population that contribute genetically toward the following generation. This parameter, which reflects the life history and demography of a population or species, may be measured using both temporal changes in allele frequencies (e.g., Waples 1989) as well as the statistical distribution of nucleotide substitutions in a population (Nei 1987). The ratio of N_e to the actual census size depends on a number of factors (Turner et al. 2002). Historical bottlenecks will reduce the measure of inbreeding effective population size (a geometric mean of the number of successfully reproducing individuals over time), and high variances in reproductive success will also lower N_e of a population. The strength of **random genetic drift**, or the stochastic variation in allele frequencies from one generation to the next, is inversely proportional to N_e; small isolated populations are subject to stronger effects of drift.

When a species' source region contains several demographically isolated populations (**population genetic structure**), the introduction of that species involves a sample from at least one—and maybe several—genetically distinct lineages. Depending on the time inferred from the phylogeny, admixture of populations in the introduced region may involve closely related populations or distinct species. **Admixture** refers to the demographic linkage of two or more historically isolated lineages in the same introduced range, and may or may not involve **hybridization**, the successful

BOX 9.1 *(continued)*

mating between members of different lineages. Hybridization allows genetic interactions between two isolated populations to occur; this may involve both dominance and epistatic interactions that affect the fitness of the hybrid individuals (Willett and Burton 2003). **Introgression** occurs only if hybrid individuals are successful in reproducing with individuals from one or both of the original lineages (Stewart et al. 2003). Methods of assessing deviations in allele frequencies from Hardy-Weinberg equilibrium provide some insight into the degree to which introgression has occurred (Anderson and Thompson 2002).

and in some cases to reconstruct directly the geographic, demographic, and ecological history of an introduction or invasion, genetic studies of introduced and invasive species can often provide unusually detailed, and sometimes replicated, chronicles of evolutionary change (Le Page et al. 2000). Moreover, to the extent that introductions and invasions involve rapid evolution, it is often possible to observe directly associated genetic and phenotypic changes. Biological introductions therefore represent potentially powerful—albeit often undesirable—experiments that may improve our understanding of how stochastic and deterministic forces shape evolutionary change.

There is little question that the strength and direction of natural selection can change rapidly during the course of an invasion (reviewed in Sakai et al. 2001; Lee 2002). It is also clear that many of the source populations for introductions and invasions possess ample additive genetic variance and may therefore endow introduced populations with the genetic variation necessary to respond to selection (e.g., Huey et al. 2000). On the other hand, if relatively few individuals found an introduced population, and the population therefore experiences an extended bottleneck before establishment and expansion, then the loss of allelic diversity and additive genetic variance could seriously limit the potential for rapid evolutionary change (reviewed in Willis and Orr 1993; but see Whitlock 1995; Whitlock and Fowler 1999). Founder effects and drift may also lead to the "conversion" of epistatic (e.g., Goodnight 1988; Cheverud 2000) and dominance (e.g., Willis and Orr 1993) variance for fitness into additive genetic variance. This conversion could subsequently fuel evolutionary change during an invasion (Reznick and Ghalambor 2001; Lee 2002). Thus, in theory, both natural selection and stochastic processes could cause (or inhibit) evolutionary change over the course of an invasion: newly founded populations could experience a loss or gain of additive genetic variance, an increase or decrease in the importance of epistatic and dominance interactions, and an enhanced or diminished response to selection.

In this chapter, we analyze patterns of genetic change that accompany introductions and invasions, emphasizing the general insights that such patterns (or their absence) provide into the roles of stochastic and deterministic processes as agents of evolutionary change. We begin by discussing the importance of genetically characterizing the full spectrum of populations that represent potential sources for introductions. We then assess the strengths and limitations of several genetic approaches widely used to identify source populations and to reconstruct the demographic history of invasion, a step that is easy to discount but critically important to our ability to make inferences of ecological and evolutionary change in introduced species. Those familiar with these methodological issues should skip to the next section, "Predicted Changes in Genetic Diversity during Introductions," in which we summarize some of the relevant theory and survey genetic studies of introduced (primarily animal) species to evaluate the contributions that drift, selection, mutation, and higher-order genetic interactions make to observed changes. The final section of the chapter evaluates the types of evolutionary forces that are likely to dominate the evolution of new phenotypes in introduced populations. (See Novak and Mack, this volume, for a comparable analysis that emphasizes plants.)

Establishing the Baseline for Evolutionary Changes

The process of establishing the evolutionary "baseline" for an introduction or invasion, as we define it here, has three key components. The first is to identify the source population(s) of an introduction. The second is to identify the genetic characteristics of that source with respect to selectively neutral markers as well as heritable traits that may be under selection. The third is to characterize the demographic history of introduction, establishment, and spread, as this history provides critical information on the size of an introduced population and hence provides critical clues to the potential action of different evolutionary forces. We emphasize the analysis of neutral markers in this section partly because such markers most accurately reflect a population's demographic history (Avise 2000). However, the genetic markers used to identify source populations and characterize the demography of introductions and invasions are unavoidably not those of greatest interest in terms of evolutionary responses of introduced species. Unfortunately, there are just a handful of studies, which we discuss later in this chapter, that characterize baseline conditions for heritable traits that are potentially subject to selection.

Identifying the source population(s)

The assertion that biological introductions and invasions often entail evolutionary change comes from observations of introduced or invasive populations exhibiting altered ecological interactions, behaviors, or phenotypes relative to source (or other introduced) populations (Sakai et al. 2001). Thus, identifying the source population(s) of an introduction reduces the potential for making false inferences about the nature and magnitude of evolutionary changes during introductions and invasions (Kolbe et al. 2004). For instance, observed divergence among introduced populations may reflect extant differences among source populations, rather than post-introduction evolution (Lee 1999). From a practical perspective, identification of the correct source population(s) for an introduction can also simplify the development of effective biological control (Bartlett et al. 2002; Roderick and Navajas 2003; Baliraine et al. 2004).

There are in effect two steps to identifying the source population(s) for an introduction or invasion. The first is to characterize the geographic region that encompasses potential source populations and to distinguish those sources from regions or populations that represent introductions; the second involves reconstructing the history of introductions and identifying the source populations that gave rise to those introductions. Ecological data, including historical descriptions of a species' range and potential range extensions, typically are the starting point for circumscribing the native range of an introduced species and for characterizing phenotypic attributes that may be related to evolutionary changes associated with the introduction. For example, historical observations show that the brown anole (*Anolis sagrei*) is native to Caribbean islands, although populations have now been introduced worldwide (Kolbe et al. 2004). However, the historical data fail to reveal the pathways by which *A. sagrei* spread beyond the Caribbean and the evolutionary changes that occurred as the species' range expanded.

Genetic information provides an approach that is complementary to ecological and historical data for distinguishing source populations, establishing baseline conditions, identifying introduced populations, and characterizing evolutionarily important changes in the genetic composition and population characteristics (such as size and connectivity) of species during the stages of introduction, establishment, and spread (Hebert and Cristescu 2002). The first component of a genetic approach to identifying the source population(s) of an introduction involves surveying multiple populations across the entire native range for variation at molecular markers. If populations in the native range are genetically homogeneous, then it will be difficult to narrow the search for a source population. In contrast, if the native region contains genetically diverse populations—as in *A. sagrei* (Kolbe et al. 2004)—and these populations exhibit significant population genetic structure, then it may be possible to identify the source population(s) most likely to have contributed to an introduction. Once the native region has been surveyed, the same molecular markers can be used in the introduced populations to help characterize (1) their sources, (2) the evolutionarily relevant demography of the species introduction, including the number of founder individuals and the frequency and timing of subsequent introductions from multiple source populations, and (3) the genetic changes that accompanied the introduction.

Genetic methods for identifying sources

There are two basic methods of obtaining some, or all, of this genetic information. One method combines genealogical and geographic information in a phylogeographic framework to reconstruct the history of introductions and invasions (Avise 2000); the other uses comparisons of allele frequencies across populations to determine the likelihood that a particular introduced population originated from an array of potential sources (see Davies et al. 1999, and Roderick and Navajas 2003 for reviews). In this section, we briefly assess the strengths and weaknesses of both of these approaches.

Phylogeographic methods establish the genealogical relationships of individuals from throughout a species' distribution. Much of this work is grounded in coalescent theory (Kingman 1982; Hudson 1990), which highlights the inherent link between the distribution of genetic variation in a population and the demographic history of that population. Greater genetic diversity is attributable to greater effective population size (N_e), and as a population grows or shrinks, or splits into smaller ones, the genetic diversity in each area reflects these demographic and distributional changes. The genetic data (such as DNA sequence data) collected for these analyses are statistically analyzed with respect to the distribution of variation within and among populations to determine whether a species' range consists of a single homogeneous set of populations linked by migration and gene flow, or whether there is significant genetic structure among populations, indicating that they have been evolutionarily and demographically isolated from one another (see Grosberg and Cunningham 2001; Roderick and Navajas 2003). The outcome is a portrait of a species' history that integrates geography with information about the shared ancestry of alleles (Beerli and Felsenstein 1999) and inferred aspects of the species' demographic history, such as effective population size and connectivity, that can strongly influence a population's evolutionary response to selection (Avise 2000; Pascual et al. 2001; Bohonak et al. 2001). Consequently, a phylogeographic analysis can be used not only to identify source locations for introduced species, but also to reconstruct the timing and mechanisms of introduction.

The second genetic approach for identifying an introduction's source population(s) uses information about allele frequencies throughout some or all of a species' range to assign an individual, based on its multilocus genotype, to the population from which it has the highest probability of origin (Paetkau et al. 1995; Davies et al. 1999; Hansen et al. 2001; Guinand et al. 2002). Methods based on allele frequencies are especially powerful for deciphering complex patterns of admixture among populations, as when multiple genetically differentiated source populations contribute propagules to an invasive population. A principal strength of frequency-based approaches lies in their use of a large number of unlinked loci to reconstruct complex introduction and invasion scenarios. In particular, estimates of genetic parameters such as effective population size, average heterozygosity, and allelic diversity based on multiple loci usually have lower variances than estimates based on sequences from one or a few loci (Wakeley and Hey 1997). Moreover, these methods are particularly useful when different classes of molecular markers are used in tandem. For example, Estoup et al. (2001) used rapidly evolving microsatellite

markers together with more slowly evolving allozyme markers to characterize the duration and intensity of population bottlenecks in introduced *Bufo
marinus* (cane toad) populations.

The power of allele frequency-based tests depends on the degree of genetic
differentiation among potential source populations, the number of potential
source populations, polymorphism at each locus, and the number of loci analyzed (Hansen et al. 2001; Guinand et al. 2002). Moreover, as Estoup et al. (2001)
emphasized, evolutionary inferences derived from frequency-based methods
are constrained by our limited understanding of how variation in many widely
used markers, such as microsatellites and other fragment length polymorphisms, evolves. It is therefore difficult to specify how alleles of different sizes
are genealogically related to one another. In contrast, the phylogeographic
approach, directly based on DNA sequences, can incorporate more realistic
models of molecular evolution. This approach can reconstruct the genealogical relationships between different sequences (and the individuals that carry
those sequences). However, it is typically more expensive to obtain sequence
data at even a single locus and more challenging to find variable sequence-
based markers. Consequently, only one or a few loci are sequenced in most
studies, seriously limiting the power of many phylogeographic studies to dis-
criminate among invasion scenarios, especially with respect to characteristics
such as admixture, the magnitude of bottlenecks, and rates of post-introduction population expansion; these limitations arise because any single gene
genealogy can vary substantially from the actual evolutionary history of the
introduction and invasion. If there have been multiple introductions through
time so that introduced populations consist of a mixture of different sources,
and/or population demography varies in space and time, then more loci will
provide a more accurate and comprehensive portrait of the evolutionary
dynamics of an introduction and invasion.

Recently, coalescent-based admixture methods have been developed to
account explicitly for the information provided by multilocus, sequence-based
genotypes. These new methods incorporate the genetic divergence between
alleles into the reconstruction, rather than relying simply on allele frequencies
or population summary statistics (e.g., Bertorelle and Excoffier 1998; Chikhi et
al. 2001; Anderson et al. 2002). Some of these methods incorporate effects of
sampling error, genetic drift, uncertainty regarding estimates of allele frequencies in source populations, and variable population sizes (Chikhi et al.
2001). These approaches appear to outperform frequency-based methods
(Choisy et al. 2004). Thus, although there are still formidable empirical and
economic obstacles to obtaining the necessary sequence data at multiple loci
for a sufficiently large sample to estimate allele frequencies with any measure
of precision, models suggest that such efforts will be necessary for reconstructing the evolutionary history of all but the simplest patterns of species
introductions.

Predicted Changes in Genetic Diversity during Introductions

In this section, we briefly review some of the basic theoretical foundations for our expectations of how the genetic diversity of introduced populations should vary, primarily with respect to heterozygosity and allelic diversity. To the extent possible, we use direct genetic evidence to examine these predictions; however, in some cases in which genetic changes cannot be explicitly measured, we use presumably correlated phenotypic effects. We first explore how the population bottlenecks that are often assumed to typify the first stages of introductions and invasions should affect heterozygosity and allelic diversity. We also consider the effects of mutation and hybridization, two mechanisms that can increase components of genetic variance. Finally, we survey the available genetic data that compare source and introduced populations to determine generally how, or if, introductions affect the genetic composition of populations. As with the characterization of source populations and genetic baselines, we emphasize the changes revealed by neutral genetic markers, because data from neutral markers more directly reflect the demographic and historical elements of a species introduction, with minimal intrusion of signals due to selection. Once again, however, this emphasis is also partly due to our relative ignorance in most introduced species about changes at loci that directly influence fitness traits. Such changes will, of course, often present very different patterns of diversity and divergence among populations (Luikart et al. 2003).

Effects of population bottlenecks and drift

The prediction that stochastic processes—notably founder effects, population bottlenecks, and genetic drift—should have major effects on evolution during introductions and invasions stems from expected changes in population demography associated with the introduction, establishment, and spread of an introduced species (reviewed by Sakai et al. 2001). Introductions typically begin with a small number of propagules derived from a much larger native source population. Compared with its source population, an introduced population should have (1) lower allelic diversity; (2) lower heterozygosity; (3) different allele frequencies due to sampling effects; and (4) lower additive genetic variance. The final prediction remains the most controversial because of the potential contribution of interactions among alleles (dominance effects) and among genes (epistatic effects) to overall genetic variance (see discussion below).

The magnitude of each of these potential effects depends strongly on the extent and duration of a population bottleneck, the rate of population expansion following introduction, and levels of ongoing migration from source populations and other invasive populations (Nei et al. 1975; Austerlitz et al. 1997; Ellstrand and Schierenbeck 2000). For example, a bottleneck of short duration may not lead to a significant loss of genetic variation (Nei et al. 1975); however, even if a population's census size rebounds quickly, N_e may still be suppressed relative to expectations at demographic and evolutionary equilibrium (Grosberg and Cunningham 2001), since N_e is largely determined by the minimum historical census size. Therefore, the relative strength of selection (directly proportional to N_e) and drift (inversely proportional to N_e) will change during the different stages of a species introduction.

Effects of mutation

The fundamental obstacle to characterizing the importance of mutation to evolutionary changes during introductions and invasions is detection. Novel alleles sampled in invasive populations may represent in situ mutations; however, limited genetic sampling of a source population may not provide a comprehensive picture of the genetic diversity present in the source region (Ewens 1972). It may be difficult, therefore, to determine whether "novel" alleles in expanding, invasive populations represent in situ mutations or whether they were simply rare and undetected in the source population (Demelo and Hebert 1994; Bastrop et al. 1998; Wares et al. 2002). Because detection becomes much simpler when full genetic characterization of source populations can be carried out, much of our understanding of how mutation contributes to the genetic and phenotypic variation of invasive populations comes from agricultural and laboratory experiments rather than natural populations (see Keightley and Lynch 2003).

Most mutations reduce fitness and should therefore be purged from populations over time (Barton 1989; Willis and Orr 1993; Keightley and Lynch 2003). However, one key difference regarding the fate of mutations—even slightly deleterious ones—distinguishes growing populations from stationary ones: new mutations have a much better chance of becoming established in rapidly expanding populations (such as invasive populations following a bottleneck) than in populations at demographic equilibrium (Carson 1968; Nei et al. 1975; Maruyama and Fuerst 1984; Otto and Whitlock 1997). Thus, mutations should more often contribute to the evolution of invasive and other growing populations than to that of populations at equilibrium. Empirical studies, however, find that mutations often represent a negligible component of the allelic diversity present in founder populations (Bohonak et al. 2001; Estoup et al. 2001; Knowles 2004). Nevertheless, novel mutations at loci under negative frequency-dependent selection, or those that are favored under unusual or unpredictable circumstances, have a higher likelihood of becoming established in rapidly expanding populations, and could therefore play a disproportionately large role in the evolution of invasive species.

Effects of admixture and hybridization

Introductions can bring into contact previously geographically isolated and genetically differentiated taxa that can hybridize. The extent of hybridization following an introduction may span a range of potential interactions, from matings between conspecific individuals from distinct source populations (admixture of multiple introductions) to hybridization between introduced and native species. Hybridization may be limited to the production of F_1 (first-generation) crosses, with little genetic exchange between the parental taxa. However, if hybrids can reproduce successfully with either of the parental taxa, there may be extensive introgression and, ultimately, the homogenization of genetic (and ecological) differences that distinguish the parental taxa. At the same time, introgression can bring novel genes into both the introduced and resident populations, increasing their potential for responding to selection, either through increased additive genetic variance or novel epistatic interactions.

Successful mating between previously isolated taxa can directly increase levels of additive genetic variance in hybrid progeny (Ellstrand and Schierenbeck 2000; Ayres et al. 1999). However, hybridization—especially between distantly related taxa—usually produces offspring with lower fitness relative to the fitness of parental taxa (see Arnold 1997; Willett and Burton 2003). The interactions between alleles and distinct loci that are brought together through hybridization (as with sexual recombination) accelerate the evolutionary process in these populations—some good combinations will thrive, and bad combinations will often be fatal or otherwise deleterious. However, hybridization and the subsequent influence of additive and higher-order interactions can sometimes produce offspring whose range of phenotypes transgresses that observed in the parental taxa (Arnold 1997; Ellstrand and Schierenbeck 2000; Rieseberg et al. 2003). Such transgressive phenotypes may have higher fitness than either of the parental taxa, especially in disturbed or novel environments. In this way, post-introduction hybridization among conspecific or heterospecific populations may be a primary source of rapid genetic and phenotypic change in non-native populations (e.g., Ayres et al. 1999).

Observed Effects of Introductions on Genetic Variation

The preceding sections imply that the genetic changes that accompany species introductions strongly depend on the magnitude and duration of founder effects and population bottlenecks. Neither of these variables is easy to measure directly in nature. Consequently, estimates of heterozygosity and allelic diversity, based on presumably neutral markers, are often used as proxies for how a species' evolutionary potential has been altered by a founder event.

In an extensive survey of the literature on the genetic characteristics of introductions and invasions, we found 29 recent studies on animals in which relevant characteristics such as the heterozygosity and allelic diversity of source and introduced populations could be compared (Figure 9.2, Table 9.1; for a discussion of plant introductions, see Novak and Mack, this volume). These comparisons revealed that the loss of heterozygosity is typically rather small (Figure 9.2A). On average, the proportional reduction in heterozygosity ($F = 1 - H_i/H_s$, where H_i is the average heterozygosity across loci in the invasive population and H_s is the same value calculated for the putative source region) is only about 17% across a broad range of animal introductions, with most values

TABLE 9.1 *Loss of genetic diversity in naturalized animal populations*

Species	F	A	Reference
Passer montanus	0.174	–0.05	St. Louis and Barlow 1988
Carduelis chloris	0	–0.025	Merilä et al. 1996
Fringilla coelebs	–0.375	–0.035	Baker 1992
Passer domesticus	0.057	0.167	Parkin and Cole 1985
Acridotheres tristis	0.193	0.210	Baker and Moeed 1987
*Sturnus vulgaris**	0	0.283	Ross 1983; Cabe 1998
*Carpodacus mexicanus**	0.036	—	Wang et al. 2003
Junco hyemalis	0.125	0.373	Rasner et al. 2004
Thymallus thymallus	0.561	0.581	Koskinen et al. 2002
Linepithema humile	0.681	0.492	Tsutsui et al. 2000
*Drosophila subobscura**	0.098	0.623	Balanyà et al. 1994; Pascual et al. 2001
Musca autumnalis	0.2	0.063	Bryant et al. 1981
Solenopsis richteri	0.33	0.25	Ross and Trager 1990
Solenopsis invicta	0.226	0.183	Ross et al. 1993
*Ceratitis capitata**	0.656	0.648	Malacrida et al. 1998
Ceratitis rosa	0.051	0.241	Baliraine et al. 2004
*Dreissena polymorpha**	–0.117	–0.077	Boileau and Hebert 1992; Stepien et al. 2002
*Carcinus maenas**	0.189	0.478	Bagley and Geller 1999
Perna perna	0.109	0.053	Holland 2001
Theba pisana	0.5	0.2	Johnson 1988
Littorina saxatilis	0.125	0	Janson 1987
Bosmina coregoni	–0.111	0	Demelo and Hebert 1994
Bythotrephes longimanus	–0.042	0	Berg et al. 2002
*Marenzellaria viridis**	0.029	0.058	Rohner et al. 1996
Crepidula fornicata	0.114	–0.098	Dupont et al. 2003
Peromyscus leucopus	0.125	0	Browne 1977
Macropus rufogriseus	0.097	0.244	Le Page et al. 2000
Anolis grahami	0.250	0.167	Taylor and Gorman 1975
Rana ridibunda	–0.067	0.187	Zeisset and Beebee 2003

Note: This table includes species for which allelic data from both source and invasive populations, along with the proportional loss of heterozygosity (F) and allelic diversity (A) from each species invasion, are available. The asterisk (*) indicates averaged effects across introductions to multiple locations or multiple studies of the same introduction. If not recognized, multiple introductions to the same region will tend to reduce the proportional loss of diversity, potentially shifting some values lower relative to the actual effects of the bottleneck. Those introduced populations that have higher diversity than putative source populations, shown here as negative values, may represent admixture of multiple propagules. Because sampling artifacts may also generate biased values of F, the "average" F (0.17) given in the text considers these negative values to be equivalent to 0; however, the average including these cases in which diversity is higher in the non-native range is similar (0.15). Some of these studies demonstrate that extremely low diversity may occur in both populations and that small changes in heterozygosity may result in large proportional changes (e.g., *Theba pisana*, $H_{source} = 0.10$, $H_{introduced} = 0.05$), which may bias F upward for such examples.

Figure 9.2 Proportional loss of (A) heterozygosity and (B) allelic diversity in invasive populations relative to their source populations. The proportional loss of heterozygosity, averaging about 0.17 across the studies listed in Table 9.1, is frequently used as a measure of the strength of a bottleneck (*F*). However, several of these studies suggest that the loss of rare alleles is more evolutionarily significant and a more sensitive indicator of what happens to invasive populations as they recover from a bottleneck; in most of the studies summarized in B, the loss of allelic diversity is also modest (< 20% loss). Note that for both A and B, several studies listed in the leftmost column (i.e., 0, < 0.2) have negative values (i.e., show a proportional gain in heterozygosity or number of alleles); these values may reflect the effects of genetic admixture or may be due to sampling artifacts (see Table 9.1 for details).

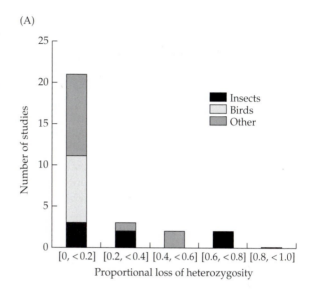

(A)

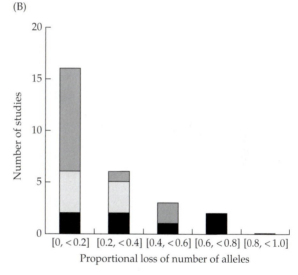

(B)

being considerably lower. Comparisons based on allelic diversity suggest that founder effects may have a slightly greater effect on the loss of rare alleles (Figure 9.2B).

Why, in terms of genetic diversity assayed through neutral markers, do introduced populations appear to differ so little from their sources? A variety of theoretical studies of colonization suggest that the effects of a bottleneck on genetic diversity can be relatively small. Consider the following example: genealogical simulations suggest that if a small founding population of 20–100 individuals rapidly expands following establishment, allelic diversity, heterozygosity, and mean time to common ancestry among individuals drop only slightly below their values in the source population (Austerlitz et al. 1997). On the other hand, if new populations are initiated by sequential colonizations of a few individuals from one population to the next, then genetic diversity can appreciably decline (also see Nei et al. 1975; Barton 1989). Studies of a natural colonization time series in an island-colonizing bird, the Australian silvereye (*Zosterops lateralis*), confirm these predictions (Clegg et al. 2002). Comparison of each recently established population with its immediate source revealed no significant differences in terms of either allelic diversity or heterozygosity; only comparisons between the earliest colonists and the most recent expansion populations showed significant differentiation. Thus, both in theory and in practice, the influence of genetic drift and founder effects on genetic diversity may have been overemphasized for many colonization events.

Given the minimal effects of species introduction on genetic diversity in introduced populations, there are only a few convincing empirical examples of drift influencing the evolution of traits that might facilitate the spread or differentiation of introduced species (Lee 2002). If levels of variation at neutral markers are strongly correlated with the additive genetic variance harbored by a population, this would suggest that the population dynamics of introduction and invasion typically do not translate into significant genetic change (Amos and Balmford 2001; Dupont et al. 2003). However, the stronger effects of bottlenecks and drift on allelic diversity (as opposed to heterozygosity per se) may be especially important when rare alleles influence fitness (Leberg 1992; Ross et al. 1993; Tsutsui et al. 2000; Box 9.2).

BOX 9.2   *Small Changes, Big Effects: The Importance of Genetic Details (and Why Additive Genetic Variance May Tell Only Part of the Story)*

The invasion of the red fire ant (*Solenopsis invicta*) incisively illustrates how small genetic changes can have profound evolutionary consequences (Lee 2002; Tsutsui and Suarez 2003). The red fire ant was originally introduced into the United States in the 1930s; however, it was not until the 1970s that the more ecologically destructive multi-queen (polygyne) colonies appeared and the species spread widely (Krieger and Ross 2002). The genetics underlying gyne number in *S. invicta* help explain this time lag between the initial introduction and spread: single-queen (monogyne) colonies are homozygous (*BB*) at the "Gp-9" locus, whereas polygyne colonies are heterozygous (*Bb*). Monogyne colonies produce new *BB* queens that disperse to found new monogyne colonies, and polygyne colonies generate *BB*, *Bb*, and *bb* queens (although *Bb* are generally the most successful).

Apparently, a secondary introduction brought the *b* allele into U.S. populations, leading to colony dynamics that fueled the spread of red fire ants (Krieger and Ross 2002).

The dramatic social and ecological changes following the addition of a single allele to a resident introduced population of *S. invicta* underscore why metrics that reflect genetic changes averaged across the entire genome (e.g., additive genetic variance), especially those based on presumably neutral markers (e.g., heterozygosity, allelic diversity), may not accurately predict how evolution may proceed during an invasion. For example, Ross et al. (1993) compared changes in diversity at putatively neutral allozyme markers (Baer 1999) with diversity at sex-determining loci in source and invasive populations of fire ants. Although the bottleneck associated with the invasion appears to have had a minimal effect on levels

So, do introduced populations possess sufficient additive genetic variance to respond rapidly to selection? The tentative answer at this point is yes. As we previously discussed, the level of additive genetic variance in the introduced population should decline in proportion to the population's inbreeding coefficient, F, which measures the strength of a bottleneck. Unless a bottleneck reduces an introduced population to just a few individuals, and the introduced population remains small for a protracted interval before (or if) it expands, there will be little loss of additive genetic variance (or genetic diversity).

Effects of Introductions and Invasions on Phenotypic Evolution, Fitness, and Population Differentiation

Although it appears that introduction bottlenecks do not typically lead to a major loss of neutral genetic variation (see Figure 9.2), it would be premature to generalize the effects of introduction on neutral markers to its effects on

---

BOX 9.2 *(continued)*

of heterozygosity at most genetic loci, estimates of the number of alleles at sex-determining loci suggest that allelic diversity may have dropped by 60%–80% because most alleles in this system are likely to be rare (Ross et al. 1993). Thus, the loss of rare alleles may be a more sensitive indicator of the strength and potential effect of a bottleneck on an invasive population (Leberg 1992; Ross et al. 1993). In addition, the loss of rare alleles may have disproportionately strong effects on the fitness of invasive species, particularly when those alleles are experiencing frequency-dependent selection and affect the expression of mating preferences, social organization, sex determination, or life history traits (Ross et al. 1993; Hiscock 2000; López-Fanjul et al. 2003).

Despite the potential, in theory, for bottlenecks and drift to influence levels of genetic variation in newly founded populations, to our knowledge there is only one well-documented example suggesting that the loss of genetic variation has affected a trait that im-

proves the fitness of members of an invasive species. Tsutsui et al. (2001) document a significant reduction in allelic variation at microsatellite loci in populations of the Argentine ant (*Linepithema humile*) in its introduced, compared with its native, range. To the extent that this reduction in genetic diversity at presumably neutral microsatellite loci in *Linepithema* corresponds to a loss in variation at loci that influence the precision of kin and nestmate recognition, a genetic bottleneck appears to have promoted a shift in colony social structure from small, mutually aggressive colonies in the species' native range to large, nonaggressive multi-queen supercolonies in the introduced range that attain high population densities (Tsutsui and Suárez 2003). Because the formation of supercolonies reduces intraspecific competition, this bottleneck appears to have facilitated the ability of introduced Argentine ants to successfully outcompete native ant species (Holway 1999).

genetic variation for fitness-related traits (McKay and Latta 2002; López-Fanjul et al. 2003). While our literature search (see Figure 9.2 and Table 9.1) suggests that the overall genetic background of an introduced population does not differ substantially from that of its source population(s), the loss of even a few alleles could, in principle, significantly constrain or facilitate evolution. However, simply placing a relatively unchanged genetic background in a different environment with different species interactions could also generate novel phenotypic responses (Coyne 1994). When a single population is the source for an introduced population, the expressed phenotypic variance of the newly founded population will often change in predictable ways related to the strength of the bottleneck and the nature of the selective regime. However, if previously isolated populations from the source range interbreed in the introduced range, interactions between distinct genetic backgrounds may generate novel phenotypes. In this section, we evaluate how small genetic changes, selection on traits tied to fitness, and novel genetic environments influence the phenotypic evolution and fitness of introduced populations.

Effects of population bottlenecks

The argument is far from settled over whether a population's failure or success following introduction depends primarily on demographic effects, such as stochastic variation in population size that increases extinction probability when census size is small and variable, or on genetic effects, such as increased inbreeding depression or the loss of additive genetic variance, which may limit the ability of a species or population to respond to further changes in the biotic and physical environment. Some insights regarding the relative importance of demographic and genetic effects might be gained by comparing the effects of population bottlenecks following a species introduction with those of bottlenecks associated with endangered species. There is, however, one key difference: introduced species that become successful invaders often pass through short-lived bottlenecks followed by rapid population expansion, whereas endangered species often remain at low effective population sizes for extended periods (Allendorf and Lundquist 2003).

A recent study by Briskie and Mackintosh (2004) clearly showed that inbreeding depression due to bottlenecks significantly affects the fitness of both endangered and invasive bird species in New Zealand. Specifically, bird populations that had passed through bottlenecks of fewer than 150 individuals experienced significantly lower egg hatching success than source populations that had not experienced a bottleneck in population size (Briskie and Mackintosh 2004). Similar negative effects of bottlenecks on competitive ability and other components of fitness occur in invasive plants and other organisms with histories of small effective population size (Saccheri et al. 1998; Knaepkens et al. 2002; Bossdorf et al. 2004). Thus, although levels of heterozygosity and allelic diversity may not change dramatically with moderate bottlenecks (Briskie and Mackintosh 2004; see also Nei et al. 1975; Amos and

Balmford 2001), the expression of deleterious recessive alleles appears to be greater in these inbred populations.

The expression of particular alleles may be modified strongly by their interaction with alleles at other genetic loci. Although the process of "conversion" of epistatic variance to additive genetic variance does not generate novel alleles, it can allow more direct expression of existing allelic variation. While overall, little allelic diversity may be lost due to a bottleneck (see Figure 9.2B), rare alleles may change in frequency or be lost (Brookes et al. 1997), and the phenotypic effect of one allele or gene could effectively increase or decrease due to reduced variation at one or more epistatically associated loci (Brodie 2000). In many ways, the fitness effects of these rare or frequency-dependent alleles are at the heart of the debate over the relative importance of drift and selection for evolutionary diversification (Coyne et al. 1997; Goodnight and Wade 2000). Some studies focus on founder events that may have eliminated alleles maintained by frequency-dependent or sexual selection (see Box 9.2). Consequently, there are circumstances under which drift *can* have a major effect on the evolutionary responses of an introduced or invasive population, even without greatly influencing overall levels of additive genetic variance.

Effects of hybridization and introgression

Most novel epistatic interactions that arise from admixture or hybridization should reduce fitness (Barton 1989; Willett and Burton 2003). However, the occasional expression of "hybrid vigor" suggests that positive interactions—which could improve the mean fitness of an invasive population—may arise when previously isolated lineages hybridize and introgress. In some cases, it appears that hybridization can lead to the formation of new species through a variety of either ecological or genetic mechanisms that reduce or eliminate backcrossing, and therefore gene flow, with parental stocks (reviewed in Dowling and Secor 1993; Ellstrand and Schierenbeck 2000; Gaskin and Schaal 2002; Rieseberg et al. 2003). Alternatively, the hybrids may backcross with native species to such an extent that they essentially "absorb" the parental stocks through introgression (Ellstrand and Schierenbeck 2000). Introgression of nonnative genes into native species (as well as established agricultural stocks), especially when it involves genetically modified organisms, is of more than academic interest (Stewart et al. 2003; Roderick and Navajas 2003) because the process may alter economically and ecologically desirable characteristics of native and domesticated species. In its extreme form, introgressive hybridization may lead to the genetic extinction of native species (Rosenfield and Kodric-Brown 2003).

Introgressive hybridization can also provide important insights into the evolution of reproductive isolation. One of the most challenging issues in the study of speciation is whether allopatric (geographically disjunct) taxa are, in fact, reproductively incompatible (the criterion for species status under the biological species concept; see Coyne and Orr 2004). For all but a handful of taxa, it is generally impossible (and unethical) to bring allopatric taxa into sympatry and examine the mechanisms that limit or promote genetic exchange between those taxa in more or less natural (compared with laboratory) surroundings. Yet this is what introductions represent, so they can be useful for analyzing the role of mating behavior and recognition cues in reproductive isolation. For instance, sexual selection for aggressive mating behavior appears to have driven the extensive introgression between native and introduced pupfishes (*Cyprinodon*) in New Mexico and Texas (Rosenfield and Kodric-Brown 2003). Female Pecos pupfish (*C. pecosensis*) prefer the aggressive mating behavior of non-native males to that of their own species, a preference that probably reflects an ancestral bias rather than a contemporary adaptation. Hybrid offspring share the aggressive mating behavior of the invasive parent species, and so continue to drive the introgression of these species. Thus, patterns of introgression, while representing a significant conservation threat, can also inform our understanding of reproductive isolation and speciation (see Rice and Sax, this volume).

Extensive introgression following hybridization is not a foregone conclusion. Successful introgression of non-native genes often requires strong selection favoring hybrid phenotypes or behaviors, or linkage to advantageous alleles that generates overdominance (Stewart et al. 2003; Rosenfield and Kodric-Brown 2003; Bernatchez and Landry 2003). Nevertheless, both direct and indirect evidence, primarily from crop plants, demonstrates that genetic markers move from crop species to native plants and vice versa (Ellstrand et al. 1999; Stewart et al. 2003). Studies of introgression also suggest that the complex interactions between genetic background and environmental variation on phenotypic variation and fitness, and their subsequent effects, often produce a hybrid genome that is a heterogeneous blend of the parental genomes (Harrison 1990; Rieseberg et al. 2003).

Additive genetic variance and phenotypic responses to selection

Many factors, some environmental, some genetic, contribute to an organism's phenotype. Consequently, it is important when measuring changes in the phenotypic composition of newly founded populations to separate the contributions of environmental, additive, and nonadditive genetic effects (López-Fanjul et al. 2003). For example, although phenotypic divergence outpaces divergence of molecular markers during the process of invasion by grayling (Koskinen et al. 2002), it remains unclear what proportion of phenotypic divergence results from selection on additive genetic variance, environmental contributions to phenotypic differentiation, or nonadditive interactions.

Grayling, of course, are not the only organisms that exhibit major phenotypic changes during introductions and invasions (reviewed in Ellstrand and Schierenbeck 2000; Lee 2002; Müller-Schärer et al. 2004). Although some of the observed changes appear to result from eco-phenotypic responses to novel environments (e.g., Sexton et al. 2002), selection in natural populations may be quite effective at generating rapid and heritable phenotypic evolution (Reznick and Ghalambor 2001). Indeed, a recent review of field selection experiments demonstrated a mean selection gradient of about 16% (Kingsolver et al. 2001). It seems likely that selection gradients acting on introduced populations could be just as steep, if not steeper (Whitlock 1997). In terms of response to these selection gradients, overwhelming evidence demonstrates that many populations can, and do, evolve in novel selective regimes at surprisingly rapid rates (Kingsolver et al. 2001; Reznick and Ghalambor 2001). While the evidence from introduced and invasive species is potentially biased because it necessarily focuses on *successful* introductions and invasions, it nevertheless suggests that rapid evolution of introduced species is possible, even for populations that have passed through significant bottlenecks. Again, the question is whether population bottlenecks sufficiently reduce additive genetic variance to the point that mechanisms other than selection in a novel environment should be invoked.

Several recent studies directly confront this question, and their results indicate that introduced populations possess sufficient additive genetic variance (despite losses due to genetic bottlenecks) to respond rapidly to selection. For example, within two decades of the introduction of *Drosophila subobscura* into the Americas, the wing length of introduced North American females had significantly increased with latitude, mirroring a previously documented pattern in native European *D. subobscura* (Huey et al. 2000; Huey et al., this volume). However, in North American populations, this phenotypic cline resulted from changes in the relative lengths of different parts of the wing than in Europe, suggesting that sufficient additive genetic variance was present in both source and invasive populations to permit genetically independent, but phenotypically equivalent, responses to a comparable environmental gradient. This necessary level of genetic variation appears to have been maintained with an introduced propagule of only about a dozen individuals (Pascual et al. 2001).

Other studies (e.g., Grant et al. 2001; Rasner et al. 2004) also suggest that the magnitude and duration of bottlenecks that accompany successful introductions are often not severe enough to prevent substantial phenotypic evolution of invading species entering novel environments and communities. Even in the most extreme cases studied—in which a single pair of individuals has founded a new population—the resulting population can respond to selection. Following colonization of the Galápagos island of Daphne Major by the large ground finch (*Geospiza magnirostris*) in 1982—an event in which only a single breeding pair successfully colonized—the population grew in size rapidly, being supplemented by only low levels of immigration (Grant et al. 2001). With these dynamics, there was no net decline in heterozygosity at 16 microsatellite loci, yet there was a dramatic heritable shift in beak shape after only 9 years. Grant et al. (2001) argue that both drift and selection were involved in the phenotypic and fitness responses of the successful colonists, as the success and/or fitness of colonists was in part determined by levels of heterozygosity in the newly founded population, as well as by the identity of the source population.

Post-introduction immigration in this and other similar circumstances may have supplemented the additive genetic variance in the founder population, and at the same time reduced the effects of inbreeding depression by limiting the expression of recessive deleterious alleles. In addition, ongoing immigration may buffer introduced populations from extinction due to demographic stochasticity.

From Population Differentiation to Speciation

What can introductions and invasions tell us about the contributions of founder effects, drift, and selection to the evolution of reproductive isolation and speciation? The available data suggest that most introduced species—examples of small populations that have become established—are capable of responding to selection and genetically diverging from their source populations. Some biologists (e.g., Mayr 1963; Carson 1968; Templeton 1980) have extended Wright's shifting balance theory and emphasis on the importance of drift in evolution by suggesting that speciation itself is promoted by strong founder events that generate novel templates for selection to act on. Following the loss of allelic diversity associated with a bottleneck, genes that had evolved in the context of a polymorphic genetic background would subsequently interact in a new, more homozygous genetic environment. These changes could then lead to selection favoring novel epistatic interactions (a "genetic revolution" *sensu* Mayr 1954) that lead to rapid evolutionary change. This outcome is largely driven by intrinsic genetic interactions; however, any introduced or recently founded population is also likely to experience a different environment and selective regime, which would confound attempts to separate the effects of genetic interactions from the effects of the environment (Coyne 1994).

In addition to the difficulty of separating the effects of genetic interactions and of novel environmental factors, there are several other reasons to question the generality of speciation theories in which drift provides the primary force driving divergence between populations (Coyne 1994). For example, Barton (1989) showed that when the inbreeding coefficient F is less than 0.2 (as with most of the examples from introduced populations in Figure 9.2), more divergence between populations can be explained by selection acting on post-bottleneck additive genetic variance than can be explained by either drift or higher-order (dominance and epistatic) genetic interactions. It is noteworthy that an important distinction between arguments favoring the primacy of Fisherian mass selection and those promoting the importance of drift-related evolutionary phenomena is whether the additive effects of alleles are more or less constant regardless of the allelic composition of the population (Goodnight and Wade 2000). Barton and Turelli (2004) argue that even if the conversion of epistatic and dominance interactions is detectably strong, the stochastic variance of this conversion process makes it less important than the overall amount of purely additive variance that passes through a bottleneck.

Overall, loss of genetic variation due to founder effects and genetic drift seems to play a minor role in the success of most species introductions, as well as natural colonization sequences (Clegg et al. 2002; Cox 2004). When multiple source populations contribute to an invasion, a relative increase in allelic diversity or novel allelic combinations in these genetic admixtures appears to increase the likelihood that new lineages will arise and succeed in an introduced range (Grant et al. 2001; Novak and Mack, this volume). Garcia et al. (1989) showed that, with respect to allele frequencies at a number of loci, invasive Californian populations of the slender wild oat *Avena barbata* are not distinct from ancestral source populations in Spain. However, the data suggest that selection in the introduced range has involved a "reorganization of the ancestral allelic ingredients of the Spanish gene pool into novel multilocus allelic combinations adapted to specific habitats" (Garcia et al. 1989). The shifting balance theory could in part explain the rise of novel multilocus allelic combinations; however, the observed changes are also easily attributable to mass selection acting on the increased additive genetic variance found in post-introduction admixtures of populations derived from distinct source regions.

The available data cannot resolve the ongoing debate on the importance of Fisherian mass selection versus Wright's shifting balance theory (and related models that emphasize the role of drift). It has been argued (Goodnight and Wade 2000; Skipper 2002) that neither model is intended to be a general description of evolution, and the relative importance of one or the other may vary from one biological example to the next. What does seem clear is that introduced species often exhibit dramatic phenotypic changes that are comparatively larger than the changes observed in the same populations at neutral loci. Thus, selection is likely to be an important force in most examples of evolutionary change in small populations (Orr and Orr 1996; Huey et al. 2000; Koskinen et al. 2002; Lee 2002; McKay and Latta 2002; Rasner et al. 2004).

Conclusions and Prospects

The extent to which natural and anthropogenically generated bottlenecks facilitate or inhibit evolutionary change has generated an enduring debate in both evolutionary and conservation biology. Evolutionary geneticists, including Mayr (1954) and Templeton (1980), championed the idea that founder effects occurring in peripheral or isolated populations could promote "genetic revolutions," resulting in striking phenotypic change and perhaps even speciation. However, Barton (1989) argued that the trade-off between the apparent reduction of genetic variation from a source to a founder population and the amount of phenotypic variance that selection in the founder population could act on limits the generality of this effect (also see Coyne 1994; Rice and Hostert 1993). The resolution of this controversy remains elusive; however, theoretical analyses of invasion scenarios and empirical comparisons of genetic variation in introduced populations with that in source population(s) offer powerful insights into the effects of founder events and isolation on evolutionary change, precisely because these introductions represent extreme examples of evolution in small isolated populations. Our synthesis shows that in general, the effect of founder events on neutral, and perhaps fitness-related, variation is not extreme: introduced populations often retain 80% or more of the genetic variation present in their source(s).

Perhaps this information will help to resolve a paradox: invasive species often displace resident species that should be better adapted to their native environments (Allendorf and Lundquist 2003). Most species are restricted to well-defined geographic distributions, even when it appears that natural dispersal mechanisms could transport them beyond their current distributions. This observation suggests that there *are* limits to a species' ability to respond to the selective forces imposed by novel environments (Fisher 1930; Antonovics 1976; Kirkpatrick and Barton 1997). Levels of genetic variation in introduced populations may be high enough, particularly if multiple source populations are involved, to respond quickly to local patterns of selection on short time scales (Huey et al. 2000; Kingsolver et al. 2001; Lee 2002; Rasner et al. 2004). The evolutionary potential of native species, on the other hand, may be constrained by local adaptation to the longer-term fluctuations in environmental conditions that historically characterize their environment (Schemske and Bierzychudek 2001; Allendorf and Lundquist 2003).

As we see it, one of the major gaps that remains to be filled lies between our growing understanding of how the demography of introductions and invasions affects neutral genetic variation, and how these same demographic processes affect the genes that control the expression of phenotypic traits that are the targets of selection. In an era of increasing availability of genomic data, it should become possible to use closely related model organisms to begin to identify those genes responsible for phenotypic changes during introductions and invasions and to determine more directly whether such changes represent the outcomes of selection. When enough homologous markers (both neutral and with known fitness consequences) can be screened, our ability to distinguish founder effects from strong selection in novel habitats will improve dramatically. At the current rates of species introduction (Sakai et al. 2001) and environmental change (Walther et al. 2002), we are likely to have plenty of opportunities to test our most fundamental models of evolution in the near future.

Acknowledgments

The authors thank the members of the NCEAS Working Group, as well as J. McKay, M. Turelli, M. McPhee, N. Tsutsui, R. Mack, T. Bell, and an anonymous reviewer, all of whom contributed substantially to the development of this chapter. This chapter was supported by grants from the Mellon Foundation, the National Science Foundation, and the University of California Agricultural Experiment Station. This work was conducted as part of the "Exotic Species:

A Source of Insight into Ecology, Evolution, and Biogeography" working group supported by the National Center for Ecological Analysis and Synthesis, a Center funded by the National Science Foundation (Grant #DEB-0072909), the University of California, and the Santa Barbara campus.

Literature Cited

Allendorf, F. W., and L. L. Lundquist. 2003. Introduction: Population biology, evolution, and control of invasive species. Conservation Biology 17:24–30.

Amos, W., and A. Balmford. 2001. When does conservation genetics matter? Heredity 87:257–265.

Anderson, E. C., and E. A. Thompson. 2002. A model-based method for identifying species hybrids using multilocus genetic data. Genetics 160:1217–1229.

Anttila, C. K. , R. A. King, C. Ferris, D. R. Ayres, and D. R. Strong. 2000. Reciprocal hybrid formation of *Spartina* in San Francisco Bay. Molecular Ecology. 9:765–770.

Antonovics, J. 1976. The nature of limits to natural selection. Annals of the Missouri Botanical Gardens 63:224–247.

Arnold, M. L. 1997. *Natural hybridization and evolution.* Oxford Series in Ecology and Evolution, Oxford University Press, Oxford.

Austerlitz, F., B. Jung-Muller, B. Godelle, and P.-H. Gouyon. 1997. Evolution of coalescence times, genetic diversity and structure during colonization. Theoretical Population Biology 51:148–164.

Avise, J. C. 2000. *Phylogeography.* Harvard University Press, Cambridge, MA.

Ayres, D. R., D. Garcia-Rossi, H. G. Davis, and D. R. Strong. 1999. Extent and degree of hybridization between exotic (*Spartina alterniflora*) and native (*S. foliosa*) cordgrass (Poaceae) in California, USA determined by random amplified polymorphic DNA (RAPDs). Molecular Ecology 8:1179–1186.

Baer, C. F. 1999. Among-Locus Variation in Fst: Fish, Allozymes, and the Lewontin-Krakauer Test Revisited. Genetics 152:653–659.

Bagley, M. J., and Geller, J. B. 1999. Microsatellite DNA analysis of native and invading populations of European green crabs. In J. Pederson, ed. *Marine bioinvasions. Proceedings of the First National Conference.*, pp. 241–244. M.I.T. Sea Grant College Program, Cambridge, MA.

Baker, A. J. 1992. Genetic and morphometric divergence in ancestral European and descendant New Zealand populations of chaffinches (*Fringilla coelebs*). Evolution 46:1784–1800.

Baker, A. J., and A. Moeed. 1987. Rapid genetic differentiation and founder effect in colonizing populations of common Mynas (*Acridotheres tristis*). Evolution 41:525–538.

Balanyà, J., C. Segarra, A. Prevosti, and L. Serra. 1994. Colonization of America by *Drosophila subobscura*: the founder event and a rapid expansion. Journal of Heredity 85:427–432.

Baliraine, F. N., M. Bonizzoni, C. R. Guglielmino, E. O. Osir, S. A. Lux, F. J. Mulaa, L. M. Gomulski, L. Zheng, S. Quilici, G. Gasperi, and A. R. Malacrida. 2004. Population genetics of the potentially invasive African fruit fly species, *Ceratitis rosa* and *Ceratitis fasciventris* (Diptera: Tephritidae). Molecular Ecology 13:683–695.

Bartlett, E., S. J. Novak, and R. N. Mack. 2002. Genetic variation in *Bromus tectorum* (Poaceae): Differentiation in the eastern United States. American Journal of Botany 89:602–612.

Barton, N. H. 1989. Founder effect speciation. In D. Otte, and J. A. Endler, eds., *Speciation and its consequences*, pp. 229–256. Sinauer Associates, Sunderland, MA.

Barton, N. H., and M. Turelli. 2004. Effects of allele frequency changes on variance components under a general model of epistasis. Evolution 58:2111–2132.

Bastrop, R., K. Jurss, and C. Sturmbauer. 1998. Cryptic species in a marine polychaete and their independent introduction from North America to Europe. Molecular Biology and Evolution 15:97–103.

Beerli, P., and J. Felsenstein. 1999. Maximum-likelihood estimation of migration rates and effective population numbers in two populations using a coalescent approach. Genetics 152:763–773.

Bertorelle, G., and L. Excoffier. 1998. Inferring admixture proportions from molecular data. Molecular Biology and Evolution 15:1298–1311.

Bohonak, A. J., N. Davies, F. X. Villablanca, and G. K. Roderick. 2001. Invasion genetics of New World medflies: testing alternative colonization scenarios. Biological Invasions 3:103–111.

Boileau, M. G., and Hebert, P.D. N. 1992. Genetics of the zebra mussel, *Dreissena polymorpha*, in populations from the Great Lakes region and Europe. In T. F. Nalepa, and D. W. Schloesser, eds. *Zebra mussels: Biology, impact, and control*, pp. 227–238. Lewis Publishers, Chelsea, MI.

Bossdorf, O., D. Prati, H. Auge, and B. Schmid. 2004. Reduced competitive ability in an invasive plant. Ecology Letters 7:346–353.

Briskie, J. V., and M. Mackintosh. 2004. Hatching failure increases with severity of population bottlenecks in birds. Proceedings of the National Academy of Sciences USA 101:558–561.

Brodie III, E. D. 2000. Why evolutionary genetics does not always add up. In J. B. Wolf, E. D. Brodie III, and M. J. Wade, eds., *Epistasis and the evolutionary process*, pp. 3–19. Oxford University Press, Oxford.

Brookes, M. I., Y. A. Graneau, P. King, O. C. Rose, C. D. Thomas, and J. L. B. Mallet. 1997. Genetic analysis of founder bottlenecks in the rare British butterfly *Plebejus argus*. Conservation Biology 11:648–661.

Browne, R. A. 1977. Genetic variation in island and mainland populations of *Peromyscus leucopus*. American Midland Naturalist 97:1–9

Bryant, E. H., Vandijk, H. and W. Vandelden. 1981. Genetic variability of the face fly, *Musca autumnalis* DeGeer, in relation to a population bottleneck. Evolution 35:872–881.

Cabe, P. R. 1998. The effects of founding bottlenecks on genetic variation in the European starling (*Sturnus vulgaris*) in North America. Heredity 80:519–525.

Carlton, J. T. 1982. The historical biogeography of *Littorina littorea* on the Atlantic coast of North America and implications for the interpretation of the structure of New England intertidal communities. Malacological Review 15:146.

Carson, H. L. 1968. The population flush and its genetic consequences. In R. C. Lewontin, ed. *Population biology and evolution*, pp. 123–138. Syracuse University Press, Syracuse, NY.

Cheverud, J. M. 2000. Detecting epistasis among quantitative trait loci. In J. B. Wolf, E. D. Brodie III, and M. J. Wade, eds. *Epistasis and the evolutionary process*, pp. 58–81. Oxford University Press, Oxford.

Chikhi, L., M. W. Bruford, and M. A. Beaumont. 2001. Estimation of admixture proportions: a likelihood-based approach using Markov chain Monte Carlo. Genetics 158:1347–1362.

Choisy, M., P. Franck, and J.-M. Cornuet. 2004. Estimating admixture proportions with microsatellites: comparison of methods based on simulated data. Molecular Ecology 13:955–968.

Clegg, S. M., S. M. Degnan, J. Kikkawa, C. Moritz, A. Estoup, and I. P. F. Owens. 2002. Genetic consequences of sequential founder events by an island-colonizing bird. Proceedings of the National Academy of Sciences USA 99:8127–8132.

Cox, G. W. 2004. *Alien species and evolution: The evolutionary ecology of exotic plants, animals, microbes, and interacting native species*. Island Press, Washington, DC.

Coyne, J. A. 1994. Ernst Mayr and the origin of species. Evolution 48:19–30.

Coyne, J. A., N. H. Barton, and M. Turelli. 1997. Perspective: a critique of Sewall Wright's shifting balance theory of evolution. Evolution 51:643–71.

Coyne, J. A., and H. A. Orr. 2004. *Speciation*. Sinauer Associates, Sunderland, MA.

Davies, N., F. X. Villablanca, and G. K. Roderick. 1999. Determining the source of individuals: multilocus genotyping in nonequilibrium population genetics. Trends in Ecology and Evolution 14:17–21.

Demelo, R., and P. D. N. Hebert. 1994. Founder effects and geographical variation in the invading cladoceran *Bosmina (Eubosmina) coregoni* Baird 1857 in North America. Heredity 73:490–99.

Dowling, T. E., and C. L. Secor. 1997. The role of hybridization in the evolutionary diversification of animals. Annual Review of Ecology and Systematics 28:593–619.

Dupont, L., D. Jollivet, and F. Viard. 2003. High genetic diversity and ephemeral drift effects in a successful introduced mollusc (*Crepidula fornicata*: Gastropoda). Marine Ecology-Progress Series 253:183–95.

Ellstrand, N. C., and K. A. Schierenbeck. 2000. Hybridization as a stimulus for the evolution of invasiveness in plants. Proceedings of the National Academy of Sciences USA 97:7043–7050.

Ellstrand, N. C., H. C. Prentice, and J. F. Hancock. 1999. Gene flow and introgression from domesticated plants into their wild relatives. Annual Review of Ecology and Systematics 30:539–563.

Estoup, A., I. J. Wilson, C. Sullivan, J. M. Cornuet, and C. Moritz. 2001. Inferring population history from microsatellite and enzyme data in serially introduced cane toads, *Bufo marinus*. Genetics 159:1671–1687.

Ewens, W. J. 1972. The sampling theory of selectively neutral alleles. Theoretical Population Biology 3:87–112.

Fisher, R. A. 1930. *The genetical theory of natural selection*. Oxford University Press, Oxford.

Garcia, P., F. J. Vences, M. Pérez De La Vega, and R. W. Allard. 1989. Allelic and genotypic composition of ancestral Spanish and colonial Californian gene pools of *Avena barbata*: Evolutionary implications. Genetics 122:687–694.

Gaskin, J. F., and B. A. Schaal. 2002. Hybrid *Tamarix* widespread in U.S. invasion and undetected in native Asian range. Proceedings of the National Academy of Sciences USA 99:11256–11259.

Goodnight, C. J. 1988. Epistasis and the effect of founder events on the additive genetic variance. Evolution 42:441–54.

Goodnight, C. J. 2000. Modeling gene interaction in structured populations. In J. B. Wolf, E. D. Brodie, and M. J. Wade, eds. *Epistasis and the evolutionary process*, pp. 58–81. Oxford University Press, Oxford.

Goodnight, C. J. and M. J. Wade. 2000. The ongoing synthesis: a reply to Coyne, Barton, and Turelli. Evolution 54:317–324.

Grant, P. R., B. R. Grant, and K. Petren. 2001. A population founded by a single pair of individuals: establishment, expansion, and evolution. Genetica 112/113:359–382.

Grosberg, R. K., and C. W. Cunningham. 2001. Genetic structure in the sea: from populations to communities. In M. D. Bertness, M. E. Hay, and S. D. Gaines, eds. *Marine community ecology,* pp. 61–84 Sinauer Associates, Sunderland, MA.

Guinand, B., A. Topchy, K. S. Page, M. K. Burnham-Curtis, W. F. Punch, and K. T. Scribner. 2002. Comparisons of likelihood and machine learning methods of individual classification. Journal of Heredity 93:260–269.

Hansen, M. M., E. E. Nielsen, D. Bekkevold, and K.-L. D.Mensberg. 2001. Admixture analysis and stocking impact assessment in brown trout (*Salmo trutta*), estimated with incomplete baseline data. Canadian Journal of Fisheries and Aquatic Sciences 58:1853–1860.

Harrison, R. G. 1990. Hybrid zones: Windows on evolutionary process. In D. J. Futuyma and J. Antonovics, eds. Oxford Surveys in Evolutionary Biology 7:69–128.

Hebert, P. D. N., and M. E. A. Cristescu. 2002. Genetic perspectives on invasions: the case of the *Cladocera.* Canadian Journal of Fisheries and Aquatic Sciences 59:1229–1234.

Hiscock, S. J. 2000. Genetic control of self incompatibility in *Senecio squalidus* L. (Asteraceae)–a successful colonising species. Heredity 84:10–19.

Holland, B. S. 2001. Invasion Without a Bottleneck: Microsatellite Variation in Natural and Invasive Populations of the Brown Mussel *Perna perna* (L). Marine Biotechnology 3:407–415.

Holway, D.A. 1999. Competitive mechanisms underlying the displacement of native ants by the invasive Argentine ant. Ecology 80:238–251.

Hudson, R. R. 1990. Gene genealogies and the coalescent process. In D. J. Futuyma, and J. Antonovics, eds. Oxford Surveys in Evolutionary Biology pp. 1–44.

Huey, R., G. W. Gilchrist, and M. Carlsen. 2000. Rapid evolution of a latitudinal cline in body size in an introduced fly. Science 287:308–309.

Janson, K. 1987. Allozyme and shell variation in two marine snails (*Littorina, Prosobranchia*) with different dispersal abilities. Biological Journal of the Linnaean Society 30:245–257.

Johnson, M. S. 1988. Founder effects and geographic variation in the land snail *Theba pisana.* Heredity 61:133–142.

Johnson, M. T. J., and A. A. Agrawal. 2003. The ecological play of predator-prey dynamics in an evolutionary theatre. Trends in Ecology and Evolution 18:549–556.

Keightley, P. D., and M. Lynch. 2003. Toward a realistic model of mutations affecting fitness. Evolution 57:683–685.

Kingman, J. F. C. 1982. On the genealogy of large populations. Journal of Applied Probability 19a:27–43.

Kingsolver, J. G., H. E. Hoekstra, J. M. Hoekstra, D. Berrigan, S. N. Vignieri, C. E. Hill, A. Hoang, P. Gibert, and P. Beerli. 2001. The strength of phenotypic selection in natural populations. American Naturalist 157:245–261.

Kirkpatrick, M. and N. H. Barton. 1997. Evolution of a species' range. American Naturalist 150:1–23.

Knaepkens, G., D. Knapen, L. Bervoets, B. Hänfling, E. Verheyen, and M. Eens. 2002. Genetic diversity and condition factor: a significant relationship in Flemish but not in German populations of the European bullhead (*Cottus gobio* L.). *Heredity* 89:280–287.

Knowles, L. L. 2004. The burgeoning field of statistical phylogeography. Journal of Evolutionary Biology 17:1–10.

Kolbe, J. J., R. E. Glor, L. R. Schettino, A. C. Lara, A. Larson, and J. B. Losos. 2004. Genetic variation increases during biological invasion by a Cuban lizard. Nature 431:177–181.

Koskinen, M. T., T. O. Haugen, and C. R. Primmer. 2002. Contemporary Fisherian life-history evolution in small salmonid populations. Nature 419:826–830.

Krieger, M. J. B., and K. G. Ross. 2002. Identification of a major gene regulating complex social behavior. Science 295:328–332.

Le Page, S. L., R. A. Livermore, D. W. Cooper, and A. C. Taylor. 2000. Genetic analysis of a documented population bottleneck: introduced Bennett's wallabies (*Macropus rufogriseus rufogriseus*) in New Zealand. Molecular Ecology 9:753–763.

Leberg, P. L. 1992. Effects of population bottlenecks on genetic diversity as measured by allozyme electrophoresis. Evolution 46:477–494.

Lee, C. E. 1999. Rapid and repeated invasions of fresh water by the copepod *Eurytemora affinis*. Evolution 53:1423–1434.

Lee, C. E. 2002. Evolutionary genetics of invasive species. Trends in Ecology and Evolution 17:386–391.

López-Fanjul, C., A. Fernández, and M. A. Toro. 2003. The effect of neutral nonadditive gene action on the quantitative index of population divergence. Genetics 164:1627–1633.

Luikart, G., P. R. England, D. Tallmon, S. Jordan, and P. Taberlet. 2003. The power and promise of population genomics: from genotyping to genome typing. Nature Reviews in Genetics 4:981–994.

Malacrida, A. R., F. Marinoni, C. Torti, L. M. Gomulski, F. Sebastiani, C. Bonvicini, G. Gasperi, and C. R. Guglielmino. 1998. Genetic aspects of the worldwide colonization process of *Ceratitis capitata.* Journal of Heredity 89:501–507.

Maruyama, T., and P. A. Fuerst. 1984. Population bottlenecks and nonequilibrium models in population genetics. I. Allele numbers when populations evolve from zero variability. *Genetics* 108:745–763.

Mayr, E. 1954. Change of genetic environment and evolution. In J. S. Huxley, A. C. Hardy, and E. B. Ford, eds. *Evolution as a process*, pp. 156–180. Allen and Unwin, London.

Mayr, E. 1963. *Animal species and evolution*. Harvard University Press, Cambridge, MA.

McKay, J. K., and R. G. Latta. 2002. Adaptive population divergence: markers, QTL and traits. Trends in Ecology and Evolution 17:285–291.

Merilä, J., M. Björklund, and A. J. Baker. 1996. The successful founder: genetics of introduced *Carduelis chloris* (greenfinch) populations in New Zealand. Heredity 77:410–422.

Müller-Schärer, H., U. Schaffner, and T. Steinger. 2004. Evolution in invasive plants and implications for biological control. Trends in Ecology and Evolution 19:417–422.

Nei, M., T. Maruyama, and R. Chakraborty. 1975. The bottleneck effect and genetic variability in populations. Evolution 29:1–10.

Nei, M. 1987. *Molecular evolutionary genetics*. Columbia University Press, New York.

Orr, H. A., and L. H. Orr 1996. Waiting for speciation: the effect of population subdivision on the time to speciation. Evolution 50:1742–1749.

Otto, S. P., and M. C. Whitlock. 1997. Fixation of beneficial mutations in a population of changing size. Genetics 146:723–733.

Paetkau, D., W. Calvert, I. Stirling, and C. Strobeck. 1995. Microsatellite analysis of population structure in Canadian polar bears. Molecular Ecology 4:347–354.

Parkin, D. T., and S. R. Cole. 1985. Genetic differentiation and rates of evolution in some introduced populations of the House Sparrow, *Passer domesticus* in Australia and New Zealand. Heredity 54:15–23.

Pascual, M., C. F. Aquadro, V. Soto, and L. Serra. 2001. Microsatellite variation in colonizing and palearctic populations of *Drosophila subobscura*. Molecular Biology and Evolution. 18:731–740.

Rasner, C. A., P. Yeh, L. S. Eggert, K. E. Hunt, D. S. Woodruff, and T. D. Price. 2004. Genetic and morphological evolution following a founder event in the dark-eyed junco, *Junco hyemalis thurberi*. Molecular Ecology 13:671–681.

Reznick, D. N., and C. K. Ghalambor. 2001. The population ecology of contemporary adaptations: what empirical studies reveal about conditions that promote adaptive evolution. Genetica 112:183–198.

Rice, W. R. and E. E. Hostert. 1993. Laboratory experiments on speciation: what have we learned in 40 years? Evolution 47:1637–1653.

Rieseberg, L. H., O. Raymond, D. M. Rosenthal, Z. Lai, K. Livingstone, T. Nakazato, J. L. Durphy, A. E. Schwarzbach, L. A. Donovan, and C. Lexer. 2003. Major ecological transitions in wild sunflowers facilitated by hybridization. Science 301:1211–1216.

Roderick, G. K., and M. Navajas. 2003. Genes in new environments: Genetics and evolution in biological control. Nature Reviews in Genetics 4:889–899.

Rohner, M., R. Bastrop, and K. Jurss. 1996 Colonization of Europe by two American genetic types or species of the genus *Marenzelleria* (Polychaeta: Spionidae). An electrophoretic analysis of allozymes. Marine Biology 127:277–287.

Rosenfield, J. A., and A. Kodric-Brown. 2003. Sexual selection promotes hybridization between Pecos pupfish, *Cyprinodon pecosensis* and sheepshead minnow, *C. variegatus*. Journal of Evolutionary Biology. 16:595–606.

Ross, H. A. 1983. Genetic differentiation of starling (*Sturnus vulgaris*: Aves) populations in New Zealand and Great Britain. Journal of the Zoological Society of London 201:351–362.

Ross, K. G., E. L. Vargo, L. Keller, and J. C. Trager. 1993. Effect of a founder event on variation in the genetic sex-determining system of the fire ant *Solenopsis invicta*. Genetics 135:843–854.

Ross, K. G. and J. C. Trager. 1990. Systematics and population genetics of fire ants (*Solenopsis saevissima* complex) from Argentina. Evolution 44:2113–2134.

Saccheri, I., M. Kuussaari, M. Kankare, P. Vikman, W. Fortelius, and I. Hanski. 1998. Inbreeding and extinction in a butterfly metapopulation. *Nature* 392:491–494.

Sakai, A. K., F. W. Allendorf, J. S. Holt, D. M. Lodge, J. Molofsky, K. A. With, S. Baughman, R. J. Cabin, J. E. Cohen, N. C. Ellstrand, D. E. Mccauley, P. O'Neil, I. M. Parker, J. N. Thompson, and S. G. Weller. 2001. The population biology of invasive species. Annual Review of Ecology and Systematics 32:305–332.

Schemske, D. W., and P. Bierzychudek. 2001. Perspective: Evolution of flower color in the desert annual *Linanthus parryae*: Wright revisited. Evolution 55:1269–1282.

Sexton, J. P., J. K. Mckay, and A. Sala. 2002. Plasticity and genetic diversity may allow saltcedar to invade cold climates in North America. Ecological Applications 12:1652–1660.

Skipper, R. A. 2002. The persistence of the R. A. Fisher–Sewall Wright controversy. Biology and Philosophy 17:341–367.

St. Louis, V. L., and J. C. Barlow. 1988. Genetic differentiation among ancestral and introduced populations of the Eurasian tree sparrow (*Passer montanus*). Evolution 42:266–276.

Stepien, C. A., C. D. Taylor, and K. A. Dabrowska. 2002. Genetic variability and phylogeographical patterns of a nonindigenous species invasion: A comparison of exotic vs. native zebra and quagga mussel populations Journal of Evolutionary Biology 15:314–328.

Stewart, C. N., M. D. Halfhill, and S. I. Warwick. 2003. Transgene introgression from genetically modified crops to their wild relatives. Nature Reviews in Genetics 4:806–817.

Stockwell, C. A., and M. V. Ashley. 2004. Rapid adaptation and conservation. Conservation Biology 18:272–273.

Taylor, C. E., and G. C. Gorman. 1975. Population genetics of a "colonizing" lizard: natural selection for allozyme morphs in *Anolis grahami*. Heredity 35:241–247.

Templeton, A. R. 1980. The theory of speciation via the founder principle. Genetics 94:1011–1038.

Tsutsui, N. D., A. V. Suarez, D. A. Holway, and T. J. Case. 2000. Reduced genetic variation and the success of an invasive species. Proceedings of the National Academy of Sciences USA 97:5948–5953.

Tsutsui, N. D., A. V. Suarez, D. A. Holway, and T. J. Case. 2001. Relationships among native and introduced populations of the Argentine ant (*Linepithema humile*) and the source of introduced populations. Molecular Ecology 10:2151–2161.

Tsutsui, N. D., and A. V. Suarez. 2003. The colony structure and population biology of invasive ants. Conservation Biology 17:48–58

Turner, T. F., J. P. Wares, and J. R. Gold. 2002. Genetic effective size is three orders of magnitude smaller than adult census size in an abundant, estuarine-dependent marine fish (*Sciaenops ocellatus*). Genetics 162:1329–1339.

Wade, M. J., and C. J. Goodnight. 1998. Perspective: The theories of Fisher and Wright in the context of metapopulations: when nature does many small experiments. Evolution 52:1537–1553.

Wakeley, J., and J. Hey. 1997. Estimating ancestral population parameters. *Genetics* 145:847–855.

Walther, G., E. Post, P. Convey, A. Menzel, C. Parmesan, T. J. C. Beebee, J.-M. Fromentin, O. Hoegh-Guldberg, and F. Bairlein. 2002. Ecological responses to recent climate change. *Nature* 416:389–395.

Waples, R. S. 1989. A generalized approach for estimating effective population size from temporal changes in allele frequency. Genetics 121:379–391.

Wang, Z., A. J. Baker, G. E. Hill, and S. V. Edwards. 2003. Reconciling actual and inferred population histories in the house finch (*Carpodacus mexicanus*). Evolution 57:2852–2864.

Wares, J. P., D. S. Goldwater, B. Y. Kong, and C. W. Cunningham. 2002. Refuting a controversial case of a human-mediated marine species introduction. Ecology Letters 5:577–584.

Whitlock, M. C. 1995. Two-locus drift with sex chromosomes: the partitioning and conversion of variance in subdivided populations. Theoretical Population Biology 48:44–64.

Whitlock, M. C. 1997. Founder effects and peak shifts without genetic drift: adaptive peak shifts occur easily when environments fluctuate slightly. Evolution 51:1044–1048.

Whitlock, M. C., and K. Fowler. 1999. The changes in genetic and environmental variance with inbreeding in *Drosophila melanogaster*. Genetics 152:345–353

Willett, C. S., and R. S. Burton. 2003. Environmental influences on epistatic interactions: viabilities of cytochrome *c* genotypes in interpopulation crosses. Evolution 57:2286–2292.

Willis, J. H., and H. A. Orr. 1993. Increased heritable variation following population bottlenecks: The role of dominance. Evolution 47:949–957.

Wolf, J. B., E. D. Brodie III, and M. J. Wade, eds. 2000. *Epistasis and the evolutionary process*. Oxford University Press, Oxford.

Wright, S. 1932. The roles of mutation, inbreeding, crossbreeding and selection in evolution. Proceedings of the 6th International Congress on Genetics 1:356–366.

Zeisset, I., and T. J. C. Beebee. 2003. Population genetics of a successful invader: the marsh frog *Rana ridibunda* in Britain. Molecular Ecology 12:639–646.

10

Theories of Niche Conservatism and Evolution

COULD EXOTIC SPECIES BE POTENTIAL TESTS?

Robert D. Holt, Michael Barfield, and Richard Gomulkiewicz

The niche of an invading species determines the range of environments in which it is initially expected to increase when rare, versus facing extinction. Given genetic variation, evolution by natural selection can lead to evolution in a species' niche. Recent theoretical studies have helped clarify when one would expect to see niche evolution, as opposed to niche conservatism or "stasis." This chapter synthesizes insights from theoretical and simulation models of coupled demographic and evolutionary change for species introduced into novel environments. Propagule pressure plays a key role in invasion "outside the niche" because the number of individuals per propagule, and the number of attempted invasions, influences the demographic opportunity and genetic potential for adaptive evolution. Infrequent colonization with propagules at low initial abundance should rarely succeed in habitats outside the ancestral niche. In spatially heterogeneous environments, after successful establishment in one habitat, evolution can either promote or impede a species' spread into other habitats. Many aspects of the basic biology and ecology of a species can influence its likelihood of niche evolution, including mating systems and temporal variation in the environment. Empirical studies of invasions potentially provide fruitful tests of this body of theory, but crucial data are often lacking because there are few records of failed invasions. Examples of lags in invasion are consistent with evolutionary adjustment to novel environments; if this is indeed the case, the magnitude of the lag should be related to the degree of difference between the novel and ancestral environments.

Introduction

Invasions by exotic species occur when organisms are transported (typically by human agency) from an original ancestral range, across dispersal barriers, to new regions where populations can then become established, persist, and potentially spread over wide areas; we refer to such species interchangeably as exotic or invading species. Because of their enormous practical importance, exotic species have received a large and growing amount of attention (Vitousek et al. 1997; Crooks and Soulé 1999; Mack et al. 2000; Olden et al. 2004). Yet, as D'Antonio et al. (2001) note, our "ability to predict establishment success and impact of nonindigenous species remains limited." There are, of course, many purely ecological reasons for this apparent failure of prediction, but beyond these, another obvious general reason why the establishment and effects of exotic species are difficult to predict might be that these species are moving targets, whose genetic and phenotypic properties change due to evolution in the novel environments in which they find themselves.

There is now a growing body of work on the genetics of exotic species. Sakai et al. (2001) and Lambrinos (2004) provide excellent overviews of empirical studies of population and evolutionary processes in exotic species, and the recent book by Cox (2004) provides numerous examples of evolution in invasions. A useful review of the evolutionary genetics of exotic species by Lee (2002) emphasizes the importance of genetic architecture for understanding the evolutionary dimension of invasions. In this chapter, we discuss exotic species from the complementary perspective of recent general theoretical studies of niche conservatism and evolution. There is a rapidly growing body of literature exploring the coevolution of demography, fitness, and species' geographic and habitat ranges that is broadly pertinent to exotic species (e.g., Holt and Gaines 1992; Kirkpatrick and Barton 1997; Ronce and Kirkpatrick 2001; Antonovics et al. 2001; Kawecki 1995, 2000, 2004; Garcia-Ramos and Rodriguez 2002; Case et al. 2005). These studies highlight the importance of the interplay of demography, landscape structure, and temporal variation as both facilitators of and constraints on adaptive evolution. We do not attempt a comprehensive literature review here, but rather provide examples from our own work and some new results. The models in question tend to be rather abstract compared with those required to address concrete management or control problems. But we suggest that these relatively simple models lead to general, qualitative insights that carry over to more complex, realistic scenarios. Moreover, we suggest that exotic species may provide fertile ground for testing and refining this body of evolutionary theory.

A particularly interesting issue is how evolution influences the likelihood of invasion in spatially heterogeneous environments. Lyford et al. (2003) note that most ecological theories that have been applied to invasions focus on the processes of dispersal, establishment, and population growth in environments that implicitly are homogeneous in space and time. Yet, in reality, invasions typically occur on landscapes with complex spatial structures and in environments that are temporally variable and often nonstationary. Sax and Brown (2000) observe that spatial variation can help explain why many introductions of potentially successful invaders actually fail. As we will see, spatial and temporal heterogeneity may also be profoundly important in governing the evolution of exotic species.

Two factors that seem to have power for predicting the success of deliberate introductions within a region (e.g., as in game bird releases) are the number of individuals per attempted introduction and the number of introduction attempts (Veltman et al. 1996; Duncan 1997; Green 1997). Levine (2000) experimentally showed that the abundance of the initial propagule had a strong influence on initial establishment (see also Byers and Goldwasser 2001; Brown and Peet 2003). D'Antonio et al. (2001; see also Lonsdale 1999) highlight the importance of propagule pressure; they suggest that the presence of a sustained source of propagules (e.g., from garden populations) also appears to facilitate the eventual spread of exotics to broader landscapes. Successful invasions are often correlated with multiple attempts at deliberate introduction (Barrett and Husband 1990); a natural analogue of this may arise when a species becomes established in one habitat patch, from which it can send repeated propagules into other habitats in the surrounding landscape. There are, of course, purely ecological reasons to expect propagule pressure (reflecting both the numbers of individuals introduced per invasion episode and the number of repeated invasion attempts observed over time) to matter in determining the likelihood of successful establishment. The evolutionary theories sketched below suggest complementary reasons for increased invasion success with increased propagule pressure.

Before discussing this body of theory, it is useful to briefly sketch some reasons why an evolutionary perspective on invasions seems almost inescapable, and in particular, how invasions are (or should be) related to analyses of species' niches. We then present theoretical results having to do with how evolution can influence invasion into homogeneous environments. With these results in hand, we turn our attention to evolution by exotic species in spatially and temporally heterogeneous landscapes.

Ecological Niches and Initial Establishment

For the purposes of this chapter, the "niche" of a species refers simply to that suite of environmental conditions within which populations of that species are expected to persist deterministically, without recurrent immigration (Hutchinson 1978; Holt and Gaines 1992). Broadly speaking, those "conditions" can include abiotic factors as well as resource availabilities and abundances of interacting species such as competitors and predators. This use of the term "niche" goes back to Grinnell (1917), who invoked the concept to characterize the factors delimiting a species' distribution. Hutchinson (1957) formalized the concept in his famous "n-dimensional hypervolume," which emphasized the multidimensional nature of organisms' interactions with their environments. The "fundamental niche" describes the requirements of a species for persistence when one discounts the negative effects of predators and competitors. Maguire (1973) later demonstrated that it could be fruitful to characterize a niche as a response surface, where the response variable was the growth rate of a species when it was rare. In effect, the niche is a mapping of deterministic population dynamics onto an abstract environmental space.

The most fundamental feature of population dynamics is extinction versus persistence, and the niche describes how this fundamental feature depends on the environment. If the niche requirements of a species are met at a given site, we expect that, on average, births should exceed deaths when the species is rare, and so introductions at that site should have a chance of becoming established, leading to populations that persist over reasonable time horizons (which can be very long if the local carrying capacity is large and the environment is constant). By contrast, if the site has conditions outside the niche, then births are fewer than deaths, and extinction is certain.

Figure 10.1A schematically depicts a simple example of a species' niche, characterized as a response surface to two abiotic factors. We assume that the combination of abiotic factors for which growth rate equals zero (the inner circle) circumscribes the fundamental niche of a species, both at the site of orig-

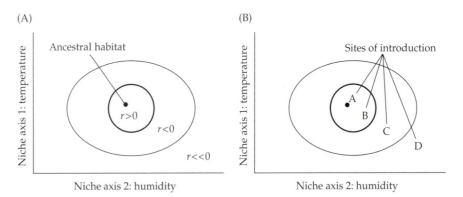

Figure 10.1 (A) A schematic representation of an ecological niche response surface. The inner circle demarcates the fundamental niche—a set of conditions that permit deterministic persistence. The outer ellipse schematically separates areas of niche space that imply gradual population decline as opposed to rapid population decline. In the optimal environment (represented by the dot), the species has a high growth rate; outside the fundamental niche, its numbers decline slowly (near the niche) or rapidly (far from the niche). (B) Different sites of introduction are likely to match a species' niche requirements to varying degrees. A and B are sites with conditions inside the niche; C and D have conditions outside the niche.

ination and at the sites where introduction occurs (at least initially), and, for the purpose of illustration, that the effects of the two abiotic factors shown outweigh those of other factors. There are, of course, many complexities in characterizing niches, most of which go beyond the intended scope of this chapter (Chase and Leibold 2003). For instance, in general, invasion is expected to depend strongly on interactions with resident resources and species, and these are likely to differ in many ways from the web of interactions within which the species evolved (Shea and Chesson 2002). The absence of coevolved natural enemies or competitors, for example, may permit a species to invade successfully into a wider range of abiotic conditions in the novel environment than in the ancestral range.

The Evolutionary Potential of Invading Species

It is likely that the environments faced by an introduced species will differ in many respects (often quite radically) from the environment harboring its ancestor. Species that leave behind their specialist natural enemies (e.g., predators, parasitoids, pathogens) (Keane and Crawley 2002) may be able to persist over a different set of abiotic conditions than that which characterizes their native range. Selection will often differ for many aspects of the phenotype relevant to coping with and utilizing the novel biotic and abiotic environment (Lewontin 1965). In a review of empirical studies of rapid adaptive evolution, Reznick and Ghalambor (2001) found that many examples involved a combination of directional selection and at least a short-term opportunity for population growth. They suggest that an opportunity for population growth may be a key attribute of rapid evolution, because otherwise directional selection might lead to population decline and even local extinction. Successful invaders, by definition, experience periods of sustained population growth, and so should be candidates for rapid evolution. Indeed, some of the clearest examples of rapid evolution come from introduced species (e.g., Quinn et al. 2000). One possible indication of evolution is that in many successful invasions there is a lag phase of very slow growth, followed by an accelerated clip of advance across space. A potential explanation for this lag is that it reflects a lag in adaptation of the exotic species to the novel environment (Crooks and Soulé 1999).

But it should not be automatically assumed that evolutionary dynamics are essential components of invasion dynamics. Indeed, an intriguing dichotomy—almost a form of schizophrenia—has arisen in the field of ecology and evolutionary biology in the last few years. On one hand, as just noted, there is a mounting body of evidence from studies of microevolution that adaptive evolution is rapid, can be readily observed (well within human life spans), and has important applied consequences for resource management, conservation, and the control of undesirable species (Hendry and Kinnison 1999; Ashley et al. 2003; Rice and Emery 2003; Stockwell et al. 2003). Stockwell and Ashley (2004) suggest that in altered environments, "rapid adaptation is the norm rather than the exception." Ashley et al. (2003) show that many instances of rapid evolution come from introduced species. Rapid evolution might appear to be the default assumption in studies of exotic species.

On the other hand, there is considerable evidence from biogeography, phylogenetics, and paleobiology that species' niches can be relatively conservative. There is a sense among many scientists that niche conservatism is ubiquitous and poorly understood (Hansen and Houle 2004). Some examples of conservatism can be found at microevolutionary time scales. Bradshaw (1991, p. 303) reviews a number of case studies in which there appears to be no evolution, despite ongoing selection, and remarks: "For a century we have been mesmerized by the successes of evolution. It is time now that we paid equal attention to its failures." Merila et al. (2001) also discuss a wide range of systems in which selection appears to be ongoing over short time scales, but stasis is observed instead of evolutionary responses. Other instances of conservatism in species' niches become apparent when species or clades are examined over broad sweeps of evolutionary history: "The presence of strong constraints on alterations in the ecological posture of species is seen in their response to long-term environmental change during the Quaternary" (Levin 2003). Many paleobiologists have commented on the prevalence of conservatism in evolutionary history (Schopf 1996). For instance, the paleobotanist Huntley (1991) argues that when climate changes, species typically migrate (or go extinct), rather than shift their niche requirements (see also Coope 1979). Indeed, some authors have argued that phylogenetic niche conservatism appears to be the norm, rather than the exception, in evolution (Peterson et al. 1999; Ackerly 2003; Peterson 2003).

A reconciliation of these two disparate perspectives may come, we suggest, from a close consideration of the demographic context in which evolution necessarily occurs. Niche conservatism is not really a species (or clade) property, but rather a reflection of the interaction of species' properties, environmental contexts, and demographic responses to environmental conditions. In the next few paragraphs, we view the problem of invasion by exotic species through the lens of evolutionary niche theory, focusing in particular on the issue of when evolution by natural selection is a necessary ingredient in successful invasion. Conversely, we expect that in the future, exotic species could provide testing grounds for assessing theories of niche conservatism and evolution. Below, we will sketch some desiderata for such tests.

As noted above, the niche of a species strongly affects the initial likelihood that a propagule will become established, leading to a potential invasion. As a simple hypothetical example (see Figure 10.1B), assume that the focal species has the fundamental niche shown in Figure 10.1A, and that the invading propagule originated in a habitat with optimal conditions like those at the dot (which we will call the ancestral habitat), and is introduced into a variety of sites (denoted by letters A, B, C, and D in Figure 10.1B) with varying local conditions. Assume for the sake of simplicity that the species is at its evolutionary optimum in its site of origination. Site A has conditions essentially similar to optimal conditions in the ancestral habitat; the species will thus face only minor differences in the selective environment, and should have a relatively high growth rate when introduced here. At site B, it is expected to become established and persist, although it will grow rather slowly; this implies that it may stay at low numbers, at which it is vulnerable to extinction due to demographic stochasticity and other factors, for a long time. Finally, at C and D, the species is predicted to decline to extinction, albeit at different rates. Thus, a first-order prediction about establishment success comes from matching a species' niche requirements against environmental conditions at the site of introduction.

This basic logic underlies the entire protocol of "climate matching" as a tool for predicting the potential success of invading species. For instance, artificial intelligence techniques are being used to develop ecological niche models in the native distributions of species, which then permit projections about the likely initial success and ultimate spatial extent of invasions by those species (Peterson 2003). However, such predictions assume that species' niches stay reasonably constant from initial establishment through the later stages of invasion.

In the region of origin, the stability of species' range boundaries and habitat distributions over evolutionary time scales is likely to reflect niche stasis. Such stasis is assumed whenever one uses a niche model developed in a site of origin to make predictions about establishment. For instance, Levin and Clay (1984) describe an experimental study of the niche conditions defining the success of introduced populations of an annual plant species, *Phlox drummondii*. In natural environments, this species inhabits sites with loose sandy loam. Levin and Clay planted seeds in adjacent habitats with denser soil. The number of seeds produced was much less than the number planted (hence, in all the introduced populations, $r < 0$, in the terms of Figure 10.1), and all the populations ultimately went extinct. A niche model with an axis of soil density would identify a soil density above which establishment is predicted to be prevented, provided the species were to remain unchanged. The best-performing population was the one placed in soil with properties closest to the soil of the ancestral environment. The theoretical results discussed below suggest that niche evolution, if it occurs at all, is most likely for that population which was not initially all that far removed from the bounds of its niche requirements.

Sites C and D in Figure 10.1B might be called "sink habitats," where conditions are outside a species' fundamental niche; by contrast, sites A and B are "source habitats." Evolution may influence a species' persistence and invasion potential in sites A and B, but is not *necessary* for invasion to occur. By contrast, if the niche itself does not evolve, invasions into sites C and D will inevitably fail. Thus, a basic question we can ask is when niche evolution can occur, converting a sink habitat into a source habitat. The issue of whether or not successful invasion into sink habitats is possible is intimately related to the potential for niche evolution.

Evolution can influence the course of establishment in several distinct ways, as shown in Figure 10.2, which plots dynamics in local population size following establishment for sites A–D in Figure 10.1B. At site A, initial growth rates are rapid, and (by assumption) the selective environment is similar to that of

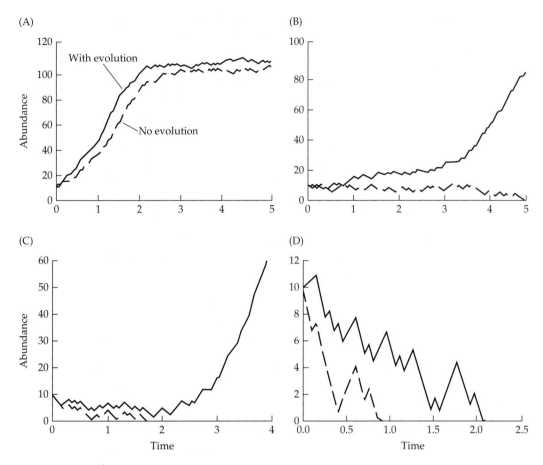

Figure 10.2   Population dynamics expected for species introduced at the niche positions shown in Figure 10.1B. A simple model was used to derive these figures for illustrative purposes. The population change was governed by $dN/dt = rN(1 - N/K)$. Evolution is simulated by linearly increasing r and K with time. In all cases with evolution, K starts at 100 and increases at a rate of 1 per time unit. Random gain or loss of one individual every 0.05 time units has been added to simulate demographic stochasticity. (A) Evolution is not needed for persistence, but can affect abundance. The initial r is 2, and it evolves at a rate of 0.5. (B) Evolution may increase growth rates, and so facilitate persistence in the face of possible extinction due to demographic stochasticity. The initial r is 0.05, and it evolves at a rate of 0.3. (C) Evolution rescues this population from extinction. The initial r is −1, and it evolves at a rate of 1. (D) Evolution occurs, but cannot save this population. The initial r is −1.5, and it evolves at a rate of 0.5. Similar patterns emerge in more complex, realistic models (see text).

the ancestral habitat. Evolution should not matter greatly in initial establishment (though there could be minor effects if selection leads to a modest acceleration of the initial rate of increase). As the population grows, density dependence should begin to limit growth. If density-dependent factors operate differently than in the ancestral environment, evolution could also subsequently act to alter population size. In the example shown, it is assumed that density-dependent selection occurs, so that as the introduced population evolves, its numbers increase. The magnitude of the change in population size depends on the local selective environment, the nature of density-dependent feedbacks, and the amount of genetic variation available for selection (either introduced in the original propagule or originating via in situ mutation). Such evolution could matter greatly in governing the transition from initial, local establishment to widespread invasion because propagule pressure across the landscape should scale with local abundance. However, this kind of evolution, though interesting and potentially important in invasion dynamics, may not greatly influence the probability of initial establishment *per se*.

Site B also has conditions within the ancestral niche, but much nearer the niche boundary. This implies that population growth rates should be lower, and hence that a population introduced at low numbers should remain at low densities for much longer than a similar introduction at site A. The population would thus face an enhanced extinction risk due to both ecological factors (e.g., demographic stochasticity, Allee effects) and genetic factors (inbreeding, mutational meltdown) (Lande 1998). Many introductions into sites with conditions just within a niche boundary should fail. If evolution could increase the initial growth rate for such introductions, it could reduce the time the population is exposed to elevated extinction risks and thus increase the probability of establishment.

The initial stages of evolution following introduction will depend on the amount of genetic variation brought in with the initial propagule (unless there is hybridization with native species: Lee 2002), which can be viewed as a "bottleneck" in the original population. In general, the rate of population growth following a bottleneck is a crucial determinant of the amount of genetic variation that is lost (Nei et al. 1975; Gaggiotti 2003). If the growth rate is rapid (as in site A), little variation will be lost. But if growth rate is slow (as in site B), considerable variation may be lost. Moreover, because numbers are low, inbreeding is likely. There is some evidence that when conditions are stressful (as is likely near the edge of the niche), inbreeding depression is more severe (Gaggiotti 2003). Saccheri et al. (1998) argue that inbreeding increased extinction risk in small populations of a butterfly, the Glanville fritillary (*Melitaea cinxia*).

Another potential cause of elevated extinction risk for a species introduced near the edge of its niche, where numbers stay low longer, is the accumulation of slightly but unconditionally deleterious mutations (Lynch et al. 1995). Natural selection is less effective at weeding out mildly deleterious mutations in small populations than in large populations. The genetic load of mutations should thus be greater for populations near the edge of a species' niche (Kawecki et al. 1997). This is a somewhat controversial prediction (Gaggiotti 2003; Whitlock et al. 2003), but the controversy has to do with the exact magnitude of the effect, rather than with its existence.

Site C has conditions just outside the ancestral niche. In the absence of evolution, the introduction will surely fail. However, if evolution can increase the growth rate of the species sufficiently rapidly, it is possible that the species will, in effect, evolve "out of the niche." If evolution can rescue a population from extinction, it will be because there has been a change in the niche of the population itself. The dynamics of such a "rescued" population, initially occupying an environment outside its niche, should usually exhibit a U-shaped trajectory (Gomulkiewicz and Holt 1995; and see below).

Site D is like site C, but considerably worse; here the population declines so rapidly that even if appropriate genetic variation is present and evolution occurs, it is insufficient to rescue the population from extinction. In effect, there is a race between a demographic process, a decline in numbers driving the introduced species to extinction, and an evolutionary process, increasing fitness in the local environment. In the example shown, evolution loses.

In both C and D, the invading propagule faces an environment much harsher than that experienced in the ancestral habitat. There is now a substantial theoretical literature on evolution in changing environments. Many papers consider environments with systematic temporal trends (due to climate change for instance), with a linear trend in a phenotypic optimum (e.g., Pease et al. 1989; Burger and Lynch 1995; Lande and Shannon 1996). Several authors have considered an alternative scenario in which there is an abrupt transition between two environmental states (e.g., Gomulkiewicz and Holt 1995). The latter models should apply equally well to analyzing the fate of a colonizing propagule plucked from an environment where its ancestors evolved in the face of one set of selective challenges and then placed in a novel environment with quite different selective challenges. In the next few paragraphs, we highlight general insights from these models, which help identify circumstances in which one might observe invasion "outside the niche."

Gomulkiewicz and Holt (1995; see also Holt and Gomulkiewicz 1997b) employed a heuristic device to examine the joint effect of population dynamics and evolution on population persistence in an environment experiencing abrupt change. Conservation biologists have suggested that there is a population size (a "critical number") below which extinction is risked for a whole suite of reasons (e.g., demographic stochasticity, Allee effects, and inbreeding). If the initial state of a population has an absolute fitness of less than 1, then without evolution, that population will certainly decline toward this critical number and face likely extinction (as at sites C and D in Figure 10.1B; see Figure 10.2C,D for numerical examples). With evolution, provided that selection can lead to a population with a positive growth rate, numbers will eventually rebound. This requires that the population have sufficient genetic variation (or can accumulate such variation rapidly) to at least persist in the novel environment. Selection in each generation increases average fitness, provided that the assumptions of Fisher's fundamental theorem approximately hold (Fisher 1958; Burt 1995). But if a population starts at low numbers, and if evolution by natural selection is slow, so that average fitness stays low for a sufficiently long time, then the population may reach low critical numbers at which it is vulnerable to extinction.

There are several qualitative messages that emerge from such models linking changes in population size with evolved changes in phenotypes (Gomulkiewicz and Holt 1995; Holt and Gomulkiewicz 2004):

- If the initial propagule number is small, then extinction is by far the most likely outcome of introduction into environments outside the niche. Upon reflection, this is not very surprising. After all, small population sizes can lead to extinction even in favorable environments, so propagules with a small number of individuals face elevated extinction risks in any case. If numbers tend to decline because niche requirements are not met, then even with appropriate genetic variation, the population is not likely to be able to evolve sufficiently rapidly to shift its niche requirements before suffering extinction.
- One effect of a large initial propagule number is to make it more likely that a species can invade environments outside its ancestral niche, because a greater number of individuals increases the demographic window of opportunity for selection to operate.
- Nonetheless, even for initial propagules with a large number of individuals, evolution is unlikely to rescue populations introduced into environments where conditions are much outside the species' ancestral niche requirements. If initial fitness is very low, the population is likely to decline rapidly and go extinct before it can adapt and grow.
- Finally, increasing genetic variability (heritability) does make evolutionary rescue more likely, but even populations with abundant genetic variability are not likely to persist in environments where initial propagule numbers and/or growth rates are very low.

To go beyond these heuristic conclusions, one needs models that describe the probability of actual extinction. Analytically describing the probability of extinction for populations evolving in tandem with changes in abundance is challenging. One approach we (Holt et al. 2003; Holt and Gomulkiewicz 2004) and others (e.g., Boulding and Hay 2001) have taken is to develop individual-based simulation models in which individuals and their genotypes are tracked and extinction probabilities calculated as a function of initial population size, initial fitnesses (reflecting the match between the local environment and a species' niche requirements), and genetic parameters. Box 10.1 describes one such model. In short, individuals are assumed to be diploid, and fitness is assumed to depend on a single quantitative trait (e.g., body size) genetically determined by n unlinked loci (in the examples shown, $n = 10$) plus an environmental effect. Genetic variation in the source population arises from a balance between mutation introducing novel variation and selection and drift weeding it out. Immigrants are drawn at random from the source population. In the introduced population, mating is random, and mutation occurs after segregation. The fitness function has a Gaussian shape, whose peak (and therefore the optimum phenotype) differs between the presumed source habitat of the population and the novel habitat, so that the initial population at the time of introduction is maladapted to the novel habitat.

BOX 10.1 *An Individual-Based Model for Analyzing Niche Evolution*

Deterministic models can illuminate the interplay of population and evolutionary dynamics (see, e.g., Gomulkiewicz and Holt 1995; Holt and Gomulkiewicz 2004; Holt et al. 2003, 2004), but a full treatment of extinction requires the incorporation of stochasticity. When populations are low in abundance, accounting for individual discreteness due to stochastic processes of mutation, birth, death, and movement becomes important. Realistic models accounting for these processes present many analytic challenges. To provide insight into the consequences of this stochasticity, we have carried out studies using individual-based simulations in which the fate of each individual is tracked (e.g., Holt et al. 2003). These simulations utilize the same basic assumptions as those of Burger and Lynch (1995), who examined adaptation to a continually changing environment for a single multilocus character. We examined both evolution in populations established by single colonizing episodes following abrupt environmental change and adaptive evolution in spatially discrete scenarios in which species successfully invaded stable source habitats in a heterogeneous landscape that were coupled by migration to sink habitats. The simulations were predicated on several key assumptions (see Holt et al. 2003).

Genetic assumptions

1. There are n additive loci, with no dominance or epistasis (each allele contributes a fixed amount to the phenotypic value, and an individual's phenotype is the sum of this quantity over n loci, plus a random term, which has a zero-mean, unit-variance normal distribution)
2. Mutational input maintains variation (the "continuum-of-alleles" model, in which mutational effects are drawn from a continuous, normal distribution)
3. There is free recombination
4. In the model for a heterogeneous landscape, the initial source habitat reaches mutation-selection-drift balance, which then determines the immigrant pool available for colonizing the sink habitats

Life history assumptions

1. Discrete, nonoverlapping generations
2. A dioecious and hermaphroditic sexual system

Ecological assumptions

1. A constant number of immigrants per generation (in spatial model)
2. "Ceiling" density dependence (i.e., density-independent growth below K)
3. A constant fecundity per mated pair
4. Offspring survival probability is a Gaussian function of phenotype

Two different mating systems assumed

1. The monogamous mating system has monogamous mating pairs (randomly formed from all adults), as in Burger and Lynch (1995) and Holt et al. (2003). This mating system has a small Allee effect (with n individuals, if n is an odd number, one individual remains unmated; this depresses average expected fecundity over all individuals, particularly when n is small).

2. The random mating system ensures that there is no Allee effect. In each generation, individuals in their female capacity are selected at random without replacement, up to the carrying capacity; in other words, below K, all individuals are chosen to be reproductive, whereas above K, only K individuals get to mate. For mating to occur, each individual selects a random individual (with replacement) from the entire population to act as a male. This protocol

BOX 10.1 *(continued)*

eliminates the small Allee effect that arises in the monogamous mating system.

We have shown elsewhere (Holt et al. 2004b) that Allee effects can influence adaptive evolution in sink habitats, so it is useful to examine scenarios in which no Allee effects exist, especially when comparing different immigration rates (or propagule sizes). With the monogamous mating system, if there are N adults, there are at most N/2 mating pairs, while with the random mating system, there are N pairs (each individual mates as a female, and, on average, once again as a male), unless limited by the carrying capacity. Therefore, to compare the two systems, there should be twice as many offspring per pair with monogamous mating to give the same average fecundity. For the simulations discussed in the text, we used four offspring per pair for the random mating system and eight offspring per pair for the monogamous mating system.

The qualitative findings reported in the text and figures appear to be robust to changes in many of these assumptions. For instance, changes in the number of loci have a relatively minor effect on the probability of adaptation (Holt and Gomulkiewicz 2004; R. D. Holt and R. Gomulkiewicz, unpublished data).

Adults were counted in each generation (N_t). After the census, in the spatial model, there is immigration, followed by random mating. The mating population is not allowed to exceed K (the carrying capacity). Individuals produce gametes with free recombination among n loci. Mutation occurs on gametes, with a stochastic mutational input per genome (distributed randomly over all n loci). Each mated pair produces f offspring, surviving to adulthood with probability

$$s(z,i) = \exp\left[\frac{-(z - \theta_i)^2}{2\omega^2}\right]$$

where z is the realized phenotype of a given offspring, θ_i is the optimum phenotype in habitat i, and ω^2 is inversely proportional to the strength of stabilizing selection. This is the life stage that experiences selection. If the average z value is sufficiently far from the optimum, mean fitness is below 1, and the population tends to decline. Individuals surviving mortality are the adults counted at the next census, N_{t+1}.

At the start of the simulation, the source population (either in the ancestral habitat or in a suitable habitat patch in a heterogeneous landscape) is allowed to reach selection-mutation-drift equilibrium. Immigrants then migrate into a sink habitat that initially has zero abundance. We should stress that in this individual-based model, stochasticity plays multiple roles. Mutation is stochastic. Gametic combinations and the genetic composition of immigrants to the sink (in the spatial model) have multilocus allelic combinations that vary due to random sampling. Finally, because survival is probabilistic, there is both genetic drift and chance variation in population size due to demographic stochasticity.

The same broad conclusions emerge from this model (illustrated in Figure 10.3) as suggested by the heuristic model of Figure 10.2. Figures 10.3A and 10.3B depict a number of population trajectories for introduced propagules drawn from similar source populations. For those populations that persist, a clear U-shaped trajectory in numbers occurs (Fig. 10.3A). The solid lines in Fig-

Figure 10.3 Sample trajectories for population size and mean genotypic state for populations introduced into a sink habitat, from a simulation using the individual-based model described in Box 10.1 and the text. (A) The trajectories for populations that adapt and persist in the novel environment. Initially, all populations shown decline in abundance, but these particular populations rebound as they adapt to the environment (a U-shaped trajectory in numbers). (B) Trajectories for populations that go extinct (solid lines) and the initial portions of trajectories for those that persist (dashed lines; the same examples as shown in part A). $K = 64$, mutational rate per haplotype $= 0.01$, mutational variance $\alpha^2 = 0.05$, strength of selection $\upsilon^2 = 1$, propagule number $= 64$; 4 births per pair. Random mating is assumed. The optimal genotype in the source scales at a phenotypic value of zero, while the sink optimum is at 2.5. Propagules are drawn at random from the source population (which is at a mutation-drift-selection equilibrium); all propagules have 64 individuals in generation 1.

(A)

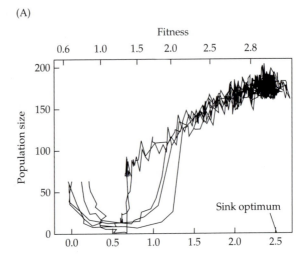

(B)

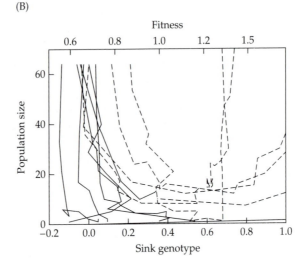

ure 10.3B are the trajectories of populations that go extinct, while the dashed lines are the trajectories of the populations shown in Figure 10.3A, which have adapted to the novel environment and are persisting. Surviving populations typically evolve more rapidly than do those that go extinct (Figure 10.3B). Often (though not always), evolutionary rescue is facilitated because, by chance, the initial propagule has a mean genotype somewhat closer to the sink habitat optimum (Figure 10.4). Once the population adapts sufficiently to persist, it then shows a period of increasing adaptation and rising population size.

A demographic signature of invasion "outside the niche" is that population numbers have a U-shaped trajectory. If the rate of initial decline is only slightly below zero and censuses are noisy (or the environment temporally variable),

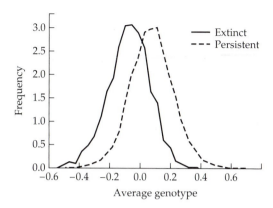

Figure 10.4 Relative frequencies of the mean genotypic state of propagules drawn from the source population (where the phenotypic optimum is assumed to be zero). Each propagule consists of 4 immigrants that were placed into a sink habitat with a phenotypic optimum of 2.5. On average, introductions that were successful (dashed line) had colonizing propagules that were somewhat closer to the new sink habitat optimum than those that went extinct (solid line). However, there is a broad overlap in the distributions. Other parameters are as in Figure 10.3. The figure illustrates the inherent uncertainty of colonization dependent on evolution in a species' niche.

then numbers may be approximately constant and then begin rising rapidly. Lags in population growth rate following introduction may be indirect indicators of periods of evolution in species' niches.

Figure 10.4 describes the initial genotypic state of propagules used in colonization attempts that either failed or succeeded (as in the examples of Figure 10.3B). In these examples, the optimal character value in the site of origination is scaled at zero, and the optimal character value in the site of introduction is a positive value (in this example, it happens to be 2.5). On average, successful introductions have a genetic composition that by chance is somewhat better adapted to the local environment. This means that they enjoy a greater demographic window during which selection can increase population fitness; they initially decline more slowly and require less evolutionary modification to enjoy a positive expected growth rate. However, note that there is a great deal of overlap in the distributions describing successful and failed invasions. Some propagules that seem to be moderately preadapted to the novel environment nonetheless fail. Conversely, some propagules that are more maladapted than the average in the source population nonetheless succeed. These results imply that there is a great deal of inherent unpredictability that arises whenever invasion depends on evolution of traits that influence a species' niche. Many invasion biologists lament the fact that it is difficult to predict invasion success. We suggest that if evolution is involved in the initial establishment of a species in a novel environment, one should not be surprised to observe a substantial degree of unpredictability in invasion success, arising from the interwoven stochastic vicissitudes of demography and evolutionary shifts.

Another issue (not addressed in the figures shown) is that the likelihood of successful invasion increases with the amount of genetic variability present in the source population from which the colonizing propagule is drawn. Novak and Mack (this volume) note that one can greatly inflate the genetic variation of a propagule by using an admixture of individuals plucked from different local populations in the ancestral range, as compared with the same number of individuals drawn from a single local population. Phylogeographic studies may be able to retrospectively identify such admixtures of source populations and identify candidate cases of rapid niche evolution in the initial stages of invasion (see Wares et al., this volume). It would be ideal to compare a number of introductions, including failures as well as successes, differing in the number of source populations included in the original immigrant gene pool. It may be difficult to achieve this ideal in practice, however, at least in retrospective analyses of accidental introductions, as generally we become aware of only those invasions that were successful.

Figure 10.5 summarizes the results of a large number of simulations using the individual-based model described in Box 10.1, which collectively document how the probability of persistence (which is also the probability of adaptation and niche evolution) varies as a function of the degree of both maladaptation and initial propagule abundance (using random mating and monogamous mating systems). These results broadly match the conclusions we drew from the heuristic model of Gomulkiewicz and Holt (1995). Overall, the more maladapted the initial propagule is in the novel environment, the less probable it is that evolution will rescue the population from extinction. Evolutionary modifications of the niche are more likely if they require only modest changes in the phenotype. Moreover, if propagules have smaller initial abundances, successful, persistent invasions into habitats outside the ancestral niche become less probable. When propagule numbers are small, the probability of successful invasion can be low even in environments within the niche; in environments with conditions outside the niche, the opportunity for evolutionary rescue is very short because extinction is expected to be rapid. In addition, note that there is a greater probability of population survival with the random mating system than with the monogamous mating system, and that the difference is greater for the smaller propagule size. This difference is probably due to the Allee effect seen with monogamous mating. The Allee

Figure 10.5 The probability of persistence and adaptation as a function of degree of initial maladaptation to the sink habitat for two different abundances of an initial colonizing propagule (I) and two mating systems. Other parameters are as in Figure 10.3, except that we assume eight births per pair for the monogamous mating system. Propagules with large abundance have a higher probability than propagules with small abundance of adapting to and thus persisting in habitats where they are initially sufficiently maladapted that extinction is expected in the absence of evolution. In addition, the random mating system (solid lines) gives a higher probability of adaptation than the monogamous mating system (dashed lines) because the latter has a small Allee effect.

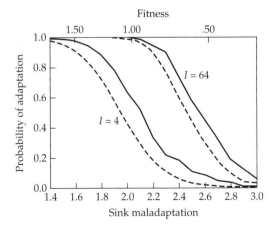

effect is quite small for a population size of 64, but, as shown in Figure 10.3, a population with this initial size can fall to much lower levels (at which the Allee effect will be more significant) before it adapts.

These theoretical exercises provide a partial justification for attempts to develop predictive niche models for assessing where invading species may gain a foothold, based on characterizations of their preexisting niches in their ancestral environments (e.g., Peterson 2003). If niches were completely plastic and could evolve very rapidly, then such models would be useless for predicting establishment. If propagule number is typically small, establishment is expected to be quite difficult (per colonizing episode) outside the ancestral niche requirements. Invasion "outside the niche" should require propagules that are larger, and should be more likely if the invading population is not too maladapted to the novel environment, so that its initial growth rate is not too far below replacement.

However, even if static models assuming no evolution can adequately characterize initial invasion success, for reasons discussed below, they should often prove less adequate in predicting the final spatial scope of widespread invasions. Such "failures" could be viewed in a positive light, as potential indicators of rapid niche evolution (e.g., as suggested for some species of New World jays: Peterson and Holt 2003). A future task will be to carefully analyze failures in predicting invasion in the light of potential evolutionary modifications in the very first stages of invasion. Again, it would be desirable to characterize failed invasions as well as successful ones, but this is rarely possible (or even attempted) for many species. In some cases, however, introductions occur quite deliberately, as in game bird or fish stocking and biological control programs. Careful comparisons of successes versus failures for such systems could reveal occasional mismatches between invasion success and predictions based on niche requirements evaluated in the sites of origination. Such mismatches could provide presumptive cases of rapid niche evolution during establishment.

To use an invading species as a crisp "test" of theories of niche conservatism versus evolution requires that one have a handle on a number of things, many of which may be absent for particular systems. Ideally, one would know (1) the site of origin of the initial propagule, (2) the degree of genetic variability in the initial propagule for ecologically relevant traits, (3) the number of individuals contained in the original propagule, and (4) the magnitude of the difference in niche requirements between the ancestral environment and the site of invasion. It is plausible that such data could be gathered for systems in which invasions are deliberate, as in biological control efforts. For accidental introductions, one must make inferences on the basis of limited information for each of these desiderata, which makes testing theory more challenging.

Another complication is that a full characterization of the conditions for evolutionary rescue via niche evolution requires that one be able to specify the genetic architecture of the traits that are under selection and can influence absolute fitness. This warrants much more study, both theoretically and empirically. For instance, there does not appear to be a consistent effect on heritability of exposure to novel environments (Hoffmann and Merila 1999). We recognize the importance of characterizing the detailed genetic underpinnings of niche traits (e.g., major vs. minor genes; genetic correlations and epistasis among distinct niche traits), but it goes beyond the scope of this chapter to treat this issue beyond this bare mention.

Evolution and Rate of Spread in Homogeneous Landscapes

Given that a population becomes established, it may then begin to spread. In a homogeneous environment, and with random movement, a general rule for spread is that the edge of the invading species' range will move across space with a velocity equal to $v = c(rD)^{1/2}$, where r is the intrinsic rate of increase when a population is rare (a mean absolute fitness), D is the root-mean-squared distance of movement (a diffusion coefficient) (Okubo and Levin 2001), and c is a proportionality constant. According to Fisher's fundamental theorem, mean fitness should increase at a rate that is proportional to the variance in fitness, or $dr/dt = \sigma^2_r$ (for simplicity, assuming a simple clonal model; Crow and Kimura 1970, p. 10).

We can use these equations to make a prediction about evolution and how it accelerates the rate of invasion: the instantaneous acceleration in invasion due to evolution by natural selection is

$$dv/dt = q(r)^{-1/2}\sigma^2_r$$

where q is a proportionality constant. The acceleration in invasion is directly proportional to variance in fitness. Moreover, the acceleration is particularly pronounced when the initial rate of increase is low. Hence, the lag observed in many empirical studies of invasion (which is tantamount to an acceleration in the rate of invasion) could well reflect the fine-tuning of adaptation to novel environments.

Anthropogenic forces can lead to directional changes in environments. Theoretical studies by Lande and Lynch (e.g., Lynch and Lande 1993) suggest that there are maximal rates of environmental change outside of which it is difficult for evolution to permit species persistence. This general point is as pertinent to invading species as to resident natives. If invasion is occurring into environments undergoing anthropogenic change, an ability to evolve rapidly may determine whether or not the invasion is ultimately successful.

The Evolutionary Ecology of Invasion into Heterogeneous Landscapes

Most invading species enter landscapes that are spatially heterogeneous, including some habitats with conditions within their niche and others with conditions outside it (With 2002). The spatial configuration of these habitats within the landscape, and how they are coupled via dispersal, potentially have very important influences on the eventual spread of the invading species (Shigesado and Kawasaki 1997). Garcia-Ramos and Rodriguez (2002) recently described a model of invasion by a species into an environment with a smooth environmental gradient, which incorporated population dynamics along with adaptation to a clinally varying phenotypic optimum. Here, we discuss complementary models that focus on invasion into landscapes with discrete and distinctly different habitats.

A species that becomes established in a source habitat with conditions well within its ancestral niche may then have the opportunity to send out a recurrent rain of propagules into other habitats that are initially not within its niche. There may be many more sites available outside the niche than within it. If the species can adapt to these sites so that it persists at one of them, it can then spread among them all. What are the constraints on adaptive evolution in these sink habitats, and how is adaptation there influenced by immigration from the source habitat? It is of particular interest to compare adaptation when sinks are mild with adaptation when they are harsh.

Theoretical studies in recent years have clarified the diverse ecological and genetic influences of dispersal on local adaptation and niche evolution in sink habitats (Holt and Gaines 1992; Hedrick 1995; Kawecki 1995, 2000, 2003; Holt 1996; Holt and Gomulkiewicz 1997a,b; Kirkpatrick and Barton 1997; Gomulkiewicz et al. 1999; Case and Taper 2000; Ronce and Kirkpatrick 2001; Tufto 2001; Kawecki and Holt 2002; Lenormand 2002; Holt et al. 2003, 2004a,b). It is now recognized that there are four distinct mechanisms involved when considering the evolutionary implications of dispersal and immigration in heterogeneous landscapes: (1) opportunities for exposure to alternative environments without risking overall extinction; (2) the provisioning of genetic variation on which natural selection can act; (3) the swamping of gene flow by recurrent immigration; and (4) shifts in population size, which can affect selection if fitnesses are density-dependent. Here we consider each of these mechanisms in turn as they pertain to invasion by an exotic species.

Testing the waters

If an exotic species can persist in one site, then emigrants from that site can repeatedly "test" alternative sites by sending out colonizing propagules. Imagine that emigration from the source habitat is sporadic. If p is the probability per colonization bout that a given propagule will persist and adapt in a particular sink habitat patch, then if there are n independent colonizing events, the probability that at least one of them will be an evolutionary and ecological success is

$$1 - (1 - p)^n$$

Figure 10.6 shows a theoretical example of the probability of eventual adaptation and invasion as a function of the initial degree of maladaptation to the sink habitat for four different numbers of independent colonizing events (n) for immigrants drawn from a specific established source population. The $n =$

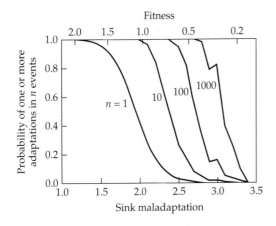

Fitness

Sink maladaptation

Figure 10.6 Propagule pressure facilitates persistence and adaptation in sink habitats. The graph shows the probability of persistence and adaptation in one or more colonizing episodes (for $n = 1$ to 1000) as a function of degree of maladaptation to the sink habitat. Other parameters are as in Figure 10.3, except that the monogamous mating system was used, assuming eight births per pair. The harsher the sink, the less likely one is to see adaptation over a long time period (1000 generations in this case). However, for more moderate sink habitats, recurrent invasion attempts can permit niche evolution to occur, leading to successful establishment in the initially unfavorable habitat. [At high values of n, the curves are close to 1 unless p is very small, which means that the initial maladaptation is very severe. Since p was determined by simulating a number of populations (up to 11,000) and dividing the number that persisted by the number simulated, there is some random variation in p, especially for small p (since in these cases very few populations persisted). This variation in p translates into a much greater variation in $1 - (1 - p)^n$ for large n. This accounts for the kinks in the $n = 100$ and $n = 1000$ curves (and to a much lesser extent in the other two curves). All result from the fact that there were slightly more populations persisting at a maladaptation of 3.0 than at 2.9 (19 versus 17 out of 11,000), going against the decreasing trend. However, the difference in these numbers is less than half the expected standard deviation of each.]

1 curve was obtained using the individual-based model to calculate the probability of adaptation and survival for a single colonizing event (giving p). The other curves were calculated using these p values and the formula for successful adaptation and establishment after n colonization attempts, $1 - (1 - p)^n$. The basic insight of this figure is that if an exotic species establishes a foothold in one habitat, it can then repeatedly send out propagules into other habitats, in effect providing repeated evolutionary "trials" in the novel environment.

Some plausible empirical examples of this phenomenon are provided by Harrison et al. (2001), who demonstrate that invasion of non-native plants into patches of serpentine soil is greatest in small patches; these authors provide suggestive evidence that the non-native species in these small patches show signs of incipient local adaptation. They argue that this enhancement of local adaptation in smaller patches reflects the greater potential rate of input of seeds from external sources into smaller patches, which in effect are "all edge." These observations could also reflect the following mechanism.

Gene flow provides adaptive genetic variation

Recurrent immigration can facilitate adaptive evolution by an invading species to a novel environment via the infusion of novel variation. Grant et al. (2001) describe how on the Galápagos island of Daphne Major, a population of the large ground finch (*Geospiza magnirostris*) was established by a single pair of individuals. Despite the extreme bottleneck at the time of establishment, heterozygosity remained roughly constant over the next 15 years, and allelic diversity doubled. Grant et al. (2001) suggest that the reason genetic variation was retained was that after its founding, the population continued to receive immigrants at a low rate from several other islands. This infusion of variation may have permitted an observed evolutionary step-function shift in beak shape due to selection. Similarly, Keller et al. (2001) described a population bottleneck in an insular population of song sparrows (*Melospiza melodia*), in which they found that genetic diversity rapidly regained pre-bottleneck levels within a few years after the crash due to a low level of immigration (about one bird every generation or two—a level of immigration that in less intensely studied populations would be almost impossible to detect). Gomulkiewicz et al. (1999) provide a model of single-locus evolution that formalizes this idea. In general, in low-density populations, immigration (which can draw on the variation maintained in abundant populations) provides a much more potent mechanism for providing variation than does mutational input.

Gene flow can hamper selection

A familiar idea in evolutionary biology is that gene flow can swamp local selection (Antonovics 1976; Hedrick 1985). For instance, Kirkpatrick and Barton (1999) analyzed a model of selection on a quantitative character along a cline in a phenotypic optimum in which local fitness determines population size. Gene flow from well-adapted, abundant populations into poorly adapted, scarce populations impedes selection in the latter and could lead to a stable distributional limit along an environmental gradient. This outcome is more likely if the gradient in fitness is sharp, so that dispersal mixes populations with substantially different optima.

Kirkpatrick and Barton's (1999) model assumed that genetic variance was fixed. If instead, genetic variance tends to increase in marginal populations (which itself might result from gene flow, as we saw above), then it becomes considerably harder for gene flow to prevent continued adaptation, and indeed, gene flow, instead of permitting a stable range margin, may lead to an evolutionary advance into novel terrain (Barton 2001; I. Filin, personal communication). Holt et al. (2003) explored an individual-based model of adaptive evolution in a sink habitat, with unidirectional migration from a source, comparable to the model described in Box 10.1. We demonstrated that an increase in the number of immigrants can increase the rate of adaptation to the sink habitat, and argued that this increase in part reflects the greater probability of drawing on genetic variation generated and maintained in the source habitat.

After the initial stage of adaptation to the new habitat, however, the perfection of local adaptation is hampered by continued migration from the source. If the new habitat is widespread, then following adaptation to any one patch of this habitat, the species could spread throughout an array of similar patches.

Over time, immigration from the initial source would become quantitatively trivial relative to the *in situ* production of these new populations, and gene flow from the source would have little or no influence on the maintenance and continued improvement of adaptation in the new habitat. With distinct habitats, stable maladaptation is also more likely if the sink habitat is relatively small relative to the source. We might predict that gene flow is less likely to hamper evolution in a widespread sink habitat to which the species is mildly maladapted than in a harsh sink because the fitness differences experienced by immigrants from the source are reduced in the former case, and more individuals are likely to be maintained there.

Immigration can alter local fitnesses

Holt and Gomulkiewicz (1995) and Gomulkiewicz et al. (1999) explored models for a "black-hole" sink, to which there was recurrent immigration from a source, but no backflow to the source from the sink. If we consider a single mutant that arises in such a sink habitat, it should increase in relative frequency if and only if it has an absolute fitness exceeding 1 (Gomulkiewicz et al. 1999).

If fitnesses are density-dependent, then changes in population size can affect the fate of favorable mutations. Immigration typically boosts population size, at least in simple models (e.g., Holt 1983); if there is negative density dependence (e.g., due to exploitation of a resource present in limited supply), then this increased abundance will lower average fitnesses. If there is negative density dependence in the sink habitat, adaptive evolution occurs most rapidly at an intermediate rate of immigration (Gomulkiewicz et al. 1999). The reason is that if immigration is very low, little variation is drawn in each generation from the source habitat, and so little variation is available for natural selection in the sink. Conversely, if immigration is high, numbers are boosted and, because of negative density dependence, fitnesses are lowered, reducing the likelihood of spread of favorable alleles. An intermediate rate of immigration appears to be optimal for adaptive evolution to harsh, novel environments, provided that density dependence is important at low initial densities.

This becomes even more likely if there are Allee effects at low densities. For instance, in very low-density sink populations, fitnesses may be further reduced because it is difficult for individuals of different sexes to find each other to set up mating pairs. An increase in immigration rate can boost local population size, alleviate Allee effects, and thereby indirectly facilitate adaptation to the sink environment (for details, see Holt et al. 2004b).

There is a temporal signature of niche evolution in a heterogeneous landscape. As discussed in more detail in Holt et al. (2003), in models of adaptation to sink habitats, given recurrent immigration from source habitats, one often observes a punctuational pattern, in which a population stays low in numbers and maladapted to the sink, often for long periods, and then undergoes a rapid increase in both adaptation and population size. This pattern could be expressed as a substantial lag in the spread of the invasion across space. A retrospective study of invasion lags could identify potential cases of niche evolution. If such evolution has occurred, then (1) one should be able to find environmental differences between the sites of original introduction and the sites in which spread later occurred, (2) there should be genetic differences between the original sites and the sites of later invasion, and (3) reciprocal transplants should show evidence of local adaptation. (Predictions 2 and 3 require that the evolution of adaptation to the novel environment not lead to significant gene flow back into the original site of introduction; Ronce and Kirkpatrick 2001.)

In invasion biology, models of evolution in sink habitats such as that sketched above could pertain to two distinct scenarios: evolution in a heterogeneous landscape, with initial establishment in a single habitat and subsequent spread to other habitats (our focus above), or repeated inputs of propagules drawn from the ancestral species range (we thank Jeb Byers for this observation). There is a potentially important difference between these two scenarios. Habitats within a heterogeneous landscape are likely to be correlated in significant ways—ways that are absent, on average, when one instead contrasts these initially unfavorable habitats with the sites of origination. For instance, an invader might be able to persist in a local site with disturbance, where certain native competitors and predators are absent or their effects are reduced. However, this site is likely to share climatic variables with a broader landscape, and also to be exposed to various natural enemies (e.g., mobile consumers that are habitat generalists). Once established in a disturbed site, the invader can then, in the fullness of evolutionary time, adapt to these other ecological factors. This in turn may make it more likely to invade (with niche evolution) sites in the landscape where colonization attempts by propagules from the ancestral range will almost surely fail. This scenario in effect permits a species to "bootstrap" its way through niche space (Holt and Gaines 1992) by gaining an original toehold in one habitat, from which it can spread as it evolves.

All these analyses become more complicated (and in interesting ways) if there is "backflow" from sinks to source habitats (see, e.g., Holt 1997; Kawecki and Holt 2002; Kawecki 2004). If adaptive trade-offs exist between habitats, then improvement in one comes at a cost of adaptation to the other. In this case, evolution in the habitat of initial colonization may make subsequent adaptation to other habitats more difficult and even lead to a kind of evolutionary "trapping" of a species in a limited habitat range (Ronce and Kirkpatrick 2001; Kisdi 2002; Holt 2003).

Conclusions

Boxes 10.2 and 10.3 summarize some key conclusions that emerge from considering invasions in the context of evolutionary theories of niche conservatism and evolution. We conclude this chapter with some speculations about which species are most likely to exhibit rapid niche evolution in the course of invasions.

BOX 10.2   *Key Insights from Evolutionary Theory*

1. The contingent details of genetic variation, evolution, and demographic stochasticity add considerable unpredictability to invasion success.

2. Propagule number influences the likelihood of invasion per colonizing episode via an interplay of adaptive evolution and demographic stochasticity. Propagules with more individuals provide more genetic variation drawn from the source, and also provide a longer window of opportunity for adaptive evolution to rescue a declining population from extinction. Consequently, predictive invasion models based purely on ecology (assessing niche requirements in the area of origin and ignoring the potential for evolution in the novel environment) should be more accurate for invading propagules containing few individuals.

3. In spatially heterogeneous environments, after the successful establishment of an invader in one habitat, evolution can then influence its spread into other habitats, either promoting or impeding such spread.

4. Migration has multiple effects on niche evolution:

   • It provides opportunities for evolution by sustaining local populations in sites outside the initial niche of the species (sink habitats).

   • It increases local abundances, potentially altering local fitnesses via density dependence and enhancing opportunities for local mutational input.

   • It introduces genetic variation from the source population.

   • It dilutes locally adapted gene pools, hampering adaptation.

The relative importance of these effects can change in a time-dependent manner through the historical process of adaptation to a novel environment.

5. Recurrent migration provides repeated evolutionary trials for adaptation to a novel environment.

6. If colonization is rare and involves only propagules small in number, invasion usually occurs only within the preexisting niche. By contrast, if colonization is frequent and propagule numbers are large, niche evolution can occur.

7. Niche evolution is more likely if the novel environment is not too different from the species' fundamental niche. By contrast, adaptation to strongly maladaptive environments should be extremely slow.

8. Evolution along gentle environmental gradients is more rapid than across abrupt environmental transitions.

9. Long periods of stasis are possible before niche evolution occurs, and often the evolutionary shift is abrupt when it occurs.

10. Temporal environmental variation can increase the probability that niche evolution will occur, particularly when the variation is strongly autocorrelated.

11. Most traditional models of community assembly assume that species have fixed properties. Because community assembly consists of repeated invasion attempts by multiple species, theories of niche conservatism help characterize when purely ecological approaches should suffice. In particular, standard ecological approaches should work best when colonization episodes are infrequent and propagules are small in number, and when fitness dif-

BOX 10.2 *(continued)*

ferences between source and target communities are either minor or large, and temporally stable.

12. Adaptation to one habitat may lead to the evolutionary "pinning" of invasions in space if there are strong adaptive trade-offs between the site of introduction (which will be the principal focus of initial adaptive evolution for an invader) and other habitats elsewhere in a landscape or geographic region.

As noted above, Holt et al. (2004b) recently argued that in populations that show Allee effects at low densities (e.g., due to the existence of distinct sexes), immigration can facilitate adaptive evolution in sink habitats because an increase in immigration typically increases population size. Given an Allee effect, at low densities, an increase in abundance *increases* individual fitness. This makes it more likely that alleles with small to moderate positive effects on fitness will yield an absolute fitness greater than unity, and so can be retained by selection. Other life history factors can also influence the probability of adaptation to novel environments. For instance, Kawecki (2003) recently suggested that in species without male parental care, female dispersal is more likely than male dispersal to lead to niche evolution. In this case, male dispersal mainly increases gene flow from source populations without enhancing local recruitment, whereas female dispersal can bolster local recruitment (with no more gene flow than for an equivalent amount of male dispersal), which can substantially affect the likelihood of adaptation to the sink. Thus, if niche evolution is required for an invading species to adapt and spread in a heterogeneous landscape, this process may be more likely to be observed for species with female-biased dispersal (or at least not strongly male-biased dispersal).

BOX 10.3   *Using Invasions as Critical Tests of Niche Evolution Theory: A Summary*

1. Few data exist to test many of the theoretical predictions for niche evolution, since what is required is a contrast between successful and failed invasions, and the latter are rarely recorded.

2. If retrospective analysis shows that many invasions stem from infrequent introductions of low-abundance propagules into harsh environments (as assessed by the niche properties in the ancestral range), then that finding would cast considerable doubt on the applicability of these theories.

3. If invasion lags involve adaptation to novel habitats, then the length of lag time should be positively correlated with the degree of difference between the novel and ancestral environments.

Figure 10.7 Probability of adaptation under constant conditions and with random variation in the degree of maladaptation to sink habitats (compared with the source optimum), with coefficient of variation (cov) = 0.2 and autocorrelation coefficients (ρ) of 0, 0.9, and approaching 1. The solid line denotes a constant environment. Other parameters and mating system are as in Figure 10.6. As noted in the text, this figure describes the probability of adaptation for a single bout of colonization. Autocorrelated variation in the degree of maladaptation facilitates adaptation to harsh sinks and makes adaptation less likely in sinks that are only slightly maladaptive (see also Holt et al. 2004a).

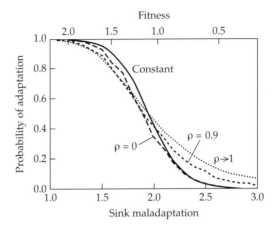

Recently we (Holt et al. 2004a) have argued that in some circumstances, autocorrelated temporal variation in a sink (e.g., in the degree of maladaptation experienced by immigrants) can facilitate adaptive evolution. If there are transient favorable periods during which the local population can grow, it can better escape the inhibitory effects of gene flow, permitting its numbers to grow even further. Temporal variation may also facilitate adaptive evolution in a sink even if there is no recurrent gene flow. In Figure 10.7 we revisit the situation explored in Figure 10.6 ($n = 1$ curve), in which we examined the probability of adaptation for one-time colonizations of empty sinks. Here, we assume that the local degree of maladaptation is described by a first-order autoregressive Gaussian process with a mean equal to the abscissa (which means that the degree of maladaptation experienced will sometimes be greater and sometimes less than the average degree of maladaptation, but the net effect will be no net change in the average degree of maladaptation experienced over time), and an autocorrelation coefficient of ρ. In sink environments that are on average quite maladapted, strongly autocorrelated variation can substantially enhance the probability of adaptation, and thus of invasion out of the initial source habitat.

An explanation for this comes from Jensen's inequality. Note that the functional form describing the probability of adaptation as a function of the degree of maladaptation in a constant environment is sigmoidal and is concave upward for harsh sinks (viz., where the initial degree of maladaptation faced by an immigrant propagule is large). If the autocorrelation coefficient approaches 1, this is in effect the same as imagining that an array of patches is available, with a fixed (Gaussian or normal) distribution among them in degrees of maladaptation. At high degrees of maladaptation, averaging over the functional form that describes the probability of adaptation over a given time horizon, as a function of the degree of initial maladaptation, leads to an expected value greater than the function evaluated at the average value. Note that the opposite is true at low degrees of maladaptation.

Thus, temporal variation in highly suboptimal sink habitats can make invasions more probable if there are at least some periods in which populations can adapt, persist, and then spread. Hoffman and Hercus (2000) argue that evolutionary patterns of diversification that show the greatest magnitude of adaptive evolution are observed in environments that are intermittently stressful. The causal mechanisms they suggest are somewhat different from those we suggest (e.g., they posit that stress could reduce gene flow), but the pattern they outline matches expectations from the theory presented here. The bottom line is that invasions into disturbed, temporally unstable environments may be facilitated for evolutionary as well as ecological reasons.

Invading species, we suggest, provide potential systems for testing many of these recent theoretical ideas about how demography can constrain (or, alternatively, at times facilitate) niche evolution. Levin (2003) has similarly suggested that invading species provide potential tests for understanding ecological speciation. Ecological speciation, he suggests, involves two stages: the colonization or invasion of a new habitat, and a subsequent process of selection on adaptation that refines the genotypic and phenotypic composition of the population to fit it better into its novel environment. Speciation can arise as a by-product of such selection.

Another issue to which we cannot do justice here is that dispersal itself is likely to evolve in a novel environment. In a landscape comprising a mix of good and bad habitats, if the latter are sufficiently bad, then selection should act on dispersal so that such habitats are avoided. Temporal variation can permit the evolutionary maintenance of dispersal, even into sink habitats. Evolutionary branching can occur so that a polymorphism in dispersal arises, with low-dispersal individuals concentrated in the habitat that initially has the higher carrying capacity and high-dispersal individuals concentrated in the lower-quality habitat (McPeek and Holt 1992; Mathias et al. 2001). This sets up an interesting situation in which, if a mutant arises that improves fitness in the lower-quality habitat, the population is preadapted to spread quickly over a much larger landscape.

Finally, our discussion has focused closely on evolutionary dynamics within invading species. More broadly, it is important for these microevolutionary processes to be embedded in a community context. Stable range limits often involve interactions with other species, including coevolutionary dynamics (Case et al. 2005). Success and failure in invasions may frequently reflect differences in the biotic environment experienced by invaders from that in their ancestral communities. A species that enjoys an escape from specialist natural enemies in a novel environment may be able to tolerate abiotic conditions from which it is excluded in their presence. If the species can persist, it can then adapt to suites of environmental factors that were originally outside its evolutionary repertoire.

In conclusion, almost all the processes studied by evolutionary biologists— the evolution of local adaptation versus generalized phenotypes, the evolution of dispersal, even mutational pressures—are potentially relevant to understanding the capacity of exotic species to persist in and invade novel environments. We have argued that a particularly pertinent area of evolutionary theory is that concerned with the evolution of the niche itself, because this abstract character describes the range of environments in which a species can persist, and thus defines the potential domain into which an exotic species can expand. An important and largely unaddressed challenge is utilizing the myriad opportunities provided by exotic species for empirically testing the emerging theory of niche conservatism and evolution. The theoretical models we have surveyed provide some expectations regarding invasion dynamics. For instance, climate-matching and other descriptive niche models (e.g., Peterson 2003) should be most successful at predicting invasion success when overall propagule pressure is low. Invasion associated with evolutionary transitions may be particularly likely in heterogeneous landscapes, where a species may in effect encounter gentle gradients in conditions relative to the ancestral environment to which it is already adapted. Finally, if it turns out that ancestral niche requirements describe not only the circumstances in which an invading species is initially established, but also the eventual range it occupies, then it is likely that there are strong constraints (e.g., due to absence of suitable genetic variation) on the evolution of that species' niche. To address these hypotheses, we suggest that broad qualitative surveys of factors correlated with successes and failures of attempted invasions may reveal patterns consistent with niche evolution being involved in invasions. Beyond this, more refined tests of the theory would require quantitatively rigorous population models that describe in detail spatial variation in demography, genetic variation in niche traits, and how the selective environment differs between ancestral locales and the sites of invasion.

An increasing number of empirical examples demonstrate that evolution frequently is associated with invasions (e.g., Thomas et al. 2001; Sexton et al. 2003; Blair and Wolfe 2004; Cox 2004; Maron et al. 2004; Muller-Scharer et al. 2004). It is an open question how often the evolution of species' niches is directly responsible for invasion success in any of these examples. It is our hope that the perspective we have presented in this chapter will foster further empirical studies of this important question.

Acknowledgments

We thank the editors, Jeb Byers, Richard Mack, and two anonymous reviewers for very helpful suggestions and thoughts. We thank the National Science Foundation and the University of Florida Foundation for support. This work was conducted as a part of the "Exotic Species: A Source of Insight into Ecology, Evolution, and Biogeography" Working Group supported by the National Center for Ecological Analysis and Synthesis, a center funded by NSF (grant DEB-0072909), the University of California, and the Santa Barbara campus.

Literature Cited

Ackerly, D. D. 2003. Community assembly, niche conservatism, and adaptive evolution in changing environments. International Journal of Plant Science 164 (Supp.): S165–S184.

Antonovics, J. 1976. The nature of limits to natural selection. Annals of the Missouri Botanical Gardens 63:224–247.

Antonovics, J., T. J. Newman, and B J. Best. 2001. Spatially explicit studies on the ecology and genetics of population margins. In J. Silvertown and J. Antonovics, eds. *Integrated ecology and evolution in a spatial context*, pp. 97–116. Blackwell, London.

Ashley, M. V., M. F. Willson, O. R. W. Pergrams, D. J. O'Dowd, S. M. Gende, and J. S. Brown. 2003. Evolutionary enlightened management. Biological Conservation 111:115–123.

Barrett, S. C. H., and B. C. Husband. 1990. The genetics of plant migration and colonization. In A. H. D. Brown, M. T. Clegg, and A. L. Kahler, eds. *Plant population genetics, breeding and resources*, pp. 254–277. Sinauer Associates, Sunderland, MA.

Barton, N. H. 2001. Adaptations at the edge of a species' range. In J. Silvertown and J. Antonovics, eds. *Integrating ecology and evolution in a spatial context*, pp. 365–392. Blackwell, London.

Blair, A. C., and L. W. Wolfe. 2004. The evolution of an invasive plant: an experimental study with *Silene latifolia*. Ecology 85:3035–3042.

Boulding, E. G., and T. Hay. 2001. Genetic and demographic parameters determining population persistence. Heredity 86:313–324.

Bradshaw, A. D. 1991. Genostasis and the limits to evolution. Philosophical Transactions of the Royal Society of London B 333:289–305.

Brown, R. L., and R. K. Peet. 2003. Diversity and invasibility of southern Appalachian plant communities. Ecology 84:32–39.

Burger, R., and M. Lynch. 1995. Evolution and extinction in a changing environment: a quantitative-genetic analysis. Evolution 49:151–163.

Burt, A. 1995. Perspective—the evolution of fitness. Evolution 49:1–8.

Byers, J. E., and L. Goldwasser. 2001. Exposing the mechanisms and timing of impact of nonindigenous species on native species. Ecology 82:1330–1343.

Case, T. J., and M. L. Taper. 2000. Interspecific competition, environmental gradients, gene flow, and the coevolution of species' borders. American Naturalist 155:583–605.

Case, T. J., R. D. Holt, and M. McPeek. 2005. The community context of species borders. Oikos 108:18–27.

Chase, J. M., and M. A. Leibold. 2003. *Ecological niches: Linking classical and contemporary approaches*. University of Chicago Press, Chicago.

Coope, G. R. 1979. Late-Cenozoic fossil Coleoptera: evolution, biogeography, and ecology. Annual Review of Ecology and Systematics. 10:247–267.

Cox, G. W. 2004. *Alien species and evolution*. Island Press, Washington, DC.

Crooks, J. A., and M. E. Soulé. 1999. Lag times in population explosions of invasive species: causes and implications. In O. T. Sandlund, P. J. Schei, and A. Viken, eds. *Invasive species and biodiversity management*, pp. 103–125. Kluwer, New York.

Crow, J. F., and M. Kimura. 1970. *An introduction to population genetics theory*. Harper and Row, New York.

D'Antonio, C., L. A. Meyerson, and J. Denslow. 2001. Exotic species and conservation. In M. E. Soulé and G. H. Orians, eds. *Conservation biology: Priorities for the next decade*, pp. 59–80. Island Press, Washington, DC.

Duncan, R. P. 1997. The role of competition and introduction effort in the success of passeriform birds introduced into New Zealand. The American Naturalist 149:903–915.

Fisher, R. A. 1958. *The genetical theory of natural selection*. Dover Publications, New York.

Gaggiotti, O. E. 2003. Genetic threats to population persistence. Annales Zoologici Fennici 40:155–168.

Garcia-Ramos, G., and D. Rodriguez. 2002. Evolutionary speed of species invasions. Evolution 56:661–668.

Gomulkiewicz, R., and R. D. Holt. 1995. When does evolution by natural-selection prevent extinction? Evolution 49:201–207.

Gomulkiewicz, R., R. D. Holt, M. Barfield. 1999. The effects of density dependence and immigration on local adaptation and niche evolution in a black-hole sink environment. Theoretical Population Biology 55:283–296.

Grant, P. R., B. R. Grant, and K. Petren. 2001. A population founded by a single pair of individuals: establishment, expansion, and evolution. Genetica 112–113:359–382.

Green, R. E. 1997. The influence of numbers released on the outcome of attempts to introduce exotic bird species to New Zealand. Journal of Animal Ecology 66:25–35.

Grinnell, J. 1917. The niche-relationships of the California thrasher. Auk 34:427–433.

Hansen, T. F., and D. Houle. 2004. Evolvability, stabilizing selection, and the problem of stasis. Evolvability, stabilizing selection, and the problem of stasis. In M. Pigliucci and K. Preston, eds. *The evolutionary biology of complex phenotypes*, pp. 130–150. Oxford University Press, Oxford.

Harrison, S., K. Rice, and J. Maron. 2001. Habitat patchiness promotes invasion by alien grasses on serpentine soil. Biological Conservation 100:45–53.

Hedrick, P. W. 1985. *Genetics of populations*. Jones and Bartlett, Boston.

Hendry, A. P., and M. T. Kinnison. 1999. The pace of modern life: measuring rates of contemporary microevolution. Evolution 53:1637–1653.

Hoffman, A. A., and M. J. Hercus. 2000. Environmental stress as an evolutionary force. Bioscience 50:217–226.

Hoffmann, A. A., and J. Merila. 1999. Heritable variation and evolution under favourable and unfavourable conditions. Trends in Ecology and Evolution 14:96–101.

Holt, R. D. 1983. Models for peripheral populations: The role of immigration. In H. I. Freedman and C. Strobeck, eds. *Lecture notes in biomathematics*, pp. 25–32. Springer-Verlag, Berlin.

Holt, R. D. 1996. Demographic constraints in evolution: Towards unifying the evolutionary theories of senescence and niche conservatism. Evolutionary Ecology 10:1–11.

Holt, R. D. 2003. On the evolutionary ecology of species ranges. Evolutionary Ecology Research 5:159–178.

Holt, R., and M. Gaines. 1992. The analysis of adaptation in heterogeneous landscapes: Implications for the evolution of fundamental niches. Evolutionary Ecology 6:433–447.

Holt, R. D., and R. Gomulkiewicz. 1997a. How does immigration influence local adaptation? A reexamination of a familiar paradigm. American Naturalist 149:563–572.

Holt, R. D. and R. Gomulkiewicz. 1997b. The evolution of species' niches: a population dynamic perspective. In H. Othmer, F. Adler, M. Lewis, and J. Dallon, eds. *Case studies in mathematical modeling: Ecology, physiology, and cell biology*, pp. 25–50. Prentice-Hall, New Jersey.

Holt, R. D. and R. Gomulkiewicz. 2004. Conservation implications of niche conservatism and evolution in heterogeneous environments. In R. Ferrière, U. Dieckmann, and D. Couvet, eds. *Evolutionary conservation biology*, pp. 244–264. Cambridge University Press, Cambridge.

Holt, R. D., R. Gomulkiewicz, and M. Barfield. 2003. The phenomenology of niche evolution via quantitative traits in a 'black-hole' sink. Proceedings of the Royal Society of London B 270:215–224.

Holt, R. D., M. Barfield, and R. Gomulkiewicz. 2004a. Temporal variation can facilitate niche evolution in harsh sink environments. American Naturalist 164:187–200.

Holt, R. D., T. M. Knight, and M. Barfield. 2004b. Allee effects, immigration, and the evolution of species' niches. American Naturalist 163:253–262.

Huntley, B. 1991. How plants respond to climate change: migration rates, individualism, and the consequences for plant communities. Annals of Botany 67 (Supp. 1): 15S–22s.

Hutchinson, G. E. 1957. Concluding remarks. Cold Spring Harbor Symposiums on Quantitative Biology 22:415–427.

Hutchinson, G. E. 1978. *An introduction to population ecology*. Yale University Press, New Haven.

Kawecki, T. J. 1995. Demography of source-sink populations and the evolution of ecological niches. Evolutionary Ecology 9:38–44.

Kawecki, T. J. 2000. Adaptation to marginal habitats: contrasting influence of the dispersal rate on the fate of alleles with small and large effects. Proceedings of the Royal Society of London, B 267:1315–1320.

Kawecki, T. J. 2003. Sex-biased dispersal and adaptation to marginal habitats. American Naturalist 162:415–426.

Kawecki, T. J. 2004. Ecological and evolutionary consequences of source-sink population dynamics. In I. Hanski and O. E. Gaggiotti, eds. *Ecology, genetics and evolution of metapopulations*, pp. 386–414. Elsevier, Burlington, MA.

Kawecki, T. J., N. H. Barton, J. D. Fry. 1997. Mutational collapse of fitness in marginal habitats and the evolution of ecological specialization. Journal of Evolutionary Biology 10:407–429.

Kawecki, T. J., and R. D. Holt. 2002. Evolutionary consequences of asymmetric dispersal rates. American Naturalist 160:333–347.

Keane, R. M., and M. J. Crawley. 2002. Exotic plant invasions and the enemy release hypothesis. Trends in Ecology and Evolution 17:164–170.

Keller, L. F., K. J. Jeffery, P. Arcese, M. A. Beaumont, W. M. Hachachka, M. N. M. Smith, and M. W. Bruford. 2001. Immigration and the ephemerality of a natural population bottleneck: evidence from molecular markers. Proceedings of the Royal Society of London B 268:1387–1394.

Kirkpatrick, M., and N. H. Barton. 1997. Evolution of a species' range. American Naturalist 150:1–23.

Kisdi, E. 2002. Dispersal: risk spreading versus local adaptation. American Naturalist 159:579–596.

Lambrinos, J. G. 2004. How interactions between ecology and evolution influence contemporary invasion dynamics. Ecology 85:2061–2070.

Lande, R. 1998. Anthropogenic, ecological and genetic factors in extinction and conservation. Res. Popul. Ecol. 40:259–269.

Lande, R., and S. Shannon. 1996. The role of genetic variation in adaptation and population persistence in a changing environment. Evolution 50:434–437.

Lee, C. E. 2002. Evolutionary genetics of invasive species. Trends in Ecology and Evolution 17:386–391.

Lenormand, T. 2002. Gene flow and the limits to natural selection. Trends in Ecology and Evolution 17:183–189.

Levin, D. 2003. Ecological speciation: lessons from invasive species. Systematic Botany 28:643–650.

Levin, D., and K. Clay. 1984. Dynamics of synthetic *Phlox drummondii* populations at the species margin. American Journal of Botany 71:1040–1050.

Levine, J. M. 2000. Species diversity and biological invasions: relating local process to community pattern. Science 288:852–854.

Lewontin, R. C. 1965. Selection for colonizing ability. In H. G. Baker and G. L. Stebbins, eds. *The genetics of colonizing species*, pp. 77–91. Academic Press, New York.

Lonsdale, W. M. 1999. Global patterns of plant invasions and the concept of invasibility. Ecology 80:1522–1536.

Lyford, M. E., S. T. Jackson, J. L. Betancourt, and S. T. Gray. 2003. Influence of landscape structure and climate variability on a late Holocene plant migration. Ecological Monographs 73:567–584.

Lynch M, J. Conery, and R. Burger. 1995. Mutational meltdowns in sexual populations. Evolution 49:1067–1080.

Lynch, M. J., and R. Lande. 1993. Evolution and extinction in response to environmental change.In P. M. Kareiva, J. G. Kingsolver, and R. B. Huey, eds. *Biotic interactions and global change*, pp. 234–250. Sinauer Associates, Sunderland, MA.

Mack, R. N., D. Simberloff, W. M. Lonsdale, H. Evans, and M. Clout. 2000. Biotic invasions: causes, epidemiology, global consequences, and control. Ecological Applications 10:698–710.

Maguire, B. J. 1973. Niche response structure and the analytical potentials of its relationship to the habitat. American Naturalist 107:213–246.

Maron, J. L., M. Vila, R. Bommarco, S. Elmendorf, and P. Beardsley. 2004. Rapid evolution of an invasive plant. Ecological Monographs 74:261–280.

Mathias, A., E. Kisdi, and I. Olivieri. 2001. Divergent evolution of dispersal in a heterogeneous landscape. Evolution 55:246–259.

McPeek, M. A., and R. D. Holt. 1992. The evolution of dispersal in spatially and temporally varying environments. American Naturalist 6:1010–1027.

Merila, J., B. C. Sheldon, and L. E. B. Kruuk. 2001. Explaining stasis: microevolutionary studies in natural populations. Genetica 112–113:199–222.

Muller-Scharer, H., U. Schaffner, and T. Steinger. 2004. Evolution in invasive plants: implications for biological control. Trends in Ecology and Evolution 19:417–422.

Nei, M., T. Maruyam, and R. Chakraborty. 1975. The bottleneck effect and genetic variability in populations. Evolution 29:1–10.

Okubo, A., and S. A. Levin. 2001. *Diffusion and ecological problems*. Springer-Verlag, New York.

Olden, J. D., N. L. Poff, M. R. Douglas, M. E. Douglas, and K. D. Fausch. 2004. Ecological and evolutionary consequences of biotic homogenization. Trends in Ecology and Evolution 19:18–24.

Pease, C. M., R. Lande, and J. Bull. 1989. A model of population-growth, dispersal and Evolution in a changing environment. Ecology 70:1657–1664.

Peterson, A. T. 2003. Predicting the geography of species' invasions via ecological niche modeling. Quarterly Review of Biology 78:419–433.

Peterson, A. T., and R. D. Holt. 2003. Niche differentiation in Mexican birds: using point occurrence data to detect ecological innovation. Ecology Letters 6:654–664.

Peterson, A. T., J. Soberon, and V. Sanchez-Cordero. 1999. Conservatism of ecological niches in evolutionary time. Science 285:1265–1267.

Quinn, T. P., M. J. Unwin, and M. T. Kinnison. 2000. Evolution of temporal isolation in the wild: genetic divergence in timing of migration and breeding by introduced chinook salmon populations. Evolution 54:1372–1385.

Reznick, D. N., and C. K. Ghalambor. 2001. The population ecology of contemporary adaptations: what empirical studies reveal about the conditions that promote adaptive evolution. Genetica 112–113:183–198.

Rice, K. J., and N. C. Emery. 2003. Managing microevolution: restoration in the face of global change. Frontiers in Ecology and the Environment 1:469–478.

Ronce, O., and M. Kirkpatrick. 2001. When sources become sinks: Migrational meltdown in heterogeneous habitats. Evolution 55:1520–1531.

Saccheri, I., M. Kuussaari, M. Kankare, P. Vikman, W. Fortelius, and I. Hanski. 1998. Inbreeding and extinction in a butterfly metapopulation. Nature 392:491–494.

Sakai, A. K., F. W. Allendorf, J. S. Holt, D. M. Lodge, J. Molofsky, K. A. With, S. Baughman, R. J. Cabin, J. E. Cohen, N. C. Ellstrand, D. E. McCauley, P. O'Neil, I. M. Parker, J. N. Thompson, and S. G. Weller. 2001. The population biology of invasive species. Annual Review of Ecology and Systematics 32:305–332.

Sax, D. F., and J. H. Brown. 2000. The paradox of invasion. Global Ecology and Biogeography 9:363–371.

Schopf, K. M. 1996. Coordinated stasis: biofacies revisited and the conceptual modeling of whole-fauna dynamics. Paleogeography, Paleoclimatology, Paleoecology 127:157–175.

Sexton, J. P. J. K. McKay, and A. Sala. 2002. Plasticity and genetic diversity may allow saltcedar to invade cold climate in North America. Ecological Applications 12:1652–1660.

Shea, K., and P. Chesson. 2002. Community ecology theory as a framework for biological invasions. Trends in Ecology and Evolution 17:170–176.

Shigesada, N., and K. Kawasaki. 1997. *Biological invasions: Theory and practice*. Oxford University Press, Oxford.

Stockwell, C. A., and M. V. Ashley. 2004. Rapid evolution and conservation. Conservation Biology 18:272–273.

Stockwell, C. A., A. P. Hendry, and M. T. Kinnison. 2003. Contemporary evolution meets conservation biology. Trends in Ecology and Evolution 18:94–101.

Thomas, C. D., E. J. Bodsworth, R. J. Wilson, A. D. Simmons, Z. G. Davies, M. Musche, and L. Conradt. 2001. Ecological and evolutionary processes at expanding range margins. Nature 411:577–581.

Tufto, J. 2001. Effects of releasing maladapted individuals: a demographic-evolutionary model. American Naturalist 158:331–340.

Veltman, C. J., and S. Nee. 1996. Correlates of introduction success in exotic New Zealand Birds. American Naturalist 147:542–557.

Vitousek, P. M., C. M. D'Antonio, L. L. Loope, and R. Westbrooks. 1996. Biological invasions as global environmental change. American Scientist 84:468–478.

Whitlock, M. C., C. K. Griswold, and A. D. Peters. 2003. Compensating for the meltdown: the critical effective size of a population with deleterious and compensatory mutations. Annales Zoologici Fennici 40:169–183.

With, K. A. 2002. The landscape ecology of invasive spread. Conservation Biology 16:1192–1203.

11

Testing Fundamental Evolutionary Questions at Large Spatial and Demographic Scales

SPECIES INVASIONS AS AN UNDERAPPRECIATED TOOL

William R. Rice and Dov F. Sax

Species invasions can serve as model systems for the study of fundamental questions in evolutionary biology, particularly by allowing studies at spatial and temporal scales that cannot readily be accomplished in laboratory or simplified field experiments. We illustrate this here by considering the opportunity species invasions provide for better understanding three topics in evolutionary biology: the process of sexual selection; the evolution of reproductive isolation; and the adaptive significance of sexual reproduction. Our understanding of sexual selection can be advanced by exploring patterns of introgression of sexually selected genes and traits following the introduction of exotic species capable of hybridizing with native species, and patterns of local adaptation in sexual signaling phenotypes following the introduction of exotics into disparate environments. Our understanding of the evolution of reproductive isolation can be advanced by exploring the timing and development of prezygotic versus postzygotic isolation in potentially interbreeding native and exotic species, the operation of reinforcement in preventing reticulation following secondary contact of related populations, and the operation and likelihood of sympatric speciation. Our understanding of the costs and benefits of sexual reproduction can be advanced by exploring the relative success of sexually and asexually reproducing congeners introduced to new re-

gions, and the relative importance of species interactions in garnering evolutionary fitness to sexual and asexual species. These explorations, taken collectively and together with previously published work on this topic, suggest that species invasions should serve as one of the most important model systems in evolutionary biology.

Introduction

The study of exotic species has two major aims: their management, and their use as tools to study basic science. Here we focus on the latter aim in the context of evolutionary biology. The purpose of this chapter is to illustrate the power and scope of exotic species as a model system for the study of fundamental questions in evolutionary biology. Many biologists have already taken advantage of exotic species to test specific evolutionary hypotheses (e.g., see Chapters 7–10 in this volume and the examples discussed below), but we consider these extant studies to be the tip of the iceberg with respect to the true, underutilized potential of this model system. To illustrate our point, suppose that a poll were taken at the annual meetings of the Society for the Study of Evolution asking participants to name the most powerful model systems in evolutionary biology. We suspect that few people would mention exotic species. But we think that exotic species should be at the top of the list, and we use this chapter to illustrate why we have come to this conclusion. Our aim is not to catalog an exhaustive set of all of the possibilities, but to use a few illustrative examples to motivate other evolutionary biologists to ask how their basic research programs would benefit by using exotic species as an extension of controlled experimentation.

The processes and mechanisms of evolution have been a subject of substantial scientific inquiry for well over two centuries (e.g., Lamarck 1809; Darwin 1859; Fisher 1930; Wright 1969; Dobzhansky 1970). However, the handshaking between theory and experimentation has been frustratingly insufficient and largely restricted to laboratory or simplified field conditions. The limitations imposed by the physical, temporal, and monetary constraints on conducting large-scale experimental work under natural field conditions have reduced the number and scope of such studies and hindered our ability to make empirical progress.

An alternative to fully controlled experimentation that may help to circumvent some of these constraints is the study of species invasions. Species introduced outside of their historical range and isolated from parental gene pools may evolve in unique ways as the selective pressures of a new environment are brought to bear on them. The study of evolutionary change in such species often holds several distinct advantages; for example, knowledge of how long a population has been isolated from its historical range, replication provided by the introduction of the same species to multiple localities, and introduction of species into environments that are novel to those species. For all of these reasons, the study of species invasions provides a unique opportunity to test predictions of fundamental evolutionary hypotheses. Further, although species introductions are not true manipulative experiments, they have a major advantage over all past experiments: they occur on a large spatial and temporal scale, and unlike laboratory experiments, they occur under natural conditions. Finally, the combination of studies of exotic species in the field with traditional controlled experiments can complement the deficiencies of each approach.

To date, significant insights into evolutionary processes have already been garnered by studying species invasions. For example, Huey et al. (2000), Gilchrist et al. (2004), and Huey et al. (this volume) have shown that genetically based clines in wing size in a European species of fruit fly were rapidly reestablished after the species was introduced into North and South America, but that the portion of the wing that was modified to accomplish this variation in size was different on each continent. On one hand, this finding suggests that selection can operate very quickly and consistently to select for general solutions to environmental pressures, while on the other hand it illustrates that the specific manner in which evolution ultimately works can be highly idiosyncratic. Work by Callaway and Aschehoug (2000), Callaway et al. (2004), Vivanco et al. (2004), and Callaway et al. (this volume) has focused on the different evolutionary trajectories or "arms races" that occur among plants by way of the chemical compounds they produce within the soils of geographically isolated plant communities. This work suggests that adaptations in an ancestral environment can become preadaptations in an invaded environment, where native species have no evolutionary experience in dealing with novel chemical compounds. Another avenue of research concerns the importance of genetic bottlenecks in limiting the genetic diversity of founder populations of exotic species (Novak and Mack, and Wares et al., this volume). This research extends past laboratory studies by assessing the effects of genetic bottlenecks on a much larger spatial scale, providing intriguing evidence that genetic bottlenecks may be much less common and severe than previously presumed following colonization or introduction events. As a final example, Ricklefs (this volume) explores how the study of species invasions can provide new insights into the taxon cycle of expansion and contraction of species' geographic ranges, abundances, and niche space.

In this chapter, we focus on three topics: the process of sexual selection, the evolution of reproductive isolation, and the adaptive significance of sexual reproduction. We use these as arbitrary examples to illustrate what we consider to be the underutilized potential to use exotic species as a model system to study fundamental questions in evolutionary biology.

Topic I: Sexual Selection

Experiments have advanced our understanding of the genetic basis of the process of sexual selection and the evolution of sexually selected traits. Consider several representative examples: (1) experimental addition of new ornaments to birds, fish, and spiders has demonstrated that female preference for a male ornament can precede the evolution of the male trait (Burley 1986; Basolo 1990; McClintock and Uetz 1996); (2) transplantation experiments have demonstrated that sexually selected traits can evolve rapidly in response to changes in predation pressure (Endler 1988); (3) field plantings of F_2 back-crossed progeny of bee- and hummingbird-pollinated flowers have demonstrated that a small number of large-effect mutations can rapidly lead to major shifts in the pollination system, and hence to rapid reproductive isolation (Schemske and Bradshaw 1999); (4) recent laboratory experiments with *Drosophila* have demonstrated that seminal fluid proteins are highly genetically variable, evolve rapidly, and make an important contribution to sexual selection (reviewed in Wolfner 2002). The small spatial and demographic scales of these experiments, in combination with the fact that all were necessarily carried out under laboratory or simplified natural conditions, represent important limitations.

Here we illustrate how studies of introduced species can be used as quasi-controlled experiments in the study of sexual selection. In contrast to manipulative experiments, many important factors cannot be fully controlled in such studies, but the large spatial and temporal context of exotic species introductions, as well as their more natural setting, provides offsetting advantages. We do not attempt to make any definitive conclusions about specific issues in the area of sexual selection in this analysis. Our aim instead is to provide insight into the potential of exotic species as a model system. We focus on two issues in this section: the evolution and introgression of sexually selected traits, and the role of sexual selection in local adaptation.

Evolution and introgression of new sexually selected genes and traits

A major problem with the use of the comparative method in evolutionary biology is that many genes influencing locally adapted traits are fixed in different populations. As a consequence, the endpoints of evolution can be measured, but the genetic dynamics leading to these ends can no longer be observed. This problem is removed when there is secondary contact between formerly allopatric populations of the same species, or between closely related taxa that are still capable of at least limited gene flow. Under these circumstances, many new genetic variants can introgress between the formerly allopatric populations, and a burst of adaptive evolution can ensue, because introgression replaces the relatively slow process of mutation in the generation of new favorable alleles. The successful introduction of an exotic species can mimic the early stages of secondary contact and therefore provide an otherwise rare opportunity for study of the genetic details of evolutionary change. Here we provide three illustrative examples of how introgression from introduced species can be used as a tool to study fundamental evolutionary principles.

SPREAD OF NEW SEXUALLY SELECTED ALLELES   One of the most basic phenomena that could be addressed by using exotic species invasions is how new beneficial traits spread within a population. Such examinations could be made with many species, but the mallard duck (*Anas platyrhynchos*) provides one of the best examples of the utility of this approach. The mallard has been introduced by Europeans to many new locations where it hybridizes with closely related species and produces fertile offspring. In addition, habitat modification in eastern North America has permitted a westward expansion of the mallard into the range of the closely related North American black duck (*A. rubripes*). In most locations, the mallard male is more highly ornamented than males of the indigenous species with which it hybridizes. In some cases, experimental evidence indicates that the plumage characteristics of the mallard provide an advantage in the context of sexual selection (Rhymer et al. 1994). In this case, the genes coding for the favored traits are analogous to new mutations that are favored by sexual selection. A similar situation occurs in the Pecos River of the southwestern United States, between the native Pecos pupfish (*Cyprinodon pecosensis*) and the invasive sheepshead minnow (*C. variegates*) (Rosenfield and Kodric-Brown 2003); here, the males of the more colorful sheepshead minnow are favored by sexual selection. In both cases, analyses of the traits that introgress into the native species, and the genes that code for them, represent critical opportunities to study the fate of new genetic variation in response to the joint action of natural and sexual selection.

Analyses of introgressed traits, however, have important pitfalls that must be appreciated. In the case of the mallard duck and its relatives, one would need corroborating evidence that the ornaments found in the mallard, but not in the taxa with which it hybridizes, are new traits that evolved independently after the mallard lineage split from the native species. An alternative would be that these traits were present in a common ancestor of all members of the clade (including the mallard and its close relatives) and that the colorful plumage of the mallard has been lost in all but the mallard lineage. In this case, the mallard's genes producing its colorful plumage traits would not be representative of new sexually selected mutations because there might be residual traits in the native females that previously evolved in response to the reintroduced trait. Nonetheless, an analysis of which mallard ornaments introgress into the native species would provide valuable information about the evolutionary genetic processes that occur during secondary contact between formerly allopatric sister taxa, as discussed below.

MEASURING SELECTION ON NEW SEXUALLY SELECTED TRAITS   A second way that introgression could be used in the context of exotic species invasions would be to study the success in sexual interactions of hybrid individuals (expressing newly introduced genetic variation) in their native environment. Again, consider the case of mallards hybridizing with their sister taxa. As an illustrative example, a survey of breeding pairs could be used to identify pairings between females that are phenotypically mallard or native species (i.e., females with a low hybrid phenotypic index, indicating that their genetic background is not hybridized, or is at least predominantly of one type) and hybrid males. By measuring the sexual display traits manifested in F_2 and backcrossed males, and their success in pair formation under natural conditions, an assessment of the influence of various male display traits on female preference could be made, in a manner analogous to the work of Schemske and Bradshaw (1999) on the importance of floral traits in F_2 hybrids on pollinator preference. The advantage of observing pairings in the wild is that mate selection would have taken place under natural environmental conditions at demographic and spatial scales that are implausible in the context of controlled experiments. In a similar way, observations of antagonistic interactions between pure-species males and hybrid males, in the context of competition over females, could be made and the outcomes correlated with the subset of new male displays in the hybrid males. Such observations would permit the analysis of the influence of male display traits on the outcome of male-male competition.

MEASURING CRYPTIC SEXUAL SELECTION   A third way in which recent hybridization may be useful is in the use of genomic analysis to test hypotheses about sexual selection. For example, the chicken genome has recently been sequenced, and with the reduced cost and increased speed of genome-wide sequencing, the mallard duck's genome should become available in the near future. We currently have substantial evidence in invertebrates that seminal fluid proteins evolve rapidly via positive Darwinian selection and that they play an important role in postcopulatory sexual selection. But, surprisingly, no experiments have been done to look for the influence of seminal fluid proteins in vertebrates. In cases in which mallards are hybridizing with their close relatives, the opportunity arises to look for introgression on a gene-class by gene-class basis. For example, an EST screen (i.e., an assay of expressed sequence tags, which are cDNA copies of the mRNA that is expressed within specific cell types) of tissues producing seminal fluid could be done to identify genes that code for seminal fluid proteins in the mallard, as was done previously in *Drosophila* species (Swanson et al. 2001). A similar EST screen could be done for another, nonreproductive tissue (e.g., liver) as a control. Next, PCR could be used to screen for introgressions, and the level of introgression of seminal fluid genes could be compared with that of control genes. If seminal fluid genes diverged between species at a faster rate, and if they also introgressed at a higher rate, this finding would support the hypothesis that seminal fluid genes are important in sexual selection in at least one group of vertebrates. By identifying which seminal fluid genes rapidly introgress from the mallard into the native species, candidates for genes involved in the process of sperm competition could be identified. Once this had been done, a variety of cellular and molecular techniques would become available to study the mechanisms by which sperm competition occurs.

Local adaptation in the context of sexual selection

Experimental work with guppies has demonstrated that sexually selected traits can evolve rapidly in response to environmental change (e.g., levels of predation: Endler 1988). Many exotic species have been purposely introduced to multiple locations throughout the world. By comparing the sexual signaling phenotypes among these isolated populations, we may be able to study the evolution of signal-receptor systems in the context of sexual selection. For example, many species of pheasants (e.g., the common pheasant, *Phasianus colchicus*) have been introduced to multiple locations across the globe (Lever 1987). Although humans influence the survival of many populations of these introduced species, there are also many populations that are locally self-sustaining, as opposed to being recurrently reintroduced from hatcheries, in which interactions with predators presumably strongly influence population persistence. In these populations, it should be possible to correlate sexual signals (e.g., the degree of plumage brightness) with predation intensity. Many other characters, such as harem size or age at first reproduction, also could be compared among populations to see which characters respond to environmental change and to look for more general associations between character changes and environmental factors. The advantage of conducting such studies with exotic species, as opposed to native ones, is that we will often have a reference point for what trait characteristics were prior to introduction. This allows for the examination of not only the direction in which traits have evolved, but also the relative speed and degree to which this has occurred—information that may be useful in evaluating the relative strength of sexual and nonsexual selective pressures.

Although it is generally seen as a problem, the fact that many species have been introduced to many locations (e.g., house finches, house sparrows, starlings), can be used to look for consistent associations between environmental conditions and patterns of sexually selected traits. For example, in socially monogamous birds, the proportion of extra-pair fertilizations (EPFs) varies widely among species. Suppose that the level of EPFs among different introduced populations of starlings was found to be negatively correlated with mortality rate of adults during the nesting period. This finding would support the hypothesis that sperm competition is weaker in populations with more intense predation. To see if this is a general phenomenon, the same pattern could be looked for in house sparrows and house finches. If the pattern was consistently found, then this result would support the generality of the relationship. The study of EPFs versus predation rate is but a single example from a long list of possibilities. Species that have been introduced to many locations represent replicated populations that are rife with opportunities for elucidating phenotype-by-environment associations in the context of sexual selection and many other areas of evolutionary study. An example of this approach can be found in the studies by Geof Hill and colleagues (e.g., Badyaev et al. 2000).

One complication of this approach is the possibility of phenotypic plasticity causing the observed variation among different populations of introduced species. Only a common garden experiment can fully resolve genetic versus environmentally induced variation. Nonetheless, comparison of phenotypes among different populations of introduced species would be the first step in identifying adaptation to differing local environmental conditions.

Topic 2: Evolution of Reproductive Isolation

The study of speciation is an area in which theory has far outstripped empirical work (Rice and Hostert 1993; Gavrilets 2003). Introduced species offer an opportunity to help correct this imbalance. The genetic foundation for speciation is the evolution of reproductive isolation. For pragmatic reasons, empirical studies of reproductive isolation have been dominated by laboratory studies (reviewed in Rice and Hostert 1993; Coyne and Orr 1998). This empirical context permits reasonable estimates of post-insemination/prezygotic reproductive isolation, as occurs when incompatibilities between sister taxa prevent proper sperm transport and fertilization. It also permits a reasonable estimate of that part of postzygotic isolation that occurs during embryonic development (e.g., hybrid inviability) or during the maturation of the gonads (e.g., hybrid sterility). However, laboratory studies are poorly suited to measure the postzygotic isolation that is associated with the reduced fitness of hybrids when intermediate phenotypes are poorly adapted to the ecological demands of the environment (Rundle et al. 2003). Another context in which most laboratory studies have been unsatisfactory is understanding the evolution of prezygotic reproductive isolation. For example, polar bears and grizzly bears freely interbreed in zoos, but in nature, differences in habitat preference during the reproductive period preclude interbreeding. This is an obvious example of why prezygotic isolation must be studied under natural conditions to properly assess its importance in speciation.

The introduction of exotic species offers a unique opportunity to study de novo secondary contact between formerly allopatric taxa and investigate, under more natural conditions, the reproductive isolation that occurs at this critical stage in the speciation process. Here we focus on how exotic species invasions can improve our understanding of the relative importance of pre- and postzygotic isolation, the process of reinforcement, and the occurrence of sympatric speciation.

Prezygotic versus postzygotic isolation

An important issue in the study of speciation is the question of whether prezygotic or postzygotic isolation typically evolves more rapidly. Most studies addressing this question have been done under unnatural laboratory conditions in which many forms of environment-dependent prezygotic isolation will not be manifested. By studying the interactions between introduced species and their sister taxa, valuable insights into this question are possible. For example, the domestic cat (*Felis catus*) has been moved to many locations as Europeans have migrated around the globe, and escaped cats frequently establish local feral populations (Lever 1985). Domestic cats have many close relatives with which they can potentially interbreed. By combining a molecular phylogeny of the clade (and estimating divergence times based on neutral variation) with information on the mechanism of reproductive isolation (prezygotic vs. postzygotic), it should be possible to map the form of reproductive isolation onto the phylogeny and thereby evaluate the relative speeds of the evolution of pre- and postzygotic isolation. Although this approach would not provide exact times of the evolution of pre- and postzygotic reproductive isolation, it would provide upper bounds on the time needed for these events to occur. In addition, when prezygotic isolation evolves, it precludes a test for postzygotic isolation, but this problem can be overcome in the laboratory with artificial insemination—an example of combining the analysis of exotics in the field with traditional laboratory experiments. Because domestic cats may have been hybridized with some of their relatives during the domestication process, special care would need to be taken whenever this possibility exists. Nonetheless, the cat example demonstrates how an exotic species that has been introduced to many new locations can be a useful tool in studying the evolution of reproductive isolation.

One problem with the cat example is that cats are promiscuous, predominantly nocturnal, and highly secretive. As a consequence, the measurement of prezygotic isolation in the field would require substantial effort. In contrast, diurnal species and those that form pair bonds would make the distinction between pre- and postzygotic reproductive isolation far easier to measure. Field studies with other widely distributed exotic species could be used to obtain a more general picture of the relative speeds at which pre- and postzygotic reproductive isolation evolve.

Reinforcement and de novo hybrid zones

Another important question in the context of speciation concerns the outcome that occurs when previously allopatric taxa come into secondary contact. There are four possible outcomes: the formerly allopatric taxa will (1) fuse into a single genetic population if little reproductive isolation has evolved during allopatry; (2) remain genetically isolated if complete reproductive isolation has evolved during allopatry, (3) complete the evolution of reproductive isolation via reinforcement if substantial but incomplete reproductive isolation has evolved during allopatry; or (4) evolve a stable hybrid zone. Hybrid zones formed by outcomes 1–3 are ephemeral, transient situations, while outcome 4 is persistent. As a consequence, extant hybrid zones are expected to represent predominantly outcome 4, and little opportunity is available to evolutionary biologists to study the other outcomes.

Species introductions that lead to new secondary contact between previously allopatric, closely related taxa represent a unique opportunity to study secondary contact during the speciation process. Again, consider the mallard duck, which has been introduced to Madagascar, where it hybridizes with the Meller's duck (*Anas melleri*); to New Zealand, where it hybridizes with the grey duck (*A. superciliosa*); to Hawaii, where it hybridizes with the Hawaiian duck (*A. wyvilliana*); and to South Africa, where it hybridizes with the yellow-billed duck (*A. undulata*). In addition, expansion of the mallard into new areas as a result of human-mediated habitat change and establishment of feral mallard populations has led to hybridization between the mallard and the Mexican duck (*A. diazi*), the mottled duck (*A. fulgvigula*), the Pacific black duck (*A. superciliosa*), and the American black duck (*A. rubripes*) (Johnsgard 1974). These replicated new secondary contacts between previously allopatric populations are a disappointment from a conservation standpoint, but given that they exist, they should be mined for evolutionary information. Besides these examples with mallards, there are many other human-initiated de novo secondary contacts between formerly allopatric populations (e.g., the ruddy duck, *Oxyura jamaicensis*, in Europe hybridizing with the more specialized white-headed duck, *O. leucocephala*) that can be profitably studied by evolutionary biologists.

One of the most important hypotheses that can be studied when an introduced species hybridizes with a resident species is the controversial issue of reinforcement—the third option mentioned above (Dobzhansky 1951; Butlin 1987). Reinforcement is the evolution of enhanced prezygotic reproductive isolation in response to the reduced fitness of hybrids (e.g., partial postzygotic reproductive isolation) that occurs when formerly allopatric taxa come into secondary contact. The empirical pattern, that prezygotic reproductive isolation is stronger in regions of sympatry than in regions of allopatry, is well established across many taxa (reviewed in Rice and Hostert 1993). This strikingly regular pattern is consistent with the hypothesis of reinforcement, but unfortunately, it is also equally consistent with other explanations. For example, laboratory studies have shown that it is common for partial prezygotic reproductive isolation to evolve as a correlated response to selection on other, seemingly unrelated traits, such as geotaxis, phototaxis, preference for certain foods, and so on (reviewed in Rice and Hostert 1993). If the presence of a new sister taxon with a similar ecology selects for character divergence in many traits simultaneously, then prezygotic isolation can evolve as a correlated response in a way that has nothing to do with the reinforcement process per se. Therefore, hybridization between exotic and native species creates a unique opportunity to study the evolutionary processes that occur during the initial stages of secondary contact. Although they will be difficult to disentangle, it should be possible in principle to use quantitative genetic techniques to test for the operation of the reinforcement process versus the more general pattern of the evolution of character displacement.

Sympatric speciation via host shifts

Initially, the idea that species could form without a physical barrier enforcing allopatry was considered unlikely (Mayr 1942). However, mathematical models indicate that speciation without allopatry is genetically possible (reviewed in Endler 1977; Gavrilets 2003), while some field studies have observed the evolution of strong but partial reproductive isolation in sympatry (Bush 1969), and some laboratory experiments have produced the evolution of complete reproductive isolation due to divergent selection in sympatry (Thoday and Gibson 1962; Rice and Salt 1988, 1990). We clearly need more empirical information, especially at large spatial and demographic scales, concerning the feasibility of sympatric speciation.

Exotic species introductions offer a unique opportunity to evaluate whether or not sympatric speciation is feasible under natural conditions, especially in the context of host shifts. Sympatric speciation can occur in both exotic and native species after an introduction. For example, the presence of an exotic species can produce divergent selection on native species when different adaptations are needed to exploit the exotic species. If these adaptations produce strong reproductive isolation via pleiotropy (when an allele favored by natural selection fortuitously produces reproductive isolation as an incidental by-product), then theory predicts that sympatric speciation will occur. A classic example of this situation occurred when the European apple (*Malus pumila*) was introduced to North America. A subpopulation of the native hawthorn fly adapted to apple. Many of these adaptations produced, through pleiotropy, reproductive isolation from the subpopulation continuing to utilize native hawthorn (*Crataegus* spp.) (Bush 1969). For example, apple and hawthorn produce fruit at different times, and mating takes place on the fruit of the trees. An adaptation causing the native hawthorn fly to emerge from diapause at a time that matched the peak in apple fruit production would make cross-breeding less likely between subpopulations adapted to apple and hawthorn.

Next, consider sympatric speciation in the exotic species. Exotic species are sometimes introduced to places where there are multiple available niches (e.g., host plants) that require different adaptations. Different host plants may be truly sympatric, when they co-occur in the same region, or they may be spatially separated, but not divided by a physical barrier to dispersal (e.g., parapatric locations). If sympatric speciation is feasible, it should commonly be observed, at least in its incipient stages, when exotic species adapt to multiple host plants (or more generally, to different niches) in a new location. By cataloging host use by introduced species over time (as well as their migration rates among host species, their timing of reproduction, their genetic differentiation, and so on), the feasibility of sympatric speciation can be tested on a scale and time frame that is unobtainable in the context of traditional laboratory and field experiments. Further, the use of herbarium records and/or insect museum collections may provide valuable historical and spatial information to document host shifts and the divergent selection that they produce. Because so many exotic species have been introduced to new locations, we have the opportunity to ask how frequently divergent selection leads to sympatric speciation, and whether there are taxonomic differences in the occurrence of this process.

Topic 3: The Costs and Benefits of Sexual Reproduction

Sexual recombination occurs during at least some part of the life cycle of most species, while a small minority of species (< 1%) reproduce exclusively by asexual means (Stebbins 1949; Williams 1975; White 1978; Maynard Smith 1978). Asexual reproduction has a large demographic advantage in species in which males do not provide resources for their offspring. In this case, consider an asexual subpopulation that is made up of 100% females and a sexual subpopulation that is composed of 50% females. All else being equal, the asexual subpopulation has a 2:1 demographic advantage because it has twice as many reproducing females, and it is expected to rapidly predominate in the population. Another advantage of asexual reproduction is genetic. In clonally reproducing populations, recombination does not break down polymorphic coadapted gene complexes. Further, when some individuals reproduce both sexually and asexually, the clonally produced offspring carry 100% of their mother's genes, instead of the 50% transmitted to sexually produced offspring. Offsetting these advantages of asexuality, the sexual subpopulation can adapt to environmental change more rapidly, it purges harmful mutations more rapidly, and if genes have the appropriate form of epistatic interaction, its equilibrium mutation load is reduced (for a review of these topics see Maynard Smith 1978; Rice 2002). The adaptive significance of sex depends on the net balance of these costs and benefits, which must be empirically determined. Laboratory studies provide evidence that the balance between these advantages and disadvantages frequently favors sexual over clonal reproduction, but larger-scale experiments in more natural settings are needed. However, only a very limited number of naturally occurring scenarios allow these issues to be examined in the field (e.g., Lively and Jokela 2002). We can utilize exotic species to provide such opportunities.

Here we discuss two ways of using exotic species to investigate the adaptive significance of sexual reproduction. First, we consider studies that examine variation in rates of geographic spread in closely related sexual and asexual species. Second, we examine the differential susceptibility of sexual and asexual species to parasites.

Differences in expansion and geographic spread of sexual and asexual exotic species

Asexual reproduction (as well as self-fertilization) has often been cited as a characteristic that should be advantageous in founding new populations (e.g., Baker 1955, 1965). This is because founding populations are often small and highly dispersed and can be initiated by single individuals only if those individuals can reproduce asexually (unless an individual has been previously fertilized or is capable of self-fertilization). This initial demographic advantage of asexual reproduction, as suggested above, should be offset by the genetic advantage of sex whereby recombination allows a sexual lineage to adapt and expand to new environments more rapidly. In theory, then, following establishment, sexually reproducing species should be more likely than obligately asexual species to spread across a landscape (or continent) and occupy a larger geographic range, as well as a greater number of habitats. Here we consider two examples of species pairs within genera (*Cortaderia* and *Spartina*) that differ in their reproductive strategies, but which have been introduced and become naturalized in the same biogeographic regions.

INTRODUCED TERRESTRIAL GRASSES   *Cortaderia* is a genus of perennial grasses distributed primarily in the southern hemisphere (Hickman 1993). Two congeners have been introduced to California: *C. selloana*, a sexually reproducing species, and *C. jubata*, an asexually reproducing one (Lambrinos 2001). Both were first recorded as having naturalized populations in the first half of the twentieth century, and both began to expand their ranges rapidly within California after 1950. By comparing herbarium records and by conducting additional surveys, Lambrinos (2001) was able to show four important differences in the patterns of range expansion of these two species. First, the sexual species, *C. selloana*, expanded its geographic range and number of known localities at over twice the rate of the asexual species. Second, the sexual species, which initially was found in ruderal sites (i.e., those associated with human dwellings or agriculture), eventually came to occupy more non-ruderal sites than ruderal ones, in contrast to the asexual species, which has remained predominately restricted to ruderal sites. Third, the sexual species has come to occupy multiple climatic zones within California, whereas the asexual species has remained restricted to a narrow climatic zone. Fourth, the sexual species has shown strong directional changes in a number of phenotypic characteristics over time, suggesting that significant adaptation to the environment of California has occurred, here too in contrast to the asexual species, which has remained relatively morphologically static since its invasion of California began. Taken as a whole, this evidence is consistent with the hypothesis that sexually reproducing species have an advantage over asexually reproducing species under natural field conditions, where evolutionary adaptability may significantly improve a species' ability to occupy new areas and new habitat types (and by extension, to continue to occupy habitats that experience environmental change). Whether the results observed in California for species of *Cortaderia* are general could be evaluated by conducting additional studies on this same pair of species, which are also naturalized together in New Zealand, South Africa, and Hawaii (Lambrinos 2001).

INTRODUCED MARSH GRASSES   Additional evidence for potential differences in the long-term success of sexual and asexual species (as measured by patterns of range expansion into novel environments) is available from other pairs of naturalized species. Some of the most intriguing evidence comes from *Spartina*, a genus of perennial, salt-tolerant grasses that predominately occupies estuaries and stream banks (Daehler and Strong 1996). Most species of *Spartina* are native to the Americas, with one species native to Europe and the United Kingdom, *S. maritima*. In the early 1800s *S. alterniflora* was introduced to the United Kingdom, where it became naturalized (Raybould et al. 1991). This species has subsequently hybridized with *S. maritima*, giving rise to two new species, *S. anglica* and *S. townsendii* (Raybould et al. 1991; Ayres and Strong 2001); note that the process of speciation via hybridization of existing species is common in nature (e.g., Stebbins 1949; White 1978). The first of these species, *S. anglica*, is sexual, producing seeds and spreading by clonal growth, whereas the second species, *S. townsendii*, is sexually sterile, producing no seeds and spreading only by clonal growth. One or both of these species have subsequently been introduced to many locations around the world: Australia, New Zealand, the Pacific coast of North America, and the mainland of Europe (Lambert 1964). Both species were introduced and became established in New Zealand during the twentieth century (Partridge 1987). The asexual species, *S. townsendii*, was apparently introduced first, becoming established at several locations in New Zealand by the 1920s (Partridge 1987). The sexual (and seed-producing) species, *S. anglica*, did not become established at multiple sites until the 1930s and 1940s, but by the 1950s it had largely displaced all *S. townsendii* populations, such that only a single known population of *S. townsendii* remains in New Zealand (Partridge 1987). This happened in spite of the extensive areas once occupied by *S. townsendii*, such as the approximately 40-ha area it once occupied in an estuary near Invercargill (Partridge 1987). *Spartina anglica* has also come to occupy many more sites than were initially occupied by *S. townsendii* (although this may have much to do with relative numbers of introduction attempts), such that *S. anglica* now occupies a much greater percentage of available habitats in New Zealand (Partridge 1987). Further evidence for competitive differences between these two species, or at least differential ability to persist and occupy new areas, could be acquired by comparing other regions where both species have been known to co-occur, such as the Wadden Sea in Europe (Dijkema and Wolff 1983). The evidence available from New Zealand, however, is once again consistent with the hypothesis that sexual reproduction provides significant advantages in colonizing new areas, as well as in maintaining viable populations in areas currently occupied. To strengthen this conclusion, however, many additional paired species invasions will need to be examined; this should be readily accomplished in the future as existing species invasions are better studied and additional species invasions occur.

Differences in the relative importance of species interactions for sexual and asexual species

Competition, predation, parasitism, and other pairwise species interactions play a significant role in delimiting species' local and geographic distributions (Brown and Lomolino 1998). These ecological interactions are thought to mediate long-term coevolutionary processes and the relative success of sexual and asexual species. In particular, the Red Queen hypothesis predicts that asexual species should be at a significant disadvantage in competing with sexual species over the long term because sexual species counter-evolve faster in response to evolution by their enemies (Van Valen 1973). If this is true, then sexual species should be able to displace asexual species (as suggested above by *Spartina* distributions in New Zealand), and sexual species should rapidly evolve in response to novel parasites, pathogens, or diseases. We explore these two predictions here by considering the details of an invasive milieu: gecko invasions on Pacific islands.

A unisexual gecko, *Lepidodactylus lugubris*, inhabits many islands of the Indian and Pacific oceans (Case et al. 1994). Its distribution may have been affected by the colonization of islands by Polynesians and Melanesians (see references in Radtkey et al. 1995). Regardless of whether this species came by canoe or via "waif" dispersal events aboard "rafts" of vegetation drifting in the sea, its asexual life history was undoubtedly an important contributor to its successful colonization of many islands (where the initial number of colonists reaching any individual island was likely to have been small). More recently, a sexual gecko species, *Hemidactylus frenatus*, has been introduced and has become established on many of these same islands (Case et al. 1994). Where these species meet, the abundance of the asexual species has been drastically reduced. For example, on islands where the sexual species is not present, the abundance of the asexual species is as much as 800% higher in some habitat types (Case et al. 1994). The explanation for the decline of the asexual gecko species in the presence of the sexual species is undoubtedly complex, but three factors appear to be particularly important. First, the sexual species appears to be a superior competitor for the clumped resources (of insects) found in some habitat types (Petren and Case 1996). Second, the fecundity of the asexual species appears to be reduced by proximity to exudates from the sexual species' femoral pores or the odor of their feces (Brown et al. 2002). Third, the two species may be differentially susceptible to parasites (Brown et al. 1995). In this case, however, in contrast to the theoretical expectation of an asexual species' increased vulnerability to the Red Queen process, the sexual species experiences a much greater diversity and prevalence of parasites, pathogens, and other diseases then does the asexual species (Brown et al. 1995). This observation is intriguing because it suggests that an exotic species is successfully invading and displacing a putatively native (or at least long-term resident) species despite a greater parasite load. The degree of harm caused by the parasites, however, would need to be quantified to determine conclusively whether they are affecting competitive interactions between these species. Nevertheless, the observation that the sexual species hosts a greater diversity of parasite species within the same geographic region as the asexual species is not consistent with the expectation that parasites are having a stronger negative effect on the asexual species. So, in summary, while this example is consistent with the general expectation that sexual species should be able to displace asexual species due to differences in the outcome of intraspecific species interactions, this example is inconsistent with the expectation that asexual species should be more vulnerable to parasites.

Clearly the patterns observed in the examples described here are not convincing evidence by themselves for an advantage of sexual reproduction because there are so many uncontrolled factors involved. However, the advantage of studying exotic species introductions is that there are so many, and undoubtedly more to come. With these numbers, we have the opportunity to look for consistent patterns that will allow us to "average over" the uncontrolled variables that are intrinsic to any single study.

Conclusions

Species introductions lead to punctuated episodes of adaptation to new environments, which provide many opportunities to study basic evolutionary questions. Exotic species introductions will not replace controlled manipulative experiments in the study of evolution, but they do expand the domain of empirical studies to larger demographic and spatial scales. By pooling inferences across multiple invasion locations and multiple taxonomic comparisons, we should be able to test for general patterns that will provide strong empirical evidence on many basic questions in evolutionary biology. Although species introductions have many unfortunate consequences, their numbers will inevitably grow. We can take advantage of this misfortune, however, if we use these introductions as a powerful model system to provide useful tests of many evolutionary hypotheses that would not otherwise be possible.

Acknowledgments

This work was conducted as a part of the "Exotic Species: A Source of Insight into Ecology, Evolution, and Biogeography" Working Group supported by the National Center for Ecological Analysis and Synthesis, a center funded by the National Science Foundation (grant #DEB-0072909), the University of California, and the Santa Barbara campus. In addition, W. Rice was supported by NSF grants #DEB-0128780 and #DEB-0410112.

Literature Cited

Ayres, D. R., and D. R. Strong. 2001. Origin and genetic diversity of *Spartina anglica* (Poaceae) using nuclear DNA markers. American Journal of Botany 88:1863–1867.

Badyaev, A.V., G. E. Hill, A. M. Stoehr, P. M. Nolan, and K. J. McGraw. 2000. The evolution of sexual dimorphism in the house finch: II. Population divergence in relation to local selection. Evolution 54:2134–2144

Baker, H. G. 1955. Self-compatibility and establishment after "long-distance" dispersal. Evolution 9:347–349.

Baker, H. G. 1965. Characteristics and modes of origin of weeds. In H. G. Baker and G. L. Stebbins, eds. *The genetics of colonizing species*, pp. 141–172. Academic Press, London.

Basolo, A. L. 1990. Female preference predates the evolution of the sword in swordtail. Science 250:808–810.

Brown, J. H., and M. V. Lomolino. 1998. *Biogeography*, 2nd Ed. Sinauer Associates, Sunderland, MA.

Brown, S. G., R. Lebrun, J. Yamasaki, and D. Ishii-Thoene. 2002. Indirect competition between a resident unisexual and an invading bisexual gecko. Behavior 139:1161–1173.

Brown, S. G., S. Kwan, and S. Shero. 1995. The parasitic theory of sexual reproduction: parasitism in unisexual and bisexual geckos. Proceedings of the Royal Society of London B 260:317–320.

Burley, N. 1986. Sexual selection for aesthetic traits in species with biparental care. American Naturalist 127:415–445.

Bush, G. L. 1969. Sympatric host race formation and speciation in frugivorous flies of genus *Rhagoletis* (Diptera, Tephritidae) Evolution 23:237–251.

Butlin, R. K. 1987. Species, speciation and reinforcement. American Naturalist 130:461–464.

Callaway, R. M., and E. T. Aschehoug. 2000. Invasive plants versus their new and old neighbors: a mechanism for exotic invasion. Science 290:521–533.

Callaway, R. M., G. C. Thelen, A. Rodriquez, and W. E. Holben. 2004. Release from inhibitory soil biota in Europe may promote exotic plant invasion in North America. Nature 427:731–733.

Case, T. J., D. T. Bolger, and K. Petren. 1994. Invasions and competitive displacement among house geckos in the tropical Pacific. Ecology 75:464–477.

Coyne, J. A., and H. A. Orr. 1998. The evolutionary genetics of speciation. Philosophical Transactions of the Royal Society of London B 353:287–305.

Daehler, C. C., and D. R. Strong. 1996. Status, prediction and prevention of introduced cordgrass *Spartina* spp. invasions in Pacific estuaries, USA. Biological Conservation 78:51–58.

Darwin, C. 1859. *On the origin of species by means of natural selection or the preservation of favored races in the struggle for life*. John Murray, London.

Dijkema, K. S., and W. J. Wolff. 1983. *Flora and vegetation of the Wadden Sea islands and coastal areas*. A. A. Balkema, Rotterdam.

Dobzhansky, T. 1951. *Genetics and the origin of species*, 3rd Edition. Columbia University Press, New York.

Dobzhansky, T. 1970. *Genetics of the evolutionary process*: Columbia University Press, New York.

Endler, J. A. 1977. *Geographic variation, speciation, and clines*. Princeton University Press, Princeton, NJ.

Endler, J. A. 1988. Sexual selection and predation risk in guppies. Nature 332:593–594.

Fisher, R. A. 1930. *The genetical theory of natural selection*. Clarendon Press, Oxford.

Gavrilets, S. 2003. Perspective: Models of speciation: what have we learned in 40 years? Evolution 57:2197–2215.

Gilchrist, G. W., R. B. Huey, J. Balanyà, M. Pascual, and L. Serra. 2004. A time series of evolution in action: a latitudinal cline in wing size in South American *Drosophila subobscura*. Evolution 58:768–780.

Hickman, J. C., ed. 1993. *The Jepson Manual: higher plants of California*. University of California Press, Berkeley.

Huey, R. B., G. W. Gilchrist, M. L. Carlson, D. Berrigan, and L. Serra. 2000. Rapid evolution of a geographic cline in size in an introduced fly. Science 287:308–309.

Johnsgard, P. A. 1974. *Hybridization*. In D. O. Hyde, ed. *Raising wild ducks in captivity*, pp. 142–146. E. P. Dutton and Company, New York.

Lamarck, J. P. B. A. 1809. Philosophie zoologique, ou, Exposition des considérations relative à l'histoire naturelle des animaux. Chez Dentu, Paris.

Lambert, J. M. 1964. The *Spartina* story. Nature 204:1136–1138.

Lambrinos, J. G. 2001. The expansion history of a sexual and asexual species of Cortaderia in California, USA. Journal of Ecology 89:88–98.

Lever, C. 1985. *Naturalized mammals of the world*. Longman, Essex.

Lever, C. 1987. *Naturalized birds of the world*. Longman, Essex.

Lively, C. M., and J. Jokela. 2002. Temporal and spatial distributions of parasites and sex in a freshwater snail. Evolutionary Ecology Research 4:219–226.

Maynard Smith, J. 1978. *The evolution of sex*. Cambridge University Press, Cambridge.

Mayr, E. 1942. *Systematics and the origin of species*. Columbia University Press, New York.

McClintock W. J., and G. W. Uetz. 1996. Female choice and pre-existing bias: visual cues during courtship in two *Schizocosa* wolf spiders (Araneae: Lycosidae). Animal Behavior 52:167–181.

Partridge, T. R. 1987. *Spartina* in New Zealand. New Zealand Journal of Botany 25:567–575.

Petren, K., and T. J. Case. 1996. An experimental demonstration of exploitation competition in an ongoing invasion. Ecology 77:118–132.

Radtkey, R. R., S. C. Donnellan, R. N. Fisher, C. Moritz, K. A. Hanley, and T. J. Case. 1995. When species collide: the origin and spread of an asexual species of gecko. Proceedings of the Royal Society of London B 259:145–152.

Raybould, A. F., A. J. Gray, M. J. Lawrence, and D. F. Marshall. 1991. The evolution of *Spartina anglica* C. E. Hubbard (Gramineae): origin and genetic variability. Biological Journal of the Linnaean Society 43:111–126.

Rice, R. R., and G. W. Salt. 1990. The evolution of reproductive isolation as a correlated character under sympatric conditions—experimental evidence. Evolution 44:1140–1152.

Rice, R. R., and G. W. Salt. 1988. Speciation via disruptive selection on habitat preference–experimental-evidence. American Naturalist 131:911–917.

Rice W. R., and E. E. Hostert. 1993. Laboratory experiments on speciation–what have we learned in 40 years. Evolution 47:1637–1653.

Rice, W. R. 2002. Experimental tests of the adaptive significance of sexual recombination. Nature Reviews Genetics 3:214–251.

Rosenfield, J. A., and A. Kodric-Brown. 2003. Sexual selection promotes hybridization between Pecos pupfish, *Cyprinodon pecosensis* and sheepshead minnow, *C. variegatus*. Journal of Evolutionary Biology 16:595–606.

Rundle, H. D., S. M. Vamosi, and D. Schluter. 2003. Experimental test of predation's effect on divergent selection during character displacement in sticklebacks. Proceedings of the National Academy of Sciences USA 100:14943–14948.

Rhymer, J. M., M. J. Williams, and M. J. Braun. 1994. Mitochondrial analysis of gene flow between New Zealand mallards (*Anas platyrhynchos*) and grey-ducks (*A. superciliosa*). Auk 111:970–978.

Schemske, D. W., and H. D. Bradshaw. 1999. Pollinator preference and the evolution of floral traits in monkeyflowers (*Mimulus*). Proceedings of the National Academy of Sciences USA 96:11910–11915.

Stebbins, G. L., Jr. 1949. Asexual reproduction in relation to plant evolution. Evolution 3:98–101.

Swanson, W. J., A. G. Clark, H. M. Waldrip-Dail, M. F. Wolfner, and C. F. Aquadro. 2001. Evolutionary EST analysis identifies rapidly evolving male reproductive proteins in *Drosophila*. Proceedings of the National Academy of Sciences USA 98:7375–7379.

Thoday, J. M., and J. B. Gibson. 1962. Isolation by disruptive selection. Nature 193:1164–1166.

Van Valen, L. 1973. A new evolutionary law. Evolutionary Theory 1:1–30.

Vivanco, J. M., H. P. Bais, F. R. Stermitz, G. C. Thelen, and R. M. Callaway. 2004. Biogeographical variation In community response to root allelochemistry: novel weapons and exotic invasion. Ecology Letters 7:285–292.

Williams, G. G. 1975. *Sex and evolution*. Princeton University Press, Princeton, NJ.

White, M. J. D. 1978. *Modes of speciation*. W. H. Freeman and Company, San Francisco.

Wright, S. 1969. *Evolution and the genetics of populations, Vol 1: Genetics and biometric foundations*. University of Chicago Press, Chicago.

Wolfner, M. F. 2002. The gifts that keep on giving: physiological functions and evolutionary dynamics of male seminal proteins in *Drosophila*. Heredity 88:85–93.

PART III

Insights into

BIOGEOGRAPHY

Julie L. Lockwood

Brown and Lomolino (1998) define biogeography as a science that attempts to document and understand spatial patterns in biodiversity. The roots of biogeography go back quite a long way, with perhaps a first flush of (printed) thought appearing during the age of European exploration (1600 to 1800). There is nothing like reading the accounts of the voyages of James Cook or Alexander von Humboldt to illustrate how rapidly our understanding of biogeography evolved during this period. The stories and specimens that came back from these and similar voyages clearly were incorporated into Darwin's thinking on the origin of species. And Darwin often resorts to the examination of non-native species to illustrate his arguments for descent through natural selection. It was the work of Alfred Russel Wallace that laid much of the foundation for modern biogeography, however, as his work more explicitly considered spatial patterns in species distribution and how these were produced. Interestingly, though, Wallace tended to ignore non-native species when relaying his observations and making his evolutionary arguments.

The on-again, off-again affair between biogeography and invasions

The modern renaissance of biogeography, perhaps culminating in the 1967 publication of MacArthur and Wilson's *The Theory of Island Biogeography*, for the most part did not consider non-native species. This is true despite the publication of Charles Elton's seminal book on species invasions, *The Ecology of Invasions by Animals and Plants*, at about the same time (1958). There is an undeniable similarity between the act of colonization as envisioned by MacArthur and Wilson and the act of non-native species establishment or geographic range expansion described by Elton. Elton plainly takes a stance on the wisdom of introducing non-native species and he highlights several ecologically and economically damaging examples. This is the time frame in which invasion ecology began to disassociate from its parent disciplines, with invasion ecologists tending to work towards the science of predicting invaders and managing their impacts rather than viewing them as reflecting (or testing) basic tenets in ecology and evolution (Davis et al. 2001; Davis 2005). Perhaps the lack of consideration of non-natives within island biogeography is due in part to the tone with which non-natives are discussed in Elton's work.

The explicit consideration of non-native species within the context of biogeography did not surface again (at least not widely) until the 1980s, with the publication of the SCOPE (Scientific Committee on Problems of the Environment) volumes on invasion ecology (e.g., Mooney and Drake 1986). These volumes were much more attuned to the interplay between biogeography and non-native species than previous work (Davis 2005). Much of the biogeographic literature from this time endeavors to describe the distribution of non-natives across regions and continents, and when differences are found, to explain them using ecological mechanisms (e.g., differences in native species richness or history of human disturbance). This situation was relatively short-lived, however. Most of today's discussions of non-native species invasions take a decidedly Eltonian approach (Davis 2005). Thus, we stand at a point in time where the affair between biogeography and invasions is decidedly "off." The chapters in this section can therefore be considered a concerted effort to rediscover the wisdom of Darwin and to bridge the gap between the "pure" studies of biogeography and the "applied" study of non-native species invasions.

What does this section do that the others don't?

Biogeography is unique in its explicit focus on biological patterns that are manifest over large spatial scales, and on processes that are believed to operate over long time frames. Thus, biogeographers find themselves synthesizing knowledge about—at the least—evolution, ecology, geography, and geology

(see Lomolino and Heany 2004). This leaves the authors of this section with the task of covering much of the same conceptual ground as the chapters that have come before while emphasizing how these concepts shape diversity at larger spatial scales, or with an explicit spatial component. In particular, these chapters tackle a fundamental series of questions related to understanding how species have come to be distributed as they are today.

Despite the continued, if often frustrating, quest to find ways to minimize the flow of harmful non-native species, we broadly understand why they are distributed where they are now. Human society wanted that particular species in that particular location, or else wanted a product that covertly carried non-native individuals in that location. Much more mysterious is how *native* species came to live where they are now, what keeps them from expanding their ranges, and what drives them toward extinction (natural and human-influenced). The question is whether we can use the latter to understand the former. The editors and contributing authors of this volume believe we can. In fact, it is worth noting that the existence of geographic patterns in species' distribution is what allows us to define what is a non-native species in the first place.

For the most part, native species obey biological "laws" that pin them to a particular location on the Earth's surface. It has long been a problem for biogeographers that we cannot manipulate these laws (or what we believe to be the laws) and see what happens. There is certainly much to be gained from looking at how species assort themselves after major paleoecological events such as biotic interchanges. Chapter 12 by Vermeij reminds us that the present distribution of native species is a product of several previous "invasion" events. The study of these events sheds light on fundamental biogeographic questions, such as what species and site characteristics favor range expansion; the long-term influence of new arrivals on community structure; and the distribution of species after large-scale species interchanges. Non-native species, also by definition, break biological "laws" that operate today and operated in the geologic past. Thus, they allow us the unprecedented opportunity to document the resultant effects of invasion on the spatial arrangement of biodiversity once these laws are broken.

The study of non-native species as the "experimental" arm of biogeography

Any student of biogeography is well aware of the tendency of the science to collect hypotheses that explain a well-substantiated diversity pattern. For example, there are 10 non-mutually exclusive hypotheses as to what produces the Island Rule (Dayan and Simberloff 1998). Similarly, there are more than 30 hypotheses that can explain the latitudinal gradient in species richness (Sax and Gaines 2005). The same is true, albeit not quite as dramatically, for Cope's Rule, Bergmann's Rule, Rapoport's Rule, and a bunch of patterns that are not yet named after a pioneer in the field. A skeptic would suggest that you have not really made an impression within biogeography if you haven't come up with an alternative explanation for a rule.

What makes biogeography so prone to hypothesis collecting? There are no doubt many contributing factors; however, one overriding issue is the inability of biogeographers to conduct experiments at the appropriate spatial scale that can unequivocally rule out one or more competing hypotheses. It is simply not possible to take 100 elephants out of Africa and place 50 of them on a set of islands with predators while placing the other 50 on another set of islands without predators, and then document whether the elephants present only on predator-free islands evolved toward smaller body sizes as predicted by one of the hypotheses for the Island Rule. Or is it? Perhaps elephants are stretch; however, human activities have done a fine job of distributing some species like house sparrows (*Passer domesticus*), house mice (*Mus musculus domesticus*), and fruit flies (*Drosophila subobscura*) across a variety of ecological and biophysical gradients. Furthermore, we know from the work of Huey and colleagues (Chapter 6) that these species have had plenty of time to evolve toward local optima in body size, or toward another biophysical fitness peak. The study of non-native species can thus provide us with considerable insight into what produces biogeographic patterns such as the Island Rule if we were only to pay attention to them. In this way, the study of non-native species can provide an "experimental" approach to understanding the processes that produce biogeographic patterns (e.g., Sax 2001).

The authors of the chapters in this section delve deeply into this notion. What limits local species richness, and how is endemism created and maintained? What regulates a species' abundance-distribution pattern? Is a species that is abundant in one habitat likely to be abundant in all of them? How does geography modify the coevolutionary dynamics of two species? These are all fundamental questions about biogeography. Some of the following chapters strive toward providing answers, while others illustrate that there are so few differences between non-native and native species in regard to these and similar questions that the study of non-natives can shed light on fundamental processes that produce biogeographic patterns.

Proceed with caution

The study of non-native species opens up a lot of doors to biogeographers, however it does not open them quite as wide as we would like, and some caution is in order. There are a lot of non-native species, and they are well distributed across the surface of the Earth. However, they are not distributed randomly (Blackburn and Duncan 2001a,b), nor are all species equally likely to have had the "opportunity" to be introduced (Lockwood 1999). Some of the questions biogeographers would like to ask depend on elephants—not house mice—actually being the species introduced. Or, a student of biogeography may have isolated the right non-native species to answer their question, but this species was not introduced to the right places.

This caution seems self-evident; however it has a sneaky influence on the science we conduct and the conclusions we draw (Lonsdale 1999). Most of these issues arise from a recognition that invasion is a multistage process (Kolar and Lodge 2002). Before a species can be labeled non-native or invasive, it must first be transported out of its native range, be released into a novel location, and establish a self-sustaining population there. Thus, biogeographers (and most ecologists and evolutionary biologists) are stepping into the picture after a lot of the "action" is over. A list of non-native species taken at one location is biased by a set of structuring forces evinced only in the early invasion stages. Similarly, each location on Earth has had a unique set of species transported to it and

released there. This non-random pattern of transport affects the overall number of non-natives across locations, as well as the fate of non-native populations of the same species that were introduced across several locations. We are often blind to these events, and thus we tend to discount them (Lockwood et al. 2005). It is important to remember, however, that the use of non-native species to explore fundamental aspects of biogeography is constrained by the fact that we are given as study species (and systems) only what human commerce has granted us. And human commerce is fickle.

Where to next?

Biogeography remains a relevant science because it calls forth facts and patterns that are only visible when viewed from a large-scale, spatially explicit perspective. The study of non-native species within biogeography brings with it an opportunity to reinvigorate dormant theories, call attention to new ways of exploring mechanisms that produce large-scale ecological and evolutionary patterns, and push beyond the stale debates of yesterday. In turn, the traditional methods of biogeography bring to invasion ecologists a fresh approach to understanding why some species cause ecological and economic harm while others do not (e.g., Daehler 1998; Vermeij, this volume). Each of the authors in this section seek to inspire such cross-fertilization between these fields of study in the expectation that the hybrid offspring produced will be more fit than their parents.

Literature Cited

Blackburn, T. M., and R. P. Duncan. 2001a. Establishment patterns of exotic birds are constrained by non-random patterns in introduction. Journal of Biogeography 28:927-939.

Blackburn. T. M., and R. P. Duncan. 2001b. Determinants of establishment success in introduced birds. Nature 414:195-197.

Brown, J. H., and M. V. Lomolino. 1998. *Biogeography*, 2nd Ed. Sinauer Associates, Sunderland, MA.

Daehler, C. 1998. The taxonomic distribution of invasive angiosperm plants: ecological insights and comparison to agricultural weeds. Biological Conservation 84:167-180.

Davis, M. A., K. Thompson, and J. P. Grime 2001. Charles S. Elton and the dissociation of invasion ecology from the rest of ecology. Diversity and Distributions 7:97-102.

Davis, M. A. 2005. Invasion biology 1958–2004: The pursuit of science and conservation. In M. W. Cadotte, S.M. McMahon, and T. Fukami, eds. *Conceptual ecology and invasions biology: Reciprocal approaches to nature*. Kluwer Academic Press, New York. In press.

Dayan, T., and D. Simberloff. 1998. Size patterns among competitors: ecological character displacement and character release in mammals, with special reference to island populations. Mammal Review 28:99-124.

Elton, C. S. 1958. *The ecology of invasion by animals and plants*. University of Chicago Press, Chicago.

Kolar, C. S., and D. M. Lodge 2002. Ecological predictions and risk assessment for alien fishes in North America. Science 298:1233-1236.

Lockwood, J. L. 1999. Using taxonomy to predict success among introduced avifauna: the importance of transport and establishment. Conservation Biology 13:560-567.

Lockwood, J. L., P. Cassey, and T. M. Blackburn. 2005. The role of propagule pressure in explaining species invasions. Trends in Ecology and Evolution 20:223-228.

Lomolino, M. V., and L. R. Heaney. 2004. *Frontiers of biogeography*. Sinauer Associates, Sunderland, MA.

Lonsdale, W. M. 1999. Global patterns of plant invasions and the concept of invasibility. Ecology 89:1522-1536.

MacArthur, R. H., and E. O. Wilson 1967. *The theory of island biogeography*. Princeton University Press, Princeton, NJ.

Mooney, H., and J. A. Drake. 1986. *Ecology of biological invasions in North America and Hawaii*. Springer-Verlag, New York.

Sax, D. F. 2001. Latitudinal gradients and geographic ranges of exotic species: implications for biogeography. Journal of Biogeography 28:139-150.

Sax, D. F., and S. D. Gaines. 2005. The biogeography of naturalized species and the species-area relationship: reciprocal insights to biogeography and invasion biology. In M. W. Cadotte, S. M. McMahon, and T. Fukami, eds. *Conceptual ecology and invasions biology: Reciprocal approaches to nature*. Kluwer Academic Press, New York.

12

Invasion as Expectation

A HISTORICAL FACT OF LIFE

Geerat J. Vermeij

Species invasion is a conspicuous phenomenon in today's human-dominated biosphere, but theory and evidence show that it is as old as the existence of species on earth. Barriers limiting the range of species appear and disappear as the configuration of oceans and landmasses changes through time and as the means and capacities of dispersal evolve. Moreover, most species undergo range expansion during their history. Invasion is therefore an expected and widespread phenomenon in the history of life, and not an aberration limited to our time. Biotic interchange—the movement of species between geographically distinct biotas following the reduction or elimination of the barrier between them—is a common form of invasion. A worldwide review of the fossil record of the past 25 million years, together with other evidence, indicates that biotic interchange has often been highly one-sided, with one biota acting as donor and the other as recipient for invaders. Donor biotas tend to be larger in extent, richer in species, and better endowed with species of high competitive, defensive, and reproductive performance than recipient biotas. Although the short-term consequences of invasion are generally disruptive to recipient ecosystems in that invasion homogenizes biotas, interferes with established patterns of interaction among native species, and causes extinction (mainly in island-like ecosystems), these effects are either muted or reversed in the long term by evolution in the recipient biota. Moreover, because biotic interchange involves the expansion of species of high performance, it therefore raises competitive standards of performance and probably speeds up recovery from mass extinction events. Human-caused invasion may not differ greatly from natural biotic interchange in its immediate or long-term consequences.

Introduction

Practically every ecosystem with which humans come into regular contact contains species that we have brought there. These alien species, introduced deliberately or by accident, have often profoundly changed the ecosystems in which they have become established. Some ecologists consider human-assisted species invasion to be one of the great environmental threats facing the biosphere (Vitousek et al. 1997). An entire discipline—invasion biology—has grown up to study and combat the problem.

There can be no doubt that many species are where they are today because of human help, but neither can there be any doubt that invasion by alien species is as old as the emergence of species. That species invasion has been part of ecological history since extremely ancient times follows from two very common circumstances. First, many species arise as small, geographically or ecologically highly circumscribed, daughter populations, which subsequently spread, invading regions and ecosystems. Second, geographic and ecological barriers that facilitate the formation of species appear, disappear, and change their positions, with the result that established species can expand their ranges into regions or ecosystems not previously occupied by those species. Knowledge of past distributions as chronicled by fossils, coupled with ancestor-descendant and sister-group studies, provides overwhelming evidence that species have dispersed to places where those species did not originate.

This rich, deep historical record must be tapped to place human-assisted species invasion in a broader context of ecological change and adaptation. Concern with human-introduced species generally centers on the harm that alien species have caused in invaded ecosystems, but by necessity we have been able to study the phenomenon of human-assisted species invasion only in the short term, on a time scale over which assimilation and adaptation of the foreign species and adaptation of the native species in the invaded community have not been fully realized. The study of past invasions in the history of life therefore offers opportunities to examine the long-term consequences of invasion and to examine the entire phenomenon of species invasion more objectively and dispassionately than is the rule in much of current invasion biology.

The comparative approach permitted by the study of past invasions leads to many questions whose answers can illuminate present-day invasions. Are there ecosystems or regions that disproportionately produce invading taxa, and if so, what do they have in common? Are there ecosystems or regions that disproportionately receive and assimilate invaders, and if so, what are their characteristics? How do invaders affect the lives and the evolutionary (selective) environments of species in recipient communities? Perhaps most pressingly, is invasion in the long run as destructive to ecosystems as it often appears to be in the short run in the biosphere today? Finally, and perhaps most crucially, are invasions of the past reliable guides to the phenomenology of human-caused invasion?

Fully satisfactory answers to these questions are not yet within reach, but I believe that enough data are available in the scattered literature to sketch out some tentative conclusions. After discussing the general features of the process of invasion and of expanding species, I turn to the phenomenology of past invasion by considering known cases from the last 65 million years of earth history, the geologic interval corresponding to the Cenozoic era. Based on this survey, I attempt some answers to the questions I posed above. Against common intuition, and more importantly, against most people's sentiments, I argue that past invasions have over the long term led to a globalization of high performance standards and have bolstered, not weakened, overall resistance to the kinds of disturbances that cause widespread extinction and ecological disruption. Whether human-assisted species introductions will continue this trend is debatable, in part because we have changed the rules for how species transcend barriers; but the long view suggests that invasion is neither a new nor necessarily a destructive phenomenon in the biosphere.

Arrival and Establishment

The process of invasion involves at least two phases: (1) dispersal from the donor region or ecosystem to the recipient site; and (2) establishment of a self-sustaining population in the recipient region or ecosystem (see Vermeij 1996). These two steps call for potentially quite different characteristics of the invading species, and they are influenced by physical and biological conditions in both the donor and recipient biotas.

Extension of a species' range occurs under two distinct classes of circumstances, whose effects on which types of species can spread are quite different: (1) the evolved ability of individuals at one or more points in the life cycle to cross an inhospitable barrier and (2) the physical elimination or reduction of a barrier that previously set limits to the species' distribution.

The first class of circumstances comprises phenotypic characteristics that enable species to spread from one favorable site to another across hostile habitats. Many terrestrial plants as well as marine species raft as adults, juvenile stages, or seeds and eggs on floating objects. Such spread has been documented for some tropical trees (e.g., Dick et al. 2003), high-latitude marine invertebrates (Highsmith 1985), and even tropical reef corals growing on floating pumice (Jokiel 1989, 1990). Others are ferried by animals such as birds and humans. For example, migrating birds are evidently responsible for the disjunct distributions of the marine intertidal littorinid gastropod *Littorina saxatilis* in Europe and southern Africa (Reid 1996) and of some freshwater mollusks in Europe and North America (Wesselingh et al. 1999). Birds flying with the wind around the southern hemisphere are probably responsible for close similarities among insular species of the plant genus *Peperomia*, whose sticky fruits are well suited to cling to feathers (Valdebenito et al. 1990). Winds transport spores, seeds, adult insects, and even marine diatoms over long distances (Polis et al. 1997; Nathan et al. 2002; Muñoz et al. 2004).

Barrier-crossing abilities fall into two categories, which may be broadly circumscribed as passive and active. Passive means involve resisting the inclement conditions of the barrier by shutting down metabolism and activity through suspended animation in a spore, seed, egg, cyst, or dormant adult. Active means include the ability to function by using up large reserves of food or by feeding during the crossing. This is what the pelagic larvae of many bottom-dwelling marine animals do; it is what humans do while they transport themselves and many other species in ships, airplanes, and land vehicles.

The second class of circumstances enabling species to spread comprises the physical removal of barriers. An example is the formation of a land bridge between North and South America during the Pliocene, which allowed many terrestrial plants and animals to expand from one continent to the other. Similarly, the establishment of a marine corridor between the Mediterranean and Red seas through the Suez Canal enabled many marine species from the Red Sea to enter the Mediterranean. As changes in sea level and movements of tectonic plates cause landmasses to collide and separate, land and sea barriers are continually created and dissolved. Given the enormous sea level fluctuations and the tectonic tumult that have characterized the earth's physical history, barriers can be expected to arise, disappear, and shift frequently. Species spreading as barriers of this kind disappear can be expected to do so as active individuals and should not be subject to the metabolic constraints that apply to passively dispersing species that are able to surmount existing barriers.

From the perspective of selectivity of invasion, it is important to note that competitively dominant terrestrial organisms with large effects on other species are generally poor passive dispersers, but good active dispersers. Prominent in this category are mammals, social insects, and large animals in general. These organisms, as well as large forest trees, are unable to cross wide hostile barriers, but can and do spread rapidly across newly available suitable terrain. This limitation does not apply to marine competitive dominants because many of these species disperse as active, pelagically feeding individuals (Vermeij 2004a).

There are interesting, but probably rare, exceptions to these generalizations. Compelling phylogenetic evidence indicates that some rainforest trees and some land mammals (rodents and primates) crossed the Atlantic between Africa and South America during the Oligocene to Early Miocene, probably by rafting (Renner et al. 2001; Dick et al. 2003). Even if invasions involving such high-energy species crossing hostile habitats are rare, they may prove to have important implications and therefore deserve further scrutiny.

The ability of a species to disperse either actively or passively to a place not previously occupied is a necessary condition for the successful spread of that species, but it is not a sufficient one. Arrivals must be able to establish a self-sustaining population—one whose persistence does not depend exclusively on continual recruitment from the source population. Establishment depends on conditions in the recipient biota, both biological and physical, as well as on the properties of arrivals. On an evolutionary time scale, the newly established foreign species have the potential to affect the selective regime of native species in the recipient biota, and the recipient biota imposes a selective regime on the new arrivals that may differ substantially from the evolutionary conditions prevailing in the donor biota (Vermeij 1996).

Attributes that enhance the successful establishment of invading species include (1) absence of powerful enemies (predators, competitors, disease agents, and parasites) in the recipient biota, either because these enemies were never present there or because other agencies had removed them before the invaders arrived (Darwin 1859; Vermeij 1991a); (2) unique ways of life or modes of exploitation in the invading species, without parallels in the recipient biota (Sax and Brown 2000); (3) the shedding of parasites and disease agents by colonizing individuals, so that invaders are free of the pests that depress their populations in the donor biota (Klironomos 2002; Wolfe 2002; Mitchell and Power 2003; Torchin et al. 2003); (4) retention of symbionts in the invading species, so that invaders can acquire resources as effectively with the aid of symbionts in the recipient biota as in the donor biota (Richardson et al. 2000); (5) higher competitive performance of the invader than of any species in the recipient biota; and (6) similar physical and chemical environments in the donor and recipient ecosystems, or modifications to the recipient environment that make the latter more compatible with the evolved physical and chemical tolerances of invading species (Brown 1989).

Although this is not the place to dwell on these six attributes in detail, one conclusion emerging from a synthesis of these attributes is worth emphasizing: the ecological roles and levels of performance of successfully established invading species typically are greater in the recipient biota than in the donor biota. If parasites disperse less easily than their hosts, and if enemies in the recipient biota individually and collectively offer less resistance to invaders than enemies in the donor biota do to the parent populations of those same invaders, then the restraints that limited the invading species in their ancestral homes no longer apply in the invaded biota. It is this higher performance level, often expressed as greater abundance, that accounts for the often destructive role and bad reputation of invaders. Furthermore, the vigor of invaders is most strongly expressed in the early stages of establishment and assimilation, before adaptation stabilizes the invader's role (see Ricklefs, this volume). Invasion in the short run is therefore a destabilizing, often destructive, phenomenon.

This perspective on the arrival and establishment of invading species in a recipient community or ecosystem is in many ways similar to the recent suggestion by Davis and colleagues (2000) that a community becomes invasible when it contains plentiful unused resources. The temporary surplus could result from prior disturbance or from an influx of nutrients that native species have been unable to absorb (see also Davis 2003). According to this view, potential invaders need not differ ecologically from native species because there would, at least temporarily, be sufficient resources to sustain populations of both the invader and its native equivalent (Davis 2003). However, when resources in the recipient biota come to be more or less fully exploited, the initial success of some invaders may give way to decline or (rarely) extinction. The important point is that the process of invasion depends both on the characteristics of potential invaders in the donor species pool and on conditions (including resource availability) in the recipient environment.

Biotic Interchanges

Most of the episodes of dispersal that the fossil record chronicles involve the more or less simultaneous spread of many species from one geographic region to another. This type of invasion is referred to as biotic interchange, or simply interchange (Simpson 1947; Vermeij 1991a), because species can spread from region A to region B as well as from region B to region A. In most instances of interchange, the biotas involved have radically different evolutionary histories, characterized by a lack of contact for millions to tens of millions of years.

The recognition of biotic interchange requires knowledge of past distributions. If a given species is hypothesized to have taken part in the invasion of a recipient biota B from a donor biota A, then fossils of that species from the time before the episode of spread should be confined to biota A, whereas the same species should occur in both biotas after the time of spread. Often, a clade or lineage will spread, giving rise to a daughter species endemic to the recipient biota. In that case, fossil evidence of distributions must be coupled with phylogenetic evidence linking the species in the recipient biota with ancestral stocks in the donor biota. Inferences based only on current distributions are unreliable and often misleading because many, if not most, species and clades have undergone significant shifts in geographic range since the time of their evolution or spread. If we knew nothing of the fossil distribution of horses, for example, we would never know that the Equidae originated in the New World and invaded the Old World (probably several times) beginning some 11 million years ago during the Late Miocene (Simpson 1947; Garcés et al. 1997).

Biotic interchange is a common, widespread, and important biogeographic phenomenon. During the last 25 million years (that is, from earliest Miocene time onward), one or more intervals of interchange have taken place in at least four regions that include continental biotas and in at least ten regions involving shallow-water, bottom-dwelling marine biotas (Table 12.1). Not counted in this tally are cases of spread from tropical to temperate zones or instances of natural invasion of islands by species from continents. Most importantly,

TABLE 12.1 *Biotic interchanges of the last 25 million years*

Event	Locations and timeframes
Continental interchanges	
Trans-Beringian	Between Asia and North America; periodically throughout interval
Great American	Between North and South America; mostly Pliocene to Recent
Trans-Indonesian	Between Southeast Asia and New Guinea-Australia; Late Miocene onward
Trans-Arabian	Between Africa and Asia; throughout interval
Marine interchanges	
Trans-Arctic	Between North Pacific and Arctic-Atlantic; early phase latest Miocene to mid-Pliocene, Pacific to Atlantic invasion mid-Pliocene onward
North Trans-Pacific	Between western and eastern North Pacific; east to west during early middle Miocene; mainly west to east from Late Miocene onward
North Trans-Atlantic	Between eastern and western North Atlantic; mid-Pliocene onward
Tropical Trans-Pacific	Between western and eastern tropical Pacific; Pleistocene onward
Tropical Trans-Atlantic	Between eastern and western tropical Atlantic; episodically throughout interval
Western Atlantic	Between Caribbean and southeastern United States; mainly Pliocene onward
Trans-South African	Between Indian and Atlantic oceans; mostly Pleistocene onward
Trans-American	Between western Atlantic and eastern Pacific; throughout interval to Late Pliocene
Trans-Suez	Between Red and Mediterranean seas; mainly since 1869
Pacific Trans-Equatorial	Between temperate western North and South America; various intervals, especially Pliocene onward
Atlantic Trans-Equatorial	Between temperate Europe and southern Africa; mainly Pliocene onward
Circum-Antarctic	Around cool-temperate to sub-Antarctic southern hemisphere; throughout interval

these tallies exclude the thousands of cases in which a species arose in a geographically small region and rapidly spread across apparently continuous habitat to achieve its current very wide distribution. I have also excluded the multitude of cases of geographic spread in response to climate change, such as occurred when temperatures rose and fell with the contraction and expansion of glaciers during the Pleistocene.

Evidence from phylogenies and the fossil record is revealing that many terrestrial groups of plants and animals, including mammals, have invaded new areas through long-distance dispersal across oceans. The predominantly New World family Cactaceae, for example, created a minor beachhead in Africa and even in South Asia (Thorne 1973). The Bromeliaceae, another mainly Neotropical plant family, has one West African species (Thorne 1973). Primates and caviomorph rodents may have crossed from Africa to South America during the Oligocene (Wood 1950). Long-distance dispersals from South America to Africa are documented for several clades of Melastomaceae (Renner et al. 2001) and Malpighiaceae (C. C. Davis et al. 2002) and for the freshwater gastropod *Biomphalaria glabrata* (Woodruff and Mulvey 1997; Morgan et al. 2001, 2002; DeJong et al. 2003). These and similar examples of long-distance dispersal are important for understanding how ecosystems are assembled and how foreign species are assimilated by them, but I shall not consider such long-distance dispersal further here.

It has proved difficult to document and characterize biotic interchanges. In order to discern the big picture of how biotas interact, we need data not just on particular clades or species, but on as large a fraction of the two interacting biotas as possible. Thus it is comparatively easy with phylogenetic evidence to show that the Equidae expanded from North America to Eurasia, but rather more challenging to show what proportion of the mammalian biota, or of the terrestrial biota as a whole, took part in this New World to Old World invasion, or to ascertain what proportion of the Asian biota expanded in the opposite direction. Similarly, an understanding of the characteristics that set invading clades apart from noninvading clades demands a systematic survey across many potential clades. We thus need (1) a survey of the biota, including characteristics of component species, in the donor region; (2) a survey of the biota, including characteristics of component species, in the recipient region; (3) a survey of conditions in the region separating the donor and recipient biotas, both before and during the time of interchange; and (4) a survey of the species taking part in the interchange, including comparisons with noninvading species and including comparative studies of the lineages spreading in opposite directions.

Few biotic interchanges have been dissected to the extent outlined in the above ideal case. The Great American interchange, the spread of species between North and South America beginning about 3 million years ago, is one of the most intensively studied cases, especially for mammals (Webb 1969, 1991; Marshall 1981; Marshall et al. 1982; Stehli and Webb 1985; Webb and Rancy 1996). On the marine front, the Trans-Arctic interchange—the movement of species between the North Pacific and Arctic-North Atlantic basins—is reasonably well understood (Durham and MacNeil 1967; Vermeij 1991b), especially for mollusks. Other cases (see Table 12.1) are less completely studied, although some of their characteristics can be discerned from the available evidence.

Biotic interchange is almost always a highly asymmetrical process. When two biotas come into contact, the predominant direction of invasion is from the larger, more diverse biota, in which antipredator and competition-related adaptations have evolved to absolutely higher performance standards, to the smaller, less diverse biota, in which maximal levels of performance are lower (Darwin 1859; Darlington 1959; Briggs 1966, 1967a,b, 1999, 2003; Vermeij 1978, 1991a). Moreover, the donor biota has suffered less extinction of species than the biota receiving the bulk of invaders (Vermeij 1991a). Disruption of recipient biotas by the elimination of incumbents thus favors the establishment of vigorous invaders from larger adjacent biotas.

Most of the biotic interchanges of the last 25 million years fit this pattern. For example, the predominant direction of species invasion in the Great American interchange is from North to South America for open-country mammals, reptiles, and plants and from South America to the southern parts of North America for rainforest plants, mammals, and birds (Webb 1969, 1991; Marshall 1981; Gentry 1982; Marshall et al. 1982; Stehli and Webb 1985; Graham 1992). When the interchange began, grassland and savanna covered much larger areas of North America than of South America, whereas rainforests showed the reverse pattern (Cristoffer and Peres 2003). Ecologically dominant North American mammals were more highly specialized herbivores and carnivores, and probably had higher metabolic rates, than their South American counterparts, which they partially replaced (Lillegraven et al. 1987; Van Valkenburgh 1999). Similarly, species have moved from the continental biota of Southeast Asia to the island-like biotas of New Guinea and Australia in the Trans-Indonesian interchange (Hand 1984; Truswell et al. 1987), from Eurasia to North America in the Trans-Beringian interchange (Simpson 1947; Webb 1985; Repenning et al. 1987; Beard 1998; Van Valkenburgh 1999; Sanmartín et al. 2001; Xiang and Soltis, 2001), from the Pacific to the Atlantic in the Trans-Arctic interchange, at least from Late Pliocene time onward (Davies 1929; Durham and MacNeil 1967; Vermeij 1991b; Marincovich and Gladenkov 1999; Marincovich 2000; Marincovich et al. 2002), from the western to the eastern tropical Atlantic in the Tropical Trans-Atlantic interchange (Vermeij and Rosenberg 1993), from western North America to eastern Asia in the early middle Miocene North Trans-Pacific interchange (Amano and Vermeij 1998, 2003; Vermeij 2001), from New Zealand to southern South America in the Late Oligocene to Early Miocene in the Circum-Antarctic interchange (Beu et al. 1997; del Río, 2004), and from Europe to eastern North America in the North Trans-Atlantic interchange (Wares, 2001; Wares and Cunningham, 2001; Vermeij, unpublished data). Since at least Early Miocene time, the large tropical Indo-West Pacific (IWP) biota, encompassing the heterogeneous regions from the Red Sea and East Africa to eastern Polynesia, has exported species to the tropical eastern Pacific and tropical Atlantic oceans (Dana 1975; Vermeij 1987; Emerson 1991; Vermeij and Rosenberg 1993; Lessios et al. 1996, 1998), but has not incorporated any invaders from other regions (Vermeij and Rosenberg 1993; Vermeij 2001; Meyer 2003). Similarly, my continuing research on tropical western Atlantic mollusks indicates that clades from the Caribbean periodically invaded and enriched the shallow-water marine biota of the subtropical southeastern United States, and that invasion in the opposite direction is undetectable.

Some biotic interchanges chronicle a reversal in the predominant direction of invasion. One possible case of such a reversal is chronicled in the Trans-Arctic interchange. Very few molluscan species participated in this interchange during its early phase, from 5.4 to 3.6 million years ago (latest Miocene to middle Pliocene). The earliest invader, the bivalve genus *Astarte*, expanded from the Arctic Ocean, across the area formerly blocked by the Bering Land Bridge, to the North Pacific (Marincovich et al. 2002). At least four or five genera invaded in the opposite direction, into the Atlantic, as early as 4.7 to 4.8 million years ago, during the Early Pliocene (Gladenkov et al. 2002; Gladenkov and Gladenkov 2004). A flood of Pacific species invaded the Atlantic beginning about 3.6 million years ago (Vermeij 1991b; Marincovich et al. 1999; Marincovich 2000). The early phenomenology of this interchange may have several explanations. First, the North Pacific sites from which the earliest Trans-Arctic invaders from the Atlantic are recorded (Alaska, the Chukotka Peninsula of Siberia, and Kamchatka) lie very close to the Bering Strait, whereas the Atlantic sites where the earliest Pacific invaders of the Atlantic are recorded (Iceland, the Netherlands, Belgium, and Virginia) are thousands of kilometers from the strait through which all invaders had to pass (Gladenkov and Gladenkov 2004). Invasion of the Atlantic would thus appear to be later than that of the Pacific even if in fact it occurred simultaneously. Second, water flow through the Bering Strait may have been predominantly southward from the time the strait opened 5.5 to 5.4 million years ago until about 4.6 million years ago, when, as a result of the shoaling of the Central American seaway, currents reversed to a predominantly northward flow (Haug and Tiedemann 1998; Marincovich 2000). The fact that some Pacific species had already entered the Atlantic by 4.7 million years ago argues strongly against this second interpretation. Moreover, perhaps the majority of species taking part in the Trans-Arctic interchange lack planktonic dispersal stages, and therefore may not be directly influenced by the direction of flow through the Bering Strait. Rafting of juveniles or adults on floating objects remains a possibility for some species, but probably not for the many sand- and mud-dwelling bivalves (including *Astarte*) that participated in the interchange. My hypothesis (Vermeij 1991b) that the predominance of invasion from the Pacific to the Atlantic in the later phases of the Trans-Arctic interchange is explained by greater extinction of species in the Atlantic remains plausible. The matter will be resolved only when more precise times of invasion and extinction become available.

Within the North Pacific, interchange between North America and Asia was almost entirely from east to west during the early middle Miocene, but it became largely west to east during the Late Miocene and Pliocene (Zullo and Marincovich 1990; Vermeij 2001). The early phase of the North Trans-Pacific interchange coincided with a warm interval, whereas at least some of the later range expansions may have occurred during cool intervals. We do not know enough about oceanographic conditions or about the extinction histories of the biotas involved to hazard a hypothesis to explain the reversal in the direction of this interchange.

One possible exception to the pattern of asymmetry in biotic interchange may be the Trans-Arabian interchange between Africa and Asia. Before 24 million years ago (earliest Miocene), a few mammals dispersed from Asia to Africa, including artiodactyls (anthracotheres) and primates (Beard 1998; Jaeger 2003).

Following the establishment of a land bridge in the Early Miocene, modern carnivores and ungulates invaded Africa from Asia, while elephants expanded in the opposite direction (Kappelman et al. 2003; Jaeger 2003). Too little is known about pre-Miocene African faunas to permit an accurate appraisal of the extent and pattern of the Trans-Arabian interchange.

Looking back in time a little further, we see that the southern hemisphere continental landmass of Gondwana was a major source of cosmopolitan land groups. During the Late Cretaceous, when connections still existed among South America, Antarctica, and Australia, and when conditions were warmer than today's, Gondwana was a huge landmass. Paleontological and phylogenetic evidence indicates that eutherian mammals, passerine and many other groups of birds, and grasses all originated in the southern hemisphere and subsequently spread to the northern hemisphere (Olson 1989; Springer et al. 1997, 2003; Madsen et al. 2001; Murphy et al. 2001a,b; Bremer 2002). In the marine realm, the most diverse warm-water fauna during the early Cenozoic was in Europe, where the region of the Paris Basin in France supported some 1850 species of mollusks during the Middle Eocene (49 to 44 million years ago). In the Late Paleocene to Middle Eocene interval, waves of invaders from this region reached the western hemisphere, including the North American Pacific coast (see Squires 2003 for a review). The New World marine tropics may be the source of several clades that, beginning in Late Oligocene time, invaded the IWP region and underwent a dramatic diversification there (Vermeij 2001; Myers 2003; Williams and Reid 2004).

These historical cases of biotic interchange show that, although invasion is a persistent phenomenon affecting the geographic distribution of species, the source biota contributing disproportionate numbers of invading species has changed over time. The spatially most extensive, biologically most diverse biotas—those in which the standards of competitive performance are highest—are most likely to act as donor biotas, whereas smaller, less diverse biotas, including those that underwent prior impoverishment, are most likely to serve as recipient biotas.

It is important to note that the spatial extent of a biota, and therefore the role of that biota as donor or recipient, depends not only on the extent of physical continuity of habitats, but also on the biological traits of resident species. All places on earth are physically, chemically, and biologically heterogeneous at many spatial scales. To species with poor dispersal capacities, this heterogeneity is "coarse-grained"—that is, the boundaries between patches of different types act as barriers—whereas for species capable of broad dispersal, those boundaries are easily crossed, and the heterogeneity is "fine-grained" (Levins 1962, 1968; Levins and MacArthur 1966). When many species are good dispersers, meaning that individuals traverse a large area and that heterogeneity occurs on large spatial scales, the biota and ecosystems of which those species are a part are effectively large. Put another way, widely dispersing species unite disparate local biotas and ecosystems into larger, but still cohesive, ecological units. Within those larger units, wide dispersers can achieve large populations, in which selection and functional specialization proceed very far. The larger units are equivalent to large nation-states, which in human affairs typically exert disproportionate influence. Asia and Africa today are large continuous landmasses whose biotas have, over Cenozoic time, contributed disproportionate numbers of invaders to other landmasses. They have held this position not only because their landmasses are large, but more importantly, because many of the competitively dominant species that have evolved there have dispersed over wide areas within these landmasses. Rather than being a series of discrete habitat islands, Asia and Africa are, at least for some species, single ecological units, or biotas. This combination of physical continuity and biological traits of species—dispersal and competitive ability, among others—has enabled Asia and Africa to serve as major donor regions. Organisms amplify or diminish chemical and physical variation, so that observed biogeography can be thought of as a collective, evolved outcome of feedbacks between species and their environments (Vermeij 2004a,c).

The Long-Term Consequences of Invasion

Ecologists confronted with the quickening pace of human-aided species invasions have pointed to at least four short-term consequences of this phenomenon: homogenization, community disassembly, extinction, and a global loss of diversity.

First, ecologists predict a homogenization of the world's biota (Vitousek et al. 1997; Lövei 1997; Myers and Knoll 2001). This homogenization is most evident in communities of opportunistic species, or weeds, and among human disease agents, but it is also one of the consequences of biotic interchange. During the Early Pliocene, for example, before Pacific species reached the North Atlantic, the rich cool-temperate faunas of the North Pacific and North Atlantic were entirely distinct and had no species in common. The faunas of the eastern and western North Atlantic were also entirely distinct. Today, about 17% of temperate North Atlantic molluscan species are morphologically indistinguishable from North Pacific populations, and species found today on both sides of the North Atlantic (almost all of which have their evolutionary origins in the North Pacific) account for 21% and 36% of eastern and western North Atlantic molluscan faunas, respectively (Vermeij 1991b). All these percentages would be higher if the deep-water faunas were excluded because most species found in both the Atlantic and the Pacific, and most species found on both sides of the Atlantic, occur at depths of 200 m or less.

In the cases of biotic interchange that have been thoroughly investigated, invading species constitute a minority of the donor biota. For example, only 2% to 11% of mammalian genera from North America, and 2% to 7% of South American genera, participated in the various phases of the Great American interchange (Marshall et al. 1982; Vermeij 1991a). Some 13% of gastropod and 16% of coral species in the Line Islands (an eastern outpost of the IWP region) have crossed the Pacific to colonize eastern Pacific shores (Vermeij 1987). Trans-Arctic invaders of the Atlantic constitute up to 46% of some North Pacific donor biotas—a high percentage, but still a minority. In other words, a majority of species in donor biotas do not participate in biotic interchange, meaning that the donor biota always contains many elements unique to it. On the other hand, some recipient communities consist almost entirely of invading species or their descendants. A majority of plant and animal species on rocky shores of Atlantic North America north of Cape Cod are of Pacific origin (Vermeij 1991b). However, other communities, notably those in mud and sand, preserve their essentially Atlantic character, with 80% or more of molluscan species belonging to Atlantic lineages. On land, communities of weeds in human-disturbed environments are most dominated by invading species. In sum, homogenization varies among communities and among regions, and is complete only under unusual circumstances.

A biota will remain homogeneous as long as the populations of its species continue to exchange genes on a large geographic scale. In the geologic record, however, evidence shows that the breaching of barriers and the spread of species is an episodic phenomenon, and that populations of newcomers in recipient biotas quickly diverge from sister and parent populations in the donor region. In the case of the Trans-Arctic interchange, for example, 47% of species of Pacific origin in the temperate North Atlantic are distinct at the species level from their North Pacific ancestors. For species of Pacific origin restricted to the European side of the Atlantic, this percentage is 100%; for species now restricted to the American side of the Atlantic, the percentage is 49%. Some 27% of species currently found on both sides of the North Atlantic and having a Pacific origin are Atlantic endemics derived from Pacific ancestors (Vermeij 1991b). These divergences could have occurred as long as 4.8 million years ago, the date of the earliest Pacific invasion of the Atlantic (Marincovich 2000). Similarly, of sixteen molluscan species inferred to have arrived in the western Atlantic from the tropical IWP region after 3.1 million years ago, the approximate time of closure of the Central American seaway, nine are specifically distinct from their IWP ancestors (Vermeij and Rosenberg 1993; Beu 2001; Vermeij and Snyder 2003). The earliest northwestern Pacific representatives of mollusks invading from the eastern Pacific during the early middle Miocene typically cannot be distinguished morphologically from their eastern Pacific counterparts, but populations in the eastern and western Pacific soon diverged, so that by the Late Miocene most invading lineages were represented by distinct eastern and western species (Amano et al. 1993; Vermeij 2001; Amano and Vermeij 2003). Mammals taking part in the Great American interchange show similar divergence following invasion. North American lineages of dogs, bears, cats, deer, camels, and especially cricetid rodents diversified markedly in South America (Berta 1988; Webb 1991; Webb and Rancy 1996; Van Valkenburgh 1999).

The point is that, although biotic interchange leads to homogenization in the short run, divergence and speciation almost always counteract this tendency over geologic time because the invaders in the recipient biota become isolated from the populations in the donor biota that gave rise to them. Whether this pattern will continue in our own time is, of course, uncertain, given that continuing human-aided rapid movements of species from region to region may prevent such isolation.

The second short-term consequence of invasion is disassembly of recipient ecosystems. Ecological interactions among native species are disrupted by the newcomers, especially if the latter are competitively vigorous species and if they affect many native species. Invaders can also indirectly change the intensities and perhaps even the nature of interactions among natives. Compelling recent examples of such invasion-related disruptions come from studies of the effects of invading ants (O'Dowd et al. 2003; Sanders et al. 2003). On Christmas Island in the Indian Ocean, invading crazy ants (*Anoplolepis gracilipes*) have eliminated most of the population of an endemic land crab, and they have indirectly changed the dynamics between a native tree and human-introduced honey-producing scale insects (O'Dowd et al. 2003). On Guam, the introduction of a new top predator (the brown tree snake, *Boiga irregularis*) and of several new potential prey (lizards, rats, and some birds) has changed the trophic structure of the forest food web from one dominated by an endemic top predator (a kingfisher) and endemic bird prey to one occupied almost entirely by human-introduced species (Fritts and Rodda 1998). It is important to note, however, that these ecosystem-wide effects of invading species are difficult to disentangle from current or prior effects of human-caused disturbance, such as deforestation and widespread wartime bombardment in the case of Guam.

Much of the early short-term success of established invaders in the recipient biota may reflect the absence of effective parasites, competitors, and predators. In the longer term, however, the initial vigor of newly established invaders is blunted by the arrival of other invaders or by adaptation of native enemies (see also Ricklefs, this volume). Introduced species, in other words, accumulate enemies in their new homes. Many birds, mammals, lizards, and land snails introduced to Guam during the twentieth century at first underwent a population explosion, but later declined sharply in abundance (Fritts and Rodda 1998). Although the reasons for these declines have not been investigated, the changing competitive landscape as new species were introduced and as invaders accommodated to one another may be responsible. In Queensland, Australia, the introduced tropical American toad *Bufo marinus* harbored few parasites in the first 2 years of invasion as the population expanded vigorously, acquired many more parasites until the population was 19 years old, and then witnessed a reduction in parasite load as well as in population size by the time the population was 47 years old (Freeland et al. 1986). Data on the accumulation of pests on sugarcane and introduced trees (Strong et al. 1977; Connor et al. 1980; Auerbach et al. 1988) imply that periods of decades to a few centuries are sufficient for introduced species to acquire as many parasites as co-occurring native species possess.

Community relationships change as immigrants become assimilated, but most members survive and adapt. This is so because interactions among species

are numerous and flexible, and because adaptations arising from these interactions are often of a very general nature and thus applicable under a wide variety of circumstances. In the early decades of experimental ecology, when
researchers dissected interactions at one or a few sites, the predominant view
was that interactions and adaptations are highly specialized. Parasites were
thought to be host-specific, predators were thought to specialize on particular
prey, and plants and their insect herbivores were thought to target each other
with specialized countermeasures. As ecologists and evolutionary biologists
expanded their experimental work to more sites, and as experimental work
came to be supplemented by comparative observations from many places, this
view of rigid and narrow specialization was gradually replaced by the realization that species play different roles in different places. Parasites may have more
than one host (Fox 1981; Fox and Morrow 1981); predators at different sites concentrate on different prey, and even their preferred methods of attack vary from
site to site according to the characteristics of the available prey and the risks
faced by the predators (Paine 1980; Vermeij et al. 1994). Interactions that are
mutually beneficial in one place may be antagonistic in another according to
the broader context of interactions in which a particular relationship between
species exists (Thompson 1994; Thompson and Cunningham 2002).

This kind of flexibility allows well-functioning communities and ecosystems to emerge even when the various member species do not share a long history of coexistence and mutual adaptation (Janzen 1985). For example, an
apparently permanent and lush tropical cloud forest composed entirely of introduced tree species has replaced a treeless, fern-dominated landscape on Green
Mountain, a peak on Ascension Island in the South Atlantic (Wilkinson 2004).

The third short-term consequence of invasion, which is related to the second, is extinction in the recipient biota. Although the fossil record does not
have the temporal resolution to conclusively evaluate the possibility that
invaders cause extinctions, it is consistent with this expectation in some cases,
most notably when the invaders are large mammals. The demise of many carnivorous and herbivorous South American mammals after the arrival of North
American carnivores and ungulates, for example, can reasonably be ascribed
to predation by and competition with the metabolically more active North
American invaders (Marshall 1981; Vermeij 1991a). Introduced predators have
certainly brought about extinctions on most oceanic islands, on island-like small
continents such as New Zealand, and in lakes. In cases involving marine organisms, land plants, and smaller terrestrial animals, evidence for invasion-caused
extinctions is weak or nonexistent, except again on islands (Brown 1983; Vermeij 1991a; Rosenzweig 2001; Mooney and Cleland 2001; Davis 2003).

Common intuition has led many ecologists to view communities and ecosystems as fragile edifices that easily crumble when members become extinct or
new species enter (Ehrlich and Ehrlich 1981; Levin 1999). In a striking, if misleading, metaphor, Ehrlich and Ehrlich (1981) likened an ecosystem to an air-
plane, which might come apart if even one rivet were removed. Notwith-
standing such rhetoric, it has become clear that ecosystems are quite robust

structures, in which the rules of engagement and the identity of players are flexible. Mathematical models of food webs have demonstrated that the presence of numerous weak links between species, together with long feedback loops among species, impart stability to trophically complex communities and ecosystems (McCann et al. 1998; Polis 1998; McCann 2000: Neutel et al. 2002; Krause et al. 2003). In such communities, there are so many controls on species and so much functional redundancy that the removal or addition of a species is unlikely to cause collateral damage great enough to lead to the extinction of other species. Only wholesale interference with production will destroy the system as a whole (Vermeij 2004b). Short-term and long-term adaptive flexibility and interactive diversity thus enable most species to adapt as species come and go.

We should not be surprised by this robustness. At every level of biological organization, disturbances from without and within disrupt activity. Given that disruption has buffeted living things and the more inclusive structures created by them from time immemorial, it would be astonishing if the great majority of species living today were not adapted to deal with such relatively frequent disturbances as species invasions. The flexibility and adaptability observable at every level of organization from cells to ecosystems are powerful, general mechanisms by which the units of life resist and respond to a regime of frequent change (Vermeij 2004c). The ability to absorb invaders is thus an expected manifestation of biological organization at the ecosystem level.

Fourth, invasion is normally thought to reduce diversity in the recipient biota. Studies of birds and land plants on islands, however, have demonstrated that, although endemic species often become extinct as a result of invasion, local diversity either remains unchanged or actually rises (Sax et al. 2002), sometimes by as much as a factor of two, following invasion. Interchange among river drainages made possible by construction of the Panama Canal similarly led to increased species richness in fish communities on both the Atlantic and Pacific slopes of the Isthmus of Panama (Smith et al. 2004). On a continental scale, Webb (1969) noted long ago that the Great American interchange enriched the South American mammalian fauna. Although the South American fauna has decreased in diversity since the late Pleistocene, this decrease can be blamed on technologically sophisticated humans, who are in a class by themselves relative to other invading species (Diamond 1984). Although experiments in microcosmic zooplankton communities indicate limits to diversity and a corresponding tendency for artificially introduced species not to enrich recipient communities (Shurin 2000), many, if not most, communities, ecosystems, and biotas exist in a state far below species saturation (Vermeij 1991a; Ricklefs 2004). As Ricklefs (2004) notes, populations are compressible, so that established invaders often increase the number of species at both the local and the regional scale.

A fifth consequence of invasion has been largely overlooked, but may be of great importance on an evolutionary time scale. If invaders are competitively and reproductively vigorous species, invasion may be a key process that spreads these traits to all parts of the world. The net effect of invasion would therefore be to raise standards of competitive performance not just in the large donor biotas where the vigorous invaders evolved, but also in the recipient biotas to which they spread. Such a process doubtless speeds recovery after major biotic crises and sets the evolutionary agenda for the post-crisis biosphere.

When a species spreads to an area it has not previously occupied, its population becomes effectively larger. If many species invade, as during biotic interchange, the biota to which these species belonged and in which they evolved also becomes effectively larger. As noted in the previous section, larger biotas support dominant species whose competitive performance, as measured in units of power, is higher than that of dominant species in smaller biotas. Invasion therefore effectively raises the performance standards in recipient biotas. Although this change is accompanied by short-term disruptions, potentially including the loss of species, ultimately it spreads competitive vigor and indirectly intensifies selection within the native recipient biota.

Comparisons between Biotic Interchange and Human-Caused Invasion

The reader may be willing to grant that species invasion has taken place throughout the history of life and that the long-term consequences of invasion have increased regional diversity as well as competition. But do such conclusions apply to human-assisted invasions of species? Are the processes and participants involved in the human introduction of species similar enough to their "natural" analogues in biotic interchange that we can use interchange to inform studies of human-assisted invasion, or are the differences so great that the principles governing interchange cannot be applied to the invasions now sweeping the biosphere?

I can think of at least four ways in which interchanges of the past might differ from present-day human-assisted invasions. First, the rate at which species invade may be faster today than in the past. Second, the extent of invasion—the proportion of invaders in both donor and recipient biotas—may be greater in the human-dominated biosphere than it was in prehuman times. In particular, invasion today may be global in scope. Third, invaders today may come from very distant biotas, whereas in the past invaders were exchanged mainly between adjacent biotas whose evolutionary histories were less divergent. Finally, the kinds of species participating in invasions today may differ from those taking part in prehuman biotic interchange. I argue that these differences either do not exist or are exaggerated, and that the lessons of past interchanges surely apply to our current predicament because the processes and consequences transcend many of the particulars of place, time, and participants.

Discussions I have had with other biologists and paleontologists reveal the widespread perception that biotic interchange is a slow, long-term process, whereas human-caused invasion is rapid and therefore in the long run a much more destructive and disruptive process. As a hypothesis, this perception is very difficult to test directly, because the temporal resolution of the fossil record is inadequate for measuring the time interval over which interchange takes place. In fact, on the scale of tens of thousands to hundreds of thousands of years, we cannot distinguish between continuous interchange and episodic interchange. All human-caused invasions—even those that involve decade-scale spread of species from one end of a continent to the other—would look instantaneous in the fossil record. Of course, one could calculate a rate of interchange by dividing the number or percentage of species known to have invaded by a duration of time known to have included interchange, but such calculations are apt to be misleading. For example, if we argued that 300 marine clades participated in the Trans-Arctic interchange over the last 5.5 million years, we would obtain a rate that I would argue is meaningless. There are two related reasons for this meaninglessness. First, the conduit through which all invaders had to spread (the Bering Strait) was closed for long intervals during Pleistocene glacial episodes, and even when it was open, conditions during the warmest Early Pliocene would have been far more favorable to interchange than, say, those during the rather brief Pleistocene interglacials. Second, calculations of rate are notoriously sensitive to the time interval over which the rate is measured (reviewed in Roopnarine 2003). The longer the interval, the lower the rate will appear to be, chiefly because there will be times when the rate is low or even zero.

We can approach the problem of measuring rates of past interchange by examining the rate of spread of species in historic times. I am not in a position to review the literature on rates of species spread, but all accounts with which I am familiar indicate that most species spread throughout an environmentally suitable new region over a period not exceeding 50 years. After its first recorded appearance in Nova Scotia in 1840, the marine periwinkle *Littorina littorea* extended its range southward, reaching southern New Jersey by 1892 (Reid 1996). I predict that the rate of spread of invading populations without human assistance will be found to be of this magnitude. If so, then there is every reason to believe that such rates of spread applied to interchanges of the past as well. In principle, therefore, human-caused and "natural" interchanges should show no substantial difference in the rate of species invasion.

But what about the extent of invasion? Today, invading species affect ecosystems globally, whereas past interchanges mainly affected neighboring biotas. More importantly, human-introduced invaders often constitute large proportions (sometimes even a majority) of species on islands, in lakes and estuaries, and in human-disturbed or human-created habitats. San Francisco Bay and grasslands in the Central Valley of California are dominated, in both species and individuals, by human-introduced invaders (Sauer 1988; Baker 1989; Brown 1989; Mack 1989; Wagner 1989; Cohen and Carlton 1998). In other habitats, however, human-introduced invaders constitute a tiny fraction of species in recipient biotas. Forests, chaparral vegetation, and wave-swept rocky intertidal habitats in California have very few established species brought by humans (Brown 1989; Vermeij 1991b). Deforested parts of New Zealand are dominated by alien species, whereas adjacent shallow marine habitats have no invaders whatsoever. These results of invasion are not unlike those of biotic interchanges, in which some habitats became nearly completely dominated by invading species while others did not (e.g., muddy versus rocky shores of the North Atlantic: Vermeij 1991b). At the scale of entire biotas (i.e., continents and oceans), human-introduced animal species typically account for less than 1% of recipient biotas (D. F. Sax, personal communication; see also Brown 1989); this percentage is much less than those seen in biotic interchanges (discussed above). For plant species, the percentage of human-introduced species on continents is somewhat higher, but still less than that seen for many groups of species (such as mollusks) that are well preserved in the fossil record. Now, it is possible that, as humans continue to transport species around the world, the proportion of alien species in recipient biotas and communities will rise, especially because we continue to disturb and fragment habitats. As forests, grasslands, and even marine habitats become increasingly island-like, they will become more prone to the successful establishment of invaders (see also Vermeij 2004b). Even so, the ecologically broad-based invasion that has characterized many instances of biotic interchange in the past does not seem to have equivalents in the human-dominated biosphere except in habitats that are island-like or that we have strongly influenced or created.

The third purported difference between invasion today and invasion in the past is that most interchanges of species in prehuman times took place between adjacent biotas, whereas humans often introduce species from one biota to other very distant ones. This difference is probably less real than it at first appears to be, but even if it were found to be valid, it would not be important. As I noted earlier in this chapter, long-distance dispersal and colonization by means of birds, currents, and wind have brought species from one biota to very distant recipient biotas for millions of years. It may be that, with the evolution of long-distance migration in flying animals, this kind of long-distance interchange became more common after Jurassic time. Rafting on tree trunks could have occurred as far back as the Devonian, at least 370 million years ago. In other words, there has been a tendency toward globalization of colonization for hundreds of millions of years before our rapid means of transport made long-distance invasion even more likely.

Biologists might argue that biotas separated by long distances have, on average, longer independent histories than "adjacent" biotas, and therefore that the effects of long-distance dispersal are more destructive than those of invasions resulting from interchange between neighboring biotas. These perceptions, however, may be incorrect. The terrestrial faunas of neighboring North and South America were separated for most of the Cenozoic era (an interval of at least 50 million years) before the Great American interchange allowed species to pass between the two continents. The cold-water biotas of the North Pacific and Arctic-North Atlantic were separated for at least 100 million years before the Bering Strait opened (Gladenkov and Gladenkov 2004). These time spans approach or even exceed the durations of independence of, say, the widely separated Californian and Mediterranean floras.

Are the kinds of species liable to become human-assisted invaders different from the kinds of species that participated in interchanges of the past? This question brings us back full circle to the beginning of this chapter, where I drew a distinction between active and passive invaders. By providing rapid and relatively safe transport, humans have enabled many species to spread that otherwise would not have extended their range on their own. These species include both actively and passively dispersing species, but human transport particularly assists actively dispersing mammals and birds, as well as disease agents, large trees, and insects. Islands and island-like environments such as lakes and small streams have received invaders at rates and with magnitudes in the human-dominated biosphere without parallels in the past. On the other hand, species participating in past interoceanic and intercontinental interchanges include both active and passive invaders, and it is hard to argue that the kinds of species participating in these interchanges are radically different from the kinds of species humans transport.

Evaluating this prediction will not be easy. Most of the species humans spread are small or weedy and unlikely to be preserved as fossils. In order to make meaningful comparisons, we should systematically document human-introduced taxa that belong to groups with prolific fossil records and compare them with naturally interchanging taxa in the same groups. Tests like this, as well as other explicit comparisons between today's invasions and those of the geologic past, should be part of the agenda of invasion biology for the coming years. Invasion biologists would do well to learn from the invasions of the past.

Concluding Remarks

There can be no doubt that species invasion is a common, important, and ancient agency of change in communities and ecosystems. In the short term, like almost all other forms of disturbance, it upsets the status quo and is therefore generally destructive. This is especially so when invasion is accompanied or preceded by other disruptions, such as human exploitation of species or human-caused environmental change. On longer time scales, adaptive flexibility blunts, and to some extent counteracts, the destructive effects of invasion. On the longest time scales, invasion speeds up recovery from biotic crises and spreads effective adaptations worldwide.

The idea that invasion is both a curse and a blessing may appear self-contradictory, but an analogous idea applies to the more familiar phenomenon of mutation. Just as invasion is an ecological disturbance, mutation is a genetic one. Most invasions, and most mutations, are harmful or relatively benign in the short run. Yet both invasion and mutation are necessary in the long run. By providing the variation on which selection acts, mutations are ultimately essential to evolution and adaptation. In the same way, invasions create new interactions among species, and therefore ultimately bring about adaptations and enhance performance among members of the ecosystem that have successfully assimilated the invaders. Invasion may be as crucial to the long-term development and vigor of communities and ecosystems as mutation is to the long-term evolution of populations and species.

This long-term optimism should not be taken as a license to introduce species. After all, the initial effects of introduction are destructive, and reasonable measures should be taken to prevent the deliberate or accidental transport of species. But neither should we descend into a siege mentality. Ecosystems have changed since the origin of life, and they do so frequently, much as species and genomes change. Ecological change and species evolution are realities that must be studied and understood, not wished away or denied. Like evolution and extinction, invasion deserves dispassionate and objective study on both short and long time scales and at large and small spatial scales.

Literature Cited

Amano, K., and G. J. Vermeij. 1998. Origin and biogeographic history of Ceratostoma (Gastropoda: Muricidae). Venus 57:209–223.

Amano, K., and G. J. Vermeij. 2003. Evolutionary adaptation and geographic spread of the Cenozoic buccinid genus *Lirabuccinum* in the North Pacific. Journal of Paleontology 77:863–872.

Amano, K., G. J. Vermeij, and K. Narita. 1993. Early evolution and distribution of the gastropod genus *Nucella*, with special reference to Miocene species from Japan. Transactions and Proceedings of the Palaeontological Society of Japan n. s. 171:237–248.

Auerbach, M., and D. Simberloff. 1988. Rapid leaf-miner colonization of introduced trees and shifts in sources of herbivore mortality. Oikos 52:41–50.

Baker, H. G. 1989. Sources of the naturalized grasses and herbs in California grasslands. In L. F. Huenneke and H. A. Mooney, eds. *Grassland structure and function: California annual grassland*, pp. 29–38. Kluwer, Dordrecht.

Beard, C. 1998. East of Eden: Asia as an important center of taxonomic origination in mammalian evolution. Bulletin of the Carnegie Museum of Natural History 34:5–39.

Berta, A. 1988. Quaternary evolution and biogeography of the large South American Canidae (Mammalia: Carnivora). University of California Publications in Geological Sciences 132:1–149.

Beu, A. G. 2001. Gradual Miocene to Pleistocene uplift of the Central American isthmus: evidence from tropical American tonnoidean gastropods. Journal of Paleontology 75:706–720.

Beu, A. G., M. Griffin, and P. A. Maxwell. 1997. Opening of Drake Passage gateway and late Miocene to Pleistocene cooling reflected in Southern Ocean molluscan dispersal: evidence from New Zealand and Argentina. Tectonophysics 281:83–97.

Bremer, K. 2002. Gondwanan evolution of the grass alliance of families (Poales). Evolution 56:1374–1387.

Briggs, J. C. 1966. Zoogeography and evolution. Evolution 20:282–289.

Briggs, J. C. 1967a. Dispersal of tropical marine shore animals: coriolis parameters or competition? Nature 216:350.

Briggs, J. C. 1967b. Relationship of the tropical shelf regions. Studies in Tropical Oceanography 5:569–578.

Briggs, J. C. 1999. Coincident biogeographic patterns: Indo-West Pacific Ocean. Evolution 53:326–335.

Briggs, J. C. 2003. Marine centres of origin as evolutionary engines. Journal of Biogeography 30:1–18.

Brown, J. H. 1989. Patterns, modes and extent of invasions by vertebrates. In J. A. Drake, H. A. Mooney, F. di Castri, R. H. Groves, F. J. Kruger, M. Rejmánek, and M. Williamson, eds. *Biological invasions: a global perspective*, pp. 85–109. John Wiley, Chichester.

Cohen, A. N., and J. C. Carlton. 1998. Accelerating invasion rate in a highly invaded estuary. Science 279:555–558.

Connor, E. F., S. H. Faeth, D. Simberloff, and P. A. Opler. 1980. Taxonomic isolation and the accumulation of herbivorous insects: comparison of introduced and native trees. Ecological Entomology 5:205–211.

Cristoffer, C., and C. A. Peres. 2003. Elephants versus butterflies: the ecological role of herbivores in the evolutionary history of two tropical worlds. Journal of Biogeography 30:1357–1380.

Dana, T. F. 1975. Development of contemporary eastern Pacific coral reefs. Marine Biology 33:355–374.

Darlington, P. J. Jr. 1959. Area, climate and evolution. Evolution 13:488–510.

Darwin, C. 1859. The origin of species by natural selection or the preservation of favoured races in the struggle for life. John Murray, London.

Davies, A. M. 1929. Faunal migrations since the Cretaceous period. Proceedings of the Geologists' Association 40:307–327.

Davis, C. C., C. D. Bell, S. Mathews, and M. J. Donoghue. 2002. Laurasian migration explains Gondwanan disjunctions: evidence from Malpighiaceae. Proceedings of the National Academy of Sciences USA 99:6833–6837.

Davis, M. A. 2003. Biotic globalization: does competition from introduced species threaten biodiversity? BioScience 53:481–489.

Davis, M. A., J. P. Grime, and K. Thompson. 2000. Fluctuating resources in plant communities: a general theory of invasibility. Journal of Ecology 88:528–534.

DeJong, R. J., J. A. T. Morgan, W. D. Wilson, M. H. Al-Jaser, C. C. Appleton, G. Coulibaly, P. S. D'Andrea, M. J. Doenhoff, W. Hass, M. A. Idris, L. A. Magahães, H. Moné, G. Mouahid, L. Mubila, J.-P. Pointer, J. P. Webster, E. M. Zanotti-Magalhães, W. L. Paraense, G. M. Mkoji, and E. S. Loker. 2003. Phylogeography of *Biomphalaria glabrata* and *B. pfeifferi*, important intermediate hosts of *Schistosoma mansoni* in the New and Old World tropics. Molecular Ecology 12:3041–3056.

Diamond, J. M. 1984. Historic extinctions: a Rosetta Stone for understanding prehistoric extinctions. In P. S. Martin and R. G. Klein, eds. *Quaternary extinctions: a prehistoric revolution,* pp. 824–862. University of Arizona Press, Tucson.

Dick, C. W., K. Abdul-Salim, and E. Bermingham. 2003. Molecular systematic analysis reveals cryptic Tertiary diversification of a widespread tropical rain forest tree. American Naturalist 162:691–703.

Durham, J. W., and F. S. MacNeil. 1967. Cenozoic migrations of marine invertebrates through the Bering Strait region. In D. M. Hopkins, ed. *The Bering land bridge,* pp. 326–349. Stanford University Press, Stanford.

Ehrlich, P. R., and A. H. Ehrlich. 1981. *Extinction: the causes and consequences of the disappearance of species.* Random House, New York.

Emerson, W. K. 1991. First records of *Cymatium mundum* (Gould) in the eastern Pacific Ocean, with comments on the zoogeography of the tropical trans-Pacific tonnacean and non-tonnacean prosobranch gastropods with Indo-Pacific faunal affinities in West American waters. Nautilus 105: 62–80.

Freeland, W. J., B. L. J. Delvinqueir, and B. Bonnin. 1986. Food and parasitism of the cane toad, *Bufo marinus,* in relation to time since colonization. Australian Wildlife Reviews 13:489–499.

Fox, L. R. 1981. Defense and dynamics in plant-herbivore systems. American Zoologist 21:853–864.

Fox, L. R., and P. A. Morrow. 1981. Specialization: species property or local phenomenon? Science 211:887–893.

Fritts, T. H., and G. H. Rodda. 1998. The role of introduced species in the degradation of island ecosystems: a case history of Guam. Annual Review of Ecology and Systematics 29:113–140.

Garcés, M., L. Cabrera, J. Agustí, and J. M. Parces. 1997. Old World first appearance datum of *"Hipparion"* horses: late Miocene large-mammal dispersal and global events. Geology 25:19–22.

Gentry, A. H. 1982. Neotropical floristic diversity: phytogeographical connections between Central and South America, Pleistocene climatic fluctuations, or an accident of the Andean orogeny? Annals of the Missouri Botanical Garden 69:557–593.

Gladenkov, A. Y., and Y. B. Gladenkov. 2004. Onset of connections between the Pacific and Arctic Oceans through the Bering Strait in the Neogene. Stratigraphy and Geological Correlation 12:174–186.

Gladenkov, A. Yu., A. E. Oleinik, L. Marincovich Jr., and K. B. Barinov. 2002. A refined age for the earliest opening of Bering Strait. Palaeogeography, Palaeoclimatology, Palaeoecology 183:321–328.

Graham, A. 1992. Utilization of the isthmian land bridge during the Cenozoic—paleobotanical evidence for timing, and the selective influence of altitudes and climate. Review of Palaeobotany and Palynology 72:119–128.

Hand, S. 1984. Australia's oldest rodents: master mariners from Malaysia. In M. Archer and G. Clayton, eds. *Vertebrate zoogeography and evolution in Australia,* pp. 905–912. Hesperian Press, Carlisle.

Haug, G. H., and R. Tiedemann. 1998. Effect of the formation of the isthmus of Panama on Atlantic Ocean thermohaline circulation. Nature 393:673–676.

Highsmith, R. C. 1985. Floating and algal rafting as potential dispersal mechanisms in brooding invertebrates. Marine Ecology Progress Series 25:169–179.

Jaeger, J.-J. 2003. Isolationist tendencies. Nature 426: 509–511.

Janzen, D. H. 1985. On ecological fitting. Oikos 45:308–310.

Jokiel, P. L. 1989. Rafting of reef corals and other organisms at Kwajalein Atoll. Marine Biology 101:483–493.

Jokiel, P. L. 1990. Transport of reef corals in the Great Barrier Reef. Nature 347:665–667.

Kappelman, J., D. T. Rasmussen, W. Sanders, M. Feseha, T. Brown, P. Copeland, J. Crabaugh, J. Fleagle, M. Glantz, A. Gordon, B. Jacobs, M. Maga, K. Muldoon, A. Pan, L. Pyne, B. Richmond, T.

Ryan, E. R. Seiffert, S. Sen, L. Todd, M. C. Wiemann, and A. Winkler. 2003. Oligocene mammals from Ethiopia and faunal exchange between Afro-Arabia and Eurasia. Nature 426:549–552.

Klironomos, J. N. 2002. Feedback with soil biota contributes to plant rarity and invasiveness in communities. Nature 417:67–70.

Krause, A., K. A. Frank, D. M. Mason, R. E. Ulanowicz, and J. W. Taylor. 2003. Compartments revealed in food-web structure. Nature 426:282–285.

Lessios, H. A., B. D. Kessing, G. M. Wellington, and A. Graybeal. 1996. Indo-Pacific echinoids in the tropical eastern Pacific. Coral Reefs 15:131–142.

Lessios, H. A., B. D. Kessing, and D. R. Robertson. 1998. Massive gene flow across the world's most potent marine biogeographic barriers. Proceedings of the Royal Society of London B 265:583–588.

Levin, S. A. 1999. *Fragile dominion: complexity and the commons.* Perseus, Reading, MA.

Levins, R. 1962. Theory of fitness in a heterogeneous environment. American Naturalist 96:363–383.

Levins, R. 1968. *Evolution in changing environments: some theoretical explorations.* Princeton University Press, Princeton, NJ.

Levins, R., and R. H. MacArthur. 1966. The maintenance of genetic polymorphism in a spatially heterogeneous environment: variations on a theme by Howard Levene. American Naturalist 100:585–589.

Lillegraven, J. A., S. D. Thompson, B. K. McNab, and J. L. Patton. 1987. The origin of eutherian mammals. Biological Journal of the Linnean Society 32:281–336.

Lövei, G. 1997. Global change through invasion. Nature 388:627–628.

Mack, R. N. 1989. Temperate grasslands vulnerable to plant invasions: characteristics and consequences. In J. A. Drake, H. A. Mooney, F. di Castri, R. H. Groves, F. J. Kruger, M. R. Rejmánek, and M. Williamson, eds. *Biological invasions: a global perspective,* pp. 155–179. John Wiley, Chichester.

Madsen, O., M. Scally, C. J. Douady, D. J. Kao, R. W. DeBry, R. Adkins, H. M. Amrine, J. J. Stanhope, W. W. de Jong, and M. S. Springer. 2001. Parallel adaptive radiations in two major clades of placental mammals. Nature 409:610–614.

Marincovich, L. Jr. 2000. Central American paleogeography controlled Pliocene Arctic Ocean molluscan migrations. Geology 28:551–554.

Marincovich, L. Jr., and Y. A. Gladenkov. 1999. Evidence for an early opening of the Bering Strait. Nature 397:149–151.

Marincovich, L. Jr., K. B. Barinov, and A. E. Oleinik. 2002. The *Astarte* (Bivalvia: Astartidae) that document the earliest opening of Bering Strait. Journal of Paleontology 76:239–245.

Marshall, L. G. 1981. The Great American interchange—an invasion induced crisis for South American mammals. In M. H. Nitecki, ed. *Biotic crises in ecological and evolutionary time,* pp. 133–229. Academic Press, New York.

Marshall, L. G., S. D. Webb, J. J. Sepkoski, Jr., and D. M. Raup. 1982. Mammalian evolution and the Great American interchange. Science 215:351–357.

McCann, K. S. 2000. The diversity-stability debate. Nature 405:228–233.

McCann, K., A. Hastings, and G. R. Huxel. 1998. Weak trophic interactions and the balance of nature. Nature 395:794–798.

Meyer, C. P. 2003. Molecular systematics of cowries (Gastropoda: Cypraeidae) and diversification patterns in the tropics. Biological Journal of the Linnaean Society 79:401–459.

Mitchell, C. E., and A. G. Power. 2003. Release of invasive plants from fungal and viral pathogens. Nature 421:625–627.

Mooney, H. A., and E. E. Cleland. 2001. The evolutionary impact of invasive species. Proceedings of the National Academy of Sciences USA 98:5446–5451.

Morgan, J. A. T., R. J. DeJong, Y. Jung, K. Khallaayoune, S. Kock, G. M Mkoji, and E. S. Loker. 2002. A phylogeny of planorbid snails, with implications for the evolution of *Schistosoma* parasites. Molecular Phylogenetics and Evolution 25:477–488.

Muñoz, J., A. M. Felicísimo, F. Cabezas, A. R. Burgaz, and I. Martínez. 2004. Wind as a long-distance dispersal vehicle in the southern hemisphere. Science 304:1144–1147.

Murphy, W. J., E. Eizirik, W. E. Johnson, Y. P. Zhang, O. A. Ryder, and S. J. O'Brien. 2001a. Molecular phylogenetics and the origins of placental mammals. Nature 409:614–618.

Murphy, W. J., E. Eizirik, S. J. O'Brien, O. Madsen, M. Scally, C. J. Douady, E. Teeling, O. A. Ryder, M. J. Stanhope, W. W. de Jong, and M. S. Springer. 2001b. Resolution of the early placental mammal radiation using Bayesian phylogenetics. Science 294:2348–2351.

Myers, N., and A. H. Knoll. 2001. The biotic crisis and the future of evolution. Proceedings of the National Academy of Sciences USA 98:5389–5392.

Nathan, R., F. G. Katul, H. S. Horn, S. M. Thomas, R. Oren, R. Avissar, S. W. Pacala, and S. A. Levin. 2002. Mechanisms of long-distance dispersal of weeds by winds. Nature 418:409–413.

Neutel, A.-M., J. A. P. Heesterbeek, and P. C. de Ruiter. 2002. Stability in real food webs: weak links in long loops. Science 296:1120–1123.

O'Dowd, D. J., P. T. Green, and P. S. Lake. 2003. Invasional "meltdown" on an oceanic island. Ecology Letters 6:812–817.

Olson, S. L. 1989. Aspects of global avifaunal dynamics during the Cenozoic. In H. Ouellet, ed. *Acta XIX, congresus internationalis ornithologici,* pp. 2023–2029, Vol. 2. University of Ottawa Press, Ottawa.

Paine, R. T. 1980. Food webs: linkage, interaction strength and community infrastructure. Journal of Animal Ecology 49:667–685.

Polis, G. A. 1998. Stability is woven by complex webs. Nature 395:744–745.

Polis, G. A., W. B. Anderson, and R. D. Holt. 1997. Toward an integration of landscape and food web ecologies: the dynamics of spatially subsidized food webs. Annual Reviews of Ecology and Systematics 28:289–316.

Reid, D. G. 1996. *Systematics and evolution of* Littorina. Ray Society, London.

Renner, S. S., G. Clausing, and K. Meyer. 2001. Historical biogeography of Melastomataceae: the roles of Tertiary migration and long-distance dispersal. American Journal of Botany 88:1290–1300.

Repenning, C. A., E. M. Brouwers, L. D. Carter, L. Marincovich Jr., and J. A. Ager. 1987. The Beringian ancestry of *Phenacomys* (Rodentia: Cricetidae) and the beginning of the modern Arctic Ocean borderland biota. United States Geological Survey Bulletin 1687:1–31.

Richardson, D. M., N. Allsopp, C. M. D'Antonio, S. J. Milton, and M. Rejmánek. 2000. Plant invasions: the role of mutualisms. Biological Reviews 75:65–93.

Ricklefs, R. E. 2004. A comprehensive framework for global patterns in biodiversity. Ecology Letters 7:1–15.

Río, C. J. del. 2004. Tertiary marine molluscan assemblages of eastern Patagonia (Argentina): a biostratigraphic analysis. Journal of Paleontology 78:1097–1122.

Roopnarine, P. D. 2003. Analysis of rates of morphologic evolution. Annual Reviews of Ecology, Evolution, and Systematics 34:605–632.

Rosenzweig, M. L. 2001. The four questions: what does the introduction of exotic species do to diversity? Evolutionary Ecology Research 3:361–367.

Sanders, N. J., N. J. Gotelli, N. E. Heller, and D. M. Gordon. 2003. Community disassembly by an invasive species. Proceedings of the National Academy of Sciences USA 100:2474–2477.

Sanmartín, I., H. Enghoff, and F. Ronquist. 2001. Patterns of animal dispersal, vicariance and diversification in the Holarctic. Biological Journal of the Linnean Society 73:345–390.

Sauer, J. D. 1988. *Plant migration: the dynamics of geographic patterning in seed plant species*. University of California Press, Berkeley.

Sax, D. F., and J. H. Brown. 2000. The paradox of invasion. Global Ecology and Biogeography 9:363–371.

Sax, D. F., S. D. Gaines, and J. H. Brown. 2002. Species invasions exceed extinctions on islands worldwide: a comparative study of plants and birds. American Naturalist 160: 766–783.

Shurin, J. B. 2000. Dispersal limitation, invasion resistance, and the structure of zooplankton communities. Ecology 81:3074–3086.

Simpson, G. G. 1947. Holarctic mammalian faunas and continental relationships during the Cenozoic. Bulletin of the Geological Society of America 58:613–687.

Smith, S. A., G. Bell, and E. Bermingham. 2004. Cross-Cordillera exchange mediated by the Panama Canal increased the species richness of local freshwater fish assemblages. Proceedings of the Royal Society of London B 271:1889–1896.

Springer, M. S., G. C. Cleven, O. Madsen, W. W. de Jong, V. G. Waddell, H. M. Amrine, and M. J. Stanhope. 1997. Endemic African mammals shake the phylogenetic tree. Nature 388:61–64.

Springer, M. S., W. J. Murphy, E. Eizirik, and S. J. O'Brien. 2003. Placental mammal diversification and the Cretaceous-Tertiary boundary. Proceedings of the National Academy of Sciences USA 100:1056–1061.

Squires, R. L. 2003. Turnovers in marine gastropod faunas during the Eocene-Oligocene transition, west coast of the United States. In D. R. Prothero, L. C. Ivany, and E. A. Nesbitt, eds. *From greenhouse to icehouse: the marine Eocene-Oligocene transition*, pp. 14–35. Columbia University Press, New York.

Stehli, F. G., and S. D. Webb, eds. 1985. *The Great American biotic interchange*. Plenum, New York.

Strong, D. R. Jr., E. D. McCoy, and J. R. Rey. 1977. Time and the number of herbivore species: the pests of sugar cane. Ecology 58:167–175.

Thompson, J. N. 1994. *The coevolutionary process*. University of Chicago Press, Chicago.

Thompson, J. N., and B. M. Cunningham. 2002. Geographic structure and dynamics of coevolutionary selection. Nature 417:735–738.

Thorne, R. F. 1973. Floristic relationships between tropical Africa and tropical America. In B. J. Meggers, E. S. Ayensu, and W. D. Duckworth, eds. *Tropical forest ecosystems in Africa and South America: a comparative review*, pp. 27–48. Smithsonian Institution Press, Washington, DC.

Torchin, M. E., K. D. Lafferty, A. P. Dobson, V. J. McKenzie, and A. M. Kuris. 2003. Introduced species and their missing parasites. Nature 421:628–630.

Truswell, E. M., A. P. Kershaw, and I. R. Sluiter. 1987. The Australian-South-East-Asian connection: evidence from the palaeobotanical record. In T. C. Whitmore, ed. *Biogeographical evolution of the Malay Archipelago*, pp. 32–49. Clarendon Press, Oxford.

Van Valkenburgh, B. 1999. Major patterns in the history of carnivorous mammals. Annual Reviews of Earth and Planetary Sciences 27:463–493.

Valdebenito, H. A., T. F. Stuessy, and D. J. Crawford. 1990. A new biogeographic connection between islands in the Atlantic and Pacific Oceans. Nature 347:549–550.

Vermeij, G. J. 1978. *Biogeography and adaptation: patterns in marine life.* Harvard University Press, Cambridge.

Vermeij, G. J. 1987. The dispersal barrier in the tropical Pacific: implications for molluscan speciation and extinction. Evolution 41:1046–1058.

Vermeij, G. J. 1991a. When biotas meet: understanding biotic interchange. Science 253:1099–1104.

Vermeij, G. J. 1991b. Anatomy of an invasion: the Trans-Arctic interchange. Paleobiology 17:281–307.

Vermeij, G. J. 1996. An agenda for invasion biology. Biological Conservation 78:3–9.

Vermeij, G. J. 2001. Community assembly in the sea: geologic history of the living shore biota. In M. D. Bertness, S. D. Gaines, and M. E. Hay, eds. *Marine community ecology*, pp. 39–60. Sinauer Associates, Sunderland, MA.

Vermeij, G. J. 2004a. Island life: a view from the sea. In M. Lomolino and L. Heany, eds. *Frontiers of biogeography: new directions in the geography of nature*, pp. 239–254. Sinauer Associates, Sunderland, MA.

Vermeij, G. J. 2004b. Ecological avalanches and the two kinds of extinction. Evolutionary Ecology Research, 6:315–337.

Vermeij, G. J. 2004c. *Nature: an economic history.* Princeton University Press, Princeton, NJ.

Vermeij, G. J., and G. Rosenberg. 1993. Giving and receiving: the tropical Atlantic as donor and recipient region for invading species. American Malacological Bulletin 10:181–194.

Vermeij, G. J., and M. A. Snyder. 2003. The fasciolariid genus *Benimakia*: new species and a discussion of Indo-Pacific genera in Brazil. Proceedings of the Academy of Natural Sciences of Philadelphia 153:15–22.

Vermeij, G. J., H. L. Lescinsky, E. Zipser, and H. E. Vermeij. 1994. Diet and mode of feeding of the muricid gastropod *Acanthinucella lugubris angelica* in the northern Gulf of California. Veliger 37:214–215.

Vitousek, P. M., C. L. D'Antonio, M. Rejmánek, and R. Westbrooks. 1997. Introduced species: a significant component of human-caused global change. New Zealand Journal of Ecology 21:1–16.

Wagner, F. H. 1989. Grazers, past and present. In L. F. Huenneke and H. A. Mooney, eds. *Grassland structure and function: California annual grassland*, pp. 151–162. Kluwer, Dordrecht.

Wares, J. P. 2001. Biogeography of *Asterias*: North Atlantic climate change and speciation. Biological Bulletin 201:95–103.

Wares, J. P., and C. W. Cunningham. 2001. Phylogeography and historical ecology of the North Atlantic intertidal. Evolution 55:2455–2469.

Webb, S. D. 1969. Extinction-origination equilibria in Late Cenozoic land mammals of North America. Evolution 23:688–702.

Webb, S. D. 1985. Late Cenozoic mammal dispersals between the Americas. In F. G. Stehli and S. D. Webb, eds. *The Great American biotic interchange*, pp. 357–386. Plenum, New York.

Webb, S. D. 1991. Ecogeography and the Great American Interchange. Paleobiology 17:266–280.

Webb, S. D., and A. Rancy. 1996. Late Cenozoic evolution of the neotropical mammal fauna. In J. B. C. Jackson, A. F. Budd, and A. G. Coates, eds. *Evolution and environment in tropical America*, pp. 335–358. University of Chicago Press, Chicago.

Wesselingh, F. P., G. C. Cadée, and W. Renema. 1999. Flying high: on the airborne dispersal of aquatic organisms as illustrated by the distribution histories of the gastropod genera *Tryonia* and *Planorbarius*. Geologie en Mijnbouw 78:165–174.

Wilkinson, D. M. 2004. The parable of Green Mountain: Ascension Island, ecosystem construction and ecological fitting. Journal of Biogeography 31:1–4.

Williams, S. T., and D. G. Reid. 2004. Speciation and diversity on tropical rocky shores: a global phylogeny of snails of the genus *Echinolittorina*. Evolution 58:2227–2251.

Wolfe, L. M. 2002. Why alien invaders succeed: support for the escape-from-enemy hypothesis. American Naturalist 160:705–711.

Wood, A. E. 1950. Porcupines, paleogeography, and parallelism. Evolution 4:87–98.

Woodruff, D. S., and M. Mulvey. 1997. Neotropical schistosomiasis: African affinities of the host snail *Biomphalaria glabrata*. Biological Journal of the Linnean Society 60:505–516.

Xiang, Q.-Y., and D. E. Soltis. 2001. Dispersal-vicariance analyses of intercontinental disjuncts: historical biogeographical implications for angiosperms in the northern hemisphere. International Journal of Plant Sciences 162:S29–S39.

Zullo, V. A., and L. Marincovich Jr. 1990. Balanoid barnacles from the Miocene of the Alaska Peninsula, and their relevance to the extant boreal barnacle fauna. Journal of Paleontology 64:128–135.

13

Evolutionary Trajectories in Plant and Soil Microbial Communities

CENTAUREA INVASIONS AND THE GEOGRAPHIC MOSAIC OF COEVOLUTION

Ragan M. Callaway, José L. Hierro, and Andrea S. Thorpe

In this chapter, we review empirical evidence for stronger allelopathic effects of exotic European invaders on North American natives than on other European species. We argue that this evidence provides critical insights into the importance of coevolutionary relationships within plant communities. We discuss how the disruption of these and other biochemical coevolutionary relationships may lead to the success of some exotic species and the decline of the natives they encounter. Understanding this coevolution within the context of John Thompson's geographic mosaic theory may provide crucial insight into biological invasions, which in turn may provide deeper insight into the geographic mosaic theory of coevolution. We also review evidence for the evolution of native plant species in response to the allelopathic effects of invaders and for mechanisms by which native species might tolerate the effects of invaders. We look at how the disruption of evolutionary trajectories in plant–soil microbe interactions may also drive invasions. Such comparisons of interactions between native and non-native species are powerful tools for understanding the role of rhizosphere biochemistry as a driver of evolutionary trajectories in plant communities, and more generally, for understanding the geographic context of coevolution.

Introduction

Coevolution is one of the major processes organizing the earth's biodiversity.

JOHN THOMPSON

One of the most important challenges in evolutionary ecology is to understand how coevolution shapes and organizes natural communities across broad geographic landscapes (Thompson 1994, 1997, 1999). Thompson argued that explicit examination of genetically differentiated populations, and of geographic variation in their interactions, is crucial for developing a framework for coevolutionary theory. This geographic population structure exists for most interspecific interactions and must be part of any theory on the organization of the earth's biodiversity. Here, we link the geographic mosaic concept to biological invasions. Invasions are inherently biogeographic processes (Hierro et al. 2005), and they provide the most spectacular examples on earth of community patterns and dynamics that arise from rapid changes in geographic population structure. Furthermore, patterns and process in invaded communities can be dramatically different from those in the original communities. Here, we argue that the geographic mosaic theory of coevolution provides theory for understanding biological invasions, and that biological invasions provide insight into the geographic mosaic theory of coevolution.

Species can undergo rapid evolutionary changes under natural conditions (Thompson 1998) and when they are introduced to new parts of the world by humans (Rice and Emery 2003; Phillips and Shine 2005; Rice and Sax, this volume). These changes provide invaluable insight into the nature of adaptation and how it may affect coexistence among species and the formation of communities. Most plant species that are introduced into new ecosystems by humans appear to act more or less like the native species they join, but some transmogrify—they become far more abundant and competitively dominant than they are in their native regions. Unusually good luck for a newcomer is generally correlated with unusually bad luck for natives, manifested in lower local diversity and large changes in community composition (see D'Antonio and Vitousek 1992; Mack et al. 2000). Species-rich communities can be replaced by virtual monocultures (Braithwaite et al. 1989; Malecki et al. 1993; Meyer and Florence 1996; Bruce et al. 1997; Ridenour and Callaway 2001; Siemann and Rogers 2001). These dramatic changes in abundance and dominance indicate unusually strong ecological processes, which may initiate strong evolutionary responses. By drastically reducing the population sizes of native plants, invasive plants have the potential to be a powerful selective force.

Despite the explosion of interest in the evolution of invaders (Blossey and Nötzold 1995; Gaskin and Schaal 2002; Hänfling and Kollman 2002; Lee 2002), the evolution of native plants and microbes in response to invasive plants has largely been ignored, probably because the paradigms for invasive success are firmly embedded in resource or trophic interactions (Elton 1958; Maron and Vilà 2001; Keane and Crawley 2002). Furthermore, much plant community theory is rooted in Gleason's (1926) individualistic theory, in which a plant community is "scarcely even a vegetational unit, but merely a coincidence" (see Lortie et al. 2004). This individualistic paradigm (reflected in current neutral theory and lottery models) does not provide a clear context in which invasive plants can drive adaptation in other plants. If the organization of plant communities is determined by the adaptation of members to a particular abiotic environment and competition for the same nutrient, light, and water resources, what mechanisms can transform an inferior competitor into an overwhelming dominant, and how do native plants adapt to the mechanisms of their new dominant neighbors?

Prevailing "individualistic" plant community theory does not have a place for coevolution among plants and may not provide an adequate paradigm for understanding exotic plant invasion. In this chapter, we explore recent evidence for novel allelopathic interactions among plants and novel biochemical interactions among plants and soil microbes (see Callaway and Ridenour 2004), focusing on *Centaurea* invasions as case examples.

We propose that shared evolutionary trajectories in different regions may mediate coexistence in the original communities of invaders where the species have been exposed to one another for a long time, while simultaneously facilitating competitive exclusion by those invaders in their new communities. We propose that when humans introduce some plant species to new regions, they may force together different biochemically driven coevolutionary trajectories from different continents and disrupt coevolved interactions. We have organized this chapter into two sections, one focusing on biochemically mediated plant interactions, allelopathy, and how these might drive the evolution of native species, and another focusing on interactions among plants and soil microbes. Both sections are organized to consider one fundamental question: Do regional coevolutionary trajectories in plant and soil microbial communities lead to the coexistence of species and community stability?

Allelopathic Interactions among Invasive and Native Plants

Some successful plant invasions appear to be based, at least in part, on biochemistry. Plants release up to 50% of their total photosynthate as organic exudates from their roots (Flores et al. 1999; Nardi et al. 2000). These exudates play mutualistic or defensive roles as a plant communicates with its environment and ultimately develops a suite of positive and negative responses to abiotic and biotic conditions in the rhizosphere (Bais et al. 2001; Walker et al. 2003; Bais et al. 2004). The role of root exudates, or chemical interactions in general among plants (allelopathy), has a murky history; ever since a highly touted early study was discredited by field experiments, other studies of allelopathy have been viewed with more skepticism than studies of resource-based interactions (see Callaway 2002). The skepticism is probably unwarranted, and the criteria for proof of allelopathy seem to far exceed criteria for other phenomena

(Williamson 1990). In this section, we briefly discuss evidence for allelopathy in exotic invasions in general, then focus on the details of the allelopathic effects of two knapweeds, *Centaurea maculosa* and *C. diffusa*. We use the term "allelopathy" to refer to the negative effect of one plant on another through the release of chemical compounds into the environment (*sensu* Muller 1966) and consider allelopathy just one of many forms of non-resource interaction among plants (Callaway 2002).

Biogeographic differences in allelopathic effects among plants: Case studies with **Centaurea**

Unlike locally based theories of coevolution, the hypothesis of geographic mosaics of coevolution predicts the routine occurrence of transient mismatches of traits, or "maladaptations," that lead to breakdowns in coevolved coexistence (Thompson 1999); perhaps these occasional mismatches drive the crashes in community composition and diversity that occur when monocultures of invaders develop. Whether long-lasting adaptations develop to promote coexistence or become maladaptive is thought to depend on three components: the existence of selection mosaics (which exist, by definition, when species are introduced around the world), coevolutionary hot spots (which *may* exist if the right combinations of species are introduced to one another), and trait remixing among populations (which will exist as exotics and natives encounter either novel environments or neighbors with novel traits). All of these components may be relevant to the biogeography of invasions.

The replacement of diverse native communities with monospecific stands of an invading species is a clue that unusually powerful mechanisms may be at work (see Hierro and Callaway 2003). In a search for such mechanisms, many studies have suggested that allelopathy may contribute to the ability of particular exotic species to become dominants in invaded plant communities. This possibility has been explored for some of the best-known plant invaders in the world, including *Eltrygia repens* (ex. *Agropyron*: Korhammer and Haslinger 1994), *Bromus tectorum* (Rice 1964), *Cirsium arvense* (Stachon and Zimdahl 1980), *Cyperus rotundus* (Quayyum et al. 2000; Agarwal et al. 2002), *Euphorbia esula* (Steenhagen and Zimdahl 1979), *Parthenium hysterophorus* (Pandey 1994), *Setaria faberii* (Bell and Koeppe 1972), and *Sorghum halepense* (Elmore 1985). As a more detailed example, *Zoysia*-dominated seminatural grasslands of Japan have been heavily invaded by *Anthoxanthum odoratum* (Yamamoto 1995), and coumarin, a compound released from this invader, inhibits the development of native *Zoysia japonica* seedlings. As another example, Czarnota et al. (2001) found that the chemical sorgoleone constituted about 85% of the total root exudates from the invasive *Sorghum halepense* (johnsongrass) and suppressed other species by docking into the QB-binding site of the photosystem II complex; soil impregnated with sorgoleone suppressed a number of different plant species.

The evidence for allelopathic effects of *Centaurea maculosa* (spotted knapweed) and *C. diffusa* (diffuse knapweed), two of North America's worst inva-

TABLE 13.1 *Summary of evidence for the effects of phytotoxic and antimicrobial root exudates from* Centaurea *species*

Exudate	Produced by	Autotoxicity	Effect on plants in invaded range	Effect on plants in native range	Effect on microbes
(±)-Catechin	*C. maculosa*	None	Strong	Weak	Strong
8-Hydroxyquinoline	*C. diffusa*	None	Strong	Moderate	Strong

ders, is strong and integrates ecological, physiological, biochemical signal transduction, and genomic approaches (Callaway and Aschehoug 2000; Bais et al. 2002, 2003; Fitter 2003; Baldwin 2003; Vivanco et al. 2004). Although these congeners are closely related, the allelopathic chemicals produced and the results they mediate differ between these two species (Table 13.1). *C. maculosa* produces (±)-catechin, which has been clearly demonstrated to have both phytotoxic and antimicrobial properties (Box 13.1). In contrast, *C. diffusa* produces 8-hydroxyquinoline, a compound with well-known abilities as a metal chelator, fungicide, and antiseptic (Vivanco et al. 2004; Merck Index 1996). In laboratory experiments, a suite of plant species showed 100% mortality after the addition of root exudates from *C. diffusa* and after application of 8-hydroxyquinoline. Further, like those of its congener *C. maculosa*, whose roots are unaffected by the allelopathic chemical it produces, the root exudates of *C. diffusa* are not autotoxic. In North America, 8-hydroxyquinoline is abundant in soil extracts from *C. diffusa*-invaded fields, but has not been found in the rhizosphere of any other plant species. In fact, 8-hydroxyquinoline had not been previously described as a natural product (Vivanco et al. 2004). Application of concentrations of 8-hydroxyquinoline much lower than those found in the field to natural soils reduces the growth of North American species (Vivanco et al. 2004).

If biochemical interactions are involved in the transmogrification of species from mild members of their native communities into competitive dominants in new communities, then it might be expected that the allelopathic effects of these species would be much weaker at home than in the communities they invade. Evidence for this is rapidly accumulating.

The root exudates from *C. maculosa* and *C. diffusa,* and the specific biochemicals in those exudates, have very different effects on plant species from communities where they are native than on plant species in communities that the weeds have invaded (Callaway and Aschehoug 2000; Bais et al. 2003; Vivanco et al. 2004). Strong biogeographic differences in the effects of allelochemicals circumstantially suggest that plants, as well as the microbes that mediate plant interactions (see below), that have coexisted for long periods have coevolved to tolerate one another's biochemistry. Furthermore, much stronger biochemical interactions in invaded systems suggest that in some cases exotic species may bring biochemicals to new communities rendered vulnerable by their naïveté—the "novel weapons" hypothesis (Rabotnov 1982; Call-

BOX 13.1 *Summary of the Evidence for Allelopathic Effects of Centaurea maculosa*

Centaurea maculosa roots produce an enantiomeric compound (a compound composed of two mirror-image entities), (±)-catechin, with clearly documented phytotoxic properties belonging only to the (–)-form of the chemical and antimicrobial properties belonging only to the (+)-form of the chemical. (±)-Catechin is present in natural soils at concentrations above that which affects other plants, and its presence is associated only with *C. maculosa* plants.

Greenhouse experiments have demonstrated the inhibitory effects of *C. maculosa* roots on the roots and overall growth of a native American grass; activated carbon added as a purification agent ameliorated these inhibitory effects (Ridenour and Callaway 2001). The concentration of (–)-catechin appears to be about twice as high in soils occupied by *C. maculosa* in North America as in similar habitats in Europe. Experiments show inhibition of the growth and germination of native species in field soils at natural concentrations of (–)-catechin. This allelochemical shows cell-specific targeting of meristematic and elonga-tion-zone cells in the roots of target plants, as evidenced by cytoplasmic condensation followed by a cascade of cell death proceeding backward up through the root stele, induction of reactive oxygen species (ROS)-related signaling that leads to rhizotoxicity in susceptible plants, a ROS-triggered Ca^{2+} signaling cascade leading to cellular pH decrease, and allelochemical-induced genome-wide changes in gene expression patterns. Injection of (±)-catechin into soil in the field strongly suppresses native species, and the degree of suppression is dose-dependent (R. M. Callaway, unpublished data). (–)-Catechin has no effect on *C. maculosa*, apparently due to the presence of a barrier/avoidance mechanism located in the root membrane that precludes the reentrance of (–)-catechin once secreted from the root of this species. The lack of autotoxicity is an answer to one of the strongest criticisms of allelopathy theory (Williamson 1990). Finally, the germination and growth of European grasses are more resistant to (–)-catechin than those of their North American counterparts.

away and Aschehoug 2000; Mallik and Pellissier 2000; Callaway and Hierro, in press; Callaway and Ridenour 2004).

In one of the earliest examples of biogeographic differences in allelochemical effects, Callaway and Aschehoug (2000) compared the inhibitory effects of *C. diffusa* on three bunchgrass species that coexist with *C. diffusa* in Eurasia with its effects on three bunchgrass species from invaded communities in North America. Each of the three grass species from North America was paired with a congener (or closely related species) from Eurasia of a similar morphology and size. *Centaurea diffusa* had much stronger negative effects on the North American species than on the Eurasian species. Correspondingly, none of the North American species had a significant competitive effect on the biomass of *C. diffusa*, but Eurasian bunchgrasses significantly reduced *C. diffusa* biomass.

The addition of activated carbon (a potent adsorbent of allelopathic compounds: Mahall and Callaway 1992) to soils had strikingly different effects on these interactions between *C. diffusa* and the grass species from the different regions. In the presence of *C. diffusa*, the overall effect of carbon on the growth of North American species was positive: activated carbon reduced the competitive strength of *C. diffusa*. In contrast, the biomass of all the Eurasian grass species growing with *C. diffusa* was reduced dramatically in the presence of activated carbon. Correspondingly, activated carbon put *C. diffusa* at a disadvantage against the North American grasses (*Centaurea* growth decreased), but conferred an advantage to *C. diffusa* when with the Eurasian grasses (*Centaurea* growth increased). Activated carbon has a high affinity for organic compounds, such as potentially allelopathic chemicals, and a weak affinity for inorganic electrolytes, such as those in nutrient solution, and has previously been shown to reduce the negative effects of root exudates from *C. maculosa* and other species (Schreiner and Reed 1907; Cheremisinoff and Ellerbusch 1978; Mahall and Callaway 1992; Ridenour and Callaway 2001). Furthermore, the strong effects of the place of origin on the competitive ability of grass species against *C. diffusa*, and the contrasting effects of activated carbon, suggest that *C. diffusa* produces allelopathic chemicals that long-term and familiar Eurasian neighbors have adapted to, but that this adaptation and tolerance has not developed in *C. diffusa*'s new North American neighbors.

Vivanco et al. (2004) further explored these biogeographic differences in the resistance or susceptibility of plant communities to *C. diffusa* by building microcosms in which North American and Eurasian plant communities were established in both North American and Eurasian soils. In full support of the previous experiments, the regional source of the plant community was by far the most important factor in resistance to interference by *C. diffusa*: Eurasian plant communities were much more resistant to invasion. Finally, North American plant species were found to be more susceptible to identical concentrations of 8-hydroxyquinoline than Eurasian species, forging a crucial link between the general effects of *C. diffusa* and the suspicious biochemical it exudes (Vivanco et al. 2004).

Mallik and Pellissier (2000) conducted experiments comparing the effects of leaves, leaf extracts, and humus from *Vaccinium myrtillis*, a widespread understory shrub in coniferous forests of Eurasia with strong allelopathic effects, on an exotic North American neighbor, *Picea mariana*, and on a native neighbor, *Picea abies*. They found that *V. myrtillus* generally had stronger biochemical effects on the exotic *P. mariana* than on the native *P. abies*. Their results also suggest that species without a common evolutionary history have stronger allelopathic interactions.

Using a biogeographic approach similar to that of Callaway and Aschehoug, Prati and Bossdorf (2004) tested the allelopathic effects of *Alliaria petiolata* (garlic mustard), an aggressive invader of the understory of forests in North America, on the germination of two congeneric species that co-occur with *Alliaria* in the field: the North American *Geum laciniatum* and the European *G. urbanum*.

They also investigated whether the allelopathic potential of *A. petiolata* varied between native European and exotic North American populations of the weed. In support of the "novel weapons" hypothesis, they found that invasive North American populations of *A. petiolata* significantly reduced the germination of "*Alliaria*-naïve" North American *G. laciniatum* seeds, but had no effect on "*Alliaria*-experienced" European *G. urbanum* seeds. European collections of *A. petiolata*, on the other hand, significantly reduced seed germination in both North American *G. laciniatum* and European *G. urbanum* in similar proportions, a result that only partially supports the "novel weapons" hypothesis. The contrasting inhibitory effects of *A. petiolata* populations from Europe and from North America on European *G. urbanum* suggest that North American *A. petiolata* has lost competitive ability against a former neighbor.

If invaders are indeed transformed from inferiors to dominants because they bring novel biochemicals to naïve native communities—and the jury is certainly still out—then the species that co-occur naturally with them may have adapted to the biochemicals they release. If species in an invader's place of origin can adapt to the biochemicals released by that invader, then the invader's new neighbors should be able to do so as well. We pursue this possibility in the following section.

Invaders and the evolution of native plants

Plants evolve rapidly in response to man-made chemical herbicides (Powles and Holtum 1994), and may evolve as rapidly to the natural allelopathic herbicides produced by invaders. The evolution of resistance to invaders by the species they encounter in their new communities would have profound consequences for ecological and evolutionary theory as well as for conservation. Such evolutionary processes would suggest that naturally intact biological communities may be, to some degree, functionally organized units (Goodnight 1990; Wilson 1997), rather than simple mixtures of species with similar adaptations to a particular abiotic environment. If natives can evolve resistance to invaders, then invaded communities may recover some aspects of their natural structure and function, and invaders and natives may eventually coexist.

As clearly presented in the geographic mosaic theory of coevolution (Thompson 1999), it is by no means necessary to view adaptive responses to allelochemicals as indicative of selective processes occurring over whole regions. Allelochemicals may drive selection in plants occurring in a particular local area, exerting strong selective pressures much like those exerted by anthropogenic herbicides.

In an experiment with five native grass species, Callaway et al. (in press) found that *C. maculosa* harmed "naïve" grass clones (individuals that had not experienced *C. maculosa* as neighbors) more than "experienced" clones (individuals that had survived *C. maculosa* invasion), suggesting the possible development of resistance to the inhibitory effect of *C. maculosa* by natives. When activated carbon was added to the sand to adsorb organic exudates, the growth of all the experienced clones of native species combined improved significantly. Analyzed separately, activated carbon significantly increased the growth of the experienced clones of only one species, *Stipa occidentalis*. In contrast, activated carbon improved the growth of the naïve clones of all native species analyzed together by 105%, and of three species when analyzed separately. *Stipa occidentalis*, *Festuca idahoensis*, and *Koeleria cristata* all showed much stronger resistance to *C. maculosa* when they had previously experienced *C. maculosa* invasion. *Centaurea*-tolerant individuals may have simply survived invasion; however, even if *Centaurea* has simply driven large shifts in genotype frequencies in surviving populations, this would be the first step toward evolution.

When in competition with *C. maculosa*, *Festuca* and *Stipa* grown from the seeds of naïve maternal plants were smaller than those grown from the seeds of experienced maternal plants, and activated carbon eliminated these differences (Callaway et al., in press). Plants grown from seeds of experienced and naïve maternal plants of *Festuca* and *Stipa* did not differ in size when grown with activated carbon in the soil, suggesting that allelopathy may have been a selective mechanism.

In another experiment (which used the allelochemical directly, instead of using the plant that produces it), the germination of seeds from both experienced and naïve native species was highly suppressed by a low concentration of (±)-catechin. However, the germination of seeds from some populations of native grasses demonstrated selection for tolerance to the allelochemical, as they did for the *C. maculosa* plants that produce the chemical. When exposed to (±)-catechin, 16% of *Stipa* seeds from experienced maternal plants germinated, whereas no seeds from naïve maternal plants germinated in (±)-catechin. Populations of *Pseudoroegneria spicata* tested in this experiment differed substantially in their responses to (±)-catechin. Germination of seeds from one population was almost completely suppressed by (±)-catechin; however, in another population, 6% of seeds from experienced maternal plants germinated, in comparison to none of the seeds from naïve maternal plants.

The enantiomer (±)-catechin elicits the production of reactive oxygen indicators (ROI) in root cells, and this production of ROI is related to the phytotoxic effects of the chemical (Bais et al. 2003). ROI production precedes cell death by 5 to 10 minutes. Seedlings grown from *Centaurea*-experienced *P. spicata* plants showed qualitatively different ROI production than those grown from seeds from *Centaurea*-naïve *P. spicata* (R. M. Callaway, unpublished data). Tolerance to (±)-catechin corresponded with much slower ROI reactions after exposure to the allelochemical.

To our knowledge, this research on the resistance of native species to *C. maculosa* and their tolerance to its root exudates is the only work that has been conducted on potential selection for resistance to invasive plants in native plants. However, other researchers have observed variation in interactions among species from different populations, suggesting that coexistence may result in ecotypic selection. For example, Martin and Harding (1981) found that when seeds of co-occurring individuals of *Erodium cicutarium* and *Erodium obtusiplicatum* were planted together, the total seed output and reproductive rates of the two species were higher than when seeds of individuals from distant communities were grown together. Evans et al. (1985) found that ecotypes of *Lolium perenne* and *Trifolium repens* that had been collected from fields in Switzerland, Italy, France, and England grew larger when paired with a "familiar" genotype of the other species—the genotype that co-occurred naturally in the same field. Joy and Laitinen (1980) found a similar response in experiments with *Phleum pratense* and *Trifolium pratense*. Turkington and Mehrhoff (1990) transplanted *Trifolium repens* and *Lolium perenne* into three fields in British Columbia, Canada, that varied in age since clearing: 0 years, 8 years, and 46 years. They found that *T. repens* genets from older fields transplanted into 46-year-old pastures grew much larger, either in competition or when competitors had been cleared away, suggesting that selection was occurring over time. In another experiment, Turkington and Harper (1979) collected ramets of *T. repens* from different sites dominated by one of four different grass species for competition experiments. When ramets were transplanted in all possible combinations with the four grasses, each clover "type" grew best with the grass species from the site where it had originally been sampled, suggesting the occurrence of microevolutionary changes in *T. repens* in response to different constraints imposed by different neighbors.

Although the empirical evidence is limited so far, the studies cited indicate that plants may be able to adapt to the presence of species-specific neighbors in ways that favor coexistence or higher productivity. So far in this chapter, we have considered the evidence that biochemical interactions drive this sort of adaptation. In other words, plants may eventually get used to the peculiar biochemistry of their neighbors, and when they do so, they may coexist. This does not mean that plants are necessarily evolving allelochemical production for competitive advantage per se. Such selection is possible, but it is just as likely that the release of particular chemicals evolves for other reasons (defense against herbivores or resource acquisition, for example) and some of these chemicals also happen to have strong effects on naïve neighbors. Biochemicals may not evolve to combat neighbors, but neighbors may have to evolve tolerance to those biochemicals anyway.

Centaurea maculosa and (±)-catechin may provide a good example of this "inadvertent" allelopathy. Catechins (and *C. diffusa's* 8-hydroxyquinoline) function as metal chelators for Al, Fe, and Ca and thereby increase the solubility of inorganic soil phosphorus. This form of P acquisition efficiency may play an important role in plant community dynamics in P-limited environments. Hydroxyl-substituted polyphenols also have the potential to increase P solubility via chelation of metals (Stevenson and Cole 1999). The high affinity of catechin for metals increases P solubility by lowering the concentration of metals available for the precipitation of insoluble metal phosphates (A. S. Thorpe et al., unpublished data). It is possible that (±)-catechin has evolved primarily to enhance P acquisition in *C. maculosa's* native habitat, and that (−)-catechin, the highly toxic and rare enantiomer, has strong allelopathic effects on North American species as a by-product.

Interactions between Plants and Soil Microbes

Microbial communities in the soil can have strong effects on plant populations (Burdon 1993; Packer and Clay 2000; van der Putten et al. 2001; Reinhart et al. 2003). The inhibitory and facilitative effects of soil microbes on plants, and the reciprocal effects of plants on soil microbes, create contrasting dynamic feedback interactions between plants and the microbial communities that develop around their roots (Bever et al. 1997; Bever 2002). Positive feedbacks occur when plant species accumulate microbes near their roots that have beneficial effects on the plants that cultivate them, such as mycorrhizal fungi and nitrogen fixers, and are thought to lead to a loss of local microbial community diversity (Bever et al. 1997; Bever 2002). Negative feedbacks occur when plant species accumulate pathogenic microbes in their rhizospheres, creating conditions that are increasingly hostile to the plants that cultivate these pathogens (van der Putten et al. 1993; Bever 1994; Klironomos 2002). Such plant-soil feedbacks have been hypothesized to be a strong selective force on both plants and microbes (van der Putten 1997). Plant–soil microbe interactions are often mediated by biochemical processes, and much as we have argued for allelopathic interactions, geographic evolutionary trajectories may also be driven by interactions among plants and soil microbes. Combining plant and microbial species without shared evolutionary trajectories may result in the unusually strong competitive interactions that appear to be characteristic of many invasions.

It is well established that different microbial communities are associated with different plant species (Bever 1994; Westover et al. 1997; Grayston and Campbell 1998; Priha et al. 1999, 2001; Grayston et al. 2001; Klironomos 2002), probably due in part to species-specific rhizosphere biochemistry. So it should be no surprise if exotic invaders also have species-specific effects on soil microbes. We are beginning to learn how invaders (and their associated soil microbiota) may produce soil conditions that differ fundamentally from those that occur in native communities and how these changes may provide a means by which invaders transmogrify into ecosystem-altering dominants.

Understanding invasive plant-microbe interactions in the context of geographic mosaic theory is complicated because microbes affect interactions among plants in more complex ways than the allelopathic interactions described above. Microbes can act as pathogens, shifting the balance of competition in favor of a less infected plant species. They can also act as mutualists, shifting the balance of competition in favor of a plant species more highly colonized by mutualistic microbes. Microbes run the nitrogen cycle and affect the availability of other nutrients through decomposition, shifting the balance of competition in ways that are hard to predict. Despite this mechanistic complexity, in this section, we continue to pursue the fundamental question addressed in this chapter: Do regional coevolutionary trajectories within plant and soil microbe communities lead to the coexistence of species and community stability?

Introduced microbial pathogens

In this chapter, we are primarily interested in what invasive plants might teach us about regional evolutionary relationships among plants and soil microbes. However, there is strong evidence for the importance of shared evolutionary trajectories for coexistence between other types of microbes and plants—as evidenced by the devastating effects of exotic microbial pathogens on plant species that are related to, but do not share a coevolutionary trajectory with, the natural hosts of the pathogens.

Some fungal species that are merely parasitic in their native range are virulently pathogenic when introduced to new ranges. *Cronartium ribicola*, the white pine blister rust fungus, is native to Asia and has been introduced into Europe and North America. Infected trees in Europe show minimal signs of infection, perhaps because of coevolutionary interactions with the fungus on the fringes of where strong coevolutionary interactions occur (Thompson 1999). However, in North America, far removed from its Eurasian coevolutionary trajectories, *Cronartium* has devastating effects on a number of *Pinus* species in the subgenus *Strobus*. Eurasian and North American evolutionary trajectories are similar enough for *Cronartium* to complete its complex life cycle, which requires *Ribes* and *Pinus* species as alternate hosts, but dissimilar (or maladaptive, in Thompson's terms) enough to allow far greater disruption of community processes than occurs in native local communities. *Cryphonectria parasitica* (syn. *Endothia*), a relatively benign parasite of *Castanea* species (chestnut) in Eurasia, was introduced into North America at the turn of the twentieth century. Within decades, a North American species, *Castanea dentata*, was transformed from a widespread ecologically dominant canopy tree to a relictually distributed understory shrub persisting solely through clonal growth. In an ironic twist, the fact that other plant species did not disappear with the near-eradication of *C. dentata* has been a primary argument *against* the occurrence of interdependent processes among members of a plant community (Woods and Shanks 1959; McCormick and Platt 1980; Johnson and Mayeux 1992). A similar but more complex case occurred when the Dutch elm fungal parasite *Ophiostoma ulmi* (syn. *Ceratocystis*) was introduced into North America. As in the other examples, a North American species, *Ulmus americana*, was harder hit than *Ulmus* species in Eurasia, where the fungus originated. However, the fungus is spread in North America by two species of bark beetles, one European and the other North American. Because of the particular feeding patterns of the latter species, the effects of the disease tend to be stronger. These disruptions of coevolved host-parasite relationships and their consequences provide evidence for the importance of a geographic context for understanding coevolution. Below, we continue to explore this geographic context for plant–soil microbial interactions and exotic plant invasions.

Effects of invasive plants on soil microbes

We know a lot about how soil microbes affect the performance and abundance of plants (Burdon 1987; Bever 1994, 2002; Bever et al. 1997; van der Putten and Peters 1997), but we know less about how invasive plants alter soil microbial communities. Kourtev et al. (2002) found that effects on microbial communities associated with the invasion of Japanese barberry (*Berberis thunbergii*) and Japanese stilt grass (*Microstegium vimineum*) extended beyond the soil in direct contact with the roots of these invaders and into the bulk soil. They suggested that these effects could be long-lasting. In field experiments conducted in the invaded range of *C. maculosa*, Callaway et al. (2004b) found that suppressing soil fungi in the presence of different native competitors caused *C. maculosa* biomass to vary from a 10-fold decrease to a 1.9-fold increase depending on the identity of the competitor. In untreated soils, *C. maculosa* grew larger in the presence of *F. idahoensis* or *K. cristata* than when alone. When fungicide was applied, these positive effects of *F. idahoensis* and *K. cristata* on *C. maculosa* did not occur. These results suggest that soil microbes may play a role in successful invasions, but that role may not be manifested in simple direct effects.

A number of plant compounds may mediate interactions between plants and microbes. One of the mechanisms driving the interaction between *C. maculosa* and microbial communities may be biochemical: the exudation of (±)-catechin from the roots of *C. maculosa*. As previously stated (see Box 13.1), (+)-catechin displays strong antimicrobial properties for at least some groups of bacteria (Bais et al. 2002, 2003). Other invasive species produce chemicals with antimicrobial activity; however, the roles of these chemicals in the plants' invasive success are generally unknown (Ehrenfeld 2003). The allelochemical produced by *C. diffusa*, 8-hydroxyquinoline, also has strong antimicrobial properties. This chemical inhibits the activity of several important plant pathogens, including the bacteria *Xanthomonas campestris*, *Pseudomonas syringae*, *Agrobacterium radiobacter*, *Erwinia carotovora*, *E. amylovora* and the fungi *Aspergillus niger*, *Rhizoctonia solani*, *Phytophthora infestans*, and *Fusarium oxysporum* (Vivanco et al. 2004). The dry mass of leaves of *Melaleuca* spp., which has invaded large areas of the coastal southeastern United States, particularly the Everglades, can be composed of up to 7% monoterpenes (Boon and Johnstone 1997). These compounds inhibit microbial colonization and decomposition of leaf litter in both the native and invaded ranges of *Melaleuca* spp. It has also been suggested that allelopathic chemicals released by some invasive species may alter nitrogen fixation in neighboring plants (Wardle et al. 1994, 1995).

The effects of invasive plants on soil microbes set the context for exploring the role of plant-microbe interactions in invasive success and the development of evolutionary trajectories on a geographic matrix. Explicit biogeographic comparisons of the effects of soil microbes in original and invaded systems are crucial tests of the hypothesis that shared evolutionary trajectories among soil microbes and plants lead to the coexistence of species and community stability.

Biogeographic evidence for regional plant–soil microbe evolutionary trajectories

In a review of 473 plant species that have become naturalized in the United States, Mitchell and Power (2003) found that 84% fewer fungi and 24% fewer virus species infected the species in their invaded ranges than in their native ranges. Furthermore, they reported that species that experienced greater release from microbial pathogens were more invasive. Klironomos (2002) found that locally rare native species consistently exhibited negative feedback interactions with the soil microbial community (a relative decrease in growth on "home" soil in which conspecifics had previously been grown), whereas common non-native species consistently exhibited positive feedback interactions with the soil community. Klironomos's results suggest that the biogeographic origin of a plant species is crucial to the direction of its interaction with soil microbes (although not clearly separated from the effects of locally low abundance), thus suggesting an important role for shared evolutionary histories. In Klironomos's research, it appeared that plant species developed geographically tight coevolutionary relationships with pathogenic fungi, but loose (less host-specific) relationships with mutualistic fungi.

The highly contrasting soil feedback effects demonstrated by Klironomos (2002) set the stage for two recent studies in which a geographic context was explicit. In an explicitly biogeographic test of the Janzen-Connell hypothesis (Janzen 1970; Connell 1971) for local diversity, Reinhart et al. (2003) compared the effects of soil microbes on the growth of *Prunus serotina* in its native and in its invaded ranges. In the native North American range, the soil microbial community occurring near *P. serotina* strongly inhibited the establishment of neighboring conspecifics and reduced seedling performance in the greenhouse. In contrast, in the non-native European range, *P. serotina* readily became established in close proximity to conspecifics, and in greenhouse experiments the soil community enhanced the growth of seedlings. Previous research had demonstrated that soil-borne *Pythium* species (Oomycota) inhibit the survival, growth, and abundance of *P. serotina* in its home range in North America (Packer and Clay 2000, 2002). Although the genus *Pythium* is found around the world, genotypes are often host-specific (Deacon and Donaldson 1993; Mills and Bever 1998). These findings suggest that the loss of the coevolved relationship between fungal consumer and host plant in North America has facilitated the invasion of the host in Europe.

Callaway et al. (2004a) found that sterilization of European soils caused a 166% increase in the total biomass of *C. maculosa*, compared with only a 24% increase when North American soils were sterilized. The stronger suppressive effects of European soil biota lend experimental support to the demonstrations of much higher richness of fungal and viral infections on plant species in their home ranges than in invaded ranges (Mitchell and Power 2003) and indicate that *C. maculosa* in North America has escaped the controlling effects of soil biota.

Callaway et al. (2004a) further examined biogeographic differences in plant–soil microbe relationships in a feedback experiment. The microbial community in soil from a selected population in France was "trained," or pre-conditioned (Bever et al. 1997; Bever 2002), by growing either *C. maculosa* or *Festuca ovina*, a perennial bunchgrass that is native to Eurasia, in the soil for several months. The microbial community in soil from a selected population in Montana was trained with *C. maculosa* or *F. idahoensis*, a bunchgrass similar to *F. ovina* but native to western North America. As would be predicted by Klironomos (2002), *C. maculosa* experienced negative feedback interactions with soil microbial communities from its native European region—*C. maculosa* plants grown alone in nonsterile French soil trained by conspecifics were significantly smaller than those grown in French soils trained by *F. ovina*. In contrast, *C. maculosa* experienced positive feedback interactions with soil microbial communities from its invaded range. *Centaurea maculosa* planted alone in Montana soils trained by conspecifics were significantly larger than those in Montana soils trained by *F. idahoensis*. Sterilization of the soils eliminated these feedbacks.

Considered together, the results of these feedback experiments suggest that *C. maculosa* is able to modify the microbial community in invaded soils to its advantage. In contrast, *C. maculosa* is inhibited by negative feedback in European soils, probably due to the accumulation of pathogens and, potentially, adaptation by inhibitory microbial populations to antimicrobial compounds produced by *C. maculosa* (Bais et al. 2003). If coevolutionary relationships between microbial pathogens and invasive plants exist, then negative feedbacks would be expected in the native range of the exotic species due to the accumulation of species-specific soil pathogens. In contrast, positive feedbacks should be observed in the invaded range, where the exotic species is largely free from species-specific soil pathogens (Klironomos 2002; Mitchell and Power 2003). If van der Putten (1997) is right about the potential for such feedbacks as a strong selective force, we should find many more examples of regionally specific plant–soil microbial relationships occurring on geographic mosaics.

In an experiment, not previously published, designed to explicitly test biochemically mediated evolutionary trajectories involving root exudates and soil microbes, we compared the effects of (±)-catechin on microbial communities extracted from soil in the rhizospheres of *C. maculosa* plants in six populations in Romania, thought to be close to the source of the invasive genotypes, and six populations in western Montana. In the controls, bacterial growth was uniformly high, but was far greater in isolates from Montana soils than in those from Romanian soils (Figure 13.1). This difference may possibly be due to the effects of transporting the soils across the Atlantic, although other explanations are possible, including potential trade-offs between (±)-catechin tolerance and absolute bacterial growth rates. Bacterial growth was reduced in both soil types by the catechin treatment, but bacterial growth in the isolates from Romanian

Figure 13.1   Growth of bacteria cultured from soil from the native range of *Centaurea maculosa* in Romania and its invaded range in Montana in control isolates and in isolates treated with (±)-catechin. Each bar represents an isolate from soil collected from the rhizosphere of one *C. maculosa* individual in one population. Note the large difference in scale between the treatments, such that the (±)-catechin treatment reduces the number of microbes by several orders of magnitude from control levels; note also that the degree of suppression by (±)-catechin is much greater for bacteria in soil from the invaded than from the native range. These results were garnered by taking 1 gram of soil from a randomly selected individual from each of six populations in each range, adding 9 ml of sterile phosphate buffered saline solution and 0.1% Tween 80, and vortexing to separate bacteria from soil particles. Isolates from each of these 12 solutions were plated onto R2A medium in two petri dishes containing 300 µg/ml cycloheximide, added to inhibit fungal growth (total *n* = 24). For each pair of petri dishes, one was treated with an ethanol control and one was treated with 100 µg/ml (±)-catechin/ethanol.

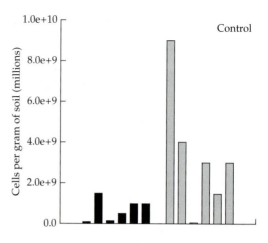

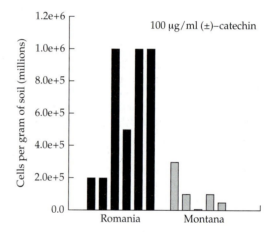

soils was over 7 times higher than in the isolates from Montana soils. Tentative identification of bacteria in the Romanian soils included *Micrococcus* sp., *Pseudomonas fluorenscens*, *Staphlococcus* sp., and *Flavobacterium* sp. It would appear that microbial communities from the native range of *C. maculosa* have developed tolerance to (±)-catechin [probably the (+) enantiomer] due to a coevolutionary history. However, we do not know whether the particular microbial species in these (±)-catechin experiments are the same ones involved in the feedback experiments described earlier (Callaway et al. 2004a).

Invaders and the evolution of soil microbes

Plant-microbe interactions and relationships are often regulated by plant biochemistry (Bais et al. 2004), but we do not know of a parallel to the allelopathic evidence we presented above for selection of resistance in native species in response to *C. maculosa* and the (±)-catechin it exudes (R. M. Callaway et al., unpublished data). Even though studies have not explicitly addressed the evolution of microbial communities in response to invasive species, it is reasonable to expect strong selection based on these interactions (van der Putten 1997). It has been suggested by van Breeman and Finzi (1998) that if the effect that a plant has on the soil alters the growth and survival of that plant, as in the feedback experiments we have described, then the soil can be considered part of the extended phenotype of the plant, and thus will be under selective pressure (see Thorpe and Callaway, in press, for a broader discussion of this idea). Microbes have short generation times and thus can respond to evolutionary pressures quickly, so they may respond relatively rapidly to pressures exerted by the invasion of exotic plants and the unique biochemicals they release. Although species-specific soil-borne pathogens are common (Agarwal and Sinclair 1997; Mills and Bever 1998; Blaney and Kotanen 2001), many soil-borne pathogens are generalists (Dix and Webster 1995). Thus, native pathogenic microbes may be able to adapt to a new invasive host relatively easily. As an example of this potential, *Sclerotinia sclerotiorum*, a generalist fungus native to North American intermountain prairies, causes very high mortality of *C. maculosa* when directly applied to the plant; however, this species does not disperse efficiently among *C. maculosa* individuals, even in dense populations of the weed (Jacobs et al. 1996; Ridenour and Callaway 2003). *Centaurea maculosa* certainly provides a large and untapped resource for consumers in North America, and perhaps selection will eventually result in an increased ability of local pathogens like *Sclerotinia* to utilize it—as suggested by the operation of the taxon cycle, in which invading species are expected to accumulate coevolved pathogens over time (Ricklefs, this volume).

We have emphasized the importance of escaping pathogens in invaded regions; however, not having a coevolutionary history with the organisms in a new region may also mean that the new species has no resistance to the pathogens that occur there. In other words, novel biochemical interactions may inhibit invasive success as well as promote it. Because most exotic introductions probably do not result in establishment, invasion, and suppression of the locals, this idea is very difficult to test. Nevertheless, future studies should consider this possibility, as it may present an important clue to a better understanding of the geographic mosaic of coevolution.

Conclusion

The idea that rhizosphere biochemistry can drive evolutionary trajectories in plant communities is derived completely from what we have learned from the study of invaders or comparisons of interactions between native and nonnative species—the contrasts in biochemical (allelopathic) effects of *Centaurea maculosa* (Bais et al. 2003), *Centaurea diffusa* (Callaway and Aschehoug 2000;

Vivanco et al. 2004), *Alliaria petiolata* (Prati and Bossdorf 2004), and *Vaccinium myrtillis* (Mallik and Pellissier 2000) on neighbors in their native communities versus neighbors in their invaded communities. These biogeographic contrasts suggest that different evolutionary trajectories may occur in different plant communities, and in particular, in communities that are isolated from one another. When different evolutionary trajectories in communities are forced together, there may be the potential for exceptionally intense biochemical interactions.

Experiments demonstrating that invasive *Centaurea* species have stronger allelopathic effects on their new neighbors in North America than on their old neighbors in Eurasia suggest the possibility that coevolutionary relationships can develop among plant species over time. Furthermore, these results imply that biochemically based evolutionary trajectories are different in different parts of the world, and that mixing species from these different trajectories increases the potential for introduction of powerful novel weapons that are inordinately effective against naïve species. Ecologists must increase the number of studies of general allelopathic interactions that utilize persuasive experimental designs, integrate biochemical and ecological approaches, take explicit biogeographic perspectives, and link competitive biochemical processes to trophic interactions (see Siemens et al. 2002).

We must also reconcile the novel weapons hypothesis and the potential evolutionary responses of natives with other theory. For example, many invasive species appear to have undergone a "lag phase," during which they spread very slowly or not at all (Kowarik 1995). If the novel weapons hypothesis is a valid explanation for some invasions, then it would seem that species with novel weapons would be highly successful from the beginning, and not undergo a lag phase. On the other hand, even exotics possessing novel weapons might be expected to experience lag phases, as these lags may occur for many reasons that are independent of the initial competitive advantage novel weapons may provide. Some potential reasons for these lags include the evolution of further increases in competitive ability after colonization has occurred (Holt et al., this volume), biological inertia exerted by the resident community on the invader (von Holle et al. 2003), and the dynamics of exponential population growth (whereby initial growth may appear to have a lag).

Another primary direction for future research would be to investigate adaptive evolution in specific microbial species in response to invasive plants. We also know very little about the effects of specific root exudates of invaders on microbial species, and to our knowledge, the experiment reported above on the comparative effects of ($\pm$)-catechin on Romanian and North American microbes is the only one of its kind. Exploring such biochemical processes will shed light on the geographic nature of plant–soil microbe coevolution and could allow integration of biochemical interactions among plants with those among plants and soil microbes. So far, biogeographic experiments on plant–soil microbe feedback (Reinhart et al. 2003; Callaway et al. 2004b) have not matched seeds from specific exotic plant populations with their co-occurring soil microbes. Until local exotic plant populations and the soil microbial communities that occur with them are studied in comparative experiments, our understanding of escape from local pathogens will continue to be limited. We encourage experiments designed to link biochemical interactions among plants with those that occur among plants and microbes. For example, some forms of catechin are consumed by microbes (e.g., *Acinetobacter calcoaceticus* utilizes catechin as its only carbon source: Arunachalam et al. 2003), so abundant and novel root exudates such as (±)-catechin produced by invasive plants constitute a potential food source for the soil microbes in their new communities, but only after the microbes have adapted to eat those exudates. Exploring microbial consumption of such root exudates in soils from native and invaded communities will be important. Finally, understanding biochemically driven evolutionary processes among invasive plants and soil microbes may open the door to a greater understanding of the regulation of populations and coexistence among species in natural communities, as well as the collapse of such regulation and coexistence that is so common in many exotic invasions.

Literature Cited

Agarwal, V.K. and J.B. Sinclair. 1997. *Principles of Seed Pathology*, 2nd edition Lewis Publishers, Boca Raton.

Agarwal, A.R., A. Gahlot, R. Verma and P.B. Rao. 2002. Effects of weed extracts on seedling growth of same varieties of wheat. *Journal Environmental Biology* 23:19–23.

Arunachalam, M., N. Mohan and A. Mahadevan. 2003. Cloning of Acinetobacter calcoaceticus chromosomal region involved in catechin degradation. *Microbial Research* 158:37–46.

Bais, H.P., V.M. Loyola-Vargas, H.E. Flores and J.M. Vivanco. 2001. Root-specific metabolism: the biology and biochemistry of underground organs. *In Vitro Cell. Development-PLANT* 37:730–741.

Bais, H.P., Walker, T.S., Stermitz, F.R., Hufbauer, R.A., Vivanco, J.M. 2002. Enantiomeric-dependent phytotoxic and antimicrobial activity of (±)-catechin. A rhizosecreted racemic mixture from spotted knapweed. *Plant Physiology* 128:1173–1179.

Bais, H.P., R. Vepachedu, S. Gilroy, R.M. Callaway, and J.M. Vivanco. 2003. Allelopathy and exotic plants: from genes to invasion. *Science* 301:1377–1380.

Bais, H.P., S. Park, T.L. Weir, R.M. Callaway and J.M. Vivanco. 2004. Root exudate- mediated rhizosphere communication. *Trends in Plant Sciences* 9:26–32.

Baldwin, I.T. 2003. At last, evidence of weapons of mass destruction. *Science STKE*, pe42.

Bell, D.T. and D.E. Koeppe. 1972. Noncompetitive effects of giant foxtail on the growth of corn. *Agronomic Journal* 64:321–325.

Bever, J.D. 1994. Feedback between plants and their soil communities in an old field community. *Ecology* 75:1965–1977.

Bever, J.D. 2002. Negative feedback within a mutualism: host-specific growth of mycorrhizal fungi reduces plant benefit. *Proceedings Royal Society Lund B* 269:2595–2601.

Bever, J.D., K.M. Westover, and J. Antonovics. 1997. Incorporating the soil community into plant population dynamics: the utility of the feedback approach. *Journal of Ecology* 85:561–573.

Blaney, C.S. and P.M. Kotanen. 2001. Effects of fungal pathogens on seeds of native and exotic plants: a test using congeneric pairs. *Journal of Applied Ecology* 38:1104–1113.

Blossey, B. and R. Nötzold. 1995. Evolution of increased competitive ability in invasive non-indigenous plants: a hypothesis. *Journal of Ecology* 83:887–889.

Boon, P.I. and L. Johnstone. 1997. Organic matter decay in coastal wetlands: an inhibitory role for essential oil from *Melaleuca alternifolia* leaves? *Archiv fur hydrobiologie* 138:428–449.

Braithwaite, R.W., W.M. Lonsdale, and J.A. Estbergs. 1989. Alien vegetation and native biota in tropical Australia: the impact of *Mimosa pigra*. *Biological Conservation* 48:189–210.

Bruce, K.A., G.N. Cameron, P.A. Harcombe, and G. Jubinsky. 1997. Introduction, impact on native habitats, and management of a woody invader, the Chinese tallow tree, *Sapium sebiferum* (L.) Roxb. *Natural Areas Journal* 17:255–260.

Burdon, J.J. 1987. Diseases and Plant Population Biology. Cambridge University Press, Cambridge, United Kingdom.

Burdon, J.J. 1993. The Structure of Pathogen Populations in Natural Plant Communities Annual Review Phytopathology 31:305–348.

Callaway, R.M. 2002. The detection of neighbors by plants. *Trends in Ecology and Evolution* 17:104–105.

Callaway, R.M., W.M. Ridenour, T. Laboski, T. Weir, and J.M. Vivanco. In press. Natural selection for resistance to the allelopathic effects of invasive plants. *Journal of Ecology.*

Callaway, R.M. and E.T. Aschehoug. 2000. Invasive plants versus their new and old neighbors: a mechanism for exotic invasion. *Science* 290:521–523.

Callaway, R.M. and J.L. Hierro. In press. Resistance and susceptibility of plant communities to invasion: revisiting Rabotnov's ideas about community homeostasis. In: Reigosa, M.J., P. Nuria and L. González, editors. Allelopathy: A Physiological Process with Ecological Implications. Kluwer Academic Publishers, The Netherlands.

Callaway, R.M. and W.M. Ridenour. 2004. Novel weapons: a biochemically based hypothesis for invasive success and the evolution of increased competitive ability. *Frontiers in Ecology and the Environment* 2:436–433.

Callaway R.M., G.C. Thelen, S. Barth, P.W. Ramsey, and J.E. Gannon. 2004b. Soil fungi alter interactions between North American plant species and the exotic invader *Centaurea maculosa* in the field. *Ecology* 85:1062–1071.

Callaway, R.M., G.C. Thelen, A. Rodriguez, and W.E. Holben. 2004a. Release from inhibitory soil biota in Europe may promote exotic plant invasion in North America. *Nature* 427:731–733.

Cheremisinoff, P.N. and F. Ellerbusch. 1978. Carbon Adsorption Handbook. Ann Arbor Science Publishers, Ann Arbor, Mich., USA

Connell J. H. 1971. Pages 298–312 *in* PJ den Boer and GR Gradwell, editors. Dynamics in Populations. Center for Agricultural Publishing and Documentation, Wageningen.

D'Antonio, C.M and P.M. Vitousek. 1992. Biological invasion by exotic grasses, the grass/fire cycle and global change. Annual Review of Ecology and Systematics 23: 63–87.

Deacon, J.W. and Donaldson, S.P. 1993. Molecular recognition in the homing responses of zoosporic fungi, with special reference to *Pythium* and *Phytophthora*. *Mycological Research* 97:1153–1171.

Dix, N. and J. Webster. 1995. *Fungal Ecology*. Chapman and Hall, London, United Kingdom.

Ehrenfeld, J. G. 2003. Effects of exotic plant invasions on soil nutrient cycling processes. Ecosystems 6: 503–523.

Elmore, C.D. 1985. Assessment of the allelopathic effects of weeds on field crops in the humid mid-south. *In* The Chemistry of Allelopathy. Ed. A.C. Thompson. pp 21–32. American Chemical Society, Washington, US.

Elton, C.S. 1958. The Ecology of invasions by animals and plants. Methuen, London, UK.

Evans, D.R., J. Hill, T.A. Williams, and I. Rhodes. 1985. Effects of coexistence in the performances of white clover-perennial rye grass mixtures. *Oecologia* 66:536–539.

Evans, R.D., R. Rimer, L. Sperry, and J. Belnap. 2001. Exotic plant invasion alters nitrogen dynamics in an arid grassland. *Ecological Applications* 11:1301–1310.

Fitter, A. 2003. Making allelopathy respectable. *Science* 301:1337–1338.

Flores, H.E., J.M. Vivanco, and V.M. Loyola-Vargas. 1999. "Radicle" biochemistry: the biology of root-specific metabolism. *Trends in Plant Science* 4:220–226.

Gaskin, J.F. and B.A. Schaal. 2002. Hybrid *Tamarix* widespread in U.S. invasion and undetected in native Asian range. *Proceedings of the National Academy of Science* 99:11256–11259.

Gleason, H.A. (1926) The individualistic concept of plant association. *Bulletin of the Torrey Botanical Club*, 53, 7–26.

Goodnight, C.J. 1990. Experimental studies of community evolution II: The ecological basis of the response to community selection. *Evolution* 44:1625–1636.

Grayston, S.J. and C.D. Campbell. 1998. Functional biodiversity of microbial communities in the rhizosphere of hybrid larch (*Larix eurolepis*) and Sitka spruce (*Picea sitchensis*). Tree Physiology 16:1031–1038.

Grayston S.J., G.S. Griffith, J.L. Mawdsley, C.D. Campbell, and R.D. Bardgett 2001. Accounting for variability in soil microbial communities of temperate upland grassland ecosystems. *Soil Biology & Biochemistry* 33:533–551.

Hänfling, B. and J. Kollman. 2002. An evolutionary perspective on invasions. *Trends in Ecology and Evolution* 17:545–546.

Hierro, J.L. and R.M. Callaway. 2003. Allelopathy and exotic plant invasion. *Plant and Soil* 256:25–39.

Hierro, J.L., J.L. Maron, and R.M. Callaway. 2005 A biogeographic approach to plant invasions: The importance of studying exotics in their introduced *and* native range. *Journal of Ecology* 93:5–15.

Jacobs, J.S., R.L. Sheley, and B.D. Maxwell. 1996. Effect of *Sclerotina sclerotiorum* on the interference between bluebunch wheatgrass (*Agropyron spicatum*) and spotted knapweed (*Centaurea maculosa*). *Weed Technology* 10:13–21.

Janzen D. H. 1970. Herbivores and the number of tree species in tropical forests. *American Naturalist* 104:501–528.

Johnson, H.B. and H.S. Mayeux. 1992. Viewpoint: A view on species additions and deletions and the balance of nature. *Journal of Range Management* 45:322–333.

Joy, P. and A. Laitinen. 1980. Breeding for co-adaptation between red clover and timothy. Hankkija's Seed Publication 13, Hankkija Plant Breeding Institute, Finland.

Keane, R.M. and M.J. Crawley. 2002. Exotic plant invasions and the enemy release hypothesis. *Trends in Ecology and Evolution* 17: 164–170.

Klironomos, J. 2002. Feedback with soil biota contributes to plant rarity and invasiveness in communities. *Nature* 417:67–70.

Korhammer S.A. and E. Haslinger. 1994. Isolation of a biologically active substance from rhizomes of quackgrass [*Elymus repens* (L.) Gould]. *Journal Agricultural Food Chemistry* 42:2048–2050.

Kourtev, P. S., J. G. Ehrenfeld, and W. Huang. 2002. Exotic species alter microbial structure and function in the soil. Ecology 85: 3152–3166.

Kowarik, I. 1995. Time lags in biological invasions with regard to the success and failure of alien species. Pages 15–38 in: Py?ek, P., K. Prach, M. Rejmánek and M. Wade, editors. Plant Invasions—General Aspects and Special Problems. SPB Academic Publishing, Amsterdam, NL.

Lee, C.E. 2002. Evolutionary genetics of invasive species. *Trends in Ecology and Evolution* 17:386–391.

Lortie, C.J., R.W. Brooker, P. Choler, Z. Kikvidze, R. Michalet, F.I. Pugnaire, and R.M. Callaway. 2004. Rethinking plant community theory. *Oikos* 107:433–438.

Mack, R.N., D. Simberloff, W.M. Lonsdale, H. Evans, M. Clout, and F.A. Bazzaz. 2000. Biotic invasions: causes epidemiology, global consequences and control. *Ecological Applications* 10:689–710.

Mahall, B.E. and R.M. Callaway. 1992. Root communication mechanisms and intracommunity distributions of two Mojave Desert shrubs. *Ecology* 73:2145–2151.

Malecki, R.A., B. Blossey, S.D. Hight, D. Schroeder, L.T. Kok, and J.R. Coulson. 1993. Biological control of purple loosestrife. *BioScience* 43:680–687.

Mallik, A.U and F. Pellissier. 2000. Effects of *Vaccinium myrtillus* on spruce regeneration: testing the notion of coevolutionary significance of allelopathy. Journal of Chemical Ecology 26:2197–2209.

Maron, J.L. and M. Vilà. 2001. Do herbivores affect plant invasion? Evidence for the natural enemies and biotic resistance hypotheses. *Oikos* 95:363–373.

Martin, M.M. and J. Harding. 1981. Evidence for the evolution of competition between two species of annual plants. *Evolution* 35:975–987.

McCormick, J. E. and Platt, R.B. 1980. Recovery of an Appalachian forest following the chestnut blight. *American Midland Naturalist* 104:264–273.

Merck Index. 1996. 8-Hydroxyquinoline. Merck and Co., Inc., Whitehouse, NJ. Page 4890.

Meyer, J-Y. and J. Florence. 1996. Tahiti's native flora endangered by the invasion of *Miconia calvescens* DC. (Melastomataceae).

Mills, K. E. and J. D. Bever. 1998. Maintenance of diversity within plant communities: soil pathogens as agents of negative feedback. *Ecology* 79:1595–1601.

Mitchell, C.G. and A.G. Power. 2003. Release of invasive plants from fungal and viral pathogens. *Nature* 421:625–627.

Muller, C.H. 1966. The role of chemical inhibition (allelopathy) in vegetational composition. *Bulletin of the Torrey Botanical Club* 93:332–351.

Nardi, S., Concheri, G., Pizzeghello, D., Sturaro, A., Rella, R., Parvoli, G. 2000. Soil organic matter mobilization by root exudates. *Chemosphere* 5: 653–658

Packer, A. and K. Clay. 2000. Soil pathogens and spatial patterns of seedling mortality in a temperate tree. *Nature* 440:278–281.

Packer, A. and K. Clay. 2002. Soil pathogens and *Prunus serotina* seedlings and sapling growth near conspecific trees. *Ecology* 84:108–199.

Pandey, D.K. 1994. Inhibition of salvinia (*Salvinia molesta* Mitchell) by parthenium (*Parthenium hysterophorus* L.). II. Relative effect of flower, leaf, stem, and root residue on salvinia and paddy. *Journal Chemical Ecology* 20:3123–3131.

Phillips and Shine. 2004. Adapting to an invasive species: toxic cane toads induce morphological change in Australian snakes. *Proceedings of the National Academy of Science* 49:17150–17155.

Powles, S.B. and J.A.M. Holtum. 1994. *Herbicide Resistance in Plants.* (Lewis Publishers, Boca Raton, FL.

Prati, D. and O. Bossdorf. 2004. Allelopathic inhibition of germination by *Alliaria petiolata* (Brassicaceae). American Journal Botany 91:285–288.

Priha O., S.J. Grayston, R. Hiukka, T. Pennanen, and A. Smolander 2001. Microbial community structure in soils under *Pinus sylvestris, Picea abies* and *Betula pendula. Biology & Fertility of Soils* 33:17–24.

Priha O., S.J. Grayston, T. Pennanen, A. and Smolander. 1999. Microbial activities related to C and N cycling, and microbial community structure in the rhizospheres of *Pinus sylvestris, Picea abies* and *Betula pendula* seedlings in an organic and mineral soil. *FEMS Microbiology and Ecology* 30:187–199.

Quayyum, H.A., A.U. Mallik, D.M. Leach and C, Gottardo. 2000. Growth inhibitory effects of nutgrass (*Cyperus rotundus*) on rice (*Oryza sativa*) seedlings. *Journal Chemical Ecology* 26:2221–2231.

Rabotnov, T.A. 1982. Importance of the evolutionary approach to the study of allelopathy. Translated from Ékologia, No. 3, May-June:5–8.

Reinhart, K.O. A. Packer, W. H. Van der Putten, and K. Clay. 2003. Plant–soil biota interactions and spatial distribution of black cherry in its native and invasive ranges. *Ecology Letters* 6:1046–1050

Rice, E.L. 1964. Inhibition of nitrogen-fixing and nitrifying bacteria by seed plants. *Ecology* 45:824–837.

Rice, K.J. and N.C. Emory. 2003. Managing microevolution: restoration in the face of global change. *Frontiers in Ecology* 1:469–478.

Ridenour, W.M. and R. M. Callaway. 2001. The relative importance of allelopathy in interference: the effects of an invasive weed on a native bunchgrass. Oecologia 126: 444–450.

Ridenour, W.L. and R.M. Callaway. 2003. Root herbivores, pathogenic fungi, and competition between *Centaurea maculosa* and *Festuca idahoensis*. *Plant Ecology* 169:161–170.

Schreiner, O. and Reed, H.S. 1907. The production of deleterious excretions by roots. *Bulletin of the Torrey Botanical Club* 34:279–303.

Siemann, E. and W.E. Rogers. 2001. Genetic differences in growth of an invasive tree species. Ecology Letters 4:514–518.

Siemens, D., S. Garner, T. Mitchell-Olds, and R.M. Callaway. 2002. The cost of defense in the context of competition. *Brassica rapa* may grow *and* defend. *Ecology* 83:505–517.

Stachon, W.J. and R.L. Zimdahl. 1980. Allelopathic activity of Canada thistle (*Cirsium arvense*) in Colorado. *Weed Science* 28:83–86.

Steenhagen, D.A. and R.L. Zimdahl. 1979. Allelopathy of leafy spurge (*Euphorbia esula*). *Weed Science* 27:1–3.

Stevenson, F.J. and M.A. Cole. 1999. Cycles of soil: carbon, nitrogen, phosphorus, sulfur, micronutrients. John Wiley and Sons, Inc., New York NY.

Thompson, J.N. 1994. The Coevolutionary Process. University of Chicago Press, Chicago.

Thompson, 1997. Evaluating the dynamics of coevolution among geographically structured populations. *Ecology* 78:1619–1623.

Thompson, J.N. 1998. Rapid evolution as an ecological process. *Trends in Ecology and Evolution* 13:329–331.

Thompson, J.N. 1999. Specific hypotheses on the geographic mosaic of coevolution. *American Naturalist* 153:1–14.

Thorpe, A.S. and R.M. Callaway. In press. Interactions between invasive plants and soil ecosystems: will feedbacks lead to stability or meltdown? Pages xxx-xxx In Cadotte, M.W., S. McMahon and T. Fukami (editors). Conceptual Ecology and Invasions Biology: Reciprocal Approaches to Nature.

Turkington, R. and L.A. Mehrhoff. 1991. The role of competition in structuring pasture communities. In Grace, J.B. and D. Tilman (editors) Perspectives on Plant Competition. Academic Press, San Diego, USA.

Turkington, R. and J.L. Harper. 1979. The growth, distribution and neighbor relationships of *Trifolium repens* in a permanent pasture. IV. Fine scale biotic differentiation. *Journal of Ecology* 67:245–254.

van Breemen, N. and A.C. Finzi. 1998. Plant-soil interactions: ecological aspects and evolutionary implications. *Biogeochemistry* 42:1–19.

van der Putten, W.H. and B.A.M. Peters. 1997. How soil-borne pathogens may affect plant competition. *Ecology* 78: 1785–1795.

van der Putten, W.H., C. Van Dijk, and B.A.M. Peters. 1993. Plant-specific soil-borne diseases contribute to succession in foredune vegetation. Nature 362:53–56.

van der Putten, W.H., L.E.M. Vet, J.A. Harvey, and F.L. Wackers. Linking above- and belowground multitrophic interactions of plants, herbivores, pathogens, and their antagonists. *Trends in Ecology and Evolution* 16, 547–554. 2001.

van der Putten, W.H. 1997. Plant-soil feedback as a selective force. *Trends in Ecology and Evolution* 12:169–170.

Vivanco, J.M., H.P. Bais, F.R. Stermitz, G.C. Thelen, and R.M. Callaway. 2004. Biogeographical variation in community response to root allelochemistry: novel weapons and exotic invasion. *Ecology Letters* 7:285–292.

Von Holle, B., H. Delcourt and D. Simberloff. 2003. The importance of biological inertia in plant community resistance to invasion. *Journal of Vegetation Science* 14:424–432.

Walker, T. S., Bais, H. P., Grotewold, E., Vivanco, J. M. 2003. Root exudation and rhizosphere biology. *Plant Physiology* 132:44–51.

Wardle, D.A., K.S. Nicholson, and A. Rahman. 1995. Ecological effects of the invasive weed species *Senecio jacobaea* L. (ragwort) in a New Zealand pasture. *Agriculture, Ecosystems, and Environment* 56:19–28.

Wardle, D.A., K.S. Nicholson, M. Ahmed, and A. Rahman. 1994. Interference effects of the invasive plant *Carduus nutans* L. against the nitrogen fixation ability of *Trifolium repens/L*. Plant and Soil 163:287–297.

Westover, K.M., A.C. Kennedy, and S.E. Kelley. 1997. Patterns of rhizosphere microbial community structure associated with co-occurring plant species. *Journal of Ecology* 85:863–873.

Williamson, G.B. 1990. Allelopathy, Koch's postulates, and the neck riddle. *In* Perspectives on Plant Competition. J.B. Grace and D. Tilman, Editors. pp. 143–162. Academic Press, Inc., San Diego, CA, US.

Wilson, D.S. 1997. Biological communities as functionally organized units. *Ecology* 78:2018–2024.

Woods, R.W. and R.E. Shanks. 1959. Natural replacement of chestnut by other species in the Great Smoky Mountains National Park. *Ecology* 40:349–361.

Yamamoto, Y. 1995. Allelopathic potential of *Anthoxanthum odoratum* for invading *Zoysia*-grassland in Japan. *Journal Chemical Ecology* 21:1365–1373.

14

Community Composition and Homogenization

EVENNESS AND ABUNDANCE OF NATIVE AND EXOTIC PLANT SPECIES

Michael L. McKinney and Julie L. Lockwood

As exotic species become increasingly widespread, many parts of the biosphere are losing their biological distinctiveness. This may be particularly true for plant communities, which have often gained many exotic species. Comparisons of native and exotic species within plant communities provide two key insights for biogeography. The first is that species abundance and evenness apparently play a surprisingly small role in homogenization. This occurs because, as with native species, each exotic species tends to vary widely among communities in its abundance. The second is that exotic species show many other patterns that are similar to those of native species. Exotic species evenness is highly correlated with native species evenness, and both exotic and native components of communities show an exponential decline in shared species with separation by distance and latitude. Further, evenness is apparently unrelated to relative number of shared species in both natives and exotics. These fine-scale similarities between natives and exotics extend previous findings at larger scales, such as similar latitudinal diversity patterns in native and exotic species. They also imply that patterns of abundance form relatively quickly via current ecological processes and without prolonged periods of coevolution.

Introduction

Biotic homogenization is often mentioned as the ultimate outcome of the widespread establishment of exotic species. Biotic homogenization occurs when exotic species become established in many places and thereby increase the compositional similarity among locations (Lockwood and McKinney 2001; Collins et al. 2002; Rahel 2000, 2002; Olden et al. 2004). Since a wide range of human activities promote biotic homogenization, the future biosphere has been described with such colorful terms as the New Pangea (Rosenzweig 2001), the Homogecene (Guerrant 1992), and the Planet of Weeds (Quammen 1998).

Although conservation biologists recognize homogenization as a basic conservation challenge (Meffe and Carroll 1998), a thorough examination of the current literature reveals how little we actually know about the process. Even though homogenization has the potential to dramatically reshape our biosphere, such basic questions as how we define the different kinds of biotic homogenization, and how we measure them, are rarely discussed.

In this chapter, we review some of these deficiencies and argue that biotic homogenization is a much more complex topic than seems to be generally realized by most biologists. We also highlight some basic related questions that address whether there are fundamental differences in the ways in which native and exotic species are distributed. Specifically, we highlight the following questions that have barely been touched on by scientific research:

1. The vast majority of homogenization studies rely on simple species richness metrics that omit abundance data when measuring homogenization. Do abundance data alter the homogenization patterns found with only presence-absence data?
2. When abundance data are included, do exotic species show similar biogeographic patterns to native species?
3. Can exotic species promote biotic differentiation as well as homogenization? If so, what factors (such as spatial scale and distance) determine these opposing outcomes?
4. Given that human activities ("disturbances") promote both native species extinction and exotic species establishment, is there evidence that increasing disturbance intensity produces increasing homogenization?

Our review of these questions is far from exhaustive. Indeed, most of them have received so little scientific attention that there is little evidence to review. Nevertheless, these issues are of tremendous applied importance. Furthermore, a better understanding of these issues can provide fundamental insights into the processes setting species distributions, abundances, and turnover. Thus, our goal is threefold: first, to describe how little we know about the processes relevant to homogenization; second, to advance our understanding of biotic homogenization through new analyses of several data sets; and third, to draw insight from data on homogenization to enhance our understanding of how ecosystems function.

Biotic Homogenization and Abundance

In the physical sense, "homogenization" typically refers to the thorough mixing of materials, liquids, or particles. It is therefore important to ask what biologists mean when they use the term. In most cases, workers studying homogenization (including ourselves) have focused on species as the "unit" of mixing and have examined the replacement of native species with exotic species. Consequently, efforts to quantify homogenization have focused on methods that simply reflect the presence or absence of species. Examples include (1) the use of species-area curves (e.g., McKinney 1998; Rosenzweig 2001; Collins et al. 2002) to measure the relative change in local and regional species richness with homogenization, (2) the use of Jaccard's and other community similarity indices (e.g., Rahel 2000; Blair 2001) to measure increases in species similarity among homogenizing communities, and (3) the use of other spatial turnover metrics (e.g., Harrison 1993; Duncan and Lockwood 2001) that demonstrate decreasing spatial turnover in species composition with homogenization.

As several workers have noted (Collins et al. 2002; Olden et al. 2004), a major limitation of using these species-counting methods to measure homogenization is that they ignore species abundance. This is potentially a huge omission, given that some exotic species are often superabundant in intensely modified landscapes, while others are rare where they are introduced. Methods that simply count species give equal weight to both rare and superabundant exotic species, despite their obvious functional dissimilarity.

To see if our views of homogenization change when species abundance is taken into account, we calculated two different similarity indices for plant communities in pairwise comparisons among 20 localities in the United States—one that ignores species abundance and one that considers it explicitly. The data come from parks with a wide range of disturbance levels, from natural areas to urban parks (listed in Table 14.1). The plant lists for these sites do not contain extinct native species, so it is likely that they underestimate the homogenizing effects of human activities. Nonetheless, we believe that these data should provide a useful approximation for considering several issues relevant to homogenization and, more generally, to patterns of species abundance and distribution. We chose the localities used in this study to reflect a wide geographic area and because published inventories of native and exotic plant species, which included ranked abundance data (1 = rare, 2 = moderately common, 3 = very common), were available. The availability of abundance information allowed us to calculate the Bray-Curtis index, which incorporates species abundance data (Clarke and Warwick 2001), in addition to Jaccard's index, which incorporates only presence-absence data.

Given two locations (or samples), j and k, Jaccard's index is calculated as

$$JI = 100\left(\frac{A}{A+B+C}\right) \tag{14.1}$$

TABLE 14.1 *"Greenspace" localities used in this study*

Locality[a]	Description	Exotic/Native[b]	Source
Big Frog Mtn, TN	Natural area	*0.04* (20/500)	Murrell and Wofford 1987
Wolf Cove, TN	Natural area	*0.05* (28/560)	Clements and Wofford 1991
Ichauway, GA	Natural area	*0.10* (93/930)	Drew et al. 1998
Passage Island, MI	National park	*0.13* (32/242)	Judziewicz 1997
Paynes Prairie, FL	State reserve	*0.13* (64/512)	Easley and Judd 1990
Myakka River, FL	State park	*0.14* (85/620)	Huffman and Judd 1998
Indian Springs, GA	State park	*0.17* (81/476)	Howell 1991
Bear Creek, TN	Natural area	*0.21* (125/610)	Carpenter and Chester 1987
Horseshoe Lake, IL	Conservation area	*0.25* (130/520)	Basinger et al. 1997
Richmond, VA	Battlefield park	*0.30* (177/590)	Hayden et al. 1989
Saratoga, NY	Battlefield park	0.34 (134/394)	Stalter et al. 1993
Mount Vernon, VA	Historic park	0.35 (71/204)	Wells and Brown 2000
Van Cortlandt, NY	City park	0.41 (100/242)	Profous and Loeb 2000
West Hills, NY	County park	0.45 (90/200)	Greller and Clemants 2001
Latimer Point, CT	Residential and park	0.47 (138/291)	Hill 1996
Hopewell Park, OH	Historic park	0.48 (141/296)	Bennett and Course 1996
Orient Beach, NY	State park	0.50 (123/246)	Lamont and Stalter 1991
Bronx Parkway, NY	Reservation	0.55 (276/505)	Frankel 1999
Chicago, IL	Urban vacant lots	1.25 (64/51)	Crowe 1979
Ellis Island, NY	Historic park	1.33 (155/116)	Stalter and Scotto 1999

[a]Localities are ranked in ascending order of the exotic/native ratio.
[b]Ratio (italicized) = exotic species richness/native species richness.

where A represents the number of species shared between sites j and k, B is the number of species present in location j but not k, and C is the number of species present in location k but not j. Jaccard's index is often scaled to 100 (as in Equation 14.1) to represent the probability that a single species picked at random will be present in both locations (Clarke and Warwick 2001).

When the species within locations j and k have associated abundance scores, it is possible to calculate similarity using the Bray-Curtis index,

$$BC = 100 \left[1 - \frac{\sum_{i=1}^{p} |y_{ij} - y_{ik}|}{\sum_{i=1}^{p} (y_{ij} + y_{ik})} \right] \tag{14.2}$$

Here, y_{ij} is the abundance of the ith species in the jth location, and y_{ik} is the abundance of the same species in the kth location. The numerator and the denominator are summed over all species in the combined species pool of richness, p. In both equations, a value of 0 represents the situation in which no species are shared, whereas a value of 100 represents the situation in which all species are shared and, in the case of the Bray-Curtis index, in which all species have the same abundance scores.

We calculated the Bray-Curtis and Jaccard's indices for the set of exotic species and again for the set of native species in our site comparisons. Thus, using only natives and then only exotics, we were able to compare the indices and see whether abundance information increased (or decreased) similarity. In addition, we were able to compare the sets of native and exotic species to see if they evinced analogous similarity patterns. Because of the large number of possible pairwise comparisons from 20 localities (see Table 14.1), we randomly selected 100 pairwise comparisons for this and other analyses that use these data.

In Figure 14.1, we plot the similarity score for each of the 100 pairwise comparisons according to the Jaccard's (*x*-axis) and Bray-Curtis (*y*-axis) indices. Our results show that there is a very strong positive relationship between Jaccard's index and the Bray-Curtis index for both exotic and native species. This result is to be expected to some degree, since each similarity index is utilizing the same underlying data structure (i.e., the species' presence and absence patterns are the same in each calculation). The indices can differ only through the added influence of abundance patterns in the Bray-Curtis index. The key question that must be addressed is how the observed linear relationship (shown in Figure 14.1) compares to the range of possible outcomes.

To further explore the relationship between Jaccard's index and the Bray-Curtis index, we analytically solved for the boundaries of the relationship between them. To do so, we defined two sites (A and B) and populated each with 100 species. To vary the similarity in species composition between the sites, we varied the number of shared species between the two sites from 5 to 100 (i.e., where 100 represents complete homogeneity of species present). We then calculated Jaccard's index on all site-by-site comparisons, considering the

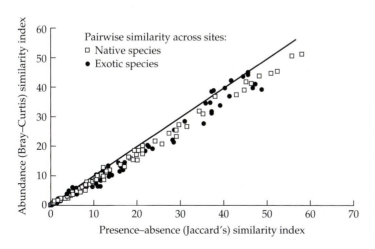

Figure 14.1 Similarity scores for pairwise comparisons between localities according to the Bray-Curtis index, which incorporates species abundance, and Jaccard's index, which does not. *N* = 100 pairwise comparisons for exotic and for native species. The line indicates equality.

species as only present or absent (0, 1). This is a relatively straightforward endeavor; for example, if 5 species are shared, Jaccard's index is 5%, whereas if 75 species are shared, the index is 75%.

The values for the Bray-Curtis index are not so straightforward, since different patterns of species abundance will alter the index value. To bound the range of possible outcomes, we considered situations that would lead to the minimum and maximum Bray-Curtis scores, as well as the expectation under a roughly random assignment of abundances to shared species. Since we could resolve abundance only to a three-tier ranking (1, 2, 3, with 3 being most abundant) in our observed data set, solving for the bounds on the Bray-Curtis index was relatively simple. The minimum Bray-Curtis score occurs when all non-shared species have an abundance of 3 and all shared species have abundances as dissimilar as possible—that is, a difference in abundance of 2 (i.e., 3 − 1). By contrast, the maximum possible Bray-Curtis score occurs when all non-shared species have the lowest possible abundances and all shared species have the highest possible abundances. Finally, we calculated the Bray-Curtis index assuming that abundance scores were close to randomly different in relation to shared and non-shared species, which we approximated by assigning all non-shared species moderate abundance (i.e., 2) while setting the average difference in abundance for shared species to 1.

Figure 14.2 illustrates our results. The expected relationship between the Jaccard's and Bray-Curtis indices is curvilinear for all three scenarios. The curvature is strongest for the maximum Bray-Curtis scenario. The minimum Bray-

Figure 14.2 The relationship between the Bray-Curtis and Jaccard's similarity indices. Differences between these indices are expected because Jaccard's index considers only presence-absence data, while the Bray-Curtis index considers presence-absence and species abundance. Given the variation in abundance scenarios described in the text, analytic solutions to these two indices show that (1) the maximum bound on this relationship occurs when all species shared between compared habitats have high abundance and all non-shared species have low abundance, (2) the null relationship (based on random expectation) occurs when abundances of shared species are moderately different (i.e., close to randomly different) and all non-shared species have moderate abundance, and (3) the minimum bound on this relationship occurs when all shared species have abundances as dissimilar as possible and all non-shared species have high abundance.

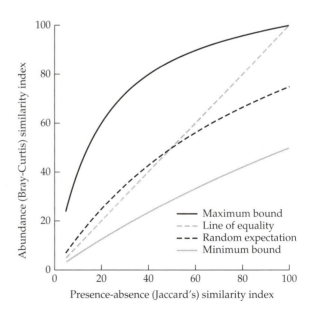

Curtis scenario approaches a linear relationship with Jaccard's, but the line is well below the line of equality for the two indices. Finally, the random Bray-Curtis scenario shows a curvilinear relationship that falls near the line of equality.

If we compare Figure 14.2 with our observed data within Figure 14.1, we can derive three insights. First, the tight linear correlation between the observed Jaccard's and Bray-Curtis indices was not a foregone conclusion. Although the two scores are always positively correlated to some extent, the tight linear correlation we found was not a necessary outcome. The observed values are constrained to a narrow range of the potential outcome space. Second, our observed relationship is below the line of equality and near the lower bound for the relationship. Thus, the abundance scores for shared species in our observed data tend to be more different among sites than random. This is true for both exotic and native species. Third, we can see the limitations of the Bray-Curtis index in relating information on similarity. Although Bray-Curtis incorporates abundance information, and in our observed data the abundances for shared species tend toward being as different as is possible, the Bray-Curtis scores are still tightly associated with the Jaccard's scores. The presence-absence component of Bray-Curtis dominates the index values, which means that the Bray-Curtis index is not telling us much more than Jaccard's index about similarity between locations, at least given the crude abundance rankings we were capable of producing (see also Clarke and Warwick 2001). True abundance values with a wider range of measures would be likely to produce a much larger effect of abundance on the index value.

Evenness of Abundance in Native and Exotic Species

Another important dimension of abundance is evenness, the extent to which abundance is distributed evenly (or skewed) among species in a community. Because of the common focus on invasive exotic species, many people assume that exotic species abundance is often more strongly skewed than native species abundance. To test this assumption, we analyzed the plant data discussed above using Hill's (1973) evenness index, which is the ratio of abundant to rare species. We used this index because, unlike most other evenness indices, it is not affected by richness (Waite 2000). Species richness was often very different between native and exotic species in our data, so this was an important criterion in comparing exotic to native patterns. In addition, because we had rank abundance data, this index provided a simple ratio by comparing the number of abundant species (rank = 3) to rare species (rank = 1).

As with similarity index patterns (see Figure 14.1), we find that exotic species tend to have evenness patterns similar to those of native species. Specifically, for all 20 communities examined, exotic species evenness tends to be highly correlated with native species evenness (Figure 14.3; $R^2 = 0.588$; $P < 0.01$). Again, these are tentative results, but they indicate that communities with skewed abundance distributions of native species also have skewed abundance dis-

Figure 14.3   Correlation between evenness in exotic and in native species. Evenness, a measure of how evenly abundance is distributed among species in a community, was calculated with Hill's evenness index (see text for details). $N = 20$ communities.

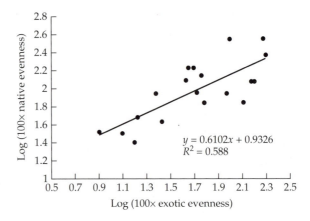

$y = 0.6102x + 0.9326$
$R^2 = 0.588$

tributions of exotic species. Whatever factors promote skewed or even abundance distributions in native species also seem to promote them in exotic species.

To see whether evenness had an effect on similarity, we regressed the presence-absence (Jaccard's) index onto the evenness difference between each pair of sites compared (absolute value of the difference between evennesses at sites 1 and 2). These calculations were performed separately for the native and exotic components of these site comparisons. We found no correlation between similarity and evenness, either for native species or for exotic species (Figure 14.4). We did the same analysis using the Bray-Curtis index, instead of Jaccard's, and obtained the same results (data not shown). Once again, native and exotics species show similar patterns.

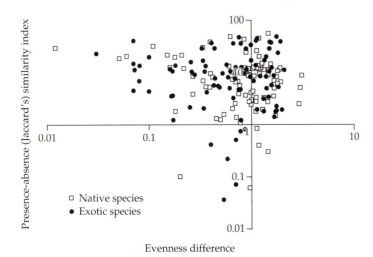

Figure 14.4   Lack of correlation between presence-absence (Jaccard's) similarity scores and differences in evenness (absolute value) between pairs of sites. These calculations were performed separately for the native and exotic species of compared sites. Both axes are log-transformed (base 10) to normalize skewed frequency distributions. $N = 100$ for both exotic and native species. $R^2 = 0.02$ for native species and 0.09 for exotic species, but neither value is statistically significant.

Exotics cause both differentiation and homogenization

While homogenization is often mentioned in the context of exotic species, it is theoretically possible that exotic species can sometimes produce the opposite effect, *reducing* the compositional similarity between biological communities. In a revealing modeling exercise, Olden and Poff (2003) review many theoretical scenarios of exotic species introductions and native species extinctions to show that exotics can have widely differing effects on species composition. They demonstrate that, while exotic introductions and native species extinctions can often increase the similarity among communities (homogenization), exotics can also have no effect on, or even reduce, community similarity, a phenomenon that Olden and Poff (2003) call "biotic differentiation" (i.e., the opposite of homogenization).

Homogenization tends to occur where the same exotic species become established in many places. This loss of distinctiveness is enhanced if introductions correspond to the extinction of unique native species. In contrast, biotic differentiation occurs where (for whatever reason) different suites of exotics occur in different places. Olden and Poff (2003) illustrate this latter case in fishes with the increasing incidence of aquarium releases, which are often idiosyncratic and thereby tend to produce biotic differentiation. In contrast, bait bucket releases tend to consist of the same fish species and thus promote homogenization.

Aside from propagule sources, a second factor influencing how exotic species affect community similarity is scale. Olden and Poff (2003) suggest that increasing the spatial extent of an analysis will increase the tendency to see patterns of homogenization because coarser scales increase the probability of recording particular exotic species as more habitats are sampled. Conversely, as spatial extent decreases, biotic differentiation may be more typically found, because at finer (local) scales the probability of recording the same exotic species in two locations is reduced. Only a very few studies have examined such effects of scale, but they seem to support these suggestions. Marchetti et al. (2001) found that the introduction of exotic fish species in California since the 1800s has produced differentiation among fish communities in local watersheds, but homogenization across geographic provinces. Rejmanek (2000) showed preliminary evidence that exotic plants in U.S. states have increased the differences between the floras of adjacent states, whereas exotics are homogenizing the floras of distant states. A similar pattern demonstrating the importance of spatial scale was shown by Sax and Gaines (2003), who reviewed evidence that exotics (particularly vascular plants) tend to increase regional species richness, and hence may increase the number of shared species between regions, whereas the effect of exotics on local (community) species richness is much more variable.

Finally, we suggest that a third factor influencing homogenization (versus differentiation) is disturbance intensity. Using the comparisons of plant community localities noted above (see Table 14.1), McKinney (2004a) examined the ratio of Jaccard's index for exotics to the same index for native species (JI_{exotic}/JI_{native}). Where (JI_{exotic}/JI_{native}) is greater than 1, we can infer that exotics have a homogenizing influence on compared sites, and where (JI_{exotic}/JI_{native}) is less than 1, we can infer that exotics are differentiating the native communities. McKinney (2004a) found that the most important factor affecting this ratio was the percentage of exotic species in the two localities being compared. There was a strong positive correlation between (JI_{exotic}/JI_{native}) and the mean percentage of exotic species in the two sites. In other words, where exotics were a small proportion of the total flora, they tended to differentiate the two communities. Where exotics were more species-rich, they tended to homogenize communities.

McKinney (2004a) presented two explanations for this observation. First, where human disturbance is low, the number of established exotics at a site may be small, but the potential pool of invading species may be large. When only a small number of exotic species are drawn from a large pool, the probability that many exotics will be shared among sites will be small. In contrast, when disturbance levels are high, the number of established exotics at a site may be large, and the likelihood that the same exotic species will be present at different sites increases purely for sampling reasons. Alternatively, if intense disturbance reduces the pool of potential exotics that can successfully colonize, which will occur if only a limited number of species can tolerate intensely disturbed sites, then the likelihood increases that many of the same exotic species will be found across those sites.

Roles of distance, latitude, and disturbance intensity

To further explore the role of disturbance in homogenization, we should account for the role of distance and latitudinal separation among sites being compared. Basic biogeographic principles imply that increasing proximity between two sites will tend to increase community similarity among their native species, because, among other things, they share more similar physical parameters and species immigration pools (Brown and Lomolino 1998). Two studies have documented an exponential decline in Jaccard's similarity index with distance between communities of native plants (Nekola and White 1999) and animals (Poulin 2003). Using the plant communities discussed above, McKinney (2004a,b) documented that, for exotic plants, Jaccard's index also showed an exponential decline with increasing distance and latitude separation between sites.

To McKinney's (2004a,b) previous studies, we add here two new analyses. The first replicates one of McKinney's studies (2004b), except that we plot the Bray-Curtis similarity score against latitude separation between localities. This analysis shows that, as with Jaccard's index, there is an exponential decline in similarity with difference in latitude (Figure 14.5). Again, both exotic and native species share similar patterns.

The second analysis uses data from inventories of native and exotic plant species for several U.S. cities, compiled by Clemants and Moore (2003). We calculated Jaccard's index among Boston, New York City, Philadelphia, Detroit, St. Louis, and other large metropolitan areas. From this pool, our specific city

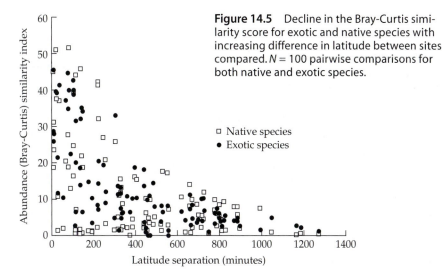

Figure 14.5 Decline in the Bray-Curtis similarity score for exotic and native species with increasing difference in latitude between sites compared. $N = 100$ pairwise comparisons for both native and exotic species.

comparisons were picked at random. Our goal was to examine the effects of distance on Jaccard's index in these highly disturbed plant communities. Cities are the most intensely modified ecosystems (McKinney 2002), and we wanted to see if this intense disturbance produced very high levels of biotic homogenization. For example, Blair (2001) found that cities increase homogenization in bird and butterfly species.

Our results show that, for any given distance separating two city sites, Jaccard's index is quite high (Figure 14.6). For comparison, the city data are plotted along with the plant data used earlier, which are mostly from parks and

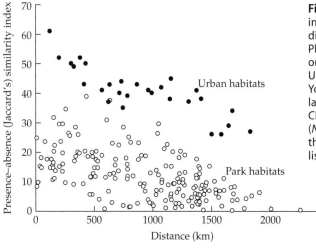

Figure 14.6 Decline in Jaccard's index in plant communities with increasing distance between localities compared. Plant communities incorporated data on both native and exotic species. Urban habitats included Boston, New York City, Detroit, St. Louis, and other large metropolitan areas listed in Clemants and Moore 2003. Park habitats ($N = 100$ pairwise comparisons) were the parks and other "greenspace" areas listed in Table 14.1.

other "greenspace" areas (see Table 14.1). The city flora similarity is consistently much higher than the similarity among greenspace communities. This analysis confirms in a systematic way (for the first time, to our knowledge) that urbanized habitats are indeed much more botanically homogenized than are more natural areas, including those parks that contain many exotic species. However, there may be a scale effect here: the greenspace sites were much smaller in size, ranging to a maximum of about 7000 ha (McKinney 2004a). In contrast, the eight urban areas compared were all over 300,000 ha. There is some evidence that larger areas are more homogenized (Marchetti et al. 2001), perhaps by spatial sampling effects (as described earlier). Therefore, these findings need more testing, with scale effects removed. Regardless of comparisons between urban and greenspace communities, a second notable pattern emerges from the urban data: there is a decline in Jaccard's index with increasing distance between urban communities (see Figure 14.6). This pattern indicates that even in these highly disturbed and homogenized environments, with enormous propagule pressure, increasing spatial separation still plays some role in reducing community similarity.

Genetic and functional homogenization and differentiation

There are several kinds of biotic homogenization. These are well described by Olden et al. (2004), who argue that, in addition to the species (taxonomic) homogenization that we have analyzed thus far, biologists also need to analyze genetic and functional homogenization. Genetic homogenization includes genetic hybridization among related species, which is clearly a conservation concern where exotic species interbreed with natives, which has occurred in many taxa (Rhymer and Simberloff 1996; Daehler and Carino 2001). Functional homogenization includes the increasing uniformity of ecological function that may occur independently of changes in taxonomic identity. While data on this phenomenon are sparse, Olden et al. (2004) describe several possible scenarios in which a decrease in functional diversity might reduce overall ecosystem functioning; this could also lead to a loss of complexity, stability, and resistance to environmental change by narrowing the available range of species-specific responses.

It is also possible that, like species (taxonomic) changes in communities, genetic and functional changes resulting from exotic species may promote differentiation. Petit (2004), for example, has noted how invasions can be studied at the gene level. In theory, a local community that is invaded by exotic genes will become genetically differentiated from adjacent communities that are not invaded. Similarly, functional differentiation could occur where exotic species drastically modify the ecosystem function of a local community relative to that of surrounding communities. Exotic earthworms, for instance, have substantially affected soil phosphorus cycling in some temperate forests (Suarez et al. 2004), while invasion of some Hawaiian ecosystems by the N-fixing plant *Myrica faya* has drastically altered their ecosystem properties (Vitousek and Walker 1989).

Conclusions

The prevalence of the term "homogenization" within scientific and popular literature has increased dramatically over the past several years. The effects of exotic species on spatial diversity patterns are increasingly recognized as a critical link to our understanding of anthropogenic changes on many levels of the biosphere: genes, community composition, and ecosystem services (Olden et al. 2004). Our analysis of homogenization has revealed two significant insights for biogeography.

One tentative insight is that, perhaps surprisingly, species abundance does not seem to play a large role in the homogenization process, or at least in our perception and measurement of that process. Certainly we need to examine more data—in particular, data with more detailed abundance measures. Still, the evidence available suggests that abundance varies widely among localities in both exotic and native species, so that information on species abundance adds little, if any, information on community similarity beyond that already provided by presence-absence metrics. In addition, evenness of abundance has no consistent effect on any metric of similarity. We suggest that the explanation for this pattern is that most species have densities that vary widely across their geographic ranges (Brown and Lomolino 1998). Such spatial variation in abundance is related to biotic and especially abiotic factors that influence habitat quality and perhaps dispersal. Our evidence indicates that this spatial abundance variation applies to both native and exotic species.

Another insight is that exotic species show many of the same biogeographic patterns as native species. Both natives and exotics show (1) a small role for abundance and evenness in homogenization patterns, (2) correlated patterns of evenness (skewness) in native and exotic species in each community, (3) and an exponential decline in shared species with distance and latitude separation. This decline is apparent even in highly urbanized areas, although at a lower rate. We can only speculate on an explanation, but these similarities imply that, as above, exotic species are influenced by many of the same abiotic and biotic variables that control the spatial distribution of native species. Sax (2001) discusses evidence that exotic species in several taxa (including mammals, birds, and plants) have latitudinal diversity patterns that are similar to those of native species. In addition, Nekola and White (1999) discuss evidence that habitat gradients and dispersal limits produce the exponential decay in community similarity with distance among native species. It seems likely that these same two factors can explain the same decay pattern in exotic species.

Our evidence here extends the parallelism between exotics and natives to a somewhat finer level, in which many patterns of abundance also show congruence. Thus, the accumulated evidence indicates that many (perhaps most) biogeographic patterns do not require long periods of species sorting or coevolution to materialize (Sax 2001). This tentative conclusion opens the opportunity to explore the mechanisms that create these large-scale patterns simply by following the fates and distributions of exotic species (Wilkinson 2004). Of course, one must be careful when making such conclusions, and some measure of the importance of the rapid evolutionary change that is possible (e.g., Holt et al., Huey et al., and Callaway et al., this volume) must also be taken into consideration.

Our tentative findings clearly need further testing with larger data sets and information from other regions and taxa. Acquiring absolute abundance data instead of our rank abundance data would significantly improve the resolution of the patterns. It is also important to extend these ideas to birds, mammals, and other animal taxa. Does abundance play a more important role in animal homogenization? Certainly, future work must expand our understanding of homogenization across taxonomic groups, as well as across broader spatial and temporal scales, if we are to gain a more mechanistic (and predictive) understanding of how the similarity of biotic assemblages is likely to vary across geographic space in the future.

Acknowledgments

This work was conducted as a part of the "Exotic Species: A Source of Insight into Ecology, Evolution, and Biogeography" Working Group supported by the National Center for Ecological Analysis and Synthesis, a center funded by the National Science Foundation (grant #DEB-0072909), the University of California, and the Santa Barbara campus.

Literature Cited

Basinger, M. A., J. Huston, R. J. Gates, and P. A. Robertson. 1997. Vascular flora of Horseshoe Lake Conservation Area, Alexander Co., Illinois. Castanea 62:82–99.

Bennett, J. P., and J. Course. 1996. The vascular flora of Hopewell Culture National Historical Park, Ross County, Ohio. Rhodora 98:146–167.

Blair, R. B. 1996. Land use and avian species diversity along an urban gradient. Ecological Applications 9:164–170.

Blair, R. B. 2001. Birds and butterflies along urban gradients in two ecoregions of the United States. In J. L. Lockwood and M. L. McKinney, eds. *Biotic homogenization*, pp. 33–56. Kluwer Academic/Plenum, New York.

Brown, J. H., and M. V. Lomolino. 1998. *Biogeography*, 2nd Ed. Sinauer Associates, Sunderland, MA.

Carlton, J. T., and J. B. Gellar. 1991. 1000 Points of invasion: rapid oceanic dispersal of coastal organisms and implications for evolutionary biology, ecology, and biogeography. American Zoologist 31:A127–A127.

Carpenter, J. S., and E. W. Chester. 1987. Vascular flora of the Bear Creek Natural Area, Stewart County, Tennessee. Castanea 52:112–128.

Case, T. J. 1996. Global patterns in the establishment and distribution of exotic birds. Biological Conservation 78:69–96.

Clarke K. R., and R. M. Warwick. 2001. *Change in marine communities: an approach to statistical analysis and interpretation*, 2nd Edition. Plymouth Marine Laboratory, Plymouth, UK.

Clemants, S. E., and G. Moore. 2003. Patterns of species diversity in eight Northeastern United States cities. Urban Habitats, 1, No. 1. (June 24), http://www.urbanhabitats.org.

Clements, R. K., and B. E. Wofford. 1991. The vascular flora of Wolf Cove, Franklin County, Tennessee. Castanea 56:268–286.

Collins, M. D., D. P. Vazquez, and N. J. Sanders. 2002. Species-area curves, homogenization and the loss of global diversity. Evolutionary Ecology Research 4:457–464.

Crowe, T. M. 1979. Lots of weeds: insular phytogeography of vacant urban lots. Journal of Biogeography 6:169–181.

Daehler, C., and D. Carino. 2001. Hybridization between native and alien plants and its consequences. In J. L. Lockwood, and M. L. McKinney, eds. *Biotic homogenization*, pp. 81–103. Kluwer Academic/Plenum, New York.

Drew, M. B., L. K. Kirkman, and A. K. Gholson. 1998. The vascular flora of Ichauway, Baker County, Georgia: a remnant longleaf pine/wiregrass ecosystem. Castanea 63:1–24.

Duncan, J. R., and J. L. Lockwood. 2001. Spatial homogenization of the aquatic fauna of Tennessee: extinction and invasion following land use change and alteration. In J. L. Lockwood, and M. L. McKinney, eds. *Biotic homogenization*, pp. 245–259. Kluwer Academic/Plenum, New York.

Easley, M. C., and W. S. Judd. 1990. Vascular Flora of the Southern Upland Property of Paynes Prairie State Reserve, Alachua County, Florida. Castanea 55:142–186.

Frankel, E. 1999. A floristic survey of vascular plants of the Bronx river Parkway Reservation in Westchester, New York: Compilation 1973–1998. Journal of the Torrey Botanical Society 126:359–366.

Greller, A. M., and S. E. Clemants. 2001. Flora of West Hills Park, Suffolk County, New York, with considerations of provenance of some long-distance disjuncts. Journal of the Torrey Botanical Society 128:76–89.

Guerrant, E. O. 1992. Genetic and demographic considerations in the sampling and reintroduction of rare plants. In P. L. Fiedler, and S. Jain, eds. *Conservation biology*, pp. 321–344. Chapman and Hall, London.

Harrison, S. 1993. Species diversity, spatial scale, and global change. In P. Kareiva, J. Kingsolver, and R. Huey, eds. *Biotic interactions and global change*, pp. 388–401. Sinauer Associates, Sunderland, MA.

Hayden, W. J., M. L. Haskins, M. F. Johnson, and J. M. Gardner. 1989. Flora of Richmond National Battlefield Park, Virginia. Castanea 54:87–104.

Hill, M. O. 1973. Diversity and evenness: unifying notation and its consequences. Ecology 54:427–432.

Hill, S. R. 1996. The flora of Latimer Point and vicinity, New London County, Connecticut. Rhodora 98:180–216.

Howell, C. L. 1991. Floristics of two state parks in the Piedmont of Georgia: Indian Springs and High Falls. Castanea 56:38–50.

Huffman, J. M., and W. S. Judd. 1998. Vascular flora of Myakka River State Park, Sarasota and Manatee Counties, Florida. Castanea 63:25–50.

Judziewicz, E. J. 1997. Vegetation and flora of Passage Island, Isle Royale National Park, Michigan. Castanea 62:27–41.

Lamont, E. E., and R. Stalter. 1991. The vascular flora of Orient Beach State Park, Long Island, New York. Bulletin of the Torrey Botanical Club 118:459–468.

Lockwood, J. L., and M. L. McKinney. 2001. *Biotic homogenization*. Kluwer Academic/Plenum, New York.

Marchetti, M. P., T. Light, J. Feliciano, T. Armstrong, Z. Hogen, J. Viers, and P. B. Moyle. 2001. Homogenization of California's fish fauna through abiotic change. In J. L. Lockwood, and M. L. McKinney, eds. *Biotic homogenization*, pp. 259–278. Kluwer Academic/Plenum, New York.

Meffe, G. K., and C. R. Carroll. 1998. *Principles of conservation biology*, 2nd Ed. Sinauer Associates, Sunderland, MA.

McKinney, M. L. 1998. On predicting biotic homogenization: species-area patterns in marine biota. Global Ecology and Biogeography Letters 7:297–301.

McKinney, M. L. 2002. Urbanization, biodiversity, and conservation. BioScience 52:883–890.

McKinney, M. L. 2004a. Measuring floristic homogenization by non-native plants in North America. Global Ecology and Biogeography 13:47–53.

McKinney, M. L. 2004b. Do exotics homogenize or differentiate communities? Roles of sampling and exotic species richness. Biological Invasions 6:495–504.

Murrell, Z. E., and B. E. Wofford. 1987. Floristics and phytogeography of Big Frog Mountain, Polk County, Tennessee. Castanea 52:262–290.

Nekola, J. C., and P. S. White. 1999. The distance decay of similarity in biogeography and ecology. Journal of Biogeography 26:867–878.

Olden, J. D., and N. L. Poff. 2003. Toward a mechanistic understanding of prediction of biotic homogenization. American Naturalist 162:442–460.

Olden, J. D., N. L. Poff, M. R. Douglas, M. E. Douglas, and K. D. Fausch. 2004. Ecological and evolutionary consequences of biotic homogenization. Trends in Ecology and Evolution 19:18–24.

Petit, R. J. 2004. Biological invasions at the gene level. Diversity and Distributions 10:159–165.

Poulin, R. 2003. The decay of similarity with geographical distance in parasite communities of vertebrate hosts. Journal of Biogeography 30:1609–1615.

Profous, G. V., and R. E. Loeb. 1984. Vegetation and plant communities of Van Cortlandt Park, Bronx, New York. Bulletin of the Torrey Botanical Club 111:80–89.

Quammen, D. 1998. Planet of weeds. Harper's Magazine 115:57–69.

Rahel, F. J. 2000. Homogenization of fish faunas across the United States. Science 288:854–856

Rahel, F. J. 2002. Homogenization of freshwater faunas. Annual Review of Ecology and Systematics 33:291–315.

Rejmanek, M. 2000. A must for North American biogeographers. Diversity and Distributions 6:208–211.

Rhymer, J. M., and D. Simberloff. 1996. Extinction by hybridization and introgression. Annual Review of Ecology and Systematics 27:83–109.

Rosenzweig M. L. 2001. The four questions: what does the introduction of exotic species do to diversity? Evolutionary Ecology Research 3:361–367.

Sax, D. F. 2001. Latitudinal gradients and geographic ranges of exotic species: implications for biogeography. Journal of Biogeography 28:139–150.

Sax, D. F., S. D. Gaines, and J. H. Brown. 2002. Species invasions exceed extinctions on islands worldwide: a comparative study of plants and birds. American Naturalist 160:766–783.

Sax D. F. and S. D. Gaines. 2003. Species diversity: from global decreases to local increases. Trends in Ecology and Evolution 18:561–566.

Scott, M. C., and G. S. Helfman. 2001. Native invasions, homogenization, and the mismeasure of integrity of fish assemblages. Fisheries 26:6–15.

Stalter, R., P. Lynch, and J. Schaberl. 1993. Vascular flora of Saratoga National Historical Park, New York. Bulletin of the Torrey Botanical Club 120:166–176.

Stalter, R., and S. Scotto. 1999. The vascular flora of Ellis Island, New York City, New York. Journal of the Torrey Botanical Society 126:367–375.

Suarez, E., D. M. Pelletier, T. Fahey, P. Groffman, P. Bohlen, and M. Fisk. 2004. Effects of exotic earthworms on soil phosphorus cycling in two broadleaf temperate forests. Ecosystems 7:28–44.

Vitousek, P. M., and L. R. Walker. 1989. Biological invasion by *Myrica faya* in Hawaii: Plant demography and nitrogen fixation ecosystem effects. Ecological Monographs 59:247–266.

Waite, S. 2000. *Statistical ecology in practice*. Prentice-Hall, Englewood Cliffs, NJ.

Wells, E. F., and R. L. Brown. 2000. An annotated checklist of vascular plants in the forest at historic Mount Vernon, Virginia: a legacy from the past. Castanea 65:242–257.

Wilkinson, D. M. 2004. The parable of Green Mountain: Ascension Island, ecosystem construction, and ecological fitting. Journal of Biogeography 31:1–4.

15

Rates of
Population Spread and
Geographic Range Expansion

WHAT EXOTIC SPECIES TELL US

Brian P. Kinlan and Alan Hastings

For the past half century, ecologists and invasion biologists have sought to identify and understand factors that control the rate and pattern of spread of introduced species. These factors are intimately related to the processes governing natural range expansions; thus, a study of modern invasions can provide insight into historical biogeographic processes. Here we review some of the major theoretical constructs that have been used to study species spread and consider empirical data on invasive species spread in marine and terrestrial ecosystems in the context of model predictions. Our recent synthesis of data on marine spread rates allows examination of the invasion process across a diverse set of life histories, environments, and dispersal mechanisms. Comparisons across and within systems suggest general features of organisms and environments that play important roles in the expansion of species across landscapes. The frequency and mode of long-distance dispersal, life history traits that affect rates of population growth at low densities (Allee effects), and feedbacks among migration, adaptation, and environmental structure emerge as critical processes influencing the dynamics of species' range expansion.

Introduction

Biogeography is rooted in the effort to describe and explain both the present distribution of organisms relative to their environment, and changes in the distribution of populations and communities over time. Inherent in these goals is the need for a rigorous mechanistic understanding of the processes determining range boundaries, not only in an equilibrium state when they are stable, but during dynamic periods when populations are actively expanding or contracting across the landscape (Hengeveld 1988; Holt et al. 2005).

An understanding of these dynamic periods is essential because all species undergo expansion (and perhaps contraction) phases in the course of their evolutionary history, regardless of whether vicariance events further modify range boundaries. Yet relatively few direct observational data exist on the dynamics of natural range fluctuations (Brown and Lomolino 1998; Parmesan et al. 2005). This is in part because range expansions and contractions of native species are viewed as relatively rare at human time scales, and in part because such fluctuations are difficult to detect without large-scale, long-term, high-resolution monitoring programs (Barry et al. 1995).

In the absence of detailed contemporary data on population expansion, biogeographers have relied heavily on paleontological and paleoenvironmental records, clues from present distribution patterns, geologic history, systematics, and more recently, population genetic structure (Avise 1992; Macdonald 1993; Brown and Lomolino 1998; Cain et al. 1998; Cox and Moore 2000). These studies have been extremely valuable in placing modern species distributions in context and in generating a rich set of hypothesized mechanisms for species origination, repeated range expansions and contractions, and eventual extinction (see Ricklefs, this volume). However, constraints on the temporal resolution of these methods and a lack of detailed knowledge of the dynamics of prehistoric populations, communities, and environments hinder a mechanistic exploration of the dynamics underlying range shifts revealed by the paleontological record.

This focus on very long time scales has also contributed to the perception of biogeography as a static feature to be mapped and explained, rather than a dynamic process (Elton 1958; Valentine 1968; Hengeveld 1988). Much early biogeographic work was aimed at the creation of static maps delineating major biogeographic provinces and understanding the persistent environmental features that define those provinces (Brown and Lomolino 1998). Even more recent approaches that consider the possibility of range shifts due to climate change often do so by assuming wholesale shifts of coherent biogeographic provinces or community assemblages (e.g., Thomas et al. 2004; but see Jackson and Williams 2004 and references cited therein for an alternative perspective). Whether biogeographic ranges really have been as stable over the past few thousand years as has been perceived, or are just too poorly sampled for dynamics to be detected, it seems almost certain that accelerating climate change will increase rates of range shifts in the near future (Barry et al. 1995; Thomas et al. 2004). In fact, a recent meta-analysis found that climate change was already implicated in an average 6.1 km per decade poleward shift in species' ranges (Parmesan and Yohe 2003). A predictive understanding of future shifts, as well as previous shifts that have led to the current biogeographic configuration, will require detailed knowledge of the processes determining species spread across landscapes (Holt et al. 2005).

Range boundaries are maintained by a complex and variable suite of interacting factors (Figure 15.1). Although the literature is replete with specific examples of mechanisms setting range boundaries, these can be broadly summa-

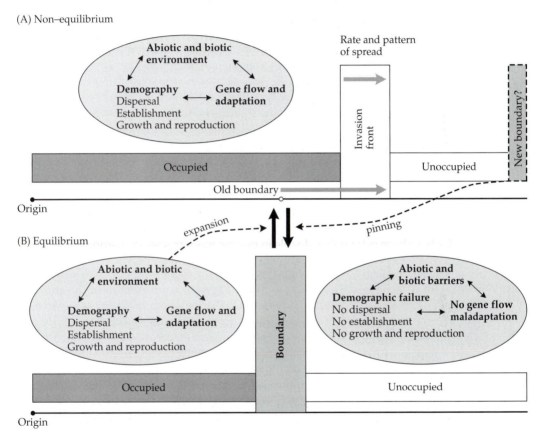

FIGURE 15.1   A conceptual model of range dynamics, summarizing the processes that influence the position of a range boundary in (A) nonequilibrium and (B) equilibrium phases. Solid arrows connect processes that interact to determine range boundaries in each phase; dashed arrows indicate processes that mediate transition between equilibrium and nonequilibrium phases (e.g., the breakdown of a biogeographic boundary) and may occur only episodically or not at all for a given species. Exotic species can inform us about processes regulating changes in species boundaries over time, including factors that can trigger the start of a nonequilibrium range expansion phase ("expansion" arrow), processes that determine the rate and pattern of spread during expansion, and mechanisms that can lead to formation of a new stable range boundary ("pinning" arrow). Note that range contraction is also possible in the nonequilibrium phase; for clarity, this process is not illustrated.

rized in terms of three interacting factors: demography, the abiotic and biotic environment, and genetic structure (Brown and Lomolino 1998; Cox and Moore 2000). Complexity arises because strong feedbacks exist between each pair of processes in this triangle. For example, genes determine which environments an organism perceives as "good" and "bad," demographic variables in turn depend on the spatial and temporal distribution of "good" places, organisms interact with and modify their abiotic and biotic environment, and both demography and environmental structure affect gene flow and the capacity for local adaptation (Levin 1992; Kirkpatrick and Barton 1997).

The same feedbacks operate whether a species' geographic range is in an equilibrium (stable) or nonequilibrium (expansion/contraction) phase, but their outcome is quite different. In the equilibrium phase, demographic, environmental, and genetic processes interact to prevent spread beyond the edge of the range (Figure 15.1B). In the nonequilibrium phase, the three fundamental processes interact to determine both the rate at which the range boundary expands (or contracts) across the landscape and the spatiotemporal pattern of expansion or contraction (Figure 15.1A). Shifts in the relative magnitude of feedbacks in this loop can result in transitions from the equilibrium to the nonequilibrium phase, and vice versa (Keitt et al. 2001; Crespi 2004; Holt et al. 2005). "Equilibrium," as used here, does not imply lack of variation. In practice, even "stable" range boundaries are rarely precise; they vary over time and are stable only in the statistical sense of being stationary in the mean. We use the term "equilibrium" to denote this kind of stability in the expected position of a range boundary, and the term "nonequilibrium" to describe situations in which the mean position of the range is changing. This view of a species range as a dynamic feature is one of the most important insights to biogeography to arise from the study of invasions.

In this chapter, we use theoretical and empirical results from the study of exotic species to examine processes influencing geographic range expansion. We focus our attention on interactions between demography, particularly dispersal and reproduction, and environmental structure that influence range expansion and boundary formation. We also discuss recent advances in theoretical and empirical work on the role of adaptation in population spread and establishment of range boundaries (Nichols and Hewitt 1994; Kirkpatrick and Barton 1997; LeCorre et al. 1997; Sakai et al. 2001; Garcia-Ramos and Rodríguez 2002), although our own data do not directly address this issue. We conclude by discussing some general insights into range expansion and range boundary formation from the past half century of invasion biology, and by identifying some exciting new areas in which exotic invasions may contribute to our knowledge of geographic range dynamics.

Exotic Species Spread as a Model of Natural Range Expansion

Two (relatively) recent phenomena have increased opportunities for the study of geographic range expansions in "real time": an accelerating effect of climate change on the world's present biotic landscape, and an explosion of exchange of non-native species across the globe. Both of these factors have, in effect, sped up the rates of biogeographic change so dramatically that it is now possible to observe in a matter of decades changes in biogeographic structure similar to shifts that may have occurred over orders-of-magnitude greater periods of time in the past (although abrupt range shifts may also have occurred during certain periods in the past: NRC 2002). This has provided an invaluable opportunity to study biogeographic-scale change on human time scales.

Expansions of exotic species by their own means from locations of anthropogenic introduction can be viewed as natural experiments, if the relevant biotic and abiotic factors can be observed and multiple similar cases can be found and compared. Our rapidly growing database of observations on invasive species spread can be married to the rich body of theory on population spread rates, yielding insights into life history and environmental features that promote or reduce the speed and likelihood of geographic range expansions. Through this process we can begin to understand sources of variation in rates of geographic range expansion, and we can parse these into *deterministic* components that are predictable from life history traits (demography and dispersal), environmental variation (abiotic and biotic), and genetic structure, as opposed to *stochastic* variation that is unpredictable except in a statistical sense. This latter source of variation refers to the fact that the rate and pattern of an invasion may differ from one "realization" of an invasion to the next, even under identical conditions; this is important, as it places limits on our ability to forecast specific outcomes.

We envision geographic range expansions as having three principal stages: (1) the period during initial establishment when population growth rates are often highly irregular and characterized by lag periods (or local extinctions and reintroductions) before rapid population growth begins, (2) a period of rapid population growth and expansion of the geographic area occupied, and (3) a period of reduced population growth, slowed spread, and an eventual end to the expansion of the geographic range, referred to hereafter as "pinning" of the geographic range (after Keitt et al. 2001). We do not focus here on the first or "lag" stage of population establishment and range expansion (for a discussion of some of the processes that may be important in this stage, see Holt et al.; Ricklefs; Novak and Mack; and Wares et al., this volume; for a general treatment of patterns of range expansion and population growth during this stage, see Shigesada and Kawasaki 1997). Instead, we focus on the second stage of the invasion process, considering insights that exotic species can provide into the spread of populations across landscapes during periods of range expansion. In so doing, we emphasize comparisons of marine and terrestrial invasions to illustrate the utility of a comparative approach for deriving general principles about the nature and pattern of range expansions. This cross-system approach allows us to compare invasion speeds across a wide range of life histories, dispersal modes, and dispersive environments (i.e., air vs. water). To a lesser extent, we also consider the third phase—that is, processes that cause slowing or halting ("pinning") of invasions—which allows us to gain insights into the formation of stable range boundaries. The term "pinning" refers to the failure of a wave to propagate due to interactions between wave-generating processes and the medium (environment); we use it here because our emphasis is on the dynamic demographic, environmental, and genetic processes that can halt the spread of invasions. In this context, we touch on recent theoretical work and discuss how data from invasive species might be used to test these models (e.g., Hastings et al. 1997; Kirkpatrick and Barton 1997; Wilson et al. 1999; McCann et al. 2000; Keitt et al. 2001; Umbanhowar and Hastings 2002).

Insights from Theory

Modeling dispersal and spatial spread of invading populations

Models and descriptions of spatial dynamics of invading ecological populations have a long history. Much of this classic theory has focused on two distinct biological questions: the rate of spread of an invasive species (e.g., Fisher 1937; Skellam 1951) and the size of a favorable habitat, embedded in a matrix of unfavorable habitat, that is required for persistence (e.g., Skellam 1951; Kierstead and Slobodkin 1953).

The initial classic results were developed using the deterministic *reaction-diffusion* modeling approach, recently reviewed in Okubo and Levin (2002) (Table 15.1). In this approach, all movement is assumed to be local, and in some sense random, and all individuals within the population are generally assumed to behave independently. Under a wide variety of further assumptions, the general character of the spread rate (after the initial lag phase) is surprisingly simple: it is linear with time, and depends on the rate of individual movement and the population growth rate. Similarly, the results for the size of favorable habitat required for population persistence are consistent across these early models. The size of the habitat required is a simple ratio, involving the individual movement rate, the population growth rate, and the habitat size. The growth rate and habitat size need to be large enough to compensate for losses through movement of individuals into unfavorable habitat, leading to death.

The results on critical habitat area from the reaction-diffusion framework carry over to another class of models, *integro-difference equations*, which operate under assumptions of discrete time and continuous space (see Table 15.1). In this description, long-range movement is allowed, and the model is phrased in terms of a dispersal kernel, $k(x,y)$, describing the probability that the offspring of an individual at location x ends up at location y. In contrast to the linear rate of spread typically seen with reaction-diffusion models, integro-difference equation models can exhibit accelerating rates of spread. This result is a consequence of the greater flexibility in dispersal behavior that can be accommodated through the explicit definition of a dispersal kernel. Because long-range movement can be incorporated into these models, certain combinations of dispersal patterns and demographic parameters can lead to accelerating rates of range expansion (Kot et al. 1996). This means that attempts to infer dispersal behavior at the individual level from rates of population spread

TABLE 15.1 *Some approaches to modeling population spread*

Modeling framework	Assumptions	Spread dynamics	Key references
Reaction-diffusion equations	Continuous space and time; local random movement; deterministic dynamics at the level of the population	Linear rate of spread under many assumptions; smooth traveling wave front	Fisher 1937; Skellam 1951; Okubo and Levin 2002
Integro-difference equations	Continuous space, discrete time; long-range movement allowed; deterministic dynamics at the population level	Rate of spread can be accelerating if long-range dispersal common enough; smooth traveling wave front	van den Bosch et al. 1990; Kot et al. 1996; van Kirk and Lewis 1997
Stratified diffusion	Dispersal occurs by a combination of local diffusion and long-distance jumps; as peripheral colonies grow in size and number, their influence increases	Nonlinear, accelerating rates of spread	Shigesada et al. 1995
Metapopulation	Discrete space, continuous time; stochastic local population dynamics and colonization; long-range movement allowed	Needs to be studied more carefully; typically not used to study spread; patchy spread front	Levins 1969; Hanski and Gilpin 1997
Stochastic point processes (interacting particle systems)	Continuous space and time; stochastic movement; stochastic dynamics	Can exhibit accelerating rates of area occupancy, patchy and variable boundaries; typically not used to study spread	Durrett and Levin 1994
Spatial moment equations (stochastic nonlinear integro-difference models)	Continuous space, discrete time; stochastic dynamics and movement	Can exhibit accelerating rates of area occupancy; patchy and variable spread	Lewis and Pacala 2000
Other approaches (e.g., spatial contact processes, cellular automata, individual-based)	Virtually any combination of discrete or continuous space, discrete or continuous time, stochastic or deterministic, linear or nonlinear, analytical or simulation		Jones et al. 1980; Mollison 1972; Wolfram 1983

of an invasive species, a goal that goes back at least as far as Skellam (1951), need to be approached with great caution. However, these differences between model results also provide a means for understanding variation in spread rates among species whose local dispersal characteristics seem similar.

A variety of other classes of models have subsequently been studied, with the general result that spread rates can be either linear or accelerating, and invasion fronts either smooth or patchy, depending on assumptions about long-range movement, demography, adaptation, and environmental structure (see Table 15.1; reviewed in Hastings et al. 2005). Individual-based models offer some of the most detailed depictions of dispersal dynamics, allowing complete relaxation of assumptions of purely local, random dispersal and incorporation of complex behaviors (e.g., Flierl et al. 1999). Such models are computationally intensive and can be difficult to generalize, but may offer the best hope for detailed prediction of specific invasions when organism dynamics and environmental structure are well described.

The development and maintenance of range boundaries

The question of what stops the spread of species is important both for understanding controls on species borders and for management of invasions. Two classic explanations for the existence and formation of range boundaries are species interactions (competition, predation, mutualism, disease) and environmental gradients (Grinnell 1917). In their most basic form, these hypotheses are not dynamic, but simply assume that carrying capacity decreases to zero at the (fixed) edge of a range due to an extrinsic abiotic or biotic factor, such as increased environmental stress, appearance of a competitor or predator, or disappearance of a mutualist. Such explanations tell us why a species fails to survive at present beyond a certain point; however, they do not reveal how things got that way. Thus, they offer only limited insight into the dynamics of species ranges. Here, we focus instead on processes that can cause the formation of stable range boundaries before all suitable habitat is occupied ("pinning"; Keitt et al. 2001). These models have the attractive property of explaining both equilibrium and nonequilibrium phases of range dynamics (see Figure 15.1). Opportunities to use these ideas to infer process from dynamic patterns are particularly exciting. Dynamic models of range boundaries also have the potential to add greater realism to the representation of environmental gradients and landscape patterns. Much of the theoretical work in the area of environmental gradients has relied on assumptions of smooth, temporally stable decreases in environmental quality and population density from the center to the edge of a species' range. This assumption remains largely untested, and more work is needed to elucidate the interplay between range-wide patterns of abundance, demography, and environmental suitability (Sagarin and Gaines 2002). Moreover, adaptation resulting from the feedback among demography, environmental structure, and genetic structure determines how a species "perceives" an environmental gradient; thus, the effective steepness or spatial pattern of an environmental gradient can itself evolve dynamically (Garcia-Ramos and Rodríguez 2002; Travis and Dytham 2002).

Recently, several models of species' spread have emerged that specifically attempt to shed light on the development and maintenance of range boundaries (Hastings et al. 1997; Wilson et al. 1999; McCann et al. 2000; Keitt et al. 2001; Umbanhowar and Hastings 2004). In particular, they highlight interactions among habitat patchiness, species interactions, and Allee effects (processes that result in negative population growth rates at low densities) that can cause range boundaries to be formed with or without the presence of large-scale environmental gradients.

First, consider a model that assumes that the underlying habitat is uniform and that species do not show an Allee effect. The classic result for this model is that population density is lowest, and per capita rate of population growth is highest, at range edges. By changing the structure of this model, we can explore processes that change this result and lead to stable range limits. Hastings et al. (1997) studied the case of a nearly immobile prey species that has a finite range, from which the species is kept from expanding by a predator. In this case, the prey population actually reaches its highest density at the edge of the range, then "jumps" to a density of zero. Since the predator species is mobile, its density changes smoothly over space—a qualitatively different pattern of abundance that can be compared against observations in nature. Although this model produces patterns that remain almost stationary over ecological time scales (Owen and Lewis 2001), if the prey movement is very small (rather than precisely zero), the range edge does shift very slowly. A model with a different form that includes interaction among predators (Harrison et al. 2005) does allow stable boundaries and shows a similar pattern of highest prey abundance at the edge of the range. Moreover, a preliminary analysis of this model shows that the area of space occupied by the prey shrinks if the carrying capacity for the prey goes down, suggesting a mechanism by which interactions among three trophic levels can play a role in setting range boundaries.

In the discrete-space analogues of both of the models discussed above, it is much easier to stabilize the edge of the species range, even in the absence of a predator (McCann et al. 2000). In this case, there is always a set of discrete patches constituting the range edge for which immigration is too low to sustain populations above the critical density needed by a species to increase when rare. A similar case exists for single-species models with Allee effects (Keitt et al. 2001). Discretization of the habitat can stop the spread of a species because, at the edge of its range, the species cannot get above the threshold for positive population growth implied by an Allee effect. These results contrast with those of continuous-space models, in which the species can always slowly build up to a population level implying positive growth as long as there is no change in the underlying habitat quality (McCann et al. 2000; Keitt et al. 2001). This pattern may have important implications for spread in patchy landscapes and could lead to "jerky" spatial spread in variable environments (where temporal variation in productivity may cause expansion to stop and start). One interesting contrast between the one- and two-species models described above is that when predators are "pinning" the edge of the range of a prey species, the highest density of the prey species is at the edge of the range, not in the interior of the range (as is the case when spread is stopped by an Allee effect).

Recent work by Umbanhowar and Hastings (2004) has found similar patterns in host-parasitoid systems.

Both spread rate and range boundary formation can also be influenced by the interplay between gene flow and environmental structure. Although a full review of this issue is beyond the scope of this chapter (the reader is referred to a number of excellent papers on this topic: Nichols and Hewitt 1994; Kirkpatrick and Barton 1997; LeCorre et al. 1997; Sakai et al. 2001; Garcia-Ramos and Rodríguez 2002), we summarize a few of the central findings of recent studies here. In the presence of an environmental gradient, gene flow from the interior of a range to the edges can cause maladaptation at peripheral regions, in turn slowing or halting the invasion (Kirkpatrick and Barton 1997; Garcia-Ramos and Rodríguez 2002; see Estoup et al. 2004 for an empirical example). This effect is dependent on the dispersal function, the pattern of population density across the range, the slope of the environmental gradient, and the strength of the genetic response (Garcia-Ramos and Rodríguez 2002). When the environmental gradient is shallow and adaptation at the edges is rapid, spread rates approach those predicted for an equivalent model considering only demography in a homogeneous favorable environment. Slowed spread, "pinning," and even contraction can result when the environmental gradient is sufficiently steep and/or adaptation is slowed by low heritability, non-stabilizing selection, or gene flow from the interior of the range. Another intriguing possibility, just beginning to be evaluated theoretically, is that life history traits affecting demography and dispersal (e.g., migration rate, reproductive ecology) may also evolve during range expansion, causing rates of spread to change over time (Lambrinos 2004). The evolution of these traits can be influenced by both habitat structure (Parvinen and Egas 2004) and Allee effects (Travis and Dytham 2002). This is an emerging area of research in which genetic data from invading populations will surely play a role.

The theoretical models discussed here provide testable predictions that allow mechanistic hypotheses about a particular range boundary to be evaluated on the basis of empirical observations. Detailed monitoring of accelerating, slowing, or stalled invasions, in addition to serving management purposes, is likely to provide insights into the mechanisms leading to natural establishment of geographic range boundaries. This work will help to illuminate differences among explanations such as Allee effects, effects of exploiter species (including parasitism and disease), genetic structure, and gradients in physiological stress or habitat suitability, and will ultimately contribute to our understanding of the dynamics of biogeographic pattern and process.

Predictability and stochasticity

The models of invasive species spread described above provide some insight into the effects of environmental variation, biotic interactions, and specific life history features on rates of spread and formation of range boundaries (reviewed in Shigesada and Kawasaki 1997 and Hastings et al. 2005). Together, this body of work offers a broad range of explanations for the wide variation in spread rates that has been observed in the natural environment. Indeed, measurements of demographic parameters and environmental characteristics have been integrated with these models to predict average rates of range expansion, with some success (e.g., Grosholz 1996).

However, a long-noted characteristic of invasions is not only variation attributable to environmental and ecological differences, but also apparent stochasticity in invasion behavior across different instances of introduction. This stochasticity results from both demographic processes (including dispersal) and environmental fluctuations.

Rather than reviewing results of stochastic invasion models here (see Lewis and Pacala 2000; Clark et al. 2003), we point out some of the principal implications of stochasticity for patterns of species spread. First, even if environmental and ecological conditions are permissive, the success of an invasion of a given species at a particular place and time is in some sense random. For example, if a small number of randomly moving individuals are introduced, they may fail to encounter mates simply by chance, fail to land in an appropriate patch of a habitat mosaic, or be wiped out by a chance environmental fluctuation. The failure of many (perhaps most) exotic species introductions is well documented (Sax and Brown 2000).

Further, intrinsic stochasticity in environmental and demographic processes can account for large spatial and temporal variation in the rate and pattern of spread, even when the long-term average spread rate is uniform. Any given realization of an invasion may be characterized by periods of apparent deceleration, acceleration, halts, and jumps. Theoretical studies of dispersal processes, in particular, often neglect this stochasticity by modeling dispersal with smooth probability distributions; in reality, the discrete and finite nature of individuals has important effects on the spatial and temporal patterns of real-world invasions (Clark et al. 2001).

Finally, even under identical ecological and environmental conditions, the ability to predict the progress of any one invasion event varies depending on the stochasticity of underlying demographic processes, and is especially sensitive to the level of long-range dispersal. Stochastic simulation and spatial moment equation models (see Table 15.1) predict that "fat-tailed" dispersal kernels will result in increased variance of spread rates across independent realizations of an invasion (Shigesada and Kawasaki 1997; Lewis and Pacala 2000).

Joint consideration of the stochastic and deterministic components of range expansion is critical for understanding the dynamics of range expansions, including the limits of predictability on individual range expansion events. Better integration of theory and data on stochastic invasion processes is important both for understanding how communities have responded during "nonequilibrium" periods of shifting habitat distribution in the past and for developing expectations about species' responses to climate change (e.g., Araujo et al. 2004).

Up to this point, we have not specifically discussed the effects of the shape of the invaded area—in particular, differences between "one-dimensional" habitats (such as coastlines and mountain ranges) and "two-dimensional" habitats (such as ocean basins and continents). In a homogeneous two-dimensional environment, where dispersal occurs symmetrically in all directions, the radial (two-dimensional) spread rate converges on the linear (one-dimensional) spread rate as the size of the invaded area grows (Murray 1993), leading to similar predictions. When Allee effects are considered, however, the situation becomes more complex. Allee effects reduce the constant traveling wave rate of spread by a similar amount in both one and two dimensions. However, inclusion of an Allee effect in two-dimensional space imposes a minimum area that must be inoculated with adults before an invasion can spread, and can reduce spread rates drastically compared with the one-dimensional case during the initial phases of spread (Lewis and Kareiva 1993). Allee effects in two-dimensional space also increase the sensitivity of initial spread rates to the size and shape of the invaded area (Lewis and Kareiva 1993). Still, with or without an Allee effect, radial spread rates of two-dimensional invasions in homogeneous environments eventually approach those of one-dimensional invasions, provided that the initial area exceeds the critical threshold for expansion. If, however, the invaded environments are not homogeneous and usable habitats are patchily distributed, further complications arise. In this case, the rate (or even success) of spread along any given axis is highly dependent on the size and spacing of patches (Lewis and Kareiva 1993; Keitt et al. 2001). This may lead to anisotropic spread; that is, different rates and patterns of spread along different axes of the invasion area. An Allee effect in conjunction with habitat heterogeneity may also prevent the radial (two-dimensional) expansion rate from ever reaching that expected in a one-dimensional patchy habitat, because the invasion can proceed only by repeated expansion from small habitat patches (Keitt et al. 2001; Holt et al. 2005). In real-world scenarios incorporating habitat heterogeneity, variation, and stochasticity, the combination of two-dimensional dispersal with Allee effects may lead to substantially slower rates of spread than when dispersal is constrained to one dimension.

An example: The interaction of demography and dispersal in a simple discrete-time integro-difference model

To illustrate how life history features that influence demography and dispersal can interact to determine spread rates, we calculated expected spread rates in an integro-difference model of population dynamics under four demography/dispersal scenarios: exponentially bounded (Gaussian) and fat-tailed dispersal kernels with and without an Allee effect (Figure 15.2). Fat-tailed dispersal kernels imply that a small proportion of propagules travel much farther than the average—that is, an increased rate of long-distance dispersal. We used the integro-difference framework because of its simplicity and the flexibility to incorporate long-distance dispersal functions. In this model, the number of adults $N_{x,t}$ at each location x along a homogeneous finite linear domain is updated at each time step Δt, equivalent to a generation, according to

FIGURE 15.2 Rates of invasion front spread during the constant-speed phase of a discrete-time, continuous-space, deterministic integro-difference model of invasion. Adults were initially introduced at the center of a one-dimensional, homogeneous, finite spatial domain (length = 2000 km). The interacting effects of two key demographic features were examined: long-distance dispersal ("fat tails") and the inability to establish and reproduce at low densities (Allee effects). Note that the presence of a strong Allee effect largely nullifies the increase in spread rate due to long-range dispersal events and causes the spread rate to be of the same order of magnitude as the average dispersal distance (measured as the standard deviation of the primary dispersal kernel). Local populations are assumed to exhibit Ricker-type density dependence and an Allee effect, reproduce in discrete generations, and disperse according to a standard Gaussian [$D = N(0, \sigma)$] or fat-tailed composite Gaussian [$D = 0.99 \times N(0, \sigma) + 0.01 \times N(0, 5 \times \sigma)$] kernel. See text for model details.

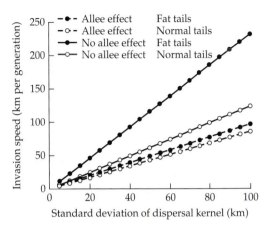

$$N_{x,t+\Delta t} = N_{x,t} - M \cdot N_{x,t} + e^{-C \cdot N_{x,t}} \cdot \int_{-\infty}^{+\infty} D_{x-x'} \cdot N_{x',t} \cdot F \cdot dx' \qquad (15.1)$$

Here M denotes constant per capita adult mortality, C denotes the strength of recruitment dependence on adult density, F denotes the net number of progeny produced per individual (i.e., the product of individual fecundity and juvenile survivorship), and D is the redistribution function (dispersal kernel). For the normal-tailed dispersal case, the kernel is defined by $D = N(0, \sigma)$; that is, the normal distribution with mean 0 and standard deviation σ. For the fat-tailed case, we used a composite Gaussian kernel with 1% of offspring having the potential to disperse distances much greater than the average (following Lewis and Pacala 2000), $D = 0.99 \times N(0, \sigma) + 0.01 \times N(0, 5\sigma)$. We chose a finite domain with absorbing boundaries (individuals dispersing past the edge of the domain are lost) to avoid artifacts caused by the infinite tails of the Gaussian distribution. Many models of spread use an infinite or periodic domain, which may lead to overestimates of spread rate, especially when long-distance dispersal processes are modeled with an unbounded distribution. We limited the range of average dispersal distances we studied to less than 1/20 of the domain size so that constant rates of spread could be observed before edge effects appeared.

Under this model, the carrying capacity at any location x is given by

$$K = \frac{\ln(F) - \ln(M)}{C} \qquad (15.2)$$

For all model runs, we assumed $F = 2$, $M = 0.1$, and $C = 0.01$, yielding $K \sim 300$. Qualitative results were not sensitive to choice of these parameters. To incorporate Allee effects, we used a modified version of the update rule in Equation 15.1,

$$N_{x,t+\Delta t} = N_{x,t} - M \cdot N_{x,t} + e^{-C \cdot N_{x,t}} \cdot \int_{-\infty}^{+\infty} D_{x-x'} \cdot N_{x',t} \cdot \frac{F \cdot N_{x',t}}{N_{x',t} + A_h} \cdot dx' \qquad (15.3)$$

where A_h denotes the density at which per capita adult fecundity reaches half of its maximum value F. A value of $A_h = 0.05K$ was assumed for all model runs incorporating an Allee effect. Initially ($t = 1$), the center of an empty domain (length = 2000 km) was inoculated with $N = K$ adults. Equation 15.3 was then evaluated numerically until invasion waves reached constant speed. The constant-speed, sigmoidal invasion fronts that can arise under this model have been described previously (see Table 15.1; reviewed in Shigesada and Kawasaki 1997). To measure spread rate, the position of the invasion front was arbitrarily defined as the position where adult density reached 10% of carrying capacity.

The results (Figure 15.2) illustrate three important points. First, a very small amount of long-range dispersal (inconsequential to the mean of the dispersal distribution) can dramatically increase spread rates; this is a general feature of models that incorporate long-distance dispersal (Kot et al. 1996; Cain et al. 2000; Neubert and Caswell 2000; Clark et al. 2003). As noted above, under certain scenarios in the integro-difference framework, long-range dispersal can even lead to rates of invasion that accelerate with time (not shown). The difficulties this presents for predicting spread rates from field dispersal data collected at local scales have been the subject of much recent research (e.g., Cain et al. 2000; Clark et al. 2003; Levin et al. 2003). Second, as noted in previous studies (e.g., Lewis and Kareiva 1993), even a modest Allee effect can substantially reduce rates of spread, regardless of the presence or absence of long-distance dispersal. Finally, there is a strong interaction between the inability of a species to increase from low densities (Allee effects) and the importance of long-distance dispersal. Allee effects can effectively nullify any effect long-range dispersal might otherwise have had on increasing the spread rate of an invasion.

Allee effects can arise from a variety of factors, including mating and social systems, reproductive biology, biotic interactions, and mobility. In general, species that are capable of asexual reproduction, are mobile, and have good sensory abilities to aid in mate finding or can mate prior to long-distance dispersal will be less likely to face Allee effects. Species that require a sexual partner to reproduce, especially if they have limited mobility, will often face strong Allee effects. Other factors that prevent population growth at low densities (including predation: Hastings et al. 1997; Owen and Lewis 2001; Harrison et al. 2005) can effectively introduce Allee-like effects that slow or halt invasions.

Another point illustrated by this modeling exercise is the importance of domain size, especially when considering long-distance dispersal. Although results are not shown for an infinite domain, comparison with previous studies reveals that the level of long-distance dispersal incorporated into our fat-tailed dispersal kernels would have led to accelerating spread rates if we had chosen an infinite domain (Kot et al. 1996; Wang et al. 2002). The use of finite (bounded) dispersal kernels can similarly limit the effect of long-distance dispersal processes on rates of spread (Clark et al. 2001). Thus, the size of the area available to be invaded can interact with long-distance dispersal and Allee effects to determine the qualitative and quantitative features of an invasion. When fat-tailed dispersal kernels are used, both Allee effects and finite domains can reduce the acceleration rate of spread, converting accelerating invasions to constant-speed invasions and/or reducing the rate of constant-speed invasions.

Other authors have found similar results in a variety of model frameworks and have presented general derivations of invasion speed in an infinite domain. For example, Lewis and Kareiva (1993) studied Allee effects and spread rate using reaction-diffusion equations, and Wang et al. (2002) examined joint interactions of Allee effects and dispersal kernel shape using integro-difference equations. Incorporation of a finite domain, and the associated decisions on how propagules behave when they encounter a boundary, will modify these general predictions and are important issues when considering real-world invasions.

Lessons from Invasions on Land and in the Sea

Value of a cross-system comparative approach

We will not attempt a comprehensive review of the vast fields of invasive species spread or geographic range dynamics here. A number of excellent reviews on these subjects are available (Hengeveld 1989; Hastings 1996; Higgins and Richardson 1996; Shigesada and Kawasaki 1997; Brown and Lomolino 1998; Cox and Moore 2000; Sakai et al. 2001; Okubo and Levin 2002; Levin et al. 2003). Instead, we focus on a recent synthesis of data on spread rates of marine and terrestrial exotic species to explore a range of factors influencing rates of spread and to illustrate some general principles of species spread that are relevant to biogeography.

Substantial effort has been devoted to modeling and predicting the spread of invading organisms. However, relatively little work has focused on modeling invasions in the marine realm, where introductions of exotic species are extremely common (Carlton and Geller 1993). This is partly due to the difficulty of obtaining direct, quantitative estimates of dispersal ability for marine species. As a substitute for direct quantitative estimates, many studies of marine dispersal employ indirect proxies of dispersal ability based on life history characteristics, such as planktonic larval duration (Siegel et al. 2003). Recent compilations of marine dispersal data (Kinlan and Gaines 2003) allow examination of relationships between life history features, dispersal, and spread rates in marine species and comparison with terrestrial data.

Together, the diversity of marine life histories and the differences between the marine and terrestrial physical environments make comparative study of spread rates in these systems especially informative. The prevalence in the marine realm of species that disperse primarily at the larval stage contrasts with many terrestrial animals, in which dispersal can occur throughout the life

TABLE 15.2 *Secondary (post-introduction) spread rates of some marine and terrestrial exotic species*

Genus and species	Type	Average spread rate (km/year)[a]	Reference
MARINE SPECIES			
Antithamnionella ternifolia	Red seaweed	64	Maggs and Stegenga 1999
Avrainvillea amadelpha	Green seaweed	0.51	Smith et al. 2002
Balanus improvisus	Barnacle	30[b]	Leppakoski and Olenin 2000; Leppakoski et al. 2002
Botrylloides violaceous	Tunicate	16	Grosholz 1996
Carcinus maenas (average, N = 3)	Crab	173	Shanks et al. 2003
Caulerpa scalpelliformis	Green seaweed	0.3	Davis et al. 1997
Caulerpa taxifolia	Green seaweed	10.9	Meinesz et al. 1993; Shanks et al. 2003
Cerithium scabridum	Snail	19.4	Por 1978
Codium fragile ssp *tomentosoides*	Green seaweed	12	Shanks et al. 2003
Dasya baillouviana	Red seaweed	40	Maggs and Stegenga 1999
Elminius modestus	Barnacle	41	Shanks et al. 2003
Ensis americanus	Clam	125	Armonies 2001
Ensis directus	Clam	111	Shanks et al. 2003
Gammarus tigrinus	Amphipod	12[c]	Gras 1971
Gracilaria salicornia	Red seaweed	0.28	Rodgers and Cox 1999
Grateloupia doryphora	Red seaweed	2	Maggs and Stegenga 1999
Hemigrapsus penicillatus	Crab	160	Shanks et al. 2003
Hemigrapsus sanguineus	Crab	33	Shanks et al. 2003
Hemimysis anomala	Shrimp	29.2	Leppakoski and Olenin 2000
Hypnea musciformis	Red seaweed	3.8	Russell and Balazs 1994
Kappaphycus alvarezii	Red seaweed	0.25	Rodgers and Cox 1999
Kappaphycus spp	Red seaweed	0.19	Smith 2002
Kappaphycus striatum	Red seaweed	0.25	Rodgers and Cox 1999
Littorina littorea	Snail	42	Shanks et al. 2003
Lutjanus kasmira	Fish	130	Shanks et al. 2003
Marenzelleria viridis	Polychaete worm	246.7	Leppakoski and Olenin 2000; Leppakoski et al. 2002

[a]Average rate of linear expansion of an invasion front measured from field surveys, unless otherwise noted. Where spread rates varied among distinct directions or time periods in a study, the maximum average rate is reported. Where multiple invasions were studied, the average rate over all invasions is reported. All spread rates represent (presumed) non-anthropogenic spread into suitable habitat.

TABLE 15.2 *(continued)*

Genus and species	Type	Average spread rate (km/year)[a]	Reference
Membranipora membranacea	Bryozoan	20	Grosholz 1996
Mytilus galloprovincialis (average, N = 2)	Mussel	97	McQuaid and Phillips 2000
Mytilus galloprovincialis	Mussel	115	Grosholz 1996
Perna perna	Mussel	235	Shanks et al. 2003
Philine auriformis	Nudibranch	80	Grosholz 1996
Portunus pelagicus	Swimming crab	8.3	Por 1978
Pranesus pinguis	Fish	13.5	Por 1978
Sargassum muticum (average, N = 3)	Brown seaweed	37.4	Leppakoski and Olenin 2000; Shanks et al. 2003
Tapes philippinarum	Clam	30	Breber 2002
Tritonia plebeian	Nudibranch	50	Grosholz 1996
Undaria pinnatifida	Brown seaweed	0.37	Fletcher and Farrell 1999
Zostera japonica	Seagrass	6	Shanks et al. 2003
TERRESTRIAL SPECIES			
Birds, spread rate			
Sturnus vulgaris	Starling	200	Grosholz 1996
Streptopelia decaocta	Collared dove	43.7	Hengeveld 1988; Grosholz 1996
Serinus serinus	Serin	40	Olsson 1971
Carpodacus mexicanus	House finch	40	Mundinger and Hope 1983
Bubulcus ibis	Cattle egret	800	Handtke and Mauersberger 1977
Mammals (N = 3),[d] spread rate		See Grosholz 1996 for data and references	
Birds (N = 76),[d] maximum observed offspring dispersal		See Sutherland et al. 2000 for data and references	
Mammals (N = 66),[d] maximum observed offspring dispersal		See Sutherland et al. 2000 for data and references	
Phytophagous insects (N = 2),[d] spread rate		See Grosholz 1996 for data and references	
Trees (N = 19),[d] spread rate, postglacial		See Kinlan and Gaines 2003 for data and references	
Trees (N = 5),[d] spread rate, modern		See Kinlan and Gaines 2003 for data and references	
Herbaceous plants (N = 47),[d] spread rate, modern		See Matlack 1994 for data and references	

[b]Minimum rate.

[c]Maximum rate

[d]N indicates the number of taxa in this group for which data were available.

cycle. Timing of dispersal within the life cycle can have a large effect on spread rates (e.g., van den Bosch et al. 1990). The marine environment, because of its fluid dynamics, may also provide more opportunities for long-distance passive transport, which probably leads to different dispersal patterns than those that result from directed individual movement (Kinlan and Gaines 2003; Carr et al. 2003; Kinlan et al. 2005). Terrestrial plants, like many marine organisms, disperse primarily in a passive propagule stage. However, seeds and reproductive structures of terrestrial plants are not nearly as buoyant in their fluid medium as those of marine plants, a factor that may limit long-distance dispersal. On the other hand, differences among marine taxa in traits such as mating system and mode of fertilization may reduce the effectiveness of long-distance dispersal in certain groups. For example, many marine taxa that are sessile or sedentary as adults—including mollusks, corals, and echinoderms—rely on external fertilization in which gametes are released into the water column. These systems are highly dependent on a critical density of adults for success (Levitan et al. 1992), a fact that may produce strong Allee effects and limit range expansion rates. Despite behaviors that can increase external fertilization success (e.g., Yund and Meidel 2003), there is no marine equivalent of the active pollinators that facilitate union of gametes over long distances in terrestrial plants. Thus, wide variation may be expected in rates of marine and terrestrial range expansion, some of which may be predictable from life history traits of taxonomic groups.

Here, we focus on a few general patterns and examples that arise from a comparison of spread rates within and among some broad taxonomic groupings of marine and terrestrial organisms. Our aim is to illustrate the utility of a cross-system, cross-taxon comparative approach for identifying general characteristics of geographic range expansions.

Patterns in marine and terrestrial spread rate data

Several recent studies have compiled and reported spread rates of invasive organisms in marine and terrestrial systems (Grosholz 1996; Kinlan and Gaines 2003; Shanks et al. 2003). We have supplemented these previous compilations with additional data on rates of spread for coastal marine organisms from the recent literature to yield a database of 38 marine and 81 terrestrial species (Table 15.2). We defined "coastal" marine species as benthic algae, fishes, and invertebrates associated with nearshore areas and generally having sedentary adults and dispersive propagules; thus, highly mobile pelagic species such as tuna, whales, and pinnipeds were excluded. For empirical estimates of spread rate, we used the mean rate of linear expansion of an advancing colonization front in km/year (*sensu* Shigesada and Kawasaki 1997).

Because the relationship between spread rate and average dispersal distance can provide insight into the effectiveness of long-distance dispersal during range expansion (Kinlan et al. 2005), we also used estimates of average propagule dispersal distance derived from two types of evidence. First, we used a simulation model developed by Palumbi (2003) to relate species' mean dispersal distances to increases in genetic differentiation with geographic distance (for detailed methods see Kinlan and Gaines 2003). Second, we used direct measures of propagule movement or known distances of settling propagules from their sources to estimate mean dispersal. These different sources of information on dispersal were employed to allay the inherent biases that any one type of evidence may have (e.g., the notorious underestimates that direct observational measures of dispersal often provide: Cain et al. 2000; Kinlan and Gaines 2003). The database considers evidence for several taxonomic groups: marine plants, marine invertebrates, terrestrial plants, and terrestrial animals.

This composite database is only now reaching the stage at which it provides a sufficient "sample" for macroecological comparative analyses to be made. In particular, two important limitations influence our analyses: first, all three types of data (spread rate, direct estimates of dispersal, and genetic estimates of dispersal) are not generally available for any single organism, preventing direct species-level comparisons, and second, the phylogenetic distribution of species covered by each data set is uneven and incomplete (Table 15.3). The first issue restricts our comparisons to large and consistent differences in patterns of spread and dispersal across environments and taxonomic groups. To address the second limitation, we adopt a Monte Carlo resampling approach to account for differences in sample size and phylogeny. When two groups differ in size, with N species in the smaller group, 1000 sets of N species are chosen at random from the larger group, and Student's t-tests are conducted using the 1000 simulated distributions. The same process is repeated replacing individual observations with mean values for each phylogenetic group (see Table 15.3). P values are reported for comparisons using randomly resampled distributions with and without taxonomic correction (Table 15.4). Data were log-transformed prior to comparison to homogenize variances. All tests were carried out using MATLAB 6.5 (The MathWorks, Inc., Natick, MA). This approach helps to account for major biases in sample size and phylogeny. However, our results still apply only to taxonomic groups that are well represented in the database; for example, several marine phyla, including nonannelid worms and ctenophores, are not included (see Table 15.3). Further, although phylogenetically independent contrast analyses could be useful here (Harvey and Pagel 1991), we have not attempted to perform them, given the lack of well-resolved phylogenies for many of the taxa we consider. Hopefully, future synthetic and systematic work will expand the data available to support more detailed analyses.

The first pattern that emerges is that spread rates of both marine plants and invertebrates tend to be larger, by up to several orders of magnitude, than those of terrestrial plants. Marine spread rates are also skewed high relative to estimates of average dispersal distance (Table 15.4; Figure 15.3). These patterns suggest that either dispersal or demographic mechanisms in the marine environment promote faster rates of spread, both in an absolute sense and relative to average dispersal distances.

TABLE 15.3 *Phylogenetic distribution of species used for comparison of spread rates and average dispersal distances*

| | Percentage of species in taxonomic group | | |
Taxonomic group	Spread rates (*n*)	Genetic dispersal estimates (*n*)	Direct dispersal estimates (*n*)
Marine plants (Division)	(*n* = 15)	(*n* = 17)	(*n* = 15)
Rhodophyta	53%	35%	13%
Chlorophyta	27%	18%	0%
Phaeophyta	13%	24%	80%
Marine angiosperms	7%	23%	7%
Marine invertebrates (Phylum)	(*n* = 20)	(*n* = 48)	(*n* = 11)
Mollusca	45%	42%	9%
Arthropoda	40%	13%	9%
Chordata	5%	0%	55%
Ectoprocta (Bryozoa)	5%	4%	9%
Annelida	5%	6%	0%
Porifera	0%	6%	0%
Echinodermata	0%	8%	0%
Cnidaria	0%	21%	18%
Marine fishes (Order: Family)	(*n* = 2)	(*n* = 25)	(*n* = 1)
Perciformes: Pomacentridae	—	24%	—
Gadiformes: ALL	—	8%	—
Perciformes: Gobiidae	—	8%	—
Perciformes: Labridae	—	8%	—
Pleuronectiformes: ALL	—	8%	—
Scorpaeniformes: ALL	—	8%	—
Perciformes: Acanthuridae	—	4%	—
Perciformes: Centropomidae	—	4%	—
Perciformes: Embiotocidae	—	4%	—
Perciformes: Kyphosidae	—	4%	—
Perciformes: Labrisomidae	—	4%	—
Perciformes: Malacanthidae	—	4%	—
Perciformes: Mullidae	—	4%	—
Perciformes: Sciaenidae	—	4%	—
Perciformes: Tripterygiidae	—	4%	—

Note: Both direct and indirect (genetic) estimates of average dispersal distance are presented for comparison with spread rates. Sources are listed in Table 15.2.

[a]Remaining 27% of species in 12 families represented by < 2% each.

[b]Remaining 36% of species in 53 families represented by < 2% each.

TABLE 15.3 *(continued)*

	Percentage of species in taxonomic group		
Taxonomic group	Spread rates (*n*)	Genetic dispersal estimates (*n*)	Direct dispersal estimates (*n*)
	Herbaceous plants/trees	Herbs and trees (combined)	
Terrestrial plants (Family)	(*n* = 47)[a]/(*n* = 24)	(*n* = 19)	(*n* = 261)[b]
Pinaceae	0%/46%	0%	4.8%
Asteraceae	11%/0%	5%	18.4%
Apiaceae	9%/0%	16%	0%
Fabaceae	2%/0%	11%	6.6%
Poaceae	0%/0%	5%	14.3%
Rosaceae	6%/0%	11%	1.1%
Betulaceae	0%/13%	0%	2.2%
Fagaceae	0%/13%	0%	0%
Rubiaceae	6%/0%	5%	0.4%
Violaceae	2%/0%	5%	3.3%
Liliaceae	9%/0%	0%	0.4%
Brassicaceae	2%/0%	5%	1.5%
Juglandaceae	0%/8%	0%	0.4%
Cyperaceae	6%/0%	0%	1.8%
Orchidaceae	2%/0%	5%	0%
Aceraceae	0%/4%	0%	2.9%
Anacardiaceae	2%/4%	0%	0.4%
Ulmaceae	2%/4%	0%	0.4%
Caprifoliaceae	6%/0%	0%	0%
Moraceae	0%/0%	5%	0.7%
Tiliaceae	0%/4%	0%	1.5%
Caesalpinieae	0%/0%	5%	0.4%
Calycophylleae	0%/0%	5%	0.4%
Meliaceae	0%/0%	5%	0.4%
Caryocaraceae	0%/0%	5%	0%
Papilionaceae	0%/0%	5%	0%
Oleaceae	0%/4%	0%	0.7%
Vitaceae	4%/0%	0%	0.7%
Berberidaceae	4%/0%	0%	0%

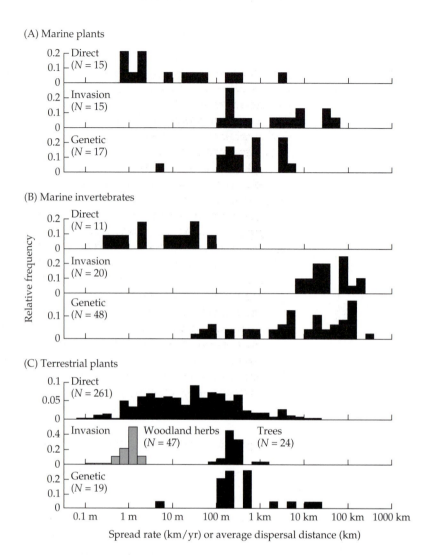

FIGURE 15.3 Comparison of marine and terrestrial spread rates ("Invasion," middle histograms of each panel) with measures of average dispersal distance inferred from two alternative methods: direct observational studies (upper histograms) and population genetic structure (lower histograms of each panel). Each histogram reflects only species for which the required data were available; thus, the species composition varies among histograms (see Table 15.3). Mean rates of linear expansion of invasion fronts were measured from field survey maps (in *kilometers per year*). For terrestrial plants, rates reported are for spread of woodland herbs from a forest boundary (shaded bars) and range expansion of trees either postglacially in North America ($n = 19$) or during modern invasions ($n = 5$) (solid bars). Genetic dispersal estimates measure the long-term average effective dispersal distance (in *kilometers per generation*) under assumptions of an isolation-by-distance model (Palumbi 2003; Kinlan and Gaines 2003). (Modified, with permission, from Kinlan and Gaines 2003. Sources for additional data compiled in the present study are given in Table 15.2.)

TABLE 15.4 *Statistical comparisons of marine and terrestrial spread rates and dispersal measures*

	P-value[b]	
Comparison[a]	**Raw data**	**Taxonomically corrected**[c]
Spread rates vs. spread rates		
Marine plants vs. terrestrial trees	0.0009	< 0.0001
vs. terrestrial herbs	< 0.0001	< 0.0001
vs. terrestrial trees (upper quartiles)	0.0002	0.007
vs. terrestrial herbs (upper quartiles)	0.0001	< 0.0001
Marine invertebrates vs. terrestrial trees	< 0.0001	< 0.0001
vs. terrestrial herbs	< 0.0001	< 0.0001
Marine plants vs. marine animals[d]	< 0.0001	0.02
Spread rates vs. measures of average dispersal distance[e]		
Marine plants, spread vs. direct	< 0.0001	0.02
vs. genetic	0.05	0.008
vs. direct (upper quartiles)	0.005	0.009
vs. genetic (upper quartiles)	0.0001	0.03
Marine invertebrates, spread vs. direct	< 0.0001	< 0.0001
vs. genetic	0.01	0.19
vs. genetic (upper quartiles)	0.22	0.18
Terrestrial trees, spread vs. direct (upper quartiles)	0.21	0.25
vs. genetic (upper quartiles)	0.12	0.21

[a] Data and taxonomic groups for comparisons are given in Table 15.2 and Table 15.3, respectively.

[b] Mean p value obtained in t-tests of 1000 resampled distributions (to account for differences in the number of species per group). Values significant at the $p < 0.05$ level are in **boldface**.

[c] Values for related species were replaced with the average value for the group.

[d] Marine animals include invertebrates and fishes

[e] Direct observational and genetic measures of average dispersal distance were employed; see text for details.

A second pattern arises from comparison of marine and terrestrial plants, groups that play the same trophic role in their respective communities and share some similar life history characteristics (e.g., propagules that have limited or no active behavior). The upper bound of marine macroalgal spread rates exceeds that of terrestrial plants by almost two orders of magnitude (Table 15.4; Figure 15.3). Moreover, the greatest macroalgal spread rates exceed the largest estimate of average dispersal distance by more than an order of magnitude, whereas estimates of terrestrial plant spread rates lie well within the bounds of observed average dispersal distances (Table 15.4; Figure 15.3). Again, these results suggest that some marine plants are capable of invading at rates far greater than their terrestrial counterparts, even when they have similar average dispersal distances. Because demographic characteristics that can enhance spread rates, such as asexual reproduction, self-compatibility, large seed sets, and rapid growth, are not unique to either environment, these two results may suggest a greater role for long-range dispersal in the marine environment. Alternatively, the observed differences could arise from differences in the spatial structure of the two environments. For example, Allee effects may be partially allayed by the concentration of dispersal and patchy habitats along an approximately one-dimensional coastline in coastal marine systems—this would apply to many of the marine invasions examined in this study, but is not a necessary component of marine invasions in general, which can occur in two or even three dimensions (for relevant theory, see the modeling example above). This issue is considered further in the summary section below.

A comparison of spread rates for marine plants and marine invertebrates provides further insight into processes that may control marine spread rates. Estimates of marine invertebrate spread rates are skewed high relative to the average dispersal distance distribution, yet, unlike those of macroalgae, they do not exceed the upper bound of the average dispersal distribution (Table 15.4; Figure 15.3). Given the wide range of average dispersal distances compared with spread rates, among marine invertebrates, it is likely that some of the fast invertebrate invaders have short average dispersal distances, similar to the macroalgal case. This conclusion is supported by the case study of *Balanus* and *Botrylloides* described below. However, invertebrates with long average dispersal distances do not appear to invade at rates that substantially exceed those average distances. This result suggests that long-distance dispersal is reduced or less effective in these species. The longer planktonic period associated with long average dispersal distances may increase the chance that long-distance travelers will be separated and settle at low density, or traits correlated with long planktonic periods may make it more difficult for these taxa to become established and reproduce at low densities. Alternatively, invertebrate larvae may use their more complex range of behaviors to avoid long-distance dispersal.

Although based on a small sample of species (with all the inherent risks of taxonomic bias) and limited by the lack of species-species correspondence across data sets, this example illustrates the ability of a cross-system comparative approach to provide clues to the mechanisms underlying variation in rates of spread across taxa and environments.

Spread rates: Three examples

More specific insights can be gained by focusing on particular taxa or taxonomic groups within the spread rate database. Three case studies, two species-specific and one more broad, highlight the types of insights that can be gained from detailed comparison of spread rates.

REPRODUCTIVE MODE: *UNDARIA PINNATIFIDA* VERSUS *CAULERPA TAXIFOLIA*   Table 15.2 includes spread rates for two species of macroalgae with very different life histories: the kelp *Undaria pinnatifida* and the green alga *Caulerpa taxifolia*. *Undaria* is unique in that it is the only known invasive kelp; by comparison, there are numerous exotic green seaweeds. *Undaria* has one of the lowest spread rates in the macroalgal data set (370 m/year: Fletcher and Farrell 1999), whereas *Caulerpa* is a notoriously rapid invader with an average spread rate of 10.9 km/year and a maximum observed spread rate of 53.7 km/year, excluding even larger jumps attributed to anthropogenic spread (Meinesz et al. 1993).

Both species grow rapidly and produce large numbers of propagules. However, they differ greatly in several important aspects of their reproductive ecology. *Undaria*, like all kelps, has a complex life cycle characterized by alternation of generations between a haploid gametophyte and a diploid sporophyte. Macroscopic diploids produce spores that settle into the benthos and develop into microscopic male and female gametophytes. Females secrete a pheromone that attracts sperm released by males. Fertilization and production of new sporophytes requires a critical gametophyte density of at least 1 per mm^2 (Reed et al. 1991). The diffusive spread of spores prior to settlement thus imposes a strong limitation on the distance over which spore dispersal can be effective. *Undaria* cannot reproduce vegetatively, and macroscopic sporophytes are not buoyant, so spores are thought to be the primary non-anthropogenic mode of dispersal.

In contrast, *Caulerpa* reproduces solely or predominantly by vegetative growth, with fragments of any part of the plant of virtually any size having a high capacity for survival and rapid growth. Fragments do not float, but are readily transported by subsurface drift. As is usually the case, multiple factors contribute to *Caulerpa*'s success as an invader and capacity for rapid spread. However, the lack of any reproductive barrier to colonization is critical to its potential for spread. Even a single isolated propagule can become established and produce a large local population over a period of months.

This example highlights the critical role of reproductive ecology in determining rates of range expansion. Any requirement that dispersive propagules be present at some critical density in order to effectively colonize will greatly limit rates of spread by reducing the effect of the longest-dispersing propagules (the "tails" of the dispersal distribution). The need for post-dispersal sexual reproduction, especially among organisms with limited mobility, is a common source of Allee effects at the edge of an expanding range. The combination of asexual reproduction and rapid growth, on the other hand, permits rapid spread.

MULTIPLE DISPERSAL MECHANISMS AND EFFECTIVENESS OF THE DISPERSAL "TAILS": *BOTRYLLOIDES VIOLACEOUS* VERSUS *BALANUS IMPROVISUS*   Sometimes the mode of dispersal that is principally responsible for spread differs from the primary, local dispersal mechanism. An obvious case is human-mediated jump dispersal, but multiple dispersal mechanisms are also a common feature of natural spread. An example of this is provided by the invasive colonial ascidian *Botrylloides violaceous* (see Table 15.2), which is capable of both sexual reproduction and asexual, clonal growth. Sexually produced tadpole larvae have a planktonic duration of only a few minutes and disperse on the order of meters. However, *Botrylloides* overgrows many substrates that are buoyant and capable of detaching and drifting for long distances, and it easily survives these periods of drift. Because of its capacity for asexual reproduction, a single such drifting adult colony can become established and begin a new local population. This secondary mode of dispersal effectively extends the "tails" of dispersal by several orders of magnitude and helps to explain the observed high spread rate of 16 km/year.

In comparison, consider the invasive barnacle *Balanus improvisus* (see Table 15.2). Despite the fact that *Balanus* larvae spend more than three orders of magnitude longer dispersing in the water column than *Botrylloides* (approximately 10–15 days), *Balanus* spreads at only a marginally higher rate (30 km/year). Like *Botrylloides*, *Balanus* is a sessile invertebrate that disperses by sexually produced larvae. However, *Balanus* is incapable of asexual reproduction, requires a mate within a few centimeters for internal fertilization, and is unlikely to survive detachment and drift (although individuals may drift on floating substrates, they cannot detach and subsequently reattach in a new location, and any larvae produced during the drift phase are subject to very high dilution effects due to small numbers of surrounding individuals). This combination of sexual reproduction, extremely limited adult mobility, and lack of any secondary dispersal mechanism causes *Balanus* to spread at a rate that is driven not by the longest dispersers, but by the bulk portion of the dispersal kernel. This is a striking example of the convergence of spread rates that is possible between organisms with very different primary dispersal distances.

ACTIVE VERSUS PASSIVE DISPERSAL: ANIMALS AND PLANTS IN MARINE AND TERRESTRIAL ENVIRONMENTS   We have already seen that terrestrial plants have quite limited spread rates compared with both marine plants and marine animals. Figure 15.4 extends this comparison by including several groups of terrestrial animals (birds, mammals, and insects). This results in a contrasting picture of the effect of passive and active dispersal mechanisms on land and in the sea.

Like marine organisms, terrestrial animals exhibit wide-ranging rates of spread, from meters to hundreds of kilometers per year. The upper bound of terrestrial animal dispersal is several orders of magnitude higher than that of terrestrial plant dispersal (~ 1000 km vs. ~ 1 km), suggesting that active dispersal provides potential for much greater rates of spread on land. Interestingly, this is not the case in marine systems. Passive dispersers in the fluid environment—even plants and other organisms that distribute the bulk of their propagules locally—are able to achieve spread rates as large as those of many fast-moving birds, mammals, and insects on land. Marine plants still exhibit significantly slower spread rates than marine animals (see Table 15.4), but there is substantial overlap in the distributions, with the upper bound of algal spread rates (64 km/year) differing from the upper bound of invertebrate/fish spread rates (247 km/year) by less than a factor of four. Marine organisms with more active dispersal mechanisms at their disposal, such as marine fishes, do not appear to exhibit greater spread rates than the average sessile invertebrate with similar planktonic larval periods (the average spread rate for the two fishes in our data set was 71 km/year).

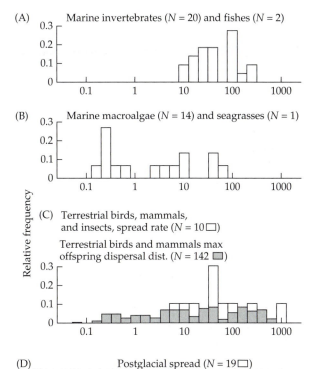

FIGURE 15.4 Distribution of observed rates of spread of invasive species in different environments (marine and terrestrial) and taxonomic groups. Mean rates of linear expansion of invasion fronts were measured from field survey maps. In part C, data on the maximum observed offspring dispersal distances of terrestrial birds and mammals are included (shaded bars; Sutherland et al. 2000) in addition to actual measured spread rates (open bars). In part D, modern (shaded bars) and paleo (open bars) spread rates of trees are plotted separately. (Sources are listed in Table 15.2.)

These accelerated spread rates of short-distance dispersers, of course, are dependent on life history characteristics that permit effective long-distance dispersal. The marine environment appears to provide more opportunities for such long-distance travel, even allowing for passive transport of large reproductive adults over long distances. General patterns of variation in the prevalence of active and passive dispersal mechanisms across systems and taxa may underlie important differences in ecology and biogeography on land and in the sea.

Summary: Life history and environmental features that promote rapid spread

A general feature that arises from examination of marine invasion data is the ability of long-distance dispersal mechanisms to promote rapid spread. Marine data also illustrate the importance of life history in determining the ultimate effect of long-distance dispersal. The combined variation in the shape of dispersal "tails" and in life history characteristics results in spread rates that are not predictable from average dispersal distances alone, and which can vary greatly among taxa. The theoretical results depicted in Figure 15.2 provide a possible explanation, summarizing the interaction between long-distance dispersal and Allee effects that can decouple spread rate from average dispersal distance.

The case study of *Balanus* and *Botrylloides*, along with comparisons between other, similar species pairs in our database, suggests reasons for the accelerated spread rates of many marine organisms with short average dispersal distances. Many marine organisms with short planktonic durations have life histories that allow one or a few individuals to colonize new areas, including secondary dispersal of reproductive adults and asexual reproduction (Strathmann 1987; Knowlton and Jackson 1993). In contrast, many organisms with long planktonic larval durations and long average dispersal distances have modes of reproduction that make them highly susceptible to Allee effects on establishment and reproduction at the invasion front. A requirement for post-dispersal sexual reproduction, especially among sedentary organisms, can render the "tails" of the dispersal distribution ineffective. This effect can occur both in organisms with internal fertilization, in which minimum spacing between individuals can influence reproductive success, and in organisms with external fertilization, in which a critical density of adults is often necessary for reproduction. Marine animals that reproduce by larvae with long planktonic periods are particularly vulnerable to these effects because the diffusive effect of the fluid environment causes settlement densities to decrease rapidly with time. The same is true of wind-dispersed seeds of land plants, although long-distance pollen dispersal may help to overcome this limitation. Pollen can survive far longer and potentially disperse much farther (Dafni and Firmage 2000) than most marine gametes, especially when carried by active pollinators (unlike larvae, marine gametes are generally short-lived; e.g., Levitan et al. 1992). A single plant that germinates far from conspecifics may still have a reasonable chance of fertilization from long-distance gamete transport (Ouborg et al. 1999). On the other hand, Allee effects on passive dispersers in patchy, two-dimensional environments, such as the habitats of many land plants, may be more pronounced; these environments contrast with the approximately linear ranges inhabited by many coastal marine species.

The potential effects of life history and long-distance dispersal are dramatically illustrated by comparison of spread rates with the length of the planktonic period available for dispersal (a proxy for average dispersal distance) in marine macroalgae and marine invertebrates (Figure 15.5). Despite the extremely short planktonic period of their primary propagules, the spread rates of macroalgae can range as high as those of invertebrates that spend 10–100 times as long in the plankton. Algal spread rates are highly variable, probably due to the combined effects of variation in reproductive ecology (as described in the macroalgal case study above) and the intrinsic stochasticity of the long-distance dispersal process (Clark et al. 2001). Invertebrate spread rates

FIGURE 15.5   Mean rate of spread (of naturalized populations) versus the average planktonic period of primary propagules (larvae, spores, etc.) for marine benthic organisms. Note that the average duration of the planktonic period, which generally correlates with average dispersal distances (Siegel et al. 2003), is a poor predictor of spread rate for marine organisms. Mean rates of linear expansion of invasion fronts were measured from field survey maps (as described in Kinlan and Gaines 2003). Larval duration was estimated from laboratory culture studies and/or field observations (as described in Siegel et al. 2003). The planktonic periods of the macroalgae and seagrasses included here are short (much less than 3 days), but are not known exactly. For comparison with invertebrates, all marine plants have been set at time in plankton = 1 day on this graph. (Species list and sources are given in Table 15.2.)

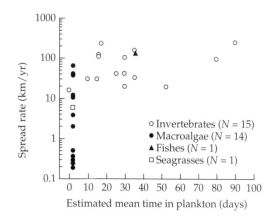

are less variable and show only a weak (nonsignificant) tendency to increase as time in the plankton increases (OLS regression, $r^2 = 0.15$, $p = 0.14$), suggesting that demographic processes reducing the effectiveness of long-distance dispersal increase even as the average dispersal distance grows larger. These results are in agreement with a recent spatially explicit model that predicted a rapid drop-off in spread rates of organisms with low mate-finding ability as dispersal distance increased (South and Kenward 2001). The result is a convergence of maximum spread rates in organisms whose primary dispersal distances range over several orders of magnitude (see Figure 15.5).

Fat-tailed dispersal kernels have important implications for the ecology and biogeography of marine species. Over ecological time scales, the interaction of dispersal kernel shape with ability to become established and reproduce at low densities determines rates of recolonization and migration in response to environmental perturbations. Over evolutionary time scales, a low level of long-distance dispersal (~ 1 migrant per generation) can be sufficient to prevent divergence of distant conspecific populations while still allowing for local adaptation and the buildup of geographic population structure. Interestingly, marine macroalgae, some of the shortest dispersers in marine benthic communities, are also capable of maintaining very large geographic ranges (S. E. Lester and B. P. Kinlan, unpublished data). Studies of range expansion that do not take long-distance dispersal into account risk seriously underestimating rates of range expansion.

In addition to the influence of life history and long-distance dispersal, our comparative analysis highlights differences in the consequences of passive and active dispersal behaviors in marine and terrestrial systems. Terrestrial plants, which are sessile and rely on passive dispersal of seeds by physical or biological agents, exhibit much slower rates of range expansion than terrestrial birds, mammals, and insects, which disperse actively as adults or juveniles (see Figure 15.4). In the marine environment, due in part to the buoyancy of the fluid medium, passive dispersers can exhibit spread rates comparable to those of active dispersers on land. However, because dispersal in land animals is often mediated by complex adult behaviors, it may be highly nonrandom and may occur episodically as a direct response to local resource availability or patch quality.

An open question in biogeography is the extent to which species have expanded their ranges in keeping with climatic fluctuations, such that present-day borders represent "equilibrium" conditions. Our results imply that sessile and sedentary marine species are capable of much more rapid range expansion than their terrestrial counterparts. This conclusion leads to two predictions. First, coastal marine species should, on average, exhibit more variable range boundaries than terrestrial plants because they are able to respond more rapidly to short-term climatic shifts. Boundaries of terrestrial birds, mammals, and insects, on the other hand, may exhibit variation similar to that of marine species because of their faster spread rates and active dispersal behaviors. Second, over millennial time scales, marine range boundaries are more likely to have kept pace with expanding regions of suitable habitat; that is, to have reached what could be termed a postglacial "equilibrium" state. Although a full evaluation of these predictions is beyond the scope of this chapter, there is some evidence to support these hypotheses. A recent study found that European trees occupy, on average, only 38% of their potential ranges (Svenning and Skov 2004). On shorter time scales, land plants tend to exhibit relatively small range extensions during brief climatic fluctuations (e.g., Ward et al. 2003). In contrast, marine fishes (Victor et al. 2001), marine invertebrates (Zacherl et al. 2003; Stachowicz et al. 2002), land birds (Patten and Marantz 1996), land mammals (Jaksic and Lima 2003), and insects (Loxdale and Lushai 1999) have been observed to undergo rapid, large range extensions during short-term climatic perturbations such as El Niño events. The ability of terrestrial plants to migrate rapidly in response to climate changes may be further impaired by the fragmented, two-dimensional nature of their habitat, which can magnify Allee effects, especially for passive dispersers. If so, this difference in range dynamics could have major implications for differences in diversity and range size patterns in marine versus terrestrial systems.

The comparative analysis presented here has led to some simple insights into mechanisms underlying variation in species spread rates, which can be understood by merging data on life history and demography with theoretical expectations. Such insights, based on syntheses of increasingly available invasive species monitoring data, contribute to a macroecological perspective on differences in range expansion processes across systems and taxa.

General Principles of Geographic Spread

The synthesis of marine and terrestrial population spread rate data that we present here is only one small piece of a large literature using exotic species as models of population spread, beginning with the seminal work of Skellam (1951) and Elton (1958). Below, we discuss five general features of range expansion that arise from this large body of work, reinforced and illustrated by the results presented in this chapter (for additional recent reviews, see Okubo and Levin 2002; Levin et al. 2003; Hastings et al. 2005).

1. Spread rates, and thus the speed of recolonization and range expansion into favorable habitat, are often dependent not on the average dispersal distance, but on the long-distance tails of the dispersal kernel. Thus, spread rate is often driven by "alternative" modes of dispersal—rare events that can carry propagules very far (Shigesada et al. 1995; Kot et al. 1996; Cain et al. 1998; Clark 1998; Higgins and Richardson 1999; Clark et al. 2003; Kinlan et al. 2005).

The role of long-distance dispersal is especially evident when contrasting the spread rates of sessile organisms with short primary propagule dispersal distances in marine versus terrestrial systems. Mechanisms that can promote rare long-distance dispersal and establishment of secondary propagules (e.g., drifting reproductive adults) are more prevalent in marine environments and appear to increase spread rates over those of terrestrial counterparts by up to several orders of magnitude. Long-distance dispersal also accelerates the spread rates of short-distance dispersers on land (Cain et al. 1998, 2000), but apparently to a lesser degree. Terrestrial and marine organisms that disperse actively as adults (e.g., mammals, birds, cetaceans, large pelagic fishes) are capable of directed, behaviorally complex long-distance dispersal events, and thus may exhibit more complicated and variable spread rates in the face of changing migration routes and resource distributions.

2. Only effective dispersal matters—that is, propagules or recruits that survive the dispersal and juvenile phases and reproduce. Thus, both life history and the spatiotemporal pattern of the environment are critical determinants of spread rate (Elton 1958; Shigesada and Kawasaki 1997; Hastings et al. 2005). Allee effects, in particular, can virtually eliminate the accelerating influence of long-distance dispersal and limit spread rates, or even halt range expansion (Lewis and Kareiva 1993; Keitt et al. 2001; Wang et al. 2002). For this reason, the ability to reproduce asexually, reproduce sexually at low densities, or spread by secondary propagules post-reproduction can be a particularly important factor influencing spread.

The ability to become established and reproduce at low densities has been a focus of invasion biology for many years (e.g., Baker and Stebbins 1965). Inclusion of marine and terrestrial species here allowed us to compare spread rates across a broad range of life histories. Our results illustrate that even species with great potential for long-distance dispersal of individual propagules (e.g., marine invertebrates with long planktonic durations) may not spread as fast as would be expected if long-distance travelers cannot survive and reproduce at low densities. In contrast, organisms such as macroalgae that produce many low-cost propagules and distribute them locally may experience more rapid spread rates than expected from their limited dispersal. The passive versus active nature of dispersal is an important determinant of the strength of Allee effects. In sexual organisms with active movement and complex behavior (e.g., birds, marine mammals), Allee effects may be reduced by the ability to actively locate mates during and after long-distance dispersal.

The genetic implications of demographic traits that increase spread, such as asexual reproduction, remain to be explored. Conceivably, there could be a long-term feedback between population genetic structure that arises from asexual spread and traits such as dispersal ability, local adaptation, vulnerability to climate change, and geographic range size (Garcia-Ramos and Rodríguez 2002; Crespi 2004; Rice and Sax, this volume).

3. Spread in real-world invasions is a stochastic process (Bolker 1999; Lewis and Pacala 2000; Cain et al. 2000; Clark et al. 2003). An important implication of this stochasticity is that spread can occur very rapidly once it begins, but initiating successful spread often requires many inoculations (e.g., Leung et al. 2004). Rates of spread (even in the same species) may also exhibit large stochastic variation across different locations and time periods (Clark et al. 2001). The more spread is driven by a small fraction of propagules that travel much farther than the mean, the less predictability will be possible across "realizations" (Lewis and Pacala 2000).

Recent studies of long-distance dispersal in both marine and terrestrial systems have suggested that even a low level of very long-range movement can increase spread rates, as well as uncertainty about patterns and rates of spread (Clark et al. 2003; Kinlan and Gaines 2003). This effect is evident in the contrast between spread rates of marine and terrestrial plants. Marine macroalgae, which have greater opportunities for long-distance dispersal of reproductive adults, exhibit spread rates that are much greater than their average dispersal distances, but are highly variable among species and across multiple invasions of a given species. Terrestrial plants, which have fewer mechanisms for effective long-distance dispersal, show lower and less variable spread rates. Again, actively dispersing organisms with highly developed locomotory and sensory capabilities may move in a more directed, deterministic fashion than taxa with passive dispersal of young, thereby reducing stochastic variation across locations and time.

4. In some cases, prehistoric range expansions may have been very rapid, like some of those observed today (e.g., Parmesan et al. 1999; Leppakoski and Olenin 2000). Spread across large expanses of suitable habitat can be understood by looking at the spread of exotic species today, especially when they have invaded multiple locations and can be observed under a range of environmental and biotic conditions.

The potential for rapid spread over many degrees of latitude (at scales of years to decades) helps to explain jumps in biogeographic distribution in coarse temporal records and argues for the continued development and application of high-temporal resolution indicators of past spread (e.g., from sediment cores). Darwin (1859) used exotic rates of spread to argue that native species would have had ample time to migrate as climate changed over glacial cycles. At millennial and longer scales, the rates of spread in our exotic species data set support this conclusion. A range expansion rate as small as 100 m/year could result in spread over nearly 10 degrees of latitude in 10,000 years.

However, recent paleoclimatic research indicates that major climatic shifts can occur much more abruptly than previously thought, perhaps over time scales as short as years to decades (NRC 2002). The ability of species to migrate over time scales of a few generations could be crucial to understanding their responses to these rapid changes. Long-distance dispersal must play a particularly important role during these periods. Even over longer periods (i.e., the Holocene), postglacial spread rates of North American flora predicted from average dispersal distances fall far short of rates of range expansion inferred from pollen records (Cain et al. 1998). Only when long-distance dispersal is incorporated do predicted and observed migration rates converge (Clark 1998). Thus, the differences in long-distance dispersal potential between marine and terrestrial environments suggested by our analyses may help to explain differences between marine and terrestrial biogeographic responses to climate change.

5. Other range expansions may have been limited in their extent by physical and biotic environmental barriers, perhaps in combination with genetic and demographic processes (Davis and Shaw 2001). The oldest and simplest incarnation of this idea is the so-called "climatic envelope" model of range shifts, which posits that species gradually shift their range to match the changing region where their tolerance for environmental (and possibly biotic) pressures allows them to survive (France 1991). However, recent studies have shown that, even across smooth environmental gradients, species spread can be slowed or halted dynamically by demographic and/or genetic factors interacting with biophysical gradients, possibly long before suitable habitat has been filled (e.g., Kirkpatrick and Barton 1997; Keitt et al. 2001; Holt et al. 2005). This may help to explain more complex and dynamic biogeographic patterns.

Although the present study focuses on spread rates into suitable habitats in the absence of barriers, meta-analyses similar to ours may be of use in identifying likely patterns of spread under different life history and environmental scenarios. To date, most predictions of species' responses to climate change have invoked the "climatic envelope" assumption that species distributions will track changes in the physical environment, including temperature and precipitation (France 1991; for a recent application, see Araujo et al. 2004). However, as the modeling studies reviewed above reveal, interactions between demographic, environmental, and genetic processes can result in more complex patterns, including slowed rates of spread and stable range boundaries along apparently smooth environmental gradients.

Thus, understanding past and future biogeographic patterns may depend on improved understanding and quantification of demography, abundance, genetic structure, and environmental variation across species ranges. For many species, especially those with limited spread rates (e.g., land plants), the dynamic processes controlling pinning and expansion may be more important in determining present and future boundaries than the "envelope" of habitat that is available based on environmental tolerance alone. By developing increasingly large databases of modern spread, we may begin to associate particular combinations of environmental, genetic, and demographic features with rates and patterns of spread.

The same approach will be useful in a paleobiogeographic context. The macroecology of modern spread rates can help to calibrate or place bounds on past rates of spread, which could in turn lead to insights on whether spread was hindered or enhanced by undetected processes and whether boundaries were historical accidents or the result of externally imposed barriers. Also essential to understanding species' responses to climate change is a better understanding of population genetics and adaptation at expanding range boundaries. Contrasts between peripheral populations of native and exotic species may yield important insights into this issue.

Conclusion

Over the past 50 years, and especially in the last decade, much progress has been made toward understanding the effects of demography, dispersal, biological interactions, environmental structure, and adaptation on the rate and pattern of species spread. Much of this work has been theoretical; recently, increased empirical data on landscape structure and dispersal have allowed a marriage of theory and observation to gain some general insights into range expansion. Exotic species have provided, and will continue to provide, an important window into the process of population spread. Exotic species range widely in their life histories and in the environments and communities they invade, and data on their rates and patterns of spread can be used to test predictions of theoretical constructs. Moreover, because invasions are ongoing and happen at human time scales, manipulative experiments can confirm what would otherwise be mere suspicions as to underlying mechanisms. An especially valuable feature of exotics that has yet to be fully exploited is the opportunity to study multiple "replicate" invasions of the same species (e.g., Grosholz and Ruiz 1996; Thornber et al. 2004). Such studies will become more common and informative as new introductions continue and invasive species data sets expand to a global scale. The next frontier for contributions of exotic species studies to our understanding of population expansion may be the opportunity they provide to test genetic models of range expansion, contraction, pinning, and evolution (e.g., Kirkpatrick and Barton 1997; Garcia-Ramos and Rodríguez 2002); such studies will be critical in connecting our knowledge of short-term range expansion processes to the dynamics of species distribution on evolutionary time scales. A greater mechanistic understanding of the causes of range boundary dynamics, in turn, will lend insight into the causes of equilibrium biogeographic patterns, such as range size and diversity gradients (Brown 1995; Sax 2001).

Acknowledgments

The authors gratefully acknowledge D. F. Sax, S. D. Gaines, D. A. Siegel, K. F. Smith, and two anonymous reviewers for discussions that aided in development of this chapter. S. Brummell provided invaluable technical assistance in preparation of the invasion data set. BPK was supported by a graduate research fellowship from the Fannie and John Hertz Foundation. This research was also supported by NSF Biocomplexity grant DEB-0083583 (to AH) and the National Center for Ecological Analysis and Synthesis, a center funded by the National Science Foundation (grant #DEB-94-21535). This is contribution number 181 from the Partnership for Interdisciplinary Studies of Coastal Oceans (PISCO), a long-term consortium funded by the David and Lucile Packard Foundation.

Literature Cited

Araujo, M. B., M. Cabeza, and W. Thuiller. 2004. Would climate change drive species out of reserves? An assessment of existing reserve-selection methods. Global Change Biology 2004:1618–1626.

Armonies, W. 2001. What an introduced species can tell us about the spatial extension of benthic populations. Marine Ecology Progress Series 209:289–294.

Avise, J. C. 1992. Molecular population structure and the biogeographic history of a regional fauna: A case history with lessons for conservation biology. Oikos 63:62–76.

Baker, H. G., and G. L. Stebbins, eds. 1965. *The genetics of colonizing species.* Academic Press, New York.

Barry, J. P., C. H. Baxter, R. D. Sagarin, and S. E. Gilman. 1995. Climate-related, long-term faunal changes in a California rocky intertidal community. Science 267:672–675.

Bolker, B. M. 1999. Analytic models for the patchy spread of plant disease. Bulletin of Mathematical Biology 61:849–874.

Breber, P. 2002. Introduction and acclimatisation of the pacific carpet clam, *Tapes philippinuarum*, to Italian waters, In E. Leppakoski, S. Gollasch, and S. Olenin, eds. *Invasive aquatic species of Europe: distribution, impacts, and management*, pp. 120–126.

Kluwer Academic Publishers, Dordrecht, Netherlands.

Brown, J. H. 1995. *Macroecology.* University of Chicago Press, Chicago.

Brown, J. H., and M. V. Lomolino. 1998. *Biogeography,* 2nd Ed. Sinauer Associates, Sunderland, MA.

Cain, M. L., H. Damman, and A. Muir. 1998. Seed dispersal and the Holocene migration of woodland herbs. Ecological Monographs 68:325–347.

Cain, M. L., B. G. Milligan, and A. E. Strand. 2000. Long-distance seed dispersal in plant populations. American Journal of Botany 87:1217–1227.

Carlton, J. T., and J. B. Geller. 1993. Ecological roulette: the global transport of nonindigenous marine organisms. Science 261:78–82.

Carr, M. H., J. E. Neigel, J. A. Estes, S. Andelman, R. R. Warner, and J. L. Largier. 2003. Comparing marine and terrestrial ecosystems: implications for the design of coastal marine reserves. Ecological Applications 13:S90–S107.

Clark, J. S. 1998. Why trees migrate so fast: confronting theory with dispersal biology and the paleorecord. American Naturalist 152:204–224.

Clark, J. S., M. Lewis, and L. Horvath. 2001. Invasion by extremes: population spread with variation in dispersal and reproduction. American Naturalist 157:537–554.

Clark, J. S., M. Lewis, J. S. McLachlan, and J. H. R. Lambers. 2003. Estimating population spread: What can we forecast and how well? Ecology 84:1979–1988.

Cox, C. B., and P. D. Moore. 2000. *Biogeography: an ecological and evolutionary approach.* Blackwell Science, Oxford.

Crespi, B. J. 2004. Vicious circles: positive feedback in major evolutionary and ecological transitions. Trends in Ecology and Evolution 19:627–633.

Dafni A., and D. Firmage. 2000. Pollen viability and longevity: practical, ecological and evolutionary implications. Plant Systematics and Evolution 222:113–132.

Darwin, C. 1859. *On the origin of species by means of natural selection.* John Murray, London.

Davis, A. R., D. E. Roberts, and S. P. Cummins. 1997. Rapid invasion of a sponge-dominated deep-reef by *Caulerpa scalpelliformis* (Chlorophyta) in Botany Bay, New South Wales. Australian Journal of Ecology 22:146–150.

Davis, M. B., and R. G. Shaw. 2001. Range shifts and adaptive responses to Quaternary climate change. Science 292:673–679.

Durrett, R., and S. A. Levin. 1994. Stochastic spatial models—a user's guide to ecological applications. Philosophical Transactions of the Royal Society of London B 343:329–350.

Elton, C. S. 1958. *The ecology of invasions by animals and plants.* Methuen, London.

Estoup, A., M. Beaumont, F. Sennedot, C. Moritz, and J. M. Cornuet. 2004. Genetic analysis of complex demographic scenarios: spatially expanding populations of the cane toad, *Bufo marinus.* Evolution 58:2021–2036.

Fisher, R. A. 1937. The waves of advance of advantageous genes. Annals of Eugenics 7:355–369.

Fletcher, R. L., and P. Farrell. 1999. Introduced brown algae in the north east Atlantic, with particular respect to *Undaria pinnatifida* (Harvey) Suringar. Helgolander Meeresunters 52:259–275.

Flierl, G., D. Grünbaum, S. Levin, and D. Olson. 1999. From individuals to aggregations: the interplay between behavior and physics. Journal of Theoretical Biology 196:397–454.

France, R. L. 1991. Empirical methodology for predicting changes in species range extension and richness associated with climate warming. International Journal of Biometeorology 34:211–216.

Garcia-Ramos, G., and D. Rodríguez. 2002. Evolutionary speed of species invasions. Evolution 56:661–668.

Gras, J. M. 1971. Range extension in the period 1968–1970 of the alien amphipod *Gammarus tigrinus* Sexton, 1939, in the Netherlands. Bulletin Zoologisch Museum 2:5–9.

Grinnell, J. 1917. The niche-relationships of the California thrasher. The Auk 34:427–433.

Grosholz, E. D. 1996. Contrasting rates of spread for introduced species in terrestrial and marine systems. Ecology 77:1680–1686.

Grosholz, E. D., and G. M. Ruiz. 1996. Predicting the impact of introduced marine species: lessons from the multiple invasions of the European green crab *Carcinus maenas.* Biological Conservation 78:59–66.

Handtke, K., and G. Mauersberger. 1977. Die Ausbreitung des Kuhreihers, *Bubulcus ibis* (Linnaeus). Mitteilungen Zoologisches Museum in Berlin (Suppl.) Band 3, Annals of Ornithology 1:1–78.

Hanski, I. A., and M. E. Gilpin. 1997. *Metapopulation biology.* Academic Press, San Diego, CA.

Harrison, S., A. Hastings, and D. R. Strong. 2005. Spatial and temporal dynamics of insect outbreaks in a complex multitrophic system: tussock moths, ghost moths, and their natural enemies on bush lupines. Annales Zoologici Fennici. In press.

Harvey, P. H., and M. Pagel. 1991. *The comparative method in evolutionary biology.* Oxford University Press, Oxford.

Hastings, A. 1996. Models of spatial spread: a synthesis. Biological Conservation 78:143–148.

Hastings, A., S. Harrison, and K. McCann. 1997. Unexpected spatial patterns in an insect outbreak match a predator diffusion model. Proceedings of the Royal Society of London B 264:1837–1840.

Hastings, A., K. Cuddington, K. F. Davies, C. J. Dugaw, S. Elmendorf, A. Freestone, S. Harrison, M. Holland, J. Lambrinos, U. Malvadkar, B. A. Melbourne, K. Moore, C. Taylor, and D. Thomson. 2005. The spatial spread of invasions: new developments in theory and evidence. Ecology Letters 8:91–101.

Hengeveld, R. 1988. Mechanisms of biological invasions. Journal of Biogeography 15:819–828.

Hengeveld, R. 1989. *Dynamics of biological invasions.* Chapman and Hall, London.

Higgins, S. I., and D. M. Richardson. 1996. A review of models of alien plant spread. Ecological Modelling 87:249–265.

Higgins, S. I., and D. M. Richardson. 1999. Predicting plant migration rates in a changing world: the role of long-distance dispersal. American Naturalist 153:464–475.

Highsmith, R. C. 1985. Floating and algal rafting as potential dispersal mechanisms in brooding invertebrates. Marine Ecology Progress Series 25:169–180.

Hobday, A. J. 2000. Persistence and transport of fauna on drifting kelp [*Macrocystis pyrifera* (L.) C. Agardh] rafts in the Southern California Bight. Journal of Experimental Marine Biology and Ecology 253:75–96.

Holt, R. D., T. H. Keitt, M. A. Lewis, B. A. Maurer, and M. L. Taper. 2005. Theoretical models of species' borders: single species approaches. Oikos 108:18–27.

Jackson, S. T., and J. W. Williams. 2004. Modern analogs in Quaternary paleoecology: here today, gone yesterday, gone tomorrow? Annual Review of Earth and Planetary Sciences 32:495–537.

Jaksic, F. M., and M. Lima. 2003. Myths and facts on ratadas: bamboo blooms, rainfall peaks and rodent outbreaks in South America. Austral Ecology 28:237–251.

Jones, R. E., N. Gilbert, M. Guppy, and V. Nealis. 1980. Long-distance movement of *Pieris rapae.* Journal of Animal Ecology 49:629–642.

Keitt, T. H., M. A. Lewis, and R. D. Holt. 2001. Allee effects, invasion pinning, and species' borders. American Naturalist 157:203–216.

Kierstead, H., and L. B. Slobodkin. 1953. The size of water masses containing plankton blooms. Journal of Marine Research 12:141–147.

Kinlan, B. P., and S. D. Gaines. 2003. Propagule dispersal in marine and terrestrial environments: a community perspective. Ecology 84:2007–2020.

Kinlan, B. P., S. D. Gaines, and S. E. Lester. 2005. Propagule dispersal and the scales of marine community process. Diversity and Distributions 11:139–148.

Kirkpatrick, M., and N. H. Barton. 1997. Evolution of a species' range. American Naturalist 150:1–23.

Knowlton, N., and J. B. C. Jackson. 1993. Inbreeding and outbreeding in marine invertebrates. In N. W. Thornhill, ed. *The natural history of inbreeding and outbreeding: theoretical and empirical perspectives*, pp. 200–249. University of Chicago Press, Chicago.

Kot, M., M. A. Lewis, and P. van den Driessche. 1996. Dispersal data and the spread of invading organisms. Ecology 77:2027–2042.

Lambrinos, J. G. 2004. How interactions between ecology and evolution influence contemporary invasion dynamics. Ecology 85:2061–2070.

LeCorre, V., N. Machon, R. J. Petit, and A. Kremer. 1997. Colonization with long-distance seed dispersal and genetic structure of maternally inherited genes in forest trees: a simulation study. Genetical Research 69:117–125.

Leppakoski, E., S. Gollasch, P. Gruszka, H. Ojaveer, S. Olenin, and V. Panov. 2002. The Baltic—a sea of invaders. Canadian Journal of Fisheries and Aquatic Science 59:1175–1188.

Leppakoski, E., and S. Olenin. 2000. Non-native species and rates of spread: lessons from the brackish Baltic Sea. Biological Invasions 2:151–163.

Leung, B., J. M. Drake, and D. M. Lodge. 2004. Predicting invasions: propagule pressure and the gravity of Allee effects. Ecology 85:1651–1660.

Levin, S. A. 1992. The problem of pattern and scale in ecology. Ecology 73:1943–1967.

Levin, S. A., H. C. Muller-Landau, R. Nathan, and J. Chave. 2003. The ecology and evolution of seed dispersal: a theoretical perspective. Annual Review of Ecology, Evolution, and Systematics 34:575–604.

Levins, R. 1969. Some demographic and genetic consequences of environmental heterogeneity for biological control. Bulletin of the Entomological Society of America 15:237–240.

Levitan, D. R., M. A. Sewell, and F. S. Chia. 1992. Kinetics of fertilization in the sea urchin *Strongylocentrotus franciscanus*: interaction of gamete dilution, age, and contact time. Biological Bulletin 181:371–378.

Lewis, M. A., and P. Kareiva. 1993. Allee dynamics and the spread of invading organisms. Theoretical Population Biology 43:141–158.

Lewis, M. A., and S. Pacala. 2000. Modeling and analysis of stochastic invasion processes. Journal of Mathematical Biology 41:387–429.

Loxdale, H. D., and G. Lushai. 1999. Slaves of the environment: the movement of herbivorous insects in relation to their ecology and genotype. Philosophical Transactions of the Royal Society of London B Biological Sciences 354:1479–1495.

Macdonald, G. M. 1993. Fossil pollen analysis and the reconstruction of plant invasions. Advances in Ecological Research 24:67–110.

Maggs, C. A., and H. Stegenga. 1999. Red algal exotics on North Sea coasts. Helgolander Meeresunters 52:243–258.

Matlack, G. R. 1994. Plant species migration in a mixed-history forest landscape in eastern North America. Ecology 75:1491–1502.

McCann, K., A. Hastings, S. Harrison, and W. Wilson. 2000. Population outbreaks in a discrete world. Theoretical Population Biology 57:97–108.

McQuaid, C. D., and T. E. Phillips. 2000. Limited wind-driven dispersal of intertidal mussel larvae: *in situ* evidence from the plankton and the spread of the invasive species *Mytilus galloprovincialis* in South Africa. Marine Ecology Progress Series 201:211–220.

Meinesz, A., J. D. Vaugelas, B. Hesse, and X. Mari. 1993. Spread of the introduced tropical green alga *Caulerpa taxifolia* in northern Mediterranean waters. Journal of Applied Phycology 5:141–147.

Mollison, D. 1972. The rate of spatial propagation of simple epidemics. Proceedings of the Sixth Berkeley Symposium on Mathematics, Statistics and Probability 3:579–614.

Mundinger, P., and S. Hope. 1983. Expansion of the winter range of the House Finch: 1947–1979. American Birds 36:347–353.

Murray, J. D. 1993. *Mathematical biology*, 2nd Ed. Springer-Verlag, New York.

NRC (National Research Council). 2002. *Abrupt climate change: inevitable surprises*. National Academies Press, Washington, DC.

Neubert, M. G., and H. Caswell. 2000. Demography and dispersal: calculation and sensitivity analysis of invasion speed for structured populations. Ecology 81:1613–1628.

Nichols, R. A., and G. M. Hewitt. 1994. The genetic consequences of long-distance dispersal during colonization. Heredity 72:312–317.

Okubo, A., and S. A. Levin. 2002. *Diffusion and ecological problems: modern perspectives*. Springer-Verlag, New York.

Olsson, V. 1971. Studies of less familiar birds 165: Serin. British Birds 64:213–223.

Ouborg, N. J., Y. Piquot, and J. M. van Groenendael. 1999. Population genetics, molecular markers and the study of dispersal in plants. Journal of Ecology 87:551–568.

Owen, M., and M. A. Lewis. 2001. Can predation slow, stall or reverse a prey invasion? Bulletin of Mathematical Biology 63:655–684.

Palumbi, S. R. 2003. Population genetics, demographic connectivity and the design of marine reserves. Ecological Applications 13:S146–S158.

Parmesan, C., N. Ryrholm, C. Stefanescu, J. K. Hill, C. D. Thomas, H. Descimon, B. Huntley, L. Kaila, J. Kullberg, T. Tammaru, W. J. Tennent, J. A. Thomas, and M. Warren. 1999. Poleward shifts in geographical ranges of butterfly species associated with regional warming. Nature 399:579–583.

Parmesan, C., and G. Yohe. 2003. A globally coherent fingerprint of climate change impacts across natural systems. Nature 421:37–42.

Parvinen, K., and M. Egas. 2004. Dispersal and the evolution of specialisation in a two-habitat type metapopulation. Theoretical Population Biology 66:233–248.

Patten, M. A., and C. A. Marantz. 1996. Implications of vagrant southeastern vireos and warblers in California. Auk 113:911–923.

Por, F. D. 1978. Lessepsian migration: The influx of red sea biota into the Mediterranean by way of the Suez Canal. Springer, Berlin.

Reed, D. C., M. Neushul, and A. W. Ebeling. 1991. Role of density in gametophyte growth and reproduction in the kelps *Macrocystis* and *Pterygophora californica*. Journal of Phycology 27:361–366.

Rodgers, S. K. U., and E. F. Cox. 1999. Rate of spread of introduced Rhodophytes *Kappaphycus alvarezii*, *Kappaphycus striatum*, and *Gracilaria salicornia* and their current distributions in Kaneo'he Bay, O'ahu, Hawai'i. Pacific Science 53:232–241.

Russell, D. J., and G. H. Balazs. 1994. Colonization by the alien marine alga *Hypnea musciformis* (Wulfen) J. Ag. (Rhodophyta:Gigartinales) in the Hawaiian Islands and its utilization by the green turtle, *Chelonia mydas* L. Aquatic Botany 47:53–60.

Sagarin, R. D., and S. D. Gaines. 2002. The "abundant centre" distribution: to what extent is it a biogeographical rule? Ecology Letters 5:137–147.

Sakai, A. K., F. W. Allendorf, J. S. Holt, D. M. Lodge, J. Molofsky, K. A. With, S. Baughman, et al. 2001. The population biology of invasive species. Annual Review of Ecology and Systematics 32:305–332.

Sax, D. F. 2001. Latitudinal gradients and geographic ranges of exotic species: implications for biogeography. Journal of Biogeography 28:139–150.

Sax, D. F., and J. H. Brown. 2000. The paradox of invasion. Global Ecology and Biogeography 9:363–371.

Shanks, A. L., B. A. Grantham, and M. H. Carr. 2003. Propagule dispersal distance and the size and spacing of marine reserves. Ecological Applications 13:S159–S169.

Shigesada, N., and K. Kawasaki. 1997. *Biological invasions: theory and practice*. Oxford University Press, Oxford.

Shigesada, N., K. Kawasaki, and Y. Takeda. 1995. Modeling stratified diffusion in biological invasions. American Naturalist 146:229–251.

Siegel, D. A., B. P. Kinlan, B. Gaylord, and S. D. Gaines. 2003. Lagrangian descriptions of marine larval dispersion. Marine Ecology Progress Series 260:83–96.

Skellam, J. G. 1951. Random dispersal in theoretical populations. Biometrika 38:196–218.

Smith, J. E., C. L. Hunter, and C. M. Smith. 2002. Distribution and reproductive characteristics of nonindigenous and invasive marine algae in the Hawaiian Islands. Pacific Science 56:299–315.

South, A. B., and R. E. Kenward. 2001. Mate finding, dispersal distances and population growth in invading species: a spatially explicit model. Oikos 95:53–58.

Stachowicz, J. J., J. R. Terwin, R. B. Whitlatch, and R. W. Osman. 2002. Linking climate change and biological invasions: ocean warming facilitates nonindigenous species invasions. Proceedings of the National Academy of Sciences USA 99:15497–15500.

Strathmann, M. F. 1987. *Reproduction and development of marine invertebrates of the northern Pacific coast: data and methods for the study of eggs, embryos, and larvae.* University of Washington Press, Seattle.

Sutherland, G. D., A. S. Harestad, K. Price, and K. P. Lertzman. 2000. Scaling of natal dispersal distances in terrestrial birds and mammals. Conservation Ecology 4:16 (June 26). http://www.consecol.org/vol14/iss11/art16.

Svenning, J. C., and F. Skov. 2004. Limited filling of the potential range in European tree species. Ecology Letters 7:565–573.

Thomas, C. D., A. Cameron, R. E. Green, M. Bakkenes, L. J. Beaumont, Y. C. Collingham, B. F. N. Erasmus et al. 2004. Extinction risk from climate change. Nature 427:145–148.

Thornber, C. S., B. P. Kinlan, M. H. Graham, and J. J. Stachowicz. 2004. Population ecology of the invasive kelp *Undaria pinnatifida* in California: environmental and biological controls on demography. Marine Ecology Progress Series 268:69–80.

Travis, J. M. J., and C. Dytham. 2002. Dispersal evolution during invasions. Evolutionary Ecology Research 4:1119–1129.

Umbanhowar, J., and A. Hastings. 2002. The impact of resource limitation and the phenology of parasitoid attack on the duration of insect herbivore outbreaks. Theoretical Population Biology 62:259–269.

Valentine, J. W. 1968. The evolution of ecological units above the population level. Journal of Paleontology 42:253–267.

van den Bosch, F., J. A. J. Metz, and O. Diekmann. 1990. The velocity of spatial population expansion. Journal of Mathematical Biology 28:529–565.

van Kirk, R. W., and M. A. Lewis. 1997. Integrodifference models for persistence in fragmented habitats. Bulletin of Mathematical Biology 59:107–137.

Victor, B. C., G. M. Wellington, D. R. Robertson, and B. I. Ruttenberg. 2001. The effect of the El Niño-Southern Oscillation event on the distribution of reef-associated labrid fishes in the eastern Pacific Ocean. Bulletin of Marine Science 69:279–288.

Wang, M. H., M. Kot, and M.G. Neubert. 2002. Integrodifference equations, Allee effects, and invasions. Journal of Mathematical Biology 44:150–168.

Ward, K. M., J. C. Callaway, and J. B. Zedler. 2003. Episodic colonization of an intertidal mudflat by native cordgrass (*Spartina foliosa*) at Tijuana Estuary. Estuaries 26:116–130.

Wilson, W. G., S. P. Harrison, A. Hastings, and K. McCann. 1999. Exploring stable pattern formation in models of tussock moth populations. Journal of Animal Ecology 68:94–107.

Wolfram, S. 1983. Statistical mechanics of cellular automata. Reviews of Modern Physics 55:601–644.

Yund, P. O., and S. K. Meidel. 2003. Sea urchin spawning in benthic boundary layers: are eggs fertilized before advecting away from females? Limnology and Oceanography 48:795–801.

Zacherl, D., S. D. Gaines, and S. I. Lonhart. 2003. The limits to biogeographical distributions: insights from the northward range extension of the marine snail, *Kelletia kelletii* (Forbes, 1852). Journal of Biogeography 30:913–924.

16

Distribution and Abundance

SCALING PATTERNS IN EXOTIC AND NATIVE BIRD SPECIES

Fabio A. Labra, Sebastián R. Abades, and Pablo A. Marquet

Scaling phenomena are at the core of a great variety of ecological processes, ranging from individual physiology to populations, communities and ecosystems, and emerge as the result of the operation of general principles governing their structure and functioning. In this chapter, we assess the generality of scaling relationships in the distribution and abundance of species by comparing exotic and native species recorded in the North American Breeding Bird Survey. We describe scaling patterns in nine exotic species compare them with those in a set of native species chosen to maximize taxonomic and ecological similarity and in a random set of native species. For each set of species, we assess the scaling of the spatial characteristics of range occupancy, the intraspecific and interspecific scaling between distribution and abundance, and the scaling of the abundance frequency distribution. Our results indicate that exotic and native species show similar scaling patterns in their distribution and abundance, which suggests that they are under the influence of similar processes, thus supporting the generality of these scaling relationships. However, exotic species do differ from natives in their ability to reach higher maximum abundances and show a more even abundance-distribution relationship, probably as a result of having broad ecological tolerances, which could be a key to their successful establishment and further spread.

Introduction

By transporting species outside their native geographic ranges, humans have been performing a long-term natural experiment in ecology and biogeography. The main focus of this book, this chapter included, is to take a close look at the results of this experiment with the aim of gaining insights into the structure and dynamics of ecological systems. We are interested here, in particular, in exploring the generality of scaling relationships in ecology.

Scaling phenomena are at the core of a great variety of ecological processes, ranging from individual physiology to populations, communities, and ecosystems. They have become a major venue of inquiry in ecology (e.g., Banavar et al. 1999; Brown and West 2000; Brown et al. 2002; Marquet et al., in press), as they probably emerge as a result of fundamental principles governing the structure and functioning of ecological systems and other complex systems (e.g., Stanley et al. 2000; Brown et al. 2004). Scaling in invasion biology has been explored in two main contexts. The first emphasizes the scaling of the invasion process itself in order to mechanistically understand the spread of exotic species in a new ecological setting. Under this approach, the most commonly reported scaling is that associated with the relationship between spatial spread (measured as the increase in the square root of area occupied from a focal introduction point) and time (Skellam 1951), which is well understood in the context of random dispersal and diffusion (Berg 1983; Okubo 1980; Lubina and Levin 1988; Andow et al., 1990; Hastings et al. 2005). This scaling pattern has also been analyzed in the context of dispersal in heterogeneous landscapes (Johnson et al. 1992) and in considering the effects of different dispersal strategies (Clark 1998; Shigesada and Kawasaki 1997). The second major approach has focused on using exotic species to test scaling relationships predicted by theory, with the aim of ascertaining their generality and the mechanisms underlying their emergence. Representative of this approach is the paper by Keitt and Marquet (1996). These authors used the exotic bird species assemblage of Hawaii to test for the existence of the phenomenon known as self-organized criticality (Bak et al. 1988) in ecological systems, a landmark of which is the existence of a power law relating the frequency or probability of an event to its size. In this case, the events were extinctions of introduced species.

In this chapter, we benefit from both approaches to analyzing scaling relationships as we compare exotic and native species in order to understand the processes that underlie the distribution and abundance of species. Exotic bird species provide an unparalleled opportunity to study the generality of scaling relationships as well as their underlying driving mechanisms. From a theoretical perspective, exotic species that establish self-sustaining populations (i.e., naturalized species) provide us with natural experiments. These experiments can help us to understand the emergence of ecological patterns, as they help to control for the effects of past historical events and evolutionary dynamics (e.g., Sax 2001). This allows us to test for generality in ecological patterns without a need to invoke the action of long-term evolution or selection, focusing instead on the action of general principles that underlie the distribution and abundance of species. On the practical side, most exotic birds have been introduced during the eighteenth and nineteenth centuries; consequently, there are good records of their distribution and abundance (see review in Duncan et al. 2003), facilitating our ability to use these natural experiments to advance ecological theory.

As pointed out above, in addition to focusing on scaling relationships in the distribution and abundance of species to assess the generality of these patterns, we also want to understand the invasion process itself, as it offers an unparalleled opportunity to understand the dynamics of geographic ranges (e.g., Gaston 2003). While the spatial characteristics of range collapse (i.e., reduction in the range of widespread species) have been the focus of several empirical investigations (e.g., Lomolino and Channell 1995; Channell and Lomolino 2000), we know little about the spatial structure of the opposite process, range expansion or buildup (but see Maurer 1994; Maurer et al. 2001; Gammon and Maurer 2002). Most studies of this process have focused on estimating and/or predicting rates of spread (e.g., Hengeveld 1989; see review in Hastings et al. 2005). Naturalized exotic species provide us with the possibility of understanding the topology of range occupancy for species whose ranges are expanding, mimicking the process that characterizes the dynamics of a species' geographic range from the point of speciation until it becomes widespread.

To understand this process, we first focus on the geometric properties of the spatial distribution of site occupancy by exotic species through time. Our null expectation is that ranges expand following the same basic processes that underlie fluctuations in the occupancy of any species native to a region. To test this hypothesis, we compare the scaling patterns observed for exotics with those characterizing native species. Further, since changes in occupancy are likely to be related to changes in abundance (e.g., Brown 1984; Holt et al. 1997; Newton 1997; Gregory 1998; Gaston et al. 2000), we also compare the intraspecific and interspecific abundance-range size scaling relationship between native and exotic species. As before, our null expectation is that naturalized exotics and natives will not differ, as has already been observed for British mammals and birds at the interspecific level (Holt and Gaston 2003). Finally, we compare the scaling relationship that describes the frequency distribution of abundance for naturalized exotics and native species. Unlike some other relationships, the distribution of abundance has been shown to be affected by factors that include habitat type, body size, phylogeny, and spatial scale (e.g., Magurran 1988; Cotgreave and Harvey 1994; Brown et al. 1995; Gregory 2000; Marquet et al. 2003); thus, we expect that the distribution of abundance may differ among native and naturalized species.

Materials and Methods

Database and general approach

To assess the potential similarities or differences between native and exotic species in their distribution and abundance, we analyzed scaling patterns using data from the Breeding Bird Survey (BBS) (Peterjohn and Sauer 1993; Peterjohn 1994; Sauer et al. 2003), a yearly sampling effort run since 1966 across North America. In Figure 16.1 we show the route coverage in North America for three different years (1966, 1985, 2002) and the total number of routes and area covered by the survey (Figure 16.1D). The number of routes and the geographic area covered have increased rapidly since 1966, but the area sampled leveled off after 1970. To minimize any potential effect of the increase in area sampled, we worked only with the data for censuses after 1970. Details on the species selection procedure are given below.

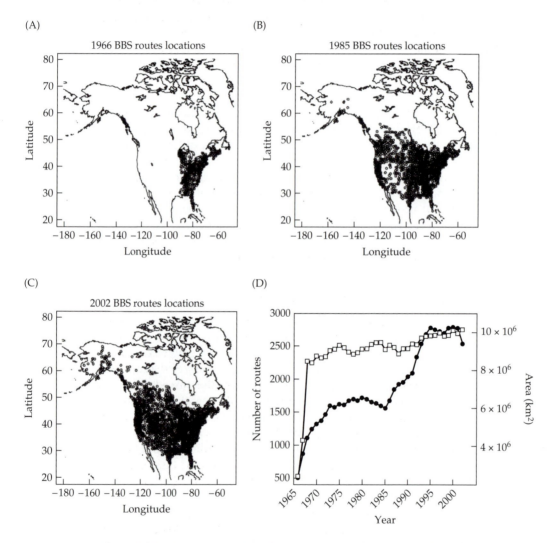

Figure 16.1 Temporal trends in BBS sampling effort. (A–C) Spatial coverage of sampling sites for the years (A) 1966, (B) 1985, and (C) 2002. (D) Temporal trends in total number of routes and in area covered.

Species included in the analysis

After identifying all exotic bird species known for the continental United States, we selected those that had adequate records (more than ten routes) in the North American Breeding Bird Survey. This procedure left us with nine species with adequate spatial and temporal coverage. We called this set of species the "exotic set." In order to gain insight into the generality of patterns in space use by exotic species, we also included in the analysis a set of ten native species selected to maximize similarity to the exotic species being examined. We selected these similar native species based on phylogenetic relatedness, ecological and life history traits, and body size; the characteristics considered are reported in Table 16.1. The ecological and life history traits used were habitat and diet, respectively. We call this set of native species the "similar set." The aim of using this group was to assess to what extent the observed patterns could be a result of phylogenetic, ecological, life history, or body size similarities.

We also studied an additional set of ten native species chosen at random from the available North American pool. We called this set of native species the "random set." As expected, the species in this set show different trends in abundance and encompass a broad spectrum of phylogenetic relatedness, life histories, and ecological attributes (see Table 16.1). The rationale for using this random set of native species was to test for the ecological generality of the observed patterns. Ideally, the patterns should be contrasted against those in a large number of random sets of species; such a task, however, would have been difficult given our computing power, and we believe that this initial set of random species should suffice in the present examination.

Assessing the geometric properties of space occupancy

To assess the geometric properties of species distribution, we used the minimum spanning tree (MST) methodology, which provides a graphic representation of the pattern of occurrences of a species. Figure 16.2A shows a set of points drawn from a uniform distribution in two-dimensional space. These points (or nodes) can be joined in many different ways by lines (or edges), forming a graph. A graph will be defined as connected if there is an edge between any pair of nodes. Figure 16.2B shows one of the many ways to form a connected graph with several circuits or loops. A connected graph containing no circuits is called a tree (Figures 16.2C,D). If a tree contains all the nodes, it is called a spanning tree. The spanning tree that minimizes the total length of connections between nodes is called the minimum spanning tree (MST, Figure 16.2D). Although the number of possible trees for a given data set may be very large, the MST is a unique configuration that reduces the "cost," or the sum of internodal distances (Gower and Ross 1969; Zahn 1971). However, MSTs will be unique only when there are complex spatial patterns. For simple spatial patterns, it is possible that more than one MST of equal length will exist. It is important to note that this unique network picks out the dominant pat-

TABLE 16.1 *Species in the exotic, similar, and random sets analyzed in this study*

Family	Common name	Scientific name	Body size (g)
EXOTIC SET			
Anatidae	Mute swan	*Cygnus olor*	11,007
Ardeidae	Cattle egret	*Bubulcus ibis*	338
Columbidae	Rock dove	*Columba livia*	355
Columbidae	Eurasian collared dove	*Streptopelia decaocto*	149
Fringillidae	House finch[c]	*Carpodacus mexicanus*	21
Passeridae	House sparrow	*Passer domesticus*	28
Phasianidae	Gray partridge	*Perdix perdix*	391
Phasianidae	Ring-necked pheasant	*Phasianus colchicus*	1,278
Sturnidae	European starling	*Sturnus vulgaris*	83
SIMILAR SET			
Anatidae	Trumpeter swan	*Cygnus buccinator*	10,701
Ardeidae	Little blue heron	*Egretta caerulea*	343
Columbidae	Inca dove	*Columbina inca*	48
Columbidae	Common ground dove	*Columbina passerina*	30
Columbidae	Mourning dove	*Zenaida macroura*	120
Fringillidae	American goldfinch	*Carduelis tristis*	13
Fringillidae	Purple finch	*Carpodacus purpureus*	25
Phasianidae	Greater sage grouse	*Centrocercus urophasianus*	2,724
Phasianidae	Blue grouse	*Dendragapus obscurus*	1,030
Icteridae	Brown-headed cowbird	*Molothrus ater*	44
RANDOM SET			
Anatidae	Northern pintail	*Anas acuta*	1,025
Anatidae	Mottled duck	*Anas fulvigula*	1,013
Anatidae	American black duck	*Anas rubripes*	1,304
Ardeidae	Tricolored heron	*Egretta tricolor*	405
Cardinalidae	Painted bunting	*Passerina ciris*	16
Columbidae	Band-tailed pigeon	*Patagioenas fasciata*	343
Odontophoridae	Gambel's quail	*Callipepla gambelii*	167
Odontophoridae	Scaled quail	*Callipepla squamata*	184
Odontophoridae	Mountain quail	*Oreortyx pictus*	233
Parulidae	Prothonotary warbler	*Protonotaria citrea*	17

Source: Diet and habitat data from the BBS and del Hoyo et al. 1994, 1997; body size from Dunning 1993.
[a] Minimum and maximum number of routes

TABLE 16.1 *(continued)*

Diet	Habitat	Distribution (routes)[a]	Abundance[b]
Plant matter, aquatic invertebrates	Wetland-open water	2–18	5–179
Insects	Wetland-open water	36–307	1,012–17,561
Seeds, plant matter, insects	Urban	280–1,262	4,961–17,858
Seeds, plant matter, insects	Urban	1–134	13–1,287
Seeds, insects	Urban	3–867	10–12,666
Seeds, insects	Urban	493–1,929	44,803–105,318
Seeds, insects	Grasslands	2–104	11–323
Plant matter, insects, worms and snails	Grasslands	127–707	1,249–10,860
Insects, fruit	Urban	507–2250	45,557–117,377
Plant matter, seeds, aquatic invertebrates	Wetland-open water	1–20	1–171
Aquatic invertebrates, fish, reptiles and amphibians	Wetland-open water	50–194	359–1,847
Seeds	Urban	5–74	19–356
Seeds, insects	Successional-scrub	28–129	188–882
Seeds	Urban	454–2,545	11,316–80,294
Seeds, insects	Successional-scrub	411–1,718	5,040–23,624
Flowers, fruit, seeds and insects	Woodland	107–444	519–2,042
Plant matter, flowers, seeds and insects	Successional-scrub	4–38	14–380
Plant matter and insects	Woodland	5–50	28–159
Seeds and insects	Open woodlands, fields	465–2,385	5,433–34,229
Plant matter, aquatic invertebrates, fish, reptiles and amphibians	Wetland-open water	4–182	7–2,099
Aquatic invertebrates, seeds, plant matter, and fish	Wetland-open water	6–33	38–455
Plant matter, seeds, aquatic invertebrates, and fish	Wetland-open water	33–75	97–313
Aquatic invertebrates, fish, reptiles and amphibians	Wetland-open water	12–56	31–535
Seeds, insects	Successional-scrub	10–199	59–2,597
Nuts, fruit, and seeds	Woodland	19–109	196–866
Seeds and insects	Successional-scrub	12–68	175–2,448
Seeds and insects	Successional-scrub	10–87	190–1,142
Seeds, plant matter, and insects	Successional-scrub	4–70	8–729
Insects and snails	Woodland	66–206	163–1,152

[b] Minimum and maximum number of individuals

[c] Note that this species was examined only in the exotic portions of its range in North America.

Figure 16.2 Examples of point, connected line, and spanning tree graphs. (A) A set of random points in space. (B) A connected line graph showing "loops." (C) A spanning tree. (D) The minimum spanning tree for the set of points shown in A.

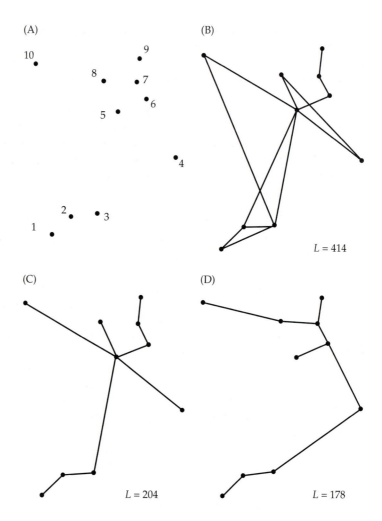

tern of connectedness among the points in a manner that emphasizes their intrinsic linear associations. This procedure has been successfully used in several disciplines, ranging from astronomy (Barrow et al. 1985; Bhavsar and Ling 1988; Adami and Mazure 1999), physics (Dussert et al. 1986; Van de Weygaert et al. 1992; Iribarne et al. 1999; Dobrin and Duxbury 2001), and pattern recognition (Zahn 1971; Hoffman and Jain 1983) to ecology (Cantwell and Forman 1993; Lockwood et al. 1993; Keitt et al. 1997; Bunn et al. 2000; Urban et al. 2001) and biology (Dussert et al. 1987; Jones et al. 1996; Wallet and Dussert 1997).

It is customary to build the MST using Prim´s algorithm (Prim 1957). This procedure starts with a fixed node. One by one, nodes that are the nearest neighbor to the subgraph already built are added to the graph. In so doing, edges that complete a circuit or loop are avoided, since a tree cannot have any circuits. The process stops when there are no further nodes to add. Thus the minimum spanning tree uniquely connects a set of N nodes with $N-1$ edges. The MST may be described in many ways, including degree, connectivity, and edge length distributions, among others (Chartrand 1977; Dussert et al. 1987). For the purposes of this work, we will describe the MST by the sum of all the edge lengths, which is a concise descriptor of the way species could connect and fill out space. It can be shown analytically that the expected total length of the MST for N randomly distributed points in a sampling window of area A scales as $(AN)^{1/2}$ (Beardwood et al. 1959; Hammersley et al. 1959; Steele 1988; Jaillet 1995). This result provides the expected pattern under the null hypothesis of random space occupancy by species through time.

Database processing and analysis

For computational purposes, and to minimize errors due to slight changes in route locations across years, we mapped the observed data onto a fixed grid of 30×30 cells per degree of latitude and longitude (with each cell having an approximate area of $15\ km^2$). This scale was chosen to be similar to the sampling area of a single BBS route (approximately $21\ km^2$). Using these data, we constructed a MST for each species in each year (from 1970 to 2002) and calculated the total edge length and number of nodes. In order to make scaling relationships comparable across species groups, we normalized MST lengths by dividing them by the square root of the area they cover. The area covered by the pattern of point occurrences in a given year was estimated using a 95% level kernel density estimate contour (Beardah et al. 1996). This procedure eliminates biases in area estimation due to runaway points separated from the densest zone of the geographic range. We call this measure the area-corrected MST. All mathematical procedures were carried out using MATLAB 6.12 (The MathWorks, Inc., Natick, MA), and the R statistical package (R Development Core Team, 2003).

Patterns in distribution and abundance

To assess the scaling relationship between distribution and abundance, we plotted the number of routes where a species was observed in a given year (a measure of distribution) versus total abundance in that year. We examined this relationship for exotic species and for the two groups of native species, both on a species-by-species basis (intraspecific scaling pattern, using total counts of individuals per year from 1966 to 2002) and for each of the three groups as an ensemble (interspecific scaling pattern, in a subset of years: 1970, 1980, 1990, and 2000). Finally, we examined the frequency distribution of abundances for exotic and native species. For each group of species, we plotted the midpoint of each abundance class versus the probability of observing a site in each abundance class (i.e., the number of sites where that abundance class was recorded divided by the total number of sites used in the analyses for that particular group) between 1970 and 2002. We used logarithmic abundance classes and plotted them versus the logarithm (base 10) of their observed probability (see Solow et al. 2003).

Results

Minimum spanning tree scaling

In Figure 16.3 we show the temporal trend in abundance and distribution for selected exotic and native species. As is apparent in these figures, there is substantial variability in both the number of BBS routes where a species was recorded and in its abundance over the study period (1970–2002). Overall, the exotic and similar sets showed an increase in the number of routes occupied by species through time (average linear regression slopes ± 1 S.E. were 13.4 ± 4.0 and 11.7 ± 5.2, respectively); however, the observed tendency in the random set, although positive, was much weaker (1.3 ± 0.27).

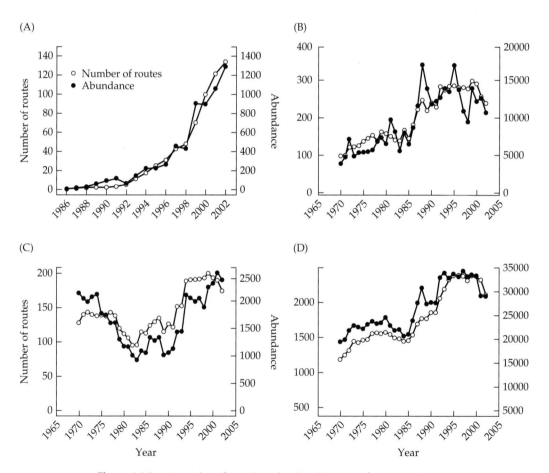

Figure 16.3   Examples of trends in the distributions of exotic and native species. The figure shows the observed dynamics in number of routes and total abundance for two exotic species, (A) Eurasian collared dove and (B) cattle egret, and two native species, (C) painted bunting and (D) brown-headed cowbird.

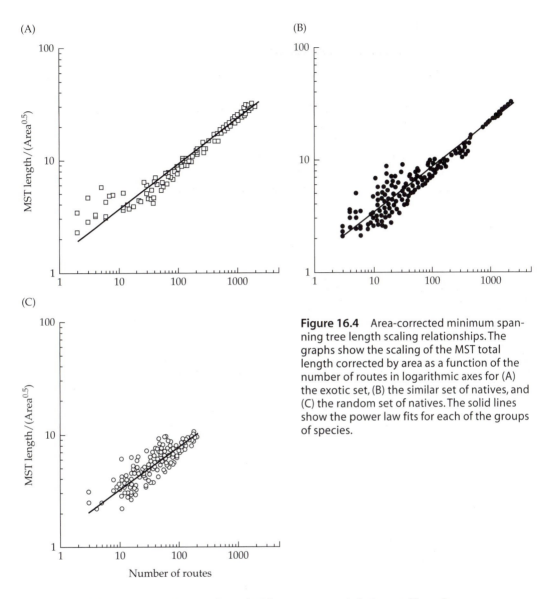

Figure 16.4 Area-corrected minimum spanning tree length scaling relationships. The graphs show the scaling of the MST total length corrected by area as a function of the number of routes in logarithmic axes for (A) the exotic set, (B) the similar set of natives, and (C) the random set of natives. The solid lines show the power law fits for each of the groups of species.

Despite this variation in the number of BBS routes occupied, the total lengths of the minimum spanning trees scale in a similar fashion within the exotic, similar, and random species sets (Figure 16.4A–C). Within each set, species show a tight scaling relationship, even though they show quite different temporal dynamics and spatial patterns of expansion (see Figure 16.3). The average scaling exponent for each of the three groups departs significantly from the null expectation of 0.5. The exponents and standard errors obtained were 0.435 ± 0.006, 0.444 ± 0.004, and 0.376 ± 0.013 ($P < 0.001$ for all relationships) for the exotic, similar, and random sets, respectively. [For this comparison, we

TABLE 16.2 *Scaling statistics for the distribution-abundance and area-corrected MST scalings fitted for all species*

| Common name | Distribution-abundance scaling | | | | |
	Intercept	S.E.	Exponent	S.E.	R^2
EXOTIC SET					
Mute swan	−0.069	0.142	0.550	0.084	0.562
Cattle egret	−0.820	0.156	0.784	0.040	0.917
Rock dove	−1.478	0.229	1.072	0.055	0.915
Eurasian collared dove	−1.377	0.166	1.122	0.071	0.943
House finch	−0.754	0.428	0.637	0.071	0.704
House sparrow	2.219	1.059	0.193	0.216	0.022
Gray partridge	−0.268	0.149	0.921	0.070	0.831
Ring-necked pheasant	−0.242	0.240	0.753	0.061	0.811
European starling	−4.201	0.466	1.492	0.094	0.878
SIMILAR SET					
Trumpeter swan	−0.071	0.122	0.601	0.091	0.591
Little blue heron	0.233	0.152	0.633	0.052	0.810
Inca dove	−0.319	0.088	0.852	0.044	0.917
Common ground dove	−0.039	0.271	0.710	0.100	0.589
Mourning dove	−1.145	0.146	0.924	0.031	0.962
American goldfinch	−0.317	0.279	0.821	0.068	0.808
Purple finch	−0.573	0.382	0.972	0.120	0.653
Greater sage grouse	−0.052	0.113	0.614	0.057	0.776
Blue grouse	−0.393	0.249	0.925	0.134	0.590
Brown-headed cowbird	−1.023	0.168	0.965	0.038	0.948
RANDOM SET					
Northern pintail	0.190	0.177	0.612	0.061	0.742
Mottled duck	−0.086	0.181	0.618	0.083	0.615
American black duck	0.498	0.156	0.553	0.070	0.643
Tricolored heron	0.289	0.110	0.524	0.052	0.747
Painted bunting	−0.291	0.128	0.754	0.040	0.910
Band-tailed pigeon	−0.375	0.277	0.810	0.103	0.653
Gambel's quail	−0.363	0.113	0.672	0.040	0.893
Scaled quail	−0.118	0.369	0.649	0.134	0.410
Mountain quail	−0.119	0.187	0.679	0.073	0.727
Prothonotary warbler	0.910	0.080	0.457	0.029	0.876

Note: The table shows the intercept and slope values with respective standard error estimates (S.E.) as well as the coefficient of determination of the fitted distribution-abundance scaling relationship. Also shown are the scaling exponents for all the species under study. Area-corrected MST scaling exponents that differ from the null expectation (of 0.5) are shown in boldfaced type. C.I. = confidence interval.

TABLE 16.2 *(continued)*

Normalized MST length scaling		
Exponent	95% Lower C. I.	95% Upper C. I.
—	—	—
0.412	0.354	0.469
0.434	0.397	0.470
0.257	0.182	0.332
0.412	0.387	0.437
0.295	0.231	0.36
0.536	0.433	0.639
0.517	0.468	0.567
0.461	0.422	0.50
0.073	−0.087	0.234
0.449	0.395	0.504
0.322	0.197	0.448
0.469	0.320	0.617
0.487	0.466	0.508
0.395	0.372	0.418
0.033	−0.108	0.174
0.229	0.063	0.553
0.308	0.191	0.425
0.472	0.437	0.507
0.412	0.317	0.507
0.298	0.136	0.460
0.307	0.069	0.545
0.340	0.239	0.441
0.369	0.314	0.425
0.371	0.251	0.491
0.460	0.378	0.541
0.422	0.311	0.530
0.468	0.398	0.537
0.339	0.262	0.416

excluded the mute swan (*Cygnus olor*) because of its restricted pattern of distribution, confined to water bodies in the eastern United States, and because it is present on fewer than 20 routes.] The random set showed a significantly lower exponent than the exotic and similar sets, which are not significantly different. This result suggests that species in the exotic and similar sets fill out space in a less aggregated way than species in the random set. Interestingly, the sampling routes for the BBS show a scaling relationship that is not significantly different from the null expectation, with an observed exponent of 0.506 ± 0.009. In Table 16.2, we report the statistics of the MST scaling for each species through time. It is interesting to note that most exponents are lower than 0.5, although in each group there are species whose MST scaling is not different from what would be expected under random occupancy.

Scaling of distribution and abundance

In general, the intraspecific pattern (i.e., for individual species among years) in the distribution-abundance relationship was positive and significant within most species (Figure 16.5; see Table 16.2). Exotic species reached the highest abundance for a given number of routes (Figure 16.5) and showed the highest scaling exponent. However, there were no significant differences among average scaling exponents for all three groups. The exponents and standard errors obtained were 0.836 ± 0.1247, 0.802 ± 0.0473, and 0.633 ± 0.033 for the exotic, similar, and random sets, respectively. Similarly, intercepts did not differ among sets. These results imply that for a given increase in occupancy, all three sets tend to increase in total abundance in a similar way.

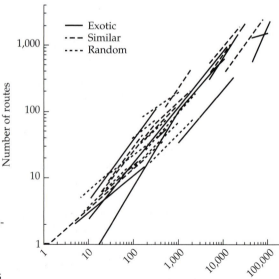

Figure 16.5 Distribution-abundance relationships. The graph shows the fitted scaling relationships for number of routes versus total abundance. Every line represents the relationship for a single species across individual years (i.e., intraspecific scaling).

At the interspecific level (Figure 16.6), the scaling of the distribution-abundance relationship shows that exotic species tend to exhibit higher abundances than natives, particularly at large numbers of routes, while at small numbers of routes they show fewer apparent differences from native species (Figure 16.6). However, there are no significant differences in the average scaling exponents among groups and across years (Table 16.3). A similar pattern is observed when we study the MST length scaling for the same groups over the same years

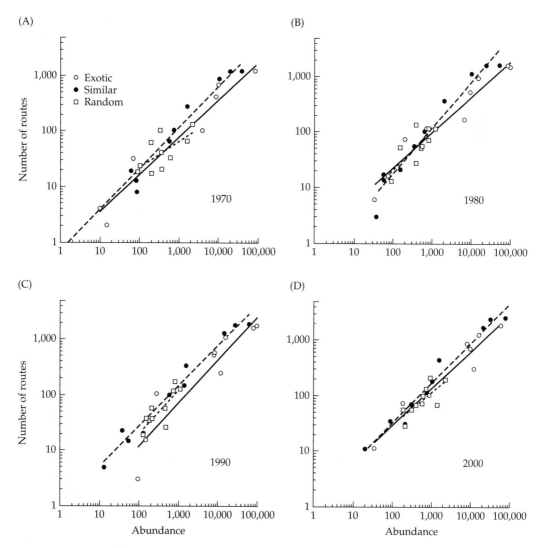

Figure 16.6 Interspecific distribution-abundance relationships for exotic and native species ensembles. Note the trend for exotic species to exhibit higher abundances than natives for a given number of routes, particularly at large number of routes.

(Figure 16.7); there are no significant differences in the scaling exponents among groups and across years (Table 16.3).

Finally, the frequency distribution of abundances for exotic and native species, considering all years and all routes, shows that species in the exotic set have a higher probability of attaining any given abundance than native

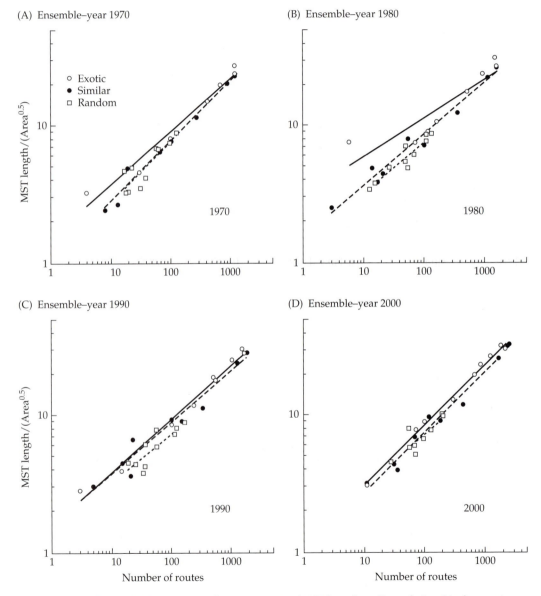

Figure 16.7 Interspecific area-corrected MST length scaling relationship for exotic and native species ensembles present for selected years.

TABLE 16.3 *Statistics for the interspecific distribution-abundance relationship and the area-corrected MST scaling relationship*

Year	Group	Slope ± S.E.[a]	R^2	P
DISTRIBUTION-ABUNDANCE RELATIONSHIP				
1970	Exotic	0.428 ± 0.141	0.606	0.023
1970	Similar	0.739 ± 0.046	0.957	0.000
1970	Random	0.471 ± 0.180	0.768	0.031
1980	Exotic	0.416 ± 0.146	0.577	0.029
1980	Similar	0.821 ± 0.064	0.954	0.000
1980	Random	0.673 ± 0.176	0.647	0.005
1990	Exotic	0.520 ± 0.201	0.488	0.036
1990	Similar	0.727 ± 0.042	0.978	0.000
1990	Random	0.767 ± 0.194	0.662	0.004
2000	Exotic	0.408 ± 0.161	0.479	0.039
2000	Similar	0.710 ± 0.043	0.972	0.000
2000	Random	0.591 ± 0.021	0.591	0.009
AREA-CORRECTED MST SCALING				
1970	Exotic	0.544 ± 0.113	0.823	0.005
1970	Similar	0.440 ± 0.022	0.983	0.000
1970	Random	0.438 ± 0.086	0.764	0.001
1980	Exotic	0.465 ± 0.075	0.884	0.002
1980	Similar	0.377 ± 0.020	0.977	0.000
1980	Random	0.378 ± 0.048	0.887	0.000
1990	Exotic	0.376 ± 0.019	0.985	0.000
1990	Similar	0.378 ± 0.027	0.961	0.000
1990	Random	0.357 ± 0.080	0.713	0.002
2000	Exotic	0.416 ± 0.020	0.986	0.000
2000	Similar	0.440 ± 0.022	0.981	0.000
2000	Random	0.370 ± 0.078	0.736	0.001

[a]S.E. = 1 standard error.

species and show a more even abundance distribution (shallower slopes; Figure 16.8), a pattern that was also observed when the analysis was repeated for years 1970, 1980, 1990, and 2000 (not shown). This pattern is correlated with the fact that only species in the exotic set are found in the highest abundance classes. In general, the scaling relationships observed in the abundance distribution correspond to a power law with an exponent that does not differ when the random and the similar species set are compared (–7.663 ± 0.513 and –7.971 ± 0.295;

Figure 16.8 Probability distribution of abundance (individuals per route) for exotic and native species sets calculated over the entire study period (1970–2002) and over all sampling routes. Note the power law scaling shown by all three groups. Regression statistics were estimated for those values in the exponentially decaying tail (i.e., probabilities greater than 0.003).

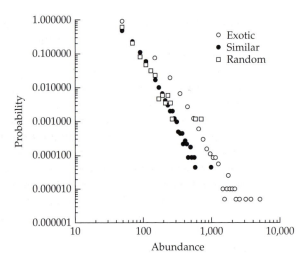

figures are estimated slopes ± 1 S.E.). However, the scaling exponent for the exotic species set (–5.714 ± 0.288) is significantly different from those of the other two sets.

Discussion

In this chapter, we have explored several scaling relationships characterizing the spatial distribution and abundance of exotic and native species. Our aim was to assess the generality of these relationships and determine whether exotic species show a distinct spatial pattern of occupancy and abundance. As discussed in detail below, on average, exotic and native species (i.e., species in the similar and random sets) do not differ in their geometry of space occupancy and in distribution-abundance scaling; however, important differences in their distributions of abundance were apparent. Further, none of the patterns herein reported were associated with the magnitude of range expansion.

Area-corrected MST scaling

We found that range dynamics for both native and exotic species, measured as the geometry of occupied sites through time, resulted in an area-corrected MST length scaling with an exponent that, on average, is different from random occupancy (i.e., lower than 0.5), implying that the distribution of these species is more clustered than that expected under a random pattern of occupancy. We claim that this result is not a consequence of the topology of the BBS route network across North America, for it follows the scaling pattern expected for a spatial distribution formed by a random process. This claim is supported by the fact that exotic species distributed across nearly all of North America (such as the rock dove, house finch, and house sparrow) show scaling exponents different from random (see Table 16.2). To further explore the potential effect of the BBS network topology, we used the observed 2002 BBS network to calculate the expected scaling pattern under the assumption that species occupy BBS sites at random. The results of simulating different occupancies (from 10% up to 100% of the total number of routes, in steps of 10%) and using ten replicates of each condition showed an exponent of 0.49 ± 0.004, $R^2 = 0.99$, which is not significantly different from 0.5. This finding reinforces the notion that the BBS network topology does not cause the observed scaling pattern.

However, there are at least two other potential sources of bias that need to be discussed. First, even if the BBS network is random at the continental scale, it might not be so at a regional scale. Nonrandomness at smaller scales could introduce a bias in the observed area-corrected MST scaling patterns for species found exclusively or predominantly in a particular region. To test for this potential effect, we repeated the analysis of the BBS network scaling, but this time we separated the continent into an eastern and a western half, using longitude 100° as the boundary between the two. The scaling exponents for the two regions were not statistically different from each other (0.458 ± 0.023 and 0.433 ± 0.036 for western and eastern halves, respectively) nor from the expected value of 0.5. Thus, the BBS network appears random, at least at the continental and large regional scales, suggesting that potential biases due to heterogeneity in the topology of the network are not a serious problem.

A second cause for concern is the potential for routes to be located near urban areas and roads in anthropogenic landscapes, which could inflate the pattern of site occupation for species (particularly exotics) associated with anthropogenic habitats (e.g., for the house finch; Gammon and Maurer 2002). The observed randomness in the BBS network suggests that this might not be a general problem. In fact, recent work by La Sorte and Boecklen (in press) has shown that BBS routes tend to be located away from urban areas and are associated with low human activity and moderately low human population densities. Nevertheless, additional work examining differences in occupancy and patterns of abundance across natural-urban gradients would be valuable.

In addition to showing that both native and exotic species do differ from random in terms of their occupancy, we found that there are differences among groups. Exotics and the similar native set showed consistently similar patterns, but both commonly differed from the random native set. Thus, given the same number of sites, the occupancy of exotic and similar native species tends to be less aggregated than the occupancy of species in the random set. This result could be due to a statistical artifact associated with the small number of occupied routes that characterized species in the random set (see Table 16.1), as such a short range in the independent variable, as compared with the other two sets, might affect the estimation of the scaling exponent through regression. However, it seems more likely (given the comparable relationships observed between the exotic and similar sets of species, which were observed despite the small sample sizes of both groups) that such differences between the random set of natives and the other two groups reflect the nonrandom characteristics of the species constituting those groups (i.e., a case in which ecological characteristics are driving the observed differences).

Another issue is how accurately the area-corrected MST scaling relationships might reflect the metapopulation dynamics associated with changes in species ranges (e.g., Gammon and Maurer 2002). One way of addressing this question is to carry out spatially explicit metapopulation models and study the relationships between the area-corrected MST scaling exponents and species extinction and colonization probabilities. In fact, preliminary data on this question (F. A. Labra, S. R. Abades, and P. A. Marquet, unpublished data) suggest that there is a connection. In a similar vein, it would be interesting to expand our analysis to a consideration of only routes above some threshold of abundance, as they are more likely to represent the source populations or density hot spots driving the dynamics of space occupancy and expansion (Brown et al. 1995; Gammon and Maurer 2002). This is probably the reason why in all three sets we found species with area-corrected MST scaling exponents that did not differ from random.

The number of sites occupied has been a common measure in ecological theory to describe species distribution. However, it fails to capture significant features of the spatial pattern, because two species with the same number of occupied sites might have very different spatial distributions. In this chapter, we have introduced a new scaling relationship (i.e., the MST length scaling) that captures some aspects of the spatial pattern in the distribution of species. However, we are not alone in these attempts. In fact, a recent paper by He and Hubbell (2003) attempts to quantify the spatial structure in the relationship between abundance and distribution. Although further work on geometric aspects of the distribution and abundance of species (see also Harte et al. 2001) is needed, we believe this work, particularly on scaling laws, will provide invaluable insights into the distribution and abundance of species.

The scaling of distribution and abundance

The distribution and abundance of species are tightly interconnected aspects of their ecology, and most of the time it is almost impossible to think of one without the other. As pointed out by Gaston et al. (2000), in the extreme, a positive interspecific correlation between both variables is almost inevitable, since it can be generated by many different patterns of space occupancy (Holt et al. 2003). This positive interspecific correlation emerges despite controlling for phylogeny, habitat, migration status, and dispersal ability (Gaston and Blackburn 2003) and occurs largely independently of intraspecific relationships (Blackburn et al. 1998).

At the intraspecific and interspecific levels (see Tables 16.2 and 16.3 and Figures 16.5 and 16.6), the average distribution-abundance scaling exponents do not differ among exotic, similar, and random sets. Further, both relationships are always positive and significant (only the house sparrow shows a nonsignificant relationship; see Table 16.2). Although our results at the intraspecific level show a consistent pattern of positive relationships, contrary to the patterns observed for British birds (e.g., Blackburn et al. 1998), it remains to be seen whether this pattern holds for the majority of the species in the BBS. Although it has been empirically shown that for some species in the BBS data set, a decline in abundance does not necessarily imply a decline in occupancy, and vice versa (e.g., Gaston and Curnutt 1998; Gaston 2003), our data show that the scaling exponent characterizing this process is similar across native and exotic species; this result points to the existence of a common process or set of processes underlying the scaling relationship between total abundance and distribution for both native and exotic species, reinforcing the conclusions reached by Holt and Gaston (2003) for British mammals and birds.

Unlike most previous studies (but see Keitt et al. 2002), we measured abundance as total number of individuals instead of average abundance. For this pattern, a scaling exponent of 1 is expected if individuals are distributed with equal probability across sites. As a null model, imagine that each population is subdivided into n equally sized, independent subpopulations, and that the number of these subpopulations depends on S, the total abundance. It has been argued by Keitt et al. (2002) that the number of subpopulations does not scale in a simple linear fashion with increasing S, but instead takes the form $n \propto S^{1-\alpha}$ (where α is the exponent of the relationship between local and regional abundance). These authors estimated $1 - \alpha$ to be 3/4, based on their analysis of the interspecific distribution-abundance relationship for the BBS in 1997. An exponent of 3/4 is not different from our observed interspecific distribution-abundance scaling exponents reported for years 1970, 1980, 1990, and 2000. Interestingly, our results also show that the same 3/4 scaling exponent holds, on average, for the intraspecific distribution-abundance relationships observed for the exotic, similar, and random sets. According to Keitt et al. (2002), this is to be expected when there is a positive relationship between regional and local abundance (e.g., Gaston and Lawton 1988). In this context, it is interesting to note that some exotic species show values of α close to zero (rock dove and Eurasian collared dove) and even negative values (European starling), which suggest that for these species, a positive relationship between local and regional abundance should not hold. As suggested by Keitt et al. (2002), the existence of a 3/4 power law scaling raises the possibility that the processes invoked to explain the emergence of 3/4 power laws in organismal physiology (e.g., West et al. 1997, 1999) could underlie the emergence of patterns in the spatial structure of metapopulations. However, as yet, there is no formal theory linking individuals to the spatial structure of populations.

Finally, our results show that, on average, exotic species do reach higher maximum abundances than native species and show a higher probability of reaching any given abundance. We are not aware of any study reporting a similar finding. Certainly, this finding deserves further scrutiny, for it points to an ecological pattern for which exotic and native species show differences that might be associated with invading species in general. One potential explanation for this pattern is the purported higher degree of ecological generalization associated with exotic species, which allows them to reach higher abundances within local communities. Although certainly not overwhelming, the existing evidence is consistent with the hypothesis that ecological generalization is related to establishment probability: McLain et al. (1999) showed that successfully introduced species had a greater tendency toward dietary and nest habit generalism; Brooks (2001) found that introduced species categorized as habitat specialists were less likely to be successfully established; and Cassey (2002) found that habitat generalism (among other variables) affected the successful establishment of introduced land bird species. Although it is not known whether those traits that favor establishment also affect spread, the observed correlation between native geographic range and the range achieved after introduction by exotic bird species in New Zealand and Australia (Duncan et al. 1999, 2001; Williamson 2001) suggests that broad ecological tolerances do indeed increase the probability of establishment and subsequent spread (Duncan et al. 2003). Thus, the observed numerical abundance and higher degree of evenness shown by exotics (see Figure 16.8) is probably associated with broad ecological tolerances.

Concluding Remarks

In theory, there are several reasons to expect that exotic species will differ from native species, both in the geometric properties of their spatial distribution and in their distribution and abundance scaling pattern. These differences, usually linked to enemy release, preadaptation to human-modified environments, ecological specialization, and life history, on one hand, make their establishment and success paradoxical (e.g., Sax and Brown 2000), and on the other, make the emergence of patterns in richness (Sax 2001; Sax et al. 2002), and in distribution and abundance, similar to those observed in native species, remarkable. Our results indicate that exotic and native species show similar scaling patterns in their distribution and abundance, which suggests that they are under the influence of similar processes. Although there is still discussion regarding what these processes and their underlying mechanisms are (e.g., Gaston et al. 2000; Harte et al. 2001), our results point to their generality in affecting range dynamics as well as the scaling of distribution and abundance relationships.

It remains to be seen whether other scaling relationships, such as those relating population density and home range size to body size, are the same when exotic and native species are compared, as theory suggests (e.g., Brown et al. 2004). In this context, it would be particularly important to assess how much energy exotic species populations use within local communities, especially considering that empirical evidence suggests that this should be approximately constant across species (i.e., the energetic equivalence rule; see review in Marquet et al., in press). Our results suggest that in at least some places within their range, exotic species reach abundances well above those of similar native species, which is reflected in the fact that exotics reach higher maximum abundances than natives, and would therefore use a disproportionate amount of available energy. This could be due to a "density compensation" effect, such that exotics occupying sites where few natives are present (for whatever reason) might then utilize a greater proportion of the available energy. Further analyses that examine the use of energy across species, the scaling relationships between distribution and abundance, and the geographic range architecture of both native and exotic species should do much to advance our understanding of the interplay between the complex factors that determine the distribution and abundance of species.

Acknowledgments

We thank the thousands of volunteers that have participated in the Breeding Bird Survey across North America, who have made this work possible. This research was funded by project FONDAP 15001-0001 program 4. This is contribution 1 to the Ecoinformatics and Biocomplexity Unit. FAL and SRA acknowledge funding by doctoral scholarship from CONICYT. PM acknowledges an international fellowship from the Santa Fe Institute during the preparation of this chapter. We are grateful for cogent criticisms and discussion provided by Roberto Comminetti, Steve Gaines, Tim Keitt, Dov Sax, and two anonymous reviewers. This work was conducted as a part of the "Exotic Species: A Source of Insight into Ecology, Evolution, and Biogeography" Working Group supported by the National Center for Ecological Analysis and Synthesis, a center funded by the National Science Foundation (grant #DEB-0072909), the University of California, and the Santa Barbara campus.

Literature Cited

Adami, C., and A. Mazure. 1999. The use of the minimal spanning tree to characterize the 2D cluster galaxy distribution. Astronomics and Astrophysics Supplement Series. 134:393–400.

Andow, D. A., P. M. Kareiva, S. A. Levin, and A. Okubo. 1990. Spread of invading organisms. Landscape Ecology 4:177–188.

Bak, P., C. Tang, and K. Wiesenfeld. 1988. Self-organized criticality. Physical Review A 38:364–374.

Banavar, J. R., J. L. Green, J. Harte, and A. Maritan. 1999. Finite size scaling in ecology. Physical Review Letters 83:4212–4214.

Barrow, J. D., S. P. Bhavsar, and D. H. Sonoda. 1985. Minimal spanning trees, filaments and galaxy clustering. Monthly Notices of the Royal Astronomical Society 216:17–35.

Beardah, C. C., and M. J. Baxter. 1996. MATLAB routines for kernel density estimation and the graphical representation of archaeological data. Analecta Prehistorica Leidensia No. 28, Leiden University, The Netherlands.

Beardwood, J., H. J. Halton, and J. M. Hammersley. 1959. The shortest path through many points. Proceedings of the Cambridge Philosophical Society 55:299–327.

Berg, H. C. 1983. *Random walks in biology*. Princeton University press, Princeton, NJ.

Bhavsar, S. P. and E. N. Ling 1988. Are the filaments real? The Astrophysical Journal 331:63–68.

Blackburn, T. M., K. J. Gaston, J. J. D. Greenwood, and R. D. Gregory. 1998. The anatomy of the interspecific abundance-range size relationship for the British avifauna: II. Temporal trends. Ecology Letters 1:47–55.

Brooks, T. 2001. Are unsuccessful avian invaders rarer in their native range than successful invaders? In J. L. Lockwood, M. L. McKinney, eds. *Biotic homogenization*. Kluwer Academic, New York.

Brown, J. H. 1984. On the relationship between abundance and distribution of species. American naturalist 124:255–279.

Brown, J. H., D. W. Mehlman, and G. C. Stevens. 1995. Spatial variation in abundance. Ecology 76:2028–2043.

Brown, J. H., and G. B. West, eds. 2000. *Scaling in biology*. Oxford University Press, Oxford.

Brown, J. H., V. K. Gupta, B.-L. Li, B. T. Milne, C. Restrepo and G. B. West. 2002. The fractal nature of nature: power laws, ecological complexity, and biodiversity. Proceedings of the Royal Society of London B 357:619–626.

Brown, J. H., J. F. Gillooly, A. P. Allen, V. M. Savage, and G. B. West. B. 2004. Toward a metabolic theory of ecology. Ecology 85:1771–1789.

Bunn, A. G., D. L. Urban, and T. H. Keitt. 2000. Landscape connectivity: a conservation application of graph theory. Journal of Environmental Management. 59:265–278.

Cantwell, M. D., and R. T. T Forman 1993. Landscale graphs: ecological modeling with graph theory to detect configurations common to diverse landscapes. Landscape Ecology 8:239–255.

Cassey, P. 2002. Life history and ecology influences establishment success of introduced land birds. Biological Journal of the Linnaean Society 76:465–480.

Channell, R., and M. V. Lomolino. 2000. Trajectories to extinction: spatial dynamics of the contraction of geographical ranges. Journal of Biogeography 27:169–179.

Chartrand, G. 1977. *Introductory graph theory*. Dover Publications, New York.

Clark, J. S. 1998. Why trees migrate so fast: confronting theory with dispersal biology and the paleo record. American Naturalist 152:204–224.

Cotgreave, P., and P. Harvey. 1994. Evenness of abundance in bird communities. Journal of Animal Ecology 63:365–374.

Del Hoyo, J., A. Elliot, and J. Sargatal. 1994. *Handbook of the birds of the world, Vol. 2. New world vultures to guineafowls*. Lynx Ediciones, Barcelona.

Del Hoyo, J., A. Elliot, and J. Sargatal. 1997. *Handbook of the birds of the world, Vol. 4. Sandgrouse to cuckoos*. Lynx Ediciones, Barcelona.

Dobrin, R., and P. M. Duxbury. 2001. Minimum spanning trees on random networks. Physical Review Letters. 86:5076–5079.

Duncan, R. P., T. M. Blackburn, and C. J. Veltman. 1999. Determinants of geographical range sizes: a test using introduced New Zealand birds. Journal of Animal Ecology 68:963–975.

Duncan, R. P., M. Bomford, D. M. Forsyth, and L. Conibear. 2001. High predictability in introduction outcomes and the geographical range size of introduced Australian birds: a role for climate. Journal of Animal Ecology 70:621–632.

Duncan, R. P., T. M. Blackburn, and D. Sol. 2003. The ecology of bird introductions. Annual Review of Ecology and Systematics 34:71–98.

Dunning, J. B. 1993. *CRC handbook of avian body masses*. CRC Press, Boca Raton, FL.

Dussert, C., G. Rasigni, M. Rasigni, J. Palmari, ad A. Llebaria. 1986. Minimal spanning tree: a new approach for studying order and disorder. Physical Review B 34:3528.

Dussert, C., G. Rasigni, M. Rasigni, J. Palmari, A. Llebaria and F. Marty. 1987. Minimal spanning tree analysis of biological structures. Journal of Theoretical Biology 125 317–323.

Gammon, D. E,. and B. A. Maurer. 2002. Evidence for non-uniform dispersal in the biological invasions of two naturalized North American bird species. Global Ecology and Biogeography 11:155–61.

Gaston, K. J., and J. H. Lawton. 1988. Patterns in the distribution and abundance of insect populations. Nature 331:709–712.

Gaston, K. J., and J. L. Curnutt. 1998. The dynamics of abundance-range size relationships. Oikos 81:38–44.

Gaston, K. J., T. M. Blackburn, J. J. D. Greenwood, R. D. Gregory, R. M. Quinn, and J. H. Lawton. 2000. Abundance-occupancy relationships. Journal of Animal Ecology 37:39–59.

Gaston, K. J., T. M. Blackburn. 2003. Dispersal and the interspecific abundance-occupancy relationships in British birds. Global Ecology and Biogegraphy 12:373–379.

Gaston, K. J. 2003. *The structure and dynamics of geographic ranges*. Oxford University Press, Oxford.

Gower, J. C., and G. J. S Ross. 1969. Minimum spanning trees and single linkage cluster analysis. Applied Statistics 18:54–64.

Gregory, R. D. 1998. An intraspecific model of species' expansion, linking abundance and distribution. Ecography 21:92–96.

Gregory, R. 2000. Abundance patterns of European breeding birds. Ecography 23:201–208.

Hammersley, J. M., J. H. Halton, and J. Beardwood 1959. The shortest path through many points. Cambridge Philosophical Society Proceedings 55:299–327.

Harte, J., T. Blackburn, and A. Ostling . 2001. Self-similarity and the relationship between abundance and range size. American Naturalist 157:374–386.

Hastings, A., K. Cuddington, K. F. Davies, C. J. Dugaw, S. Elmendorf, A. Freestone, S. Harrison, M. Holland, J. Lambrinos, U. Malvadkar, B. A. Melbourne, K. Moore, C. Taylor and D. Thomson. 2005. The spatial spread of invasions: new developments in theory and evidence. Ecology Letters 8:91–101.

He, F., and S. P. Hubbell. P. 2003. Percolation theory for the distribution and abundance of species. Physical Review Letters 91:198103.

Hengeveld, R. 1989. *Dynamics of biological invasions.* Chapman and Hall, London.

Hoffman, R. and A. K. Jain 1983. A test of randomness based on the minimal spanning tree. Pattern Recognition Letters 1:175–180.

Holt, A. R., K. J. Gaston, and F. He. 2003. Occupancy-abundance relationships and spatial distribution: a review. Basic and Applied Ecology 3:1–13.

Holt, A. R. and K. J. Gaston. 2003. Interspecific abundance–occupancy relationships of British mammals and birds: Is it possible to explain the residual variation? Global Ecology and Biogeography 12:37–46

Holt, R. D., J. H. Lawton, K. J. Gaston and T. M. Blackburn. 1997. On the relationship between range size and local abundance: back to basics. Oikos 78:183–190.

Iribarne, C., M. Rasigni, and G. Rasigni. 1999. Minimal spanning tree and percolation on mosaics: graph theory and percolation. Journal of Physics A–Mathematical and General 32:2611–2622.

Jaillet, P. 1995. On properties of geometric random problems in the plane. Annals of Operations Research 61:1–20.

Johnson, A. R., B. T. Milne, and J. A. Wiens 1992. Diffusion in fractal landscapes—simulations and experimental studies of tenebrionid beetle movements. Ecology 73:1968–1983.

Jones, C. L., G. T. Lonergan, and D. E. Mainwaring. 1996. Minimal spanning tree analysis of fungal spore spatial patterns. Bioimages 4:91–98.

Keitt, T. H., and P. Marquet. 1996. The introduced Hawaiian avifauna reconsidered: evidence for self-organized criticality? Journal of Theoretical Biology 182:161–167.

Keitt, T. H., D. L. Urban, and B. T. Milne. 1997. Detecting critical scales in fragmented landscapes. Conservation Ecology 1, no. 1. http://www.consecol.org/vol1/iss1/art4.

Keitt T. H., L. A. N. Amaral, S. V. Buldyrev, and H. E. Stanley. 2002. Scaling in the growth of geographically subdivided populations: invariant patterns from a continent-wide biological survey. Philosophical Transactions of the Royal Society of London B 357:627–633.

La Sorte, F. A., and W. J. Boecklen. In Press. Changes in the diversity structure of avian assemblages in North America. Global Ecology and Biogeography.

Lockwood, J. L., M. P. Moulton and S. K. Anderson 1993. Morphological assortment and the assembly of communities of introduced passeriforms on oceanic Islands: Tahiti versus Oahu. American Naturalist 141:398–408.

Lomolino, M. V., and R. Channell. 1995. Splendid isolation: patterns of geographic range collapse in endangered mammals. Journal of Mammology 76:335–347.

Lubina, J., and S. A. Levin. 1988. The spread rate of a reinvading organism: range expansion of the California sea otter. American Naturalist 131:526–543.

Magurran, A. E. 1988. *Ecological diversity and its measurement.* Princeton University Press, Princeton, NJ.

Marquet, P. A., J. Keymer, and H. Cofre. 2003. Breaking the stick in space: of niche models, metacommunities, and patterns in the relative abundance of species. In T. M. Blackburn, and K. J. Gaston, eds. *Macroecology: concepts and consequences,* pp. 64–84. Blackwell Scientific Publications, Oxford.

Marquet, P. A., R. A. Quiñones, S. A. Abades, F. Labra, M. Tognelli, M. Arim and M. Rivadeneira. In press. Scaling and power-laws in ecological systems. Journal of Experimental Biology.

Maurer, B. A. 1994. *Geographical population analysis: tools for the analysis of biodiversity.* Blackwell Scientific Publications, Oxford.

Maurer, B. A., E. T. Linder, and D. Gammon. 2001. A geographical perspective on the biotic homogenization process: implications from the macroecology of North American birds. In J. Lockwood, ed. *Biotic homogenization,* pp. 157–178. Kluwer Academic, Plenum Publishers, New York.

Mclain, D. K., M. P. Moulton, J. G. Sanderson. 1999. Sexual selection and extinction: The fate of plumage-dimorphic and plumage-monomorphic birds introduced onto islands. Evolutionary Ecology Research 1:549–565.

Newton, I. 1997. Links between the abundance and distribution of birds. Ecography 20:137–145.

Okubo, A. 1980. *Diffusion and ecological problems: mathematical models.* Springer-Verlag, Berlin.

Peterjohn, B. G., and J. R. Sauer. 1993. North American Breeding Bird Survey Annual Summary 1990–1991. Bird Populations 1. 52–67.

Peterjohn, B. G. 1994. The North American Breeding Bird Survey. Birding 26:386–398.

Prim, R. C. 1957. Shortest connection network and some generalizations. Bell System Technical Journal 36:1389–1401.

R Development Core Team 2003. R: A language and environment for statistical computing. R Foundation for Statistical Computing, Vienna, Austria. http://www.R-project.org.

Sauer, J. R., J. E. Hines, and J. Fallon. 2003. *The North American Breeding Bird Survey, results and analysis 1966–2002.* Version 2003. 1, USGS Patuxent Wildlife Research Center, Laurel, MD.

Sax, D. F. 2001. Latitudinal gradients and geographic ranges of exotic species: implications for biogeography. Journal of Biogeography 28:139–150.

Sax, D. F., S. D. Gaines, and J. H. Brown. 2002. Species invasions exceed extinctions on islands worldwide: A comparative study of plants and birds. American Naturalist 160:766–783.

Shigesada, N., and K. Kawasaki. 1997. *Biological invasions: theory and practice.* Oxford University Press, Oxford.

Skellam, J. G. 1951. Random dispersal in theoretical populations. Biometrika 38:196–218.

Solow, A. R., C. J. Costello, and M. Ward. 2003. Testing the power law model for discrete size data. American Naturalist 162:685–689.

Stanley, H. E., L. A. N. Amaral, P. Gopikrishnan, P. Ch. Ivanov, T. H. Keitt, and V. Plerou. 2000. Scale Invariance and Universality: Organizing Principles in Complex Systems. Physica A 281:60–68.

Steele, J. M. 1988. Growth rates of Euclidean minimal spanning trees with power weighted edges, Annals of Probability, 16 1767–1787.

Urban, D. L. and T. H. Keitt. 2001. Landscape connectedness: a graph theoretic perspective. Ecology, 82:1205–1218.

Van de Weygaert, R., B. J. T. Jones and V. J. Martinez, 1992. The minimal spanning tree as an estimator for generalized dimensions. Physics Letters 169, 145–150.

Wallet, F. and C. Dussert. 1997. Multifactorial Comparative study of spatial point pattern analysis methods. Journal of Theoretical Biology 187:437–447.

West, G. B. 1999. The origin of universal scaling laws in biology. Physica A 263:104–113.

West, G. B., J. H. Brown, and B. J. Enquist. 1997. A general model for the origin of allometric scaling laws in biology. Science 276:122–126.

Williamson, M. 2001. Can the impacts of invasive species be predicted? In R. H. Groves, F. D. Panetta, and J. G. Virtue, eds. *Weed risk assessment*, pp. 20–33. Collingwood, Australia.

Zahn, C. T. 1971. Graph-theoretical methods for detecting and describing gestalt clusters. IEEE Transactions on Computers C 20:68–86.

17

The Dynamics of Species Invasions

INSIGHTS INTO THE MECHANISMS THAT LIMIT SPECIES DIVERSITY

Dov F. Sax, James H. Brown, Ethan P. White, and Steven D. Gaines

Here we examine the effect of multiple species introductions on net changes in species richness. To do this we borrow from population dynamics theory to explore species dynamics. One extension of population theory we explore is the existence of a "species capacity," where the ultimate number of species that an area can support is constrained by some maximum value. We contrast the concept of a species capacity with existing conceptual frameworks of species diversity. To explore these issues, we examine dynamic changes in species richness for birds, vascular plants, and freshwater fishes following species introductions. This allows us, first, to examine the nature and variation of changes in species richness that have occurred to date following species introductions; and second, to explore the relative merits of different conceptual frameworks for understanding species dynamics. We show that species richness has generally increased over the past few centuries for most taxa in most areas, and only rarely has it declined. In light of these data and theory, we discuss possible future changes in species richness of invaded systems. We also discuss limitations of this present study and outline the types of data needed to make larger advances in our understanding of species dynamics. Finally, we believe this work shows the great potential the study of species invasions has for understanding the mechanisms and processes that limit species diversity.

Introduction

Most studies of biological invasions focus on the impacts of individual species. These leave the impression that a single colonizing species often causes or contributes to the extinction of multiple native species, leading to an overall decline in species richness. Often cited examples of species believed to cause multiple extirpations and extinctions include the introduced cichlid, *Chicla ocellaris*, in Lake Gatun (Zaret and Paine 1973); the brown tree snake, *Boiga irregularis*, in Guam (Fritts and Rodda 1998); and the Nile perch, *Lates nilotica*, in Lake Victoria (Goldschmidt et al. 1993). There are several reasons, however, to question whether this phenomenon—that of individual species invasions causing multiple native species extinctions—is the general rule. First, those exotic species with large ecological and economic impacts in their introduced ranges garner disproportionate attention by scientists, whereas species that are not suspected to have any significant impacts on native biota go largely unstudied (Simberloff 1981; see also Bruno et al., this volume). Second, invaded habitats where extinctions have occurred have generally experienced other human-caused changes that may have contributed to the observed extinctions of native species (e.g., habitat destruction on oceanic islands; see Atkinson and Cameron 1993; Gurevitch and Padilla 2004). Third, recent studies that have quantified changes in species richness of ecological communities and geographic regions over the last few decades to centuries have often found net increases in richness, with more exotic species becoming established (i.e., naturalized) than native species going extinct (Atkinson and Cameron 1993; Gido and Brown 1999; Sax et al. 2002; Sax and Gaines 2003).

The net changes in species richness that have accompanied recent human-caused environmental change should be of interest to both basic and applied scientists concerned with biodiversity. We refer to these fluctuations in species richness caused by the opposing effects of colonization and extinction as *species dynamics.* There are obvious parallels between species dynamics and population dynamics, where increases in abundance of individuals due to the effects of birth and immigration are opposed by decreases in abundance caused by death and emigration. Studies of species dynamics associated with invasions and extinctions can enhance our understanding of the fundamental mechanistic processes that generate and regulate species richness. Such studies can evaluate the correlates and consequences of historical events and ecological conditions on spatial patterns and temporal changes in species richness. In so doing, they can address how susceptibility to invasion varies among different kinds of environments and different taxonomic or functional groups of species, and they can ask whether there is something akin to a carrying capacity for species (a "species capacity"). The answers to such basic ecological, evolutionary, and biogeographic questions are directly relevant to accurately informing policymakers and lay people about the current status and trends in biodiversity; these answers are also critical to our ability to design and implement management practices that preserve or enhance biodiversity.

In this chapter we first consider the conceptual basis for species dynamics by examining various theories that account for such dynamics, as well as the predictions they make. We then consider how population theory can be extended to provide a conceptual framework for exploring species dynamics. Next we examine empirical patterns of species dynamics and consider whether these patterns support any particular conceptual models of biodiversity. Finally, we consider the implications and limitations of this work, suggest future work that would be valuable in further elucidating the mechanisms of species dynamics, and discuss the likelihood of future changes in species richness.

Conceptual Basis for Species Dynamics

Species dynamics refer to changes in species richness and composition that occur following the colonization or introduction of species into novel systems. These dynamics are occurring today, and have occurred over the past few hundred years, at rates that greatly exceed presumed background rates (Brown and Sax 2005). On the one hand, current accelerated rates of species dynamics provide significant conservation challenges, particularly in maintaining native biodiversity. On the other hand, rapid changes in species dynamics provide significant opportunities to advance our understanding of the mechanisms that ultimately determine and limit the number of species in any given region. Here we explore two of the conceptual frameworks commonly used to describe or explain species richness and consider the predictions they make about how richness may change in the future.

Patterns and processes of species richness were a major focus of the evolutionary ecology that developed between the late 1950s and early 1970s (e.g., Hutchinson 1957, 1959; MacArthur 1972; Cody and Diamond 1975). Particularly relevant was MacArthur and Wilson's 1967 island biogeography theory (IBT), which hypothesized that species richness on islands—and by implication in many other environments—represents a dynamic equilibrium between opposing rates of colonization and extinction. This equilibrium theory discounted the influence of historical events on geographic variation in species richness. Instead, it emphasized the influence of environmental factors, such as island size and isolation, which determine the balance between contemporary rates of colonization and extinction (Box 17.1). Island biogeography theory suggests that net species richness is effectively bounded at equilibrium points, with sustainable changes in richness possible only by altering rates of colonization and/or extinction.

A different conceptual framework can be seen in stochastic niche theory (SNT), a hybrid of neutral and niche theories that views patterns of invasion and coexistence to be primarily a function of niche partitioning (Tilman 2004). SNT makes predictions about which species are likely to coexist; it suggests that the probability of a species successfully becoming established will decrease with increasing assemblage richness, so long as natives and invaders are drawn from a similar trade-off surface (see Stachowicz and Tilman, this volume). In

BOX 17.1 *Island Biogeography Theory*

Island Biogeography Theory (IBT) was initially developed by Robert H. MacArthur and E. O. Wilson (1963, 1967). It postulates that the number of species on an island (S) are determined by the dynamic equilibrium between independent rates of colonization and extinction, such that the point of intercept between colonization and extinction rates determines the equilibrium number of species. Colonization rates are assumed to be higher on islands that are near (C_N) than islands that are far (C_F) from the source of potential colonists. Extinction rates are assumed to higher on islands that are small (E_S) than on islands that are large (E_L). Therefore, S will be greater on islands that are near than islands that are far if island size is similar. The figure illustrates an example in which both islands are large. IBT also predicts variation in rates of species turnover (T)—i.e., the replacement of one species with another. Turnover is scaled on the y-axis. In the example illustrated, turnover is predicted to be higher for large islands that are near the source popu-

lation than for large islands far from the source.

One critical aspect of classical island biogeography theory is that colonization rates reach zero when the complete pool (P) of species that could colonize from the source population has arrived; colonization curves are assumed to converge on P.

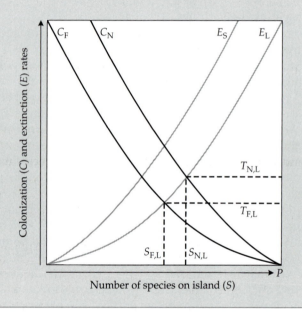

such a framework, the number of species that can be supported in a given area may not be fixed with any hard bound.

The extension of these two ideas, IBT and SNT, to address the impacts of exotic species leads to several expectations for the effects of human-assisted dispersal and the impacts of exotic species on net species richness. IBT (Box 17.1) predicts that when colonization rate is raised to a new level (as through human transport of species), a new equilibrium point with the extinction rate will be reached, such that net species richness increases to a new equilibrium level. Note, however, that the rate at which species are added could conceivably slow as net richness converges on a new equilibrium point. SNT predicts that elevated rates of species introduction should lead to continuing increases in net species richness over time, but that rates at which species addition occurs should decrease over time. Note that while SNT does not predict any fixed bound on species richness, it is conceivable that if probabilities of species addition become low enough, net richness may reach a point that appears (over short time scales) to be relatively stable.

These conceptual frameworks also make predictions about species turnover. IBT predicts that elevating colonization rates, and hence elevating the equilibrium number of species, should also lead to increased rates of species turnover (see Box 17.1); this turnover should cause a loss of some preexisting (native) species. In contrast, while SNT predicts that elevated rates of colonization should also lead to a net increase in species richness, it does not necessarily imply that rates of species turnover will increase or that preexisting (native) species will be lost. Indeed, SNT predicts that resident species, once established, should be difficult to displace.

While neither of these conceptual frameworks (IBT or SNT) suggests that introducing species should lead to decreases in species richness, there are nevertheless other reasons to suspect that introducing exotic species could lead to net decreases in total species richness, particularly at local ecological scales. Strongly interacting species such as keystone predators (e.g., Paine 1966) and other species that cause trophic cascades (Carpenter et al. 1985) allow for the possibility that adding one or more species might trigger the decline to extinction of a great number of other species. However, there is nothing about the direct and indirect interactions caused by such species that would suggest that decreases in total richness should necessarily occur, as increases in richness due to these same interactions are also possible. Indeed, the introduction of top predators or keystone species might promote the buildup of species richness, particularly if the introduction of predators leads to increased resource availability, which could facilitate the invasion of additional species (Davis et al. 2000). Similarly, ideas of "invasional meltdown" (Simberloff and Von Holle 1999), in which invading species facilitate the invasion of additional exotics, does not imply that there should necessarily be any net decreases in species richness. In summary, averaged over large number of invaders, there is little *a priori* reason to suspect a net reduction in species richness as a result of ecological interactions.

Empirical data on changes in species richness remain relatively rare, particularly at local ecological scales. Recent reviews of the available local-scale data suggest that both increases and decreases in total net species richness occur, but that the introduction of exotic species tend to lead to increases in richness much more frequently than decreases (Crooks 2002; Sax et al. 2005). At regional scales such as the scales of large watersheds, counties, states, countries and continents, far more data are available, particularly for individual taxonomic groups. At these scales, most taxonomic groups (mammals, fishes, vascular plants) show net increases in species richness, some groups (land birds) show relatively unchanged levels of richness, while some other groups (freshwater mollusks) show net decreases in richness (Neves 1999; Sax et al. 2002; Sax and Gaines 2003; Sax and Gaines 2005).

Of central interest to us in this chapter is whether the patterns of invasion and the variation in net changes in species richness can help us improve our conceptual understanding of the mechanisms that limit species richness; in turn we are interested in whether an improved conceptual understanding of these mechanisms can help us to better forecast how species richness is likely to change in the future. We believe that one of the key requirements for making progress on this front is our ability to structure and interpret the empirical evidence currently available. To do this, we need an analytical framework that can adequately characterize changes in species richness that occur over the short-term—that is, over time scales of interest to contemporary ecologists and biogeographers.

Analytical Framework

One simple and informative way to quantify several important features of species dynamics is to draw an analogy with population dynamics and calculate the per species change in species richness over some specified time interval:

$$\frac{1}{S}\Delta S = \frac{S_F - S_O}{S_O} = \frac{C - E}{S_O}$$

where S_O and S_F are the initial and final number of species, respectively, and C and E are the number of colonizations and extinctions, respectively, during the time interval. In terms of the impacts of exotic species on species richness, it is informative to measure $(1/S)(\Delta S)$ in terms of the number of exotic species that have established (C) and the number of native species that have gone extinct (E).

For the purposes of this paper, it is practical and meaningful to measure $(1/S)(\Delta S)$ over the period of the last few centuries. This is a relevant time interval for the major impact of human-caused invasions, and there are reasonably good data available for native and exotic species over that time period for a variety of taxa and geographic regions. As a consequence of using disparate data sets, the time intervals in the following analyses vary somewhat with the area and taxon being studied, but in most cases the interval is on the order of the last 200 years. This is sufficient to assess the role of exotic species that have colonized as a consequence of the spread of European people, including examinations of some locations that had been free of humans prior to the arrival of Europeans. Included as colonizing exotics are all species that have naturalized since the arrival of humans: species intentionally introduced, species unintentionally transported by humans, and species that colonized under their own power (whose establishment may have been facilitated by anthropogenic environmental change). This estimate of $(1/S)(\Delta S)$, therefore, includes as natives all species that were present prior to the arrival of Europeans. As such, it does not include extinctions or colonizations that might have been caused by aboriginal peoples prior to the arrival of Europeans. The one exception to this is data examined on oceanic islands, where estimates of extinctions and colonizations caused by aboriginal peoples have been included (for details see Sax et al. 2002).

In the following section, we implement this framework to analyze empirical species dynamics within the last few centuries, when there have been large effects of introduced exotic species and other human-caused environmental changes. We present graphs in which $(1/S)(\Delta S)$ is plotted as a function of the log of S_O (historic native species richness). This is analogous to the practice of analyzing population dynamics by plotting the per capita change in abundance—$(1/N)(\Delta N)$—over some time interval as a function of the log of initial population density, N_O.

One potentially useful analogy to draw from population dynamics to species dynamics is the concept of a carrying capacity. As in population dynamics, where any given area can only support some limited or maximum number of individuals, it is also conceivable that any given area may also be limited in the total number of species it can support. Such a limit is not a necessary outcome of current theory on species richness (discussed above). We refer to this idea of a fixed limit to the total number of species that an area can support as a "species capacity." Such a species capacity might exist, for example, if an area can support a fixed number of individuals, and if species must have some minimum number of individuals to maintain viable populations. However, we do not ascribe any particular mechanism to this capacity at this time, but instead list it here as a heuristic concept—one that we believe may be valuable to consider, particularly in contrast to the expectations of IBT and SNT.

If such a species capacity exists, then there would be some level of species richness, S_{max}, above which $(1/S)(\Delta S)$ is always negative (Figure 17.1). Additionally, the existence of a species capacity would dictate a constraint on the relationship between S_O and $(1/S)(\Delta S)$ whenever sites compared are more or less ecologically equivalent (with respect to area,

Figure 17.1 Species dynamics (i.e., changes in species richness) graphed as analogous to population dynamics. The function $(1/S)(\Delta S)$ represents the per species change in species richness from initial or original species richness (S_O) prior to anthropogenic influences; note that these net changes in richness are independent of changes in species composition. The minimum bound on $(1/S)(\Delta S)$ is −1, the point at which 100% of initial species richness is lost. The maximum bound on $(1/S)(\Delta S)$ is potentially unlimited (see text for examples). However, if there is a "species capacity" (that is, a maximum number of species that any one area can support), and if sites being compared are more or less equivalent ecologically (with respect to available energy, amount of area, etc.), then there should be a upward bound that limits $(1/S)(\Delta S)$ relative to S_O. This upward bound can be viewed as a "constraint line" that is set by $(S_{max}/S_O) − 1$.

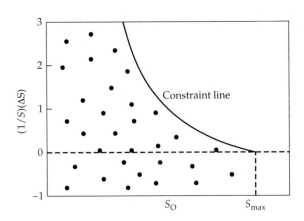

available energy, and other limiting resources), such that the closer S_O is to S_{max}, the smaller $(1/S)(\Delta S)$ is expected to be (Figure 17.1). Such a constraint line, where increases in S become limited as S_{max} is approached, is a necessary outcome of a species capacity model, where the total number of species in any one site is ultimately limited. A secondary consequence of the constraint line is that $(1/S)(\Delta S)$ should be a declining function of S_O, especially as S approaches S_{max}.

A species capacity, however, is not necessary to generate a declining relationship between $(1/S)(\Delta S)$ and S_O, as there are several other reasons to expect such a relationship. First, a null model based purely on propagule pressure could explain such a pattern. For example, if we were to assume that all species introduced became established and that an equal number of species were introduced across sites, ΔS would be the same for all values of S, and $(1/S)(\Delta S)$ would decline with increasing S_O. Second, IBT could explain such a pattern. For example, if sites that formerly differed in their rates of colonization (because of relative differences in their degree of isolation) were all elevated to a new standardized colonization rate, then sites that had previously been particularly isolated (and hence had low levels of species richness) would increase in richness more than those that had been less isolated, generating a decline in $(1/S)(\Delta S)$ with increasing S_O. Third, SNT could explain such a pattern. For example, if the number of introductions across sites is fairly constant, but the rate of successful establishment decreases with increasing levels of preexisting richness, than there should be a sharp decline in $(1/S)(\Delta S)$ with increasing S_O. In summary, there are many conceivable reasons why $(1/S)(\Delta S)$ might decrease with increasing S_O and additional evidence would be needed to support a species capacity model—an issue we return to after first considering the empirical data.

Data and interpretation

Figure 17.2 presents plots of $(1/S)(\Delta S)$ versus S_O for three taxonomic/functional groups (birds, vascular plants, and freshwater fishes) in diverse biogeographic settings ranging from continents to oceanic islands. Inspection of these graphs reveals several patterns, with important distinctions present for each of the taxonomic groups examined.

Birds

For birds, there are relatively few extreme changes in S. There is only one proportional change greater than 1 and three greater than 0.4 out of 69 data points (Figure 17.2A). Observed values cluster closely around $\Delta S = 0$ (i.e., no net change in species richness). This is especially true for land birds within the continental United States, but is also true for birds on oceanic islands, where the majority of species richness values have changed proportionally by less than 0.2. This relative consistency in the total number of bird species is remarkable

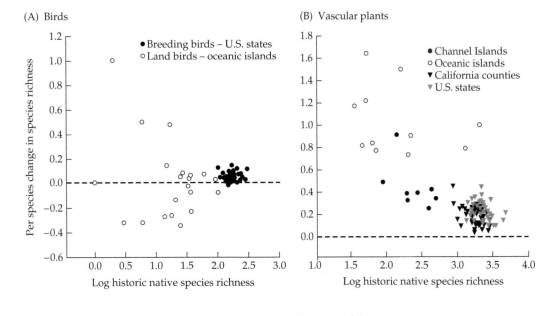

(A) Birds

(B) Vascular plants

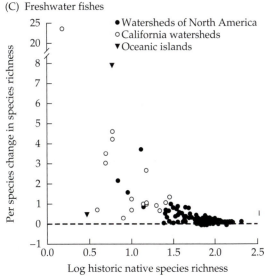

(C) Freshwater fishes

Figure 17.2 Species dynamics, where per species change in species richness [$(1/S)(\Delta S)$] is plotted against historic richness levels (S_O). These data represent change in richness over the past few centuries. With the exception of oceanic islands, they do not represent changes in richness that may have occurred as a consequence of "native" peoples. (A) Per species change in richness for breeding birds in U.S. states and for land birds on oceanic islands (data sources cited in Sax and Gaines 2003). (B) Per species change in richness for vascular plants on the Channel Islands of California (data from Junak et al. 1995); oceanic islands (data from Sax et al. 2002); California counties (data from Calflora, www.calflora.org in 2004); and U.S. states (data sources cited in Sax and Gaines 2003). Note that data for California counties and U.S. states underestimate the actual increase in richness, as only exotic species that are completely foreign to all California counties or U.S. states, respectively, were included as species additions in these counts. Also, plant data were incomplete for three U.S. states (Massachusetts, Minnesota, and North Dakota), and those states were therefore excluded from analysis. (C) Per species change in richness for freshwater fishes for watersheds of North America (data from Gido and Brown 1999), 19 California watersheds (data from Moyle 2002) and oceanic islands (data sources cited in Sax and Gaines 2003). Note that Hawaii is excluded from all counts of U.S. states, but is included as an oceanic island.

TABLE 17.1 *Per species change in species richness (S) and apparent species turnover of birds on oceanic islands*

Island	Per species change in S	Apparent turnover[a]
Ascension Island	1	1
Chatham Islands	−0.08	0.57
Cook Islands	−0.36	0.46
Easter Island	−0.33	1
Fiji Islands	0.07	0.15
Galápagos Islands	0.06	0.03
Guam Island	−0.28	0.61
Hawaiian Islands	−0.09	0.60
Lord Howe Island	0.13	0.63
Marquesas Islands	−0.14	0.39
Mauritius Island	0.07	0.57
New Zealand Islands	0.02	0.42
Norfolk Island	0.47	0.48
Reunion Island	−0.03	0.66
Rodriquez Island	−0.29	0.75
Saint Helena Island	0.50	0.87
Samoan Islands	0.03	0.07
Society Islands	0.04	0.45
Tonga Islands	−0.24	0.45
Tristan da Cunha Island	−0.33	0.60
Wake Island	0	1

Source: Data from Sax et al. 2002.

[a] Apparent turnover of species is calculated as the sum of species colonizations and extinctions divided by the sum of historic and current species richness.

given the extensive proportional turnover of species, which ranges from approximately 0.03 to 1, with the majority of islands having greater than 0.50 turnover (Table 17.1). Thus, even though some islands and continental sites were susceptible to invasion, as measured by naturalization of exotic species, these colonizations have been approximately offset by extinctions of native species; it should be noted and emphasized, however, that many of the extinctions of native bird species occurred prior to the introduction of exotic bird species. Thus, the correlation between native species lost and exotic species gained (which is highly significant) is not a causal relationship (for a discussion see Sax et al. 2002), unless the extinctions played a role in enhancing the likelihood of subsequent invasions. Nonetheless, the result is that overall land bird species richness seems to have been maintained fairly close to the historic number of species, approximated by S_O, where net richness of sites has changed on average by 0.05 on continents and 0.01 on oceanic islands.

Vascular plants

Vascular plants present a strikingly contrasting pattern (Figure 17.2B). All of the 123 points are above the $\Delta S = 0$ line. This indicates that the overwhelming effect of colonizing exotic species has been to increase overall plant species richness. Invasions have been quite numerous, and extinctions have been quite rare. This is true at all spatial scales represented here, from relatively small islands and California counties to very large islands and entire U.S. states. Additionally, there appears to be a large effect of historic biogeographic isolation. The most isolated oceanic islands have the largest $(1/S)(\Delta S)$ values, ranging from 0.80–1.60; the landbridge California Channel islands have smaller values, ranging from 0.25–0.80; and continental sites have the smallest values, ranging from 0.10–0.30 (Table 17.2). An implication of these data is that most vascular plant assemblages have been quite susceptible to invasion. Within the last two centuries, invading exotic species have approximately doubled the richness of isolated oceanic islands (see also Sax et al. 2002; Sax and

Gaines 2003), while in many areas of the North American continent, at spatial scales ranging from counties to states, colonizing exotics have increased richness by more than 20%. These data are therefore consistent with the notion that dispersal of plants is sufficiently limited that local habitats and larger geographic areas were historically far from reaching any ultimate capacity to hold species. Note, however, that the degree of increase declines with increasing S_O, both when all sites, regardless of type (e.g., island versus mainland), are considered (Figure 17.2B), and when sites are separated into more homogeneous clusters (Figure 17.3).

TABLE 17.2 *Per species increase in species richness of plants*

Geographic units	Mean	95% CI
California counties	0.17	0.15–0.19
U.S. states	0.23	0.21–0.26
Channel Islands	0.47	0.27–0.61
Oceanic islands	1.04	0.83–1.25

Note: On average, geographic units that are most isolated from continental sources of species (i.e., oceanic islands) have increased in species richness the most, while relatively non-isolated continental regions have increased in richness the least.

Freshwater fishes

The data for freshwater fishes present still a different pattern (Figure 17.2C). These fishes show by far the widest range of $(1/S)(\Delta S)$, from a few slightly negative values to several positive values in excess of 3 (i.e., to increases in richness in excess of 300%). This implies that fish faunas differ enormously in their susceptibility to invasion (or potentially in their respective number of introduction attempts). Historically, it has been difficult for freshwater fish species to disperse across terrestrial and marine barriers. For example, all of the freshwater fishes native to Hawaii have secondarily colonized freshwater environments from saltwater ones (Resh and De Szalay 1995). As a consequence, oceanic islands and some long-isolated river basins on continents (such as the Colorado, Rio Grande, and Sacramento-San Joaquin river basins in southwestern North America) have historically had depauperate fish faunas (Gido and Brown 1999; Sax and Gaines 2003). It is perhaps not surprising, therefore, that many areas have proven to be highly susceptible to invasion by introduced fish species. There is substantial variation, however, in rates of species turnover (and the loss of native species) and in the addition of exotic species (for more detail see Gido and Brown 1999).

Despite the dramatic increases in species richness in many fish assemblages, a substantial number of areas are very close to the $(1/S)(\Delta S) = 0$ line when native richness is high, which is consistent with the notion that these areas may be relatively close to their species capacity. This interpretation is confounded, however, by the composition of the pool of available invaders. In contrast to the exotic plants plotted for the United States in Figure 17.2, all of which were exotic to the United States, most "exotic" fish species within watersheds of North America are native to other watersheds in North America, with a much smaller number being exotic to the entire continent. This means that there are fewer possible species in the total pool of "exotics" that can invade watersheds with high native species richness, since many of those species are already present in these watersheds as natives. To show this, we present data from freshwater fishes in U.S.

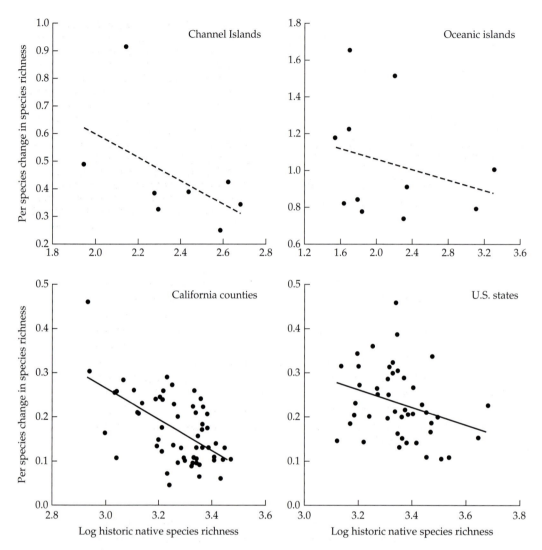

Figure 17.3 Per species change in richness [$(1/s)(\Delta S)$] of vascular plants plotted against historic richness levels (S_O). These data are identical to those in Figure 17.2B, but each group of locations is plotted individually. Note that all four groups show a trend towards decreasing change in richness with increasing historic richness, but that only the correlations for California counties ($R^2 = 0.34, P < 0.0001$) and U.S. states ($R^2 = 0.10, P < 0.04$) are significant.

states, which show a strong negative correlation between native richness and the number of exotic species from other U.S. states, but no relationship between exotics from outside the United States and native richness (Figure 17.4). These patterns are consistent with the notion that the overall decline in the value of $(1/S)(\Delta S)$ with increasing native richness for fishes in North America is not due

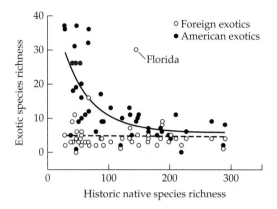

Figure 17.4 Exotic freshwater fishes in U.S. states split into two categories: species exotic to the entire United States (foreign exotics), and those native to at least one U.S. state, but introduced and established in others (American exotics). Foreign exotics show no relationship with increasing native richness (dashed line), while American exotics show a negative relationship with increasing native richness (solid line; 3-parameter exponential decay regression, Adj. $R^2 = 0.45$, $P < 0.0001$). Note that removing the outlying point for the state of Florida, which has a large number of foreign exotics, does not affect the relationship. (Data from Fuller et al. 1999 and Stein et al. 1999. The states of Hawaii and Ohio were excluded from this analysis.)

to an increased difficulty of invading with increasing richness (since exotics from outside the United States don't show this pattern), but instead to a constraint on the size of the pool of exotic species. Increased difficulty in invading species-rich regions could also be occurring, but without knowing more about the relative introduction effort across states or watersheds, it is impossible to make firm conclusions from these data. Further, the relatively small number of fishes introduced from outside the United States reduces our power to detect much of a correlation, even if one did exist. Thus, for fishes it is difficult to distinguish the relative importance of artifacts in the species invader pool from the possibility that species may indeed be near some maximal capacity or that areas with many native species are more difficult to invade.

Discussion

Plots of net change in species richness, $(1/S)(\Delta S)$, as a function of the original number of native species, S_O, provide a useful quantitative view of the net effect, to date, of exotic species on species richness. We have presented examples of such plots for land birds, vascular plants, and freshwater fish. The results run the gamut from virtually no net change in richness of birds on either islands or continents, to large increases in richness of plants, to wide variation in the responses of fish. Large increases in richness have been typical in geographically isolated and historically depauperate habitats, such as oceanic islands for both plants and fish, and geologically isolated watersheds for fish. In addition to demonstrating how net richness has changed, the plots of $(1/S)(\Delta S)$ as a function of S_O also raise several interesting questions.

1. *How many native species have gone extinct?* Since these plots show only net changes in species richness, they do not indicate this, although the data re-

quired to construct these plots do allow quantification of extinction, and examples for birds, plants, and fish have been presented elsewhere (Gido and Brown 1999; Sax et al. 2002). For plants and fishes the number of native species extinctions has generally been modest. For example, in New Zealand, where over 2000 colonizing plants have become naturalized, fewer than 10 native plant species are believed to have gone extinct; similarly, in Hawaii, where approximately 40 freshwater fish species have become naturalized, none of the 5 native species of fish have become extinct. In contrast, for birds rates of extinction have generally been higher, particularly on islands. For example, in Hawaii, at least 64 bird species have gone extinct (Sax et al. 2002). Care must be exercised, however, in interpreting the cause of native species extinction. These extinctions may not have been caused entirely or even largely by colonizing exotics, although there are clear cases when this has occurred (e.g., see Fritts and Rodda 1998). Most invaded assemblages have experienced many kinds of environmental change, both human-caused and "natural," and these changes, in addition to the potential impacts of colonizing exotics, must be evaluated to assess the causes of extinctions. For example, change in habitat structure (or outright habitat destruction) has been a particularly noteworthy contributor to native species extinction (Gurevitch and Padilla 2004; Blackburn and Gaston, this volume).

2. *Has the rate of species establishment (i.e. naturalization) decreased, increased, or remained about the same over recent decades?* In some cases there are sufficiently detailed data to address this question, although we will not try to do so here. We will, however, emphasize the importance, both practical and conceptual, of addressing it. On the one hand, human travel and commerce is continually increasing, and purposeful introductions (e.g., of garden plants) are continuing, raising the opportunity for accidental or deliberate introduction of additional exotic species. On the other hand, there have been increasing legislative, inspection, and public education efforts to prevent the importation of exotic species. A multispecies approach is required to assess the effectiveness of these activities. Some invasions are still occurring, but isolated reports of the establishment of a few spectacular and potentially damaging colonizations may lead to misleading assessments of the overall effectiveness of prevention measures. In particular, it is difficult to estimate what the rate of invasion would be if the preventative measures were not in effect.

3. *How many invaders might different environments ultimately acquire, and how many native species might ultimately be lost to extinction?* Answers to these questions vary with the conceptual model employed. If colonization rates remain elevated then SNT would allow for continuing increases in species richness (i.e., the continued addition of exotic species to these invaded systems, albeit at potentially declining rates even in the face of high propagule pressure). In contrast, the species capacity and IBT models presume that species richness will not continue to increase (despite elevated colonization rates) and will become

fixed at some set point. In one case (the species capacity model), this fixed point will be impossible to increase further and in the other case (IBT) the point could be increased further if colonization rates were elevated to an even higher level. In all cases, rates of increases in species richness are expected to decline over time, but the mechanisms by which that rate declines differ dramatically among the models.

Declines in the rate of species richness increase could be driven by extinctions that match the rate of introductions (e.g., in IBT) or through ever increasing resistance of the system to subsequent successful invasion (e.g., in SNT). These alternatives have strikingly different implications. For example, consider the case of vascular plants on oceanic islands. In these systems, there are now twice as many plant species established as there were historically. The striking consistency in this pattern, with 96% of the variation in naturalized plant richness explained by native plant richness (Sax et al. 2002), raises the possibility that oceanic islands may now be approaching their capacity to hold plant species. If the rate of colonization (i.e., exotic propagule introductions) remains at the same elevated level, then depending on which conceptual framework best matches reality, we could see elevated rates of extinctions of native species (if IBT or a species capacity model are correct) or rarely see successful colonizations, and presumably have relatively fewer native species extinctions (if SNT is correct).

4. *What would happen if colonization rates were returned to historic levels (i.e., if we were able to prevent the introduction of any additional exotic species)?* Among these models, neither SNT nor a species capacity model would predict a corresponding decline in species richness. By contrast, IBT would predict a concordant decline in species richness as colonization rates were decreased. Therefore, these models make very different predictions about the likelihood of retaining a substantial number of exotic species over the long run, if we were able to restrict the entry of additional exotic species.

In practice, it may be extremely difficult to prevent future naturalizations from occurring. This may be particularly true for some groups like vascular plants. In New Zealand, for example, there are more than 20,000 exotic plant species already present, growing in gardens around the country (Sax et al. 2002). These species effectively form a large "bank" or reserve of species that are capable of becoming naturalized in the future. This exotic species bank exists irrespective of the success in preventing still further species from entering the country. Similar banks occur in most, if not all, countries around the world, not just for plants, but for other taxonomic groups as well, such as aquarium fishes.

5. *Is species richness for all taxa governed by similar forces, or do the conceptual models discussed here vary in their applicability among different taxonomic groups?* For birds, the small $(1/S)(\Delta S)$ values relative to plants and fishes (see Figure 17.2) together with the large species turnover (see Table 17.1) are consistent with the

operation of a species capacity, in which historic levels of species richness were already at their capacity prior to human influences on colonization and extinction—a view that is further supported by the failure of many islands to exceed historic levels of richness in spite of a large pool of potential colonists. For example, since humans arrived on Reunion Island, net bird species richness has changed by only one species, since the 22 extinctions of native species have been almost perfectly matched by 21 successful naturalizations by exotic species (Sax et al. 2002). For plants, $(1/S)(\Delta S)$ shows a consistent trend of converging towards zero with increasing S_O. However, there is no case in which $(1/S)(\Delta S)$ values have actually reached zero (the point of no change in species richness) (see Figure 17.3). This suggests that if there is in fact a species capacity, or new equilibrium point, for plant richness that it has not yet been reached, and that additional plant species can be added to these systems.

For fishes, there is some evidence in continental systems to suggest the presence of a species capacity, but these patterns are confounded with the sampling artifact of a limited species invasion pool for the most diverse watersheds, and limited information on propagule pressure; therefore these data are difficult to interpret. On islands, there is scant evidence to interpret invasion patterns in fishes, but available evidence suggests that these systems are far from reaching a species capacity and that if one exists, it has not yet been reached. For none of these groups can we ascribe the clear operation of any one of these conceptual models: IBT, SNT, or a species capacity. This is unfortunate, as ultimately these models make very different predictions about how species richness is likely to change in the future, both overall and for particular taxonomic groups.

Final Thoughts

Part of the difficulty in distinguishing among these conceptual models is not just the limited data available, but also the importance of a number of other factors. For example, human impacts, independent from any effects of exotic species addition, can increase species extinction rates; in contrast, humans may also increase the capacity of systems to support species by changing habitat heterogeneity and/or the amount of available resources. Invaders themselves might alter the habitat in such a way as to change a community's species capacity (Crooks 2002). Any of these outcomes would severely complicate the interpretation of empirical patterns. Similarly, it is important to note that many of the areas compared are not equivalent ecologically (e.g., with respect to abiotic conditions), which may cause the probability of species establishment to differ among regions. Further, many of the areas compared differ in size, which may effect the total number of species likely to be supported (Sax and Gaines 2005). One of the most significant difficulties in interpreting current empirical patterns, however, is the complication posed by potential lags in species extinction, which could create an "extinction debt" *sensu* Tilman et al. 1994. Such a debt would greatly complicate the interpretation of any current empirical patterns, particularly as lags in extinction could conceivably vary in magnitude among taxonomic groups and among regions. Finally, it is important to recognize that introduction pressure has undoubtedly not been equivalent across areas compared; the degree to which this has varied is unknown, but such variation increases the difficulty that exists in interpreting these patterns. It is conceivable that variation in introduction pressure of exotic propagules explains much of the variation we have observed in changes in species richness to date.

Our difficulty in distinguishing among the different conceptual frameworks presented here raise a number of issues and point to potentially profitable directions for future research. First, it is clear that the conceptual frameworks discussed here could be explored in greater depth. For example, by altering the shape of potential colonization and extinction rates a great complexity of results could be generated from IBT. Similarly, the heuristic concept of a "species capacity" that we have presented could be further developed and tied to particular mechanisms. Second, in this chapter we have discussed only a very limited number of the potential conceptual frameworks for species richness that are currently described in the literature. A more formal review that contrasts the predictions of a larger number of frameworks was beyond the scope of this chapter, but would be extremely valuable. In particular, it would be valuable to more formally explore the predictions of the Neutral/Symmetrical Model proposed by Hubbell (2001), in light of species introductions and extinctions. Similarly, we believe that species-energy theory (see Wright 1983; Brown 1981, 1988; Currie 1991; Wright et al. 1993; Brown et al. 2001) would also be valuable to consider in this light. Third, it is clear that to make substantial progress with this type of approach that we will need access to detailed data on rates of species introductions; these data would be valuable even over limited time periods or for a limited number of areas. Fourth, we must improve our understanding of potential lags in species extinctions and how such lags may vary across taxonomic groups and geographic areas. Finally, and perhaps most critically, the work discussed here highlights just how poorly species dynamics are understood. Ecology, evolutionary biology, and biogeography all currently lack a general theory or framework for understanding species dynamics that is widely accepted—one that the scientific community feels confident can robustly forecast future changes in biodiversity. We believe developing such a framework is particularly important in light of current patterns of global change. Accomplishing this will undoubtedly require a long and sustained effort from scientists working on many aspects of these issues. We believe, however, that patterns of species introductions can help considerably in this endeavor, and that their study will ultimately help to provide much insight into the mechanisms and processes that set and limit patterns of species richness.

Acknowledgments

This chapter benefited from comments from Jay Stachowicz, Dave Tilman, and an anonymous reviewer. This work was conducted as a part of the "Exotic Species: A source of insight into ecology, evolution, and biogeography" Working Group supported by the National Center for Ecological Analysis and Synthesis, a center funded by the National Science Foundation (Grant #DEB-0072909), the University of California, and the Santa Barbara campus. This is contribution number 176 from PISCO, the Partnership for Interdisciplinary Studies of Coastal Oceans funded primarily by the Gordon and Betty Moore Foundation and David and Lucile Packard Foundation.

Literature Cited

Atkinson, I. A. E., and E. K. Cameron. 1993. Human influence on the terrestrial biota and biotic communities of New Zealand. Trends in Ecology and Evolution 8:447–451.

Brown, J. H. 1981. Two decades of homage to Santa Rosalia: toward a general theory of diversity. American Zoologist 21:877–888.

Brown, J. H. 1988. Species diversity. In A. A. Myers and P. S. Giller, eds. *Analytical biogeography*, pp. 57–89. Chapman and Hall, London.

Brown, J. H., and D. F. Sax. 2005. Biological invasions and scientific objectivity: Reply to Cassey et al. (2005). Austral Ecology 30:481–483.

Brown, J. H., S. K. M. Ernest, J. M. Parody, and J. P. Haskell. 2001. Regulation of diversity: maintenance of species richness in changing environments. Oecologia 126:321–332.

Carpenter, S. R., J. F. Kitchell, and J. R. Hodgson. 1985. Cascading trophic interactions and lake productivity. Bioscience 35:634–639.

Cody, M. L., and J. M. Diamond. 1975. *Ecology and evolution of communities*. Harvard University Press, Cambridge, MA.

Crooks, J. A. 2002. Characterizing ecosystem-level consequences of biological invasions: the role of ecosystem engineers. Oikos 97:153–166.

Currie, D. J. 1991. Energy and large-scale patterns of animal-species and plant-species richness. American Naturalist 137:27–49.

Davis, M. A., J. P. Grime, and K. Thompson. 2000. Fluctuating resources in plant communities: a general theory of invisibility. Journal of Ecology 88:528–534.

Fritts, T. H., and G. H. Rodda. 1998. The role of introduced species in the degradation of island ecosystems: a case history of Guam. Annual Review of Ecology and Systematics 29:113–140.

Fuller, P. L., L. G. Nico, and J. D. Williams. 1999. Nonindigenous fishes introduced into the inland waters of the United States. American Fisheries Society, Special Publication 27, Bethesda, MD.

Gido, K. B., and J. H. Brown. 1999. Invasion of North American drainages by alien fish species. Freshwater Biology 42:387–399.

Goldschmidt, T., F. Witte, and J. Wanink. 1993. Cascading effects of the introduced Nile perch on the detritivorous phytoplanktivorous species in the sublittoral areas of Lake Victoria. Conservation Biology 7:686–700.

Gurevitch, J., and D. K. Padilla. 2004. Are invasive species a major cause of extinctions? Trends in Ecology and Evolution 19:470–474.

Hubbell, S. P. 2001. *The unified neutral theory of biodiversity and biogeography*. Princeton University Press, Princeton, NJ.

Hutchinson, G. E. 1957. Concluding remarks. Population studies: animal ecology and demography. Cold Spring Harbor Symposia on Quantitative Biology 22:415–427.

Hutchinson, G. E. 1959. Homage to Santa Rosalia; or, Why are there so many kinds of animals? American Naturalist 93:145–159.

Junak, S., T. Ayers and R. Scott. 1995. A flora of Santa Cruz Island. Santa Barbara Botanic Gardens, Santa Barbara, CA.

MacArthur, R. H. 1972. *Geographical ecology*. Harper and Row, New York.

MacArthur, R. H., and E. O. Wilson. 1963. Equilibrium theory of insular zoogeography. Evolution 17:373–387.

MacArthur, R. H., and E. O. Wilson. 1967. *The theory of islands biogeography*. Princeton University Press, Princeton, NJ.

Moyle, P. B. 2002. Inland fishes of California, revised and expanded edition. University of California Press, Berkeley, CA.

Neves, R. J. 1999. Conservation and commerce: management of freshwater mussel (Bivalvia: Unionoidea) resources in the United States. Malacologia 41:461–474.

Paine, R. T. 1966. Food web complexity and species diversity. American Naturalist 100:65–75.

Resh, V. H., and F. A. De Szalay. 1995. Streams and rivers of Oceania. In C. E. Cushing, K. W. Cummins, and G. W. Minshall, eds. *River and stream ecosystems*, pp. 717–736. Elsevier, Amsterdam.

Sax, D. F., and S. D. Gaines. 2003. Species diversity: from global decreases to local increases. Trends in Ecology and Evolution 18:561–566.

Sax, D. F., S. D. Gaines, and J. H. Brown. 2002. Species invasions exceed extinctions on islands worldwide: A comparative study of plants and birds. American Naturalist 160:766–783.

Sax, D. F., B. P. Kinlan, and K. F. Smith. 2005. A conceptual framework for comparing species assemblages in native and exotic habitats. Oikos 108:457–464.

Simberloff, D. 1981. Community effects of introduced species. In M. H. Nitecki, ed. *Biotic crises in ecological and evolutionary time*, pp. 53–81. Academic Press, New York.

Simberloff, D., and B. Von Holle. 1999. Positive interactions of nonindigenous species: Invasional meltdown? Biological Invasions 1:21–32.

Stein, B. A., L. S. Kutner, and J. S. Adams. 2000. *Precious heritage: the status of biodiversity in the United States*. Oxford University Press, New York.

Tilman, D. 2004. A stochastic theory of resource competition, community assembly and invasions. Proceedings of the National Academy of Sciences USA 101:10854–10861.

Tilman, D., R. M. May, C. L. Lehman, and M. A. Nowak. 1994. Habitat destruction and the extinction debt. Nature 371:65–66.

Wright, D. H. 1983. Species-energy theory—an extension of species-area theory. Oikos 41:496–506.

Wright, D. H., D. J. Currie, and B. A. Maurer. 1993. Energy supply and patterns of species richness on local and regional scales. In R. E. Ricklefs and D. Schulter, eds. *Species diversity in ecological communities*, pp. 66–74. University of Chicago Press, Chicago, IL.

Zaret, T. M., and R. T. Paine. 1973. Species introduction in a tropical lake. Science 182:449–455.

CAPSTONE

Where Do We Go from Here?

Dov F. Sax, John J. Stachowicz, and Steven D. Gaines

Species invasions embody a series of apparent contradictions and paradoxes. They have taken place commonly and pervasively throughout the Earth's history, yet their current extent and frequency are unnatural (Brown and Sax 2005; Cassey et al. 2005). They present chronic economic challenges to modern economies, and yet are integral to the very foundation on which many of these same economies developed (see Pimentel et al. 2000; Sax et al., Introduction to this volume). They threaten native species with extinction and endanger native communities and ecosystems (e.g., Ebenhard 1988; Fritts and Rodda 1998), but at the same time they provide an unparalleled opportunity to study and understand the natural world, providing insights into fundamental research issues in the life sciences—insights that, somewhat ironically, can enhance our efforts to conserve native biota and biological systems.

The balance of attention on invasions has been decidedly on the side of solving applied issues (Sax et al., Introduction to this volume). Curiously, this is a somewhat incomplete view of the manner in which Darwin (1859), Grinnell (1919), Baker and Stebbins (1965), and other prominent biologists of the nineteenth and twentieth centuries viewed invasions. They looked beyond just the impacts and saw the opportunities that species invasions presented to understand basic research issues. Since the time of Elton's (1958) seminal work, however, the field of invasion biology has been dominated by a more narrow and primarily applied focus (Davis 2005). We believe that the time is now ripe to reinvigorate a more balanced approach in the study of species invasions, with increased emphasis on the search for insights into basic research issues.

Fortunately, this process is well underway. Certainly, the field of ecosystem ecology has been drawing basic research insights from species invasions for many years (see D'Antonio and Hobbie, this volume). More widespread attention in ecology, evolution and biogeography is also becoming more common (e.g., Huey et al. 2000; Sax 2001; Levin 2003). Hopefully, the work presented in this book, as well as in the forthcoming volume by Cadotte et al. (2005), will further propel the use of species invasions to study basic research issues into the mainstream.

We will not attempt to summarize here the insights derived from each of the individual chapters; we believe the work presented speaks for itself. Instead, we consider two other lines of thought. First, we explore the insights that emerge from reading "between the chapters"—that is, to consider insights that arise not from any single chapter, but from the work as a whole. Second, we explore some of the logical next steps for research on these topics.

Reading between the Chapters: Emergent Insights from the Study of Species Invasions

Any two individuals reading the chapters in this book would likely differ in the lines of thought, study, and insights they considered most intriguing. Here, we discuss four general topics that we believe the work in this book bears on in interesting ways.

Genetic bottlenecks: How common are they?

The odds that any propagule will cross some significant geographical barrier and arrive in a location that is suitable for its persistence are exceedingly low. Consider the probability of a single dandelion seed being blown in the wind from Japan, traveling across the Pacific, eventually to reach the Hawaiian Islands. Further consider that, even once on an island, most places that seed might land would be entirely unsuitable for its survival—a marsh, a stream, a coral reef. It is these low probabilities of successful arrival and establishment that presumably account for the relatively depauperate flora of Hawaii and other oceanic islands (Sax and Gaines 2005). On rare occasions, however, propagules do arrive and successfully establish. Presumably, such successful invasions are typically achieved by just a few individuals at any one time and place, creating potential founder effects and population bottlenecks.

Population bottlenecks should be a fairly common feature in the establishment of exotic species, just as they are in the establishment of isolated populations of native species. Further, such population bottlenecks should frequently create corresponding *genetic* bottlenecks, exemplified by reductions in allelic diversity and heterozygosity within newly established populations. Such reductions in genetic diversity could reduce the evolutionary potential of newly established populations and hinder their ability to successfully invade and occupy new regions. Clearly, however, population bottlenecks are not an absolute barrier to successful invasion. As a case in point, consider the invasion of *Drosophila subobscura* in South America. It is believed (based on genetic evidence) that this introduction started with fewer than a dozen individuals, yet in just a few decades *D. subobscura* has come to occupy a large range in South America, where it shows pronounced geographic clines in genetically based traits (Huey et al. 2000, and this volume).

Just how common are genetic bottlenecks in exotic species? Are genetic bottlenecks the norm? When they occur, are they typically severe? These are questions that are difficult to answer at this time. However, as the number of well-documented genetic studies of species invasions grows, the answers may become clearer. Currently available evidence suggests that genetic bottlenecks may be much less frequent and less severe than previously thought. Two of the chapters in this volume (Novak and Mack; Wares et al.) draw together a large number of the existing studies on plants and animals, respectively, that indicate many invading species show little evidence of pronounced bottlenecks. Other recent studies of individual species reach the same conclusion (e.g., Kolbe et al. 2004; Voisin et al. 2005). In fact, many invading populations show the exact opposite pattern: they are actually *more* genetically diverse (within populations) in their naturalized range than in their native range (Kolbe et al. 2004; Voisin et al. 2005). This "genetic enrichment" of naturalized populations is presumably a result of the introduction of individuals from multiple, genetically differentiated, locations (Novak and Mack, this volume). However, before we can conclude anything about the frequency of genetic bottlenecks (or genetic enrichment), it is important to bear in mind that genetic studies are usually only performed on those invaders that are ultimately successful. Perhaps many invasions ultimately fail precisely because pronounced genetic bottlenecks do occur. This is undoubtedly a very difficult question to address, and will require creative approaches and considerable data to answer. Still, given the evidence available, for species that have been successful in invading new regions, it appears that severe genetic bottlenecks may be less common than previously thought.

Evaluating this proposition opens the door to a number of related questions. For instance, how might the severity of a genetic bottleneck or the degree of genetic enrichment be related to invasion success? More specifically, how might the level of genetic diversity within an invading population affect its vulnerability or resistance to competitors, predators, and pathogens? How might genetic diversity predict the eventual size of the geographic area a species is able to occupy, or determine the abundance it maintains within an invaded region, or predict the ecological impacts it manifests? These are difficult and important questions that draw firmly at the union between fundamental issues in evolutionary biology, ecology, and biogeography.

Are exotic species increasing speciation rates?

Although there is some debate in the literature over the magnitude of the threat exotic species pose to native biodiversity (Gurevitch and Padilla 2004), and whether extinctions mediated by exotics are more likely to occur via competitive or predatory interactions (e.g., Davis 2003), there is little question that exotic species are causing or contributing to at least some native species extinctions (e.g., Fritts and Rodda 1998; Blackburn and Gaston, this volume). Indeed, exotic species are often listed as one of the principal threats to regional and global biodiversity (e.g., Wilcove et al. 1998; Myers and Norman 2001).

Recent synthetic considerations raise the possibility, however, that exotic species might also increase the likelihood of speciation (Mooney and Cleland 2001; Myers and Norman 2001; Tilman and Lehman 2001). Examples of speciation and other evolutionary diversification events being facilitated by exotic species have been documented in the literature for many years, and sometimes provide textbook examples of evolutionary diversification. One example is the speciation event that resulted when a native cordgrass, *Spartina maritima*, and an exotic, *S. alterniflora*, hybridized to form a new species, *S. anglica* (Raybould et al. 1991). Another example is the evolutionary diversification (through a potentially incipient speciation event) that has occurred since populations of native hawthorn flies (*Rhagoletis pomonella*) colonized introduced apple trees (Bush 1969). More recent work has provided many additional examples of the variety of evolutionary changes that occur following species invasions (Cox 2004) and of speciation events in particular (see examples in Ellstrand and Schierenbeck 2000).

Intimately related to the probability of speciation events are the occurrence of founder events, which are mediated through population and genetic bottlenecks. Founder events have long been hypothesized to increase the likelihood of speciation events (see Mayr 1954). Some researchers, however, have postulated that founder events should be unlikely to lead to speciation events when there are prolonged population bottlenecks, ones where newly established populations are not able to rapidly expand into newly invaded environments (e.g., Templeton 1980; Templeton et al. 2001). The work on genetic bottlenecks discussed above, however, suggests that, among successfully naturalizing species, pronounced bottlenecks may occur much less frequently than was previously thought. Therefore, founder events associated with successfully invading species should, at the very least, not reduce the likelihood of subsequent speciation events and may instead promote such events. Nonetheless, it remains unclear how commonly speciation events (or incipient speciation events) should occur as a consequence of anthropogenically facilitated invasions.

While we cannot resolve the question of how commonly such speciation (or incipient speciation) events are occurring, we can suggest a few of the processes related to species invasions that are likely to contribute to these events. First, species invasions are creating disjunct, allopatric, isolated populations. This is occurring commonly and repeatedly. For example, consider the house sparrow (*Passer domesticus*), which has naturalized populations in North America, Australia, New Zealand, and many other localities (Lever 1987). Most of these populations are likely isolated from each other, with little to no gene flow currently taking place among them. While such isolated populations might be expected to diverge from each other very slowly over time due to genetic drift, change is probably occurring much more quickly due to differences in the selective regimes present in each of the regions they now occupy.

Second, species invasions may lead to speciation events when invasions promote hybridization between taxa. Very rapid speciation events can occur following hybridizations that involve chromosomal rearrangements and subsequent genetic isolation from parent populations. Such events are believed to account for a significant proportion of native diversity, particularly among plant species (Coyne and Orr 2004). Exotic introductions have greatly increased the potential for such hybridization events, and there are several clear cases of speciation events having occurred as a consequence of human-mediated species invasions (see examples in Ellstrand and Schierenbeck 2000). Although the total number of documented cases of speciation via hybridization following exotic species invasion is modest, undetected cases may be much more common.

Third, species invasions may lead to speciation events when invasions are associated with increased genetic diversity. These increases in genetic diversity can occur when multiple source populations contribute to individual invasions, resulting in within-population genetic diversity that is higher in the naturalized than the native range (Kolbe et al. 2004; Voisin et al. 2005; Novak and Mack, this volume). Such "super invaders" should conceivably have an increased potential to speciate, particularly if environmental conditions exist that are novel for these species (Novak and Mack, this volume).

Fourth, species invasions may lead to speciation events because of the novel evolutionary pressures and opportunities associated with invasion. While novel environmental conditions should promote speciation by interacting with genetically rich and isolated populations of invaders (discussed above), so too should novel biotic interactions. In particular, we refer to the opportunities for coevolution between species that were previously isolated. Such novel interactions may lead to host-switching with respect to herbivory, parasitism, predation, mutualism, and commensalism, or to the use of newly created habitat, vegetation, or substrate types.

This brief outline of some factors and pressures that may lead to increased rates of speciation is not meant to be exhaustive. The complete set of factors that lead to and allow for speciation are complex. Further, many of the processes that limit or promote speciation may have little to do with species invasions, but may instead be mediated more strongly by other processes, like habitat fragmentation (Myers and Norman 2001; Rosenzweig 2001). Still, we believe it is apparent that species invasions may ultimately have large impacts on future rates of speciation and that this issue deserves significant attention.

Are communities saturated with species?

A container of water of a given temperature can become saturated with salt: at some point, the net number of salt molecules in a solution can no longer increase, unless the quantity or temperature of the water is also increased. This simple concept from chemistry provides the basis for an analogy frequently found in ecology, that of species saturation within a given area. Just as a specific set of factors limit how many molecules can be present in a solution, so too might specific factors, such as total area and available resources, limit how many species can be supported in any given location. Of course, simple analogies can be misleading if taken too literally; however, they can provide jumping-off points for the consideration and development of appropriate theory (e.g., Srivastava 1999; and Bruno et al.; Sax et al.; Stachowicz and Tilman, this volume).

Three important and somewhat independent bodies of literature in ecology and biogeography have considered the issue of species saturation. The first of these is the species-area literature, and the island biogeography literature that developed from it. Species-area relationships differ among taxonomic groups and geographic regions (particularly their intercept and slope when plotted in log-log space; for a discussion of this topic, see Sax and Gaines 2005). This means that species richness can vary tremendously from site to site in spite of similar environmental conditions and quantity of area, presumably due to the influence of isolation. This fact has been taken to suggest that many sites that are species-poor are not "saturated" and could contain many more species if they weren't so isolated.

A second line of research has more explicitly searched for evidence of species saturation by exploring the relationship between local and regional diversity (see references in Shurin and Srivastava 2005). This research defines *saturation* as the point at which local diversity levels off, or asymptotes, relative to increasing regional diversity. While there is some contention in the literature about how to best interpret such patterns, recent meta-analyses suggest that many species-assemblages do not appear to be saturated (or at least not completely saturated) and thus are capable of containing additional species (Shurin and Srivastava 2005). A third line of research examines the process of species additions, generally at spatial scales of smaller than a few square or cubic meters. At these scales, species-rich treatments are generally more difficult to invade, but they are also almost always invasible, in that at least some additional species can be added to even the most species-rich assemblages (e.g., Tilman 1997; Stachowicz et al. 1999; Shurin 2000; Kennedy et al. 2002).

Each of these lines of research suggests that species assemblages are not saturated. Still, this evidence is not completely convincing because the data are either observational, and therefore difficult to interpret conclusively; or experimental, but performed at small spatial scales, from which findings may not scale up.

Exotic invasions clearly provide the evidence we need to test the concept of species saturation. If there is one thing that the species invasion literature tells us, it is that species assemblages are invasible, and that rarely if ever is there evidence to suggest that no additional species could be added to any given system. This view is bolstered most strongly by recent studies showing that most invasions lead to net increases in diversity, at least at the landscape scale and larger (e.g., Atkinson and Cameron 1993; Gido and Brown 1999; Sax et al. 2002; Sax and Gaines 2003; Sax et al., Chapter 17, this volume). This finding suggests that few ecological systems are saturated with species—a view that is supported by several of the chapters in this book (see especially chapters by Bruno et al.; Lafferty et al.; Sax et al.; Stachowicz and Tilman; Vermeij).

However, it is unclear to what extent this concept is useful, particularly if no assemblages are saturated. Perhaps no assemblage can be saturated? This would certainly be valuable to determine. And, if saturation sometimes does occur, then it would be valuable to determine whether the concept of saturation is falsified for certain taxonomic groups, certain hyperdiverse species assemblages, and at certain spatial scales. Further, it would be valuable to determine how close different assemblages are to some saturation point, and to determine what factors predict this state. Perhaps human-mediated introductions have increased local richness to a saturation point for some taxa in some places—an outcome that could strongly impact the likelihood of future native species extinctions (Sax et al., Chapter 17, this volume). Finally, understanding whether and how saturation relates to the probability of invasion would also be valuable, as degree of saturation could conceivably provide a metric that could help to make invasion (at least at large scales) a more predictive discipline. Such information could help to test fundamental concepts of the processes and mechanisms that limit species diversity.

Are exotic species "different"?

There is a natural tendency to think of invasive species as being somehow different from natives. On the one hand, this view seems to have merit, as there is significant evidence to suggest that invading species are in some ways more robust in their new, naturalized habitat than in their native range. For example, individual exotic species have been shown to be heavier, taller, and in higher density in their naturalized range (see references in Lafferty et al., this volume). On the other hand, there is growing evidence that exotics as a group (i.e., when analyzed across multiple species) are fairly similar to native species. Many of the chapters in this volume bear out this latter view with respect to species interactions (Bruno et al.), patterns of range occupancy (Labra et al.), patterns of evenness within communities, and patterns of abundance across communities (McKinney and Lockwood). Many other recent examples from the literature also exist—for example, similarities in native versus exotic gradients of geographic range size, body size, and other types of ecotypic variation (e.g., Johnston and Selander 1964, 1967, 1971, 1973; Huey et al. 2000; Sax 2001). Some similarities are not unexpected, because many of the same constraints that influence the distribution and abundance of native species should similarly affect exotic species. For example, the importance of temperature in limiting length of growing season, the action of predators in limiting species abundance, and so on. Presumably, this is why so many large- and small-scale aspects of species distributions appear to be found common to both native and exotic species (op. cit.).

As a counterpoint, however, just as the same constraints limit both natives and exotics, freedom of exotics from such constraints is perhaps the most justifiable basis for the existence of differences in native and exotic performance. Predators and parasites can limit species abundance (e.g., Paine 1966), and the partial freedom from predation and parasitism that sometimes accompanies invasion (see discussion in Lafferty et al., this volume) can allow invading species to become larger and more competitive, or can increase their relative performance in a variety of other ways. Similarly, as Callaway et al. (this volume) show, species can gain competitive advantages simply by having traits (such as the production of particular secondary plant compounds) that are novel to native species. So, when exotic species do outperform native species, there is often either a shared constraint that has been lost or an approach that is evolutionarily novel relative to the native species in question (Sax and Brown 2000).

Caution is needed, however, when equating unique or impressive measures of exotic species performance with an evaluation of those species being somehow "different" from natives. Certainly many exotic species—particularly those labeled as invasive or as pest species—are renowned for their potential to damage native ecosystems. Consider the growth of kudzu (*Pueraria montana*) in the southeastern United States, where this exotic plant covers whole forests (and even buildings); or the performances of cheatgrass (*Bromus tectorum*) in the western United States, the cactus *Opuntia stricta* in Australia, and the marine alga *Caulerpa taxifolia* in the Mediterranean Sea (Dodd 1959; Mack 1981; Meinesz and Messe 1991; Forseth and Innis 2004). Indeed, many exotic species are accused of forming monocultures, or of otherwise decreasing native diversity (e.g., Farnsworth and Ellis 2001). Certainly, this does happen. But many native species have equally large impacts on species diversity (e.g., Houlahan and Findlay 2004). For example, consider the redwood forests of California, where the shade and litter produced by giant redwood trees (*Sequoia sempervirens*) severely limits plant growth, resulting in ecosystems with extremely low vascular plant diversity (Ornduff 2003).

Clearly, native species also have tremendous influences on species diversity, as well as on the structure and functioning of ecosystems. This rather obvious point is important to bear in mind when considering the performance of exotic species, as it is only some tiny fraction of exotic species that actually act in an invasive way. It would be interesting to know whether and how the fraction of exotic species that cause ecologically "extreme" impacts differs from the fraction of native species that do likewise. This question is in no way meant to imply that we shouldn't be concerned about the impacts caused by exotics—we should be! The question is simply meant to raise the point that our understanding of how exotics differ from native species, and the degree to which this difference is real and not just a function of observer bias, is unclear at this time and is in need of much greater attention.

Species Invasions: Where Do We Go from Here?

The study of species invasions is a large-scale endeavor. Researchers from many countries devote considerable attention to the management and control of invading species. It is certainly beyond our ability to predict how this field will continue to develop and whether the current attention afforded species invasions will be maintained. We believe that the study of species invasions has much to offer to fundamental research questions and that insights gained from such processes can be fed back into improved management strategies. While there are many issues that researchers could tackle next, we outline a few issues that we believe would be particularly valuable to consider and that collectively represent a series of next steps that we believe would be profitable to take.

Interaction of ecology and evolution is determining patterns of species diversity

One of the most powerful insights species invasions might provide is a better understanding of how ecological and evolutionary processes interact to determine patterns of species diversity. There are well developed theories for patterns of evolutionary diversification (see Coyne and Orr 2004). Similarly, there are well developed theories for patterns of species diversity, such as MacArthur and Wilson's island biogeography theory (1967). What we lack are robust and well developed theories that integrate ecological and evolutionary drivers of species diversity.

The shortcomings of current theory are particularly apparent in light of recently characterized patterns of species invasion, which show that net species richness generally increases following species invasions. For example, net richness of vascular plant species has approximately doubled on oceanic islands in the past few hundred years as many exotic species have been introduced and become naturalized while few native species have gone extinct (Sax et al. 2002). This result suggests that these islands historically have contained many fewer species than they are ultimately capable of supporting. So why haven't more species evolved on these islands to take advantage of these opportunities? How do ecological forces manifested through particular levels of species richness limit rates of speciation? Is the process of speciation inherently limited by slow rates, or is it instead the case that speciation rates become slow only after moderate levels of species richness have developed (or been achieved via colonization)?

Answers to parts of these questions are already well developed in the ecological, evolutionary, biogeographical and paleontological literature. Still, a broad and coherent synthesis and further development of these issues seems to be needed. While these questions have historically been very difficult to address, we believe that the study of species invasions can provide tremendous insights into these questions for two principal reasons. First, species invasions provide the species-addition "experiments" needed to help determine the capacity or ability of different areas to support species. Second, evolutionary diversification catalyzed by invasion, of both native and invading species, can potentially help to resolve which factors may limit speciation processes. We hope that work on these issues can lead to a more comprehensive theory of species diversity and help to advance a conceptual unification of ecological and evolutionary processes.

Toward a conceptual framework for invasion biology

The current study of invasion biology seems to be driven in many ways by a focus on single-species studies, with the emphasis on understanding the impacts of particular invading species. Certainly many of the most influential and valuable papers in the field of invasion biology, and indeed within the field of ecology, have come from such studies—consider, for example, the study of *Myrica faya* invasion in Hawaii (Vitousek and Walker 1989), which advanced our understanding of the influence a single plant species could have on ecosystem dynamics. More recently macroecological approaches of studying multiple species invasions have also become prominent in their attempts to identify general unifying themes of species invasions (e.g., Lonsdale 1999; Blackburn and Duncan 2001; Cassey 2002). Synthetic studies and attempts to understand invasions go back much further, at least to the Asilomar Conference and the resulting 1965 volume edited by Baker and Stebbins. What we believe is lacking, however, is a broadly powerful conceptual framework for understanding species invasions, their impacts, and their management.

Valuable attempts have been made to understand aspects of such a framework. For example, there have been attempts at a conceptual framework for understanding variation in invasiveness of different taxa (e.g., Baker 1965; Rejmánek 1996), and for assessing variation in ecosystem susceptibility to invasion (e.g., Davis et al. 2000). However, merging these efforts into a comprehensive framework has been the bugaboo of invasion biology. Perhaps it is not possible, or not possible at this time given our rudimentary understanding of ecological and evolutionary processes. Still, we believe that developing such a framework is a useful goal to work towards, and encourage greater effort on this front.

One potentially encouraging basis for the development of such a framework may be the work laid out by Ricklefs (this volume) on comparisons between the "taxon cycle" and patterns of species invasion. Taxon cycle theory provides a basis for interpreting the success of species invasions, as well as predicting their eventual decline (from particularly problematic invaders to those whose ecological footprints are much less pronounced). It incorporates both ecological and evolutionary patterns, together with biogeographical patterns of abundance and distribution. Whether the further development and extension of this theory can provide the basis for a conceptual unification of invasion theory remains to be seen, but we believe that it is one avenue that should be more thoroughly explored.

Merging research agendas in applied and basic science

Species invasions provide a wonderful opportunity to merge applied and basic research agendas. Many of the applied questions that would be valuable to understand—what makes certain species good invaders, what makes certain environments prone to invasion, what determines whether an impacted native species will eventually go extinct—can be coupled with more fundamental questions, such as what processes allow for the evolution of invasive species, what ecological and evolutionary processes create variation in susceptibility to invasion, and what processes and mechanisms limit the number of species any one area supports. We believe that there is ample room for creativity in such couplings, particularly by drawing on other applied problems in conservation. For example, by better understanding range collapse in endangered native species can we perhaps learn how to induce range collapse in invading exotic pest species (M. V. Lomolino, personal communication)? We believe that much more could and should be done on this front. It is our hope that the coupling between applied and basic research will become much stronger in the years ahead.

Concluding Remarks

There is clearly much room for the continued advancement in our understanding of species invasions and in what they can tell us about the natural world. Pursuing these advances should, and undoubtedly will, form the basis for research careers in science for at least the next several generations. We hope that the work in this volume can help in a small way to continue to push the fields of ecology, evolution, and biogeography forward, particularly towards greater synthesis among fields and greater synthesis between applied and basic research questions.

As a final thought, it is probably worth mentioning that one of the most intriguing aspects of species invasions is just how humbling they can be to the scientist who studies them. They remind us of how little we currently know or understand about the basic ways that nature works on the planet. They provide us with impetus and perhaps the inspiration to continue to advance this understanding.

Acknowledgments

This work was conducted as a part of the "Exotic Species: A Source of Insight into Ecology, Evolution, and Biogeography" Working Group supported by the National Center for Ecological Analysis and Synthesis, a Center funded by the National Science Foundation (Grant #DEB-0072909), the University of California, and the Santa Barbara campus. This manuscript benefited tremendously from the interactions at NCEAS, and from specific comments by J. Wares.

Literature Cited

Atkinson, I. A. E., and E. K. Cameron. 1993. Human influence on the terrestrial biota and biotic communities of New Zealand. Trends in Ecology and Evolution 8:447–451.

Baker, H. G. 1965. Characteristics and modes of origin of weeds. In H. G. Baker and G. L. Stebbins, eds. *The genetics of colonizing species*, pp. 147–168. Academic Press, New York.

Baker, H. G., and G. L. Stebbins, eds. 1965. *The genetics of colonizing species*. Academic Press, New York.

Blackburn, T. M., and R. P. Duncan. 2001. Determinants of establishment success in introduced birds. Nature 414:195–197.

Brown, J. H., and D. F. Sax. 2005. Biological invasions and scientific objectivity: Reply to Cassey et al. (2005). Austral Ecology 30:481–483.

Cassey, P. 2002. Life history and ecology influences establishment success of introduced land birds. Biological Journal of the Linnaean Society 76:465–480.

Cassey, P., T. M. Blackburn, R. P. Duncan, and S. L. Chown. 2005. Concerning invasive species: reply to Brown and Sax. Austral Ecology 30:475–480.

Cadotte, M. W., S. M. McMahon, and T. Fukami. 2005. *Conceptual ecology and invasions biology: reciprocal approaches to nature*. Kluwer Academic, New York. In press.

Coyne, J. A., and H. A. Orr. 2004. *Speciation*. Sinauer Associates, Sunderland, MA.

Cox, G. W. 2004. *Alien species and evolution*. Island Press, Washington, DC.

Darwin, C. 1859. *On the origin of species*. John Murray, London.

Davis, M. A., J. P. Grime, and K. Thompson. 2000. Fluctuating resources in plant communities: a general theory of invisibility. Journal of Ecology 88:528–534.

Davis, M. A. 2003. Does competition from new species threaten long-term residents with extinction? BioScience 53:481–489.

Davis, M. A. 2005. Invasion biology 1958–2004: the pursuit of science and conservation. In M. W. Cadotte, S. M. McMahon, and T. Fukami, eds. *Conceptual ecology and invasions biology: reciprocal approaches to nature*. Kluwer Academic, New York. In press.

Dodd, A. P. 1959. The biological control of prickly pear in Australia. In A. E. M. Nairn, ed. *Biogeography and ecology in Australia*, pp. 565–577. Dr. W. Junk BV, The Hague, Netherlands.

Ebenhard, T. 1988. Introduced birds and mammals and their ecological effects. Swedish Wildlife Research 13:1–107.

Ellstrand, N. C., and K. A. Schierenbeck. 2000. Hybridization as a stimulus for the evolution of invasiveness in plants? Proceedings of the National Academy of Sciences USA 97:7043–7050.

Elton, C. S. 1958. *The ecology of invasions by animals and plants*. Methuen & Co., London.

Farnsworth, E. J., and D. R. Ellis. 2001. Is purple loosestrife (*Lythrum salicaria*) an invasive threat to freshwater wetlands? Conflicting evidence from several ecological metrics. Wetlands 21:199–209.

Forseth, I. N., and A. F. Innis. 2004. Kudzu (*Pueraria montana*): History, physiology, and ecology combine to make a major ecosystem threat. Critical Reviews in Plant Sciences 23:401–413.

Fritts, T. H., and G. H. Rodda. 1998. The role of introduced species in the degradation of island ecosystems: a case history of Guam. Annual Review of Ecology and Systematics 29:113–140.

Gido, K. B., and J. H. Brown. 1999. Invasion of North American drainages by alien fish species. Freshwater Biology 42:387–399.

Grinnell, J. 1919. The English House Sparrow has arrived in Death Valley: an experiment in nature. American Naturalist 53:468–472.

Gurevitch, J., and D. K. Padilla. 2004. Are invasive species a major cause of extinctions? Trends in Ecology and Evolution 19:470–474.

Houlahan, J. E., and C. S. Findlay. 2004. Effect of invasive plant species on temperate wetland plant diversity. Conservation Biology 18:1132–1138.

Johnston, R. F., and R. K. Selander. 1964. House Sparrows: rapid evolution of races in North America. Science 144:548–550.

Johnston, R. F., and R. K. Selander. 1967. Evolution of the House Sparrow. I. Intrapopulation variation in North America. The Condor 69:217–258.

Johnston, R. F., and R. K. Selander. 1971. Evolution of the House Sparrow. II. Adaptive differentiation in North American populations. Evolution 25:1–28.

Johnston, R. F., and R. K. Selander. 1973. Evolution of the House Sparrow. III. Variation in size and sexual dimorphism in Europe and North and South America. American Naturalist 107:373–390.

Huey, R. B., G. W. Gilchrist, M. L. Carlson, D. Berrigan, and L. Serra. 2000. Rapid evolution of a geographic cline in size in an introduced fly. Science 287:308–309.

Kennedy T. A., S. Naeem, K. M. Howe, J. M. H. Knops, D. Tilman, and P. Reich. 2002. Biodiversity as a barrier to ecological invasion. Nature 417:636–638.

Kolbe, J. J., R. E. Glor, L. R. Schettino, A. C. Lara, A. Larson, and J. B. Losos. 2004. Genetic variation increases during biological invasion by a Cuban lizard. Nature 431:177–181.

Lever, C. 1987. *Naturalized birds of the world*. Longman, Essex.

Levin, D. A. 2003. Ecological speciation: lessons from invasive species. Systematic Botany 28:643–650.

Lonsdale, W. M. 1999. Global patterns of plant invasions and the concept of invasibility. Ecology 80:1522–1536.

MacArthur, R. H., and E. O. Wilson. 1967. *The theory of island biogeography*. Princeton University Press, Princeton, NJ.

Mack, R. N. 1981. Invasion of *Bromus tectorum* L. into western North America, an ecological chronicle. Agro-Ecosystems 7:145–165.

Mayr, E. 1954. Change of genetic environment and evolution. In J. S. Huxley, A. C. Hardy and E. B. Ford, eds. *Evolution as a process*, pp. 156–180. Allen & Unwin, London.

Meinesz, A., and B. Hesse. 1991. Introduction of the tropical alga *Caulerpa taxifolia* and its invasion of the northwestern Mediterranean. Oceanologica Acta 14:415–426.

Mooney, H. A., and E. E. Cleland. 2001. The evolutionary impact of invasive species. Proceedings of the National Academy of Sciences USA 98:5446–5451.

Myers, N., and A. H. Knoll. 2001. The biotic crisis and the future of evolution. Proceedings of the National Academy of Sciences USA 98:5389–5392.

Ornduff, R. [Revised by P. M. Faber and T. Keeler-Wolf]. 2003. *Introduction to California plant life*. University of California Press, Berkeley.

Paine, R. T. 1966. Food web complexity and species diversity. American Naturalist 100:65–75.

Pimentel, D., L. Lach, R. Zuniga, and D. Morrison. 2000. Environmental and economic costs of nonindigenous species in the United States. Bioscience 50:53–65.

Raybould, A. F., A. J. Gray, M. J. Lawrence, and D. F. Marshall. 1991. The evolution of *Spartina anglica* C. E. Hubbard (Gramineae): origin and genetic variability. Biological Journal of the Linnaean Society 43:111–126.

Rejmánek, M. 1996. A theory of seed plant invasiveness, the first sketch. Biological Conservation 78:171–181.

Sax, D. F. 2001. Latitudinal gradients and geographic ranges of exotic species: implications for biogeography. Journal of Biogeography 28:139–150.

Sax, D. F., and J. H. Brown. 2000. The paradox of species invasion. Global Ecology and Biogeography 9:363–372.

Sax, D. F., and S. D. Gaines. 2003. Species diversity: from global decreases to local increases. Trends in Ecology and Evolution 18:561–566.

Sax, D. F., S. D. Gaines, and J. H. Brown. 2002. Species invasions exceed extinctions on islands worldwide: A comparative study of plants and birds. American Naturalist 160:766–783.

Sax, D. F., and S. D. Gaines. 2005. The biogeography of naturalized species and the species-area relationship: reciprocal insights to biogeography and invasion biology. In M. W. Cadotte, S. M. McMahon, and T. Fukami, eds. *Conceptual ecology and invasions biology: reciprocal approaches to nature*. Kluwer Academic, New York. In press.

Srivastava, D. S. 1999. Using local-regional richness plots to test for species saturation: pitfalls and potentials. Journal of Animal Ecology 68:1–16.

Shurin, J. B. 2000. Dispersal limitation, invasion resistance, and the structure of pond zooplankton communities. Ecology 81:3074–3086.

Shurin, J. B., and D. S. Srivastava. 2005. New perspectives on local and regional diversity: beyond saturation. In M. Holyoak, M. A. Leibold, and R. D. Holt, eds. *Metacommunities*. University of Chicago Press, Chicago. In press.

Stachowicz, J. J., R. B. Whitlatch, and R. W. Osman. 1999. Species diversity and invasion resistance in a marine ecosystem. Science 286:1577–1579.

Templeton, A. R. 1980. The theory of speciation via the founder principle. Genetics 94:1011–1038.

Templeton, A. R., R. J. Robertson. J. Brisson, and J. Strasburg. 2001. Disrupting evolutionary processes: the effect of habitat fragmentation on collared lizards in the Missouri Ozarks. Proceedings of the National Academy of Sciences USA 98:5426–5432.

Tilman, D. 1997. Community invasibility, recruitment limitation, and grassland biodiversity. Ecology 78:81–92.

Tilman, D., and C. Lehman. 2001. Human-caused environmental change: Impacts on plant diversity and evolution. Proceedings of the National Academy of Sciences USA 98:5433–5440.

Vitousek, P. M., and L. R. Walker. 1989. Biological invasion by *Myrica faya* in Hawaii: plant demography, nitrogen fixation, ecosystem effects. Ecological Monographs 59:247–265.

Voisin, M., C. R. Engel, and F. Viard. 2005. Differential shuffling of native genetic diversity across introduced regions in a brown alga: Aquaculture vs. maritime traffic effects. Proceedings of the National Academy of Sciences USA 102:5432–5437.

Wilcove, D. S., J. Dubow, A. Phillips, and E. Losos. 1998. Quantifying threats to imperiled species in the United States. BioScience 48:607–615.

Index